Lecture Notes in Computer Science   1280

Edited by G. Goos, J. Hartmanis and J. van Leeuwen

Advisory Board: W. Brauer   D. Gries   J. Stoer

Springer
Berlin
Heidelberg
New York
Barcelona
Budapest
Hong Kong
London
Milan
Paris
Santa Clara
Singapore
Tokyo

Xiaohui Liu   Paul Cohen
Michael Berthold   (Eds.)

Advances in Intelligent Data Analysis

Reasoning about Data

Second International Symposium, IDA-97
London, UK, August 4-6, 1997
Proceedings

 Springer

Series Editors

Gerhard Goos, Karlsruhe University, Germany

Juris Hartmanis, Cornell University, NY, USA

Jan van Leeuwen, Utrecht University, The Netherlands

Volume Editors

Xiaohui Liu
Birkbeck College, University of London, Department of Computer Science
Malet Street, London WC1E 7HX, UK
E-mail: hui@dcs.bbk.ac.uk

Paul Cohen
University of Massachusetts, Amherst, Department of Computer Science
Amherst, MA 01003-4610, USA
E-mail: cohen@cs.umass.edu

Michael Berthold
Universität Karlsruhe, IRF
Am Zirkel 2, D-76128 Karlsruhe, Germany
E-mail: berthold@ira.uka.de

Cataloging-in-Publication data applied for

Die Deutsche Bibliothek - CIP-Einheitsaufnahme

Advances in intelligent data analysis : reasoning about data ;
second international symposium ; proceedings / IDA-97, London,
UK, August 4 - 6, 1997. X. Liu ... (ed.). - Berlin ; Heidelberg ; New
York ; Barcelona ; Budapest ; Hong Kong ; London ; Milan ; Paris ;
Santa Clara ; Singapore ; Tokyo : Springer, 1997
  (Lecture notes in computer science ; Vol. 1280)
  ISBN 3-540-63346-4

CR Subject Classification (1991): H.3, I.2, G.3, I.5.1, I.4.5, J.2, J.1, J.3

ISSN 0302-9743
ISBN 3-540-63346-4 Springer-Verlag Berlin Heidelberg New York

© Springer-Verlag Berlin Heidelberg 1997
Printed in Germany

Typesetting: Camera-ready by author
SPIN 10547800   06/3142 – 5 4 3 2 1 0   Printed on acid-free paper

Foreword

This volume is a collection of papers presented at the Second International Symposium on Intelligent Data Analysis (IDA-97) held at Birkbeck College, University of London, August 4-6, 1997. Our community is growing fast: The first Symposium, held in Baden-Baden in 1995, received 69 extended abstracts for review; this time we received 107 full papers. We accepted 50, divided equally between oral and poster presentations. Each paper was considered by at least two independent reviewers and the results were discussed in a meeting of program committee members.

IDA-97 was a single-track conference consisting of oral and poster presentations, invited speakers, demonstrations, and exhibitions. The conference Call for Papers introduced a theme, "Reasoning About Data", and many papers complement this theme, but other, exciting topics have emerged, including exploratory data analysis, data quality, knowledge discovery, and data-analysis tools, as well as the perennial technologies of classification and soft computing. A new and exciting theme involves analyzing time series data from physical systems, such as medical instruments, environmental data, and industrial processes.

We are very grateful to Professor David Hand, of the Open University, and Dr. Larry Hunter, of the National Library of Medicine in Washington D.C., for agreeing to give keynote talks. Each has made major contributions and has helped to define the interdisciplinary field of intelligent data analysis.

The symposium owes much to many hard-working individuals. Michael Berthold handled publicity and publications, and is responsible for this volume. The program committee and additional volunteer reviewers read the papers with great care and wrote excellent, informative, helpful reviews. Terrie Korpita and Peggy Weston at the University of Massachusetts coordinated the international reviewing process, a huge job performed flawlessly. The organization at Birkbeck has been tremendous. Phil Gregg was responsible for the IDA-97 homepage, which served to keep the community abreast of developments and was responsible in no small part for the increasing size of the symposium. Betty Walters, Sylvie Jami, Trevor Fenner, Claude Gierl, and Steven Swift were responsible for daily operations, enquiries, registrations, and the details that grew exponentially as the conference date drew nearer. The Systems group in Computer Science at Birkbeck College provided valuable computing support for the reviewing process and conference exhibitions. The symposium would not exist were it not for the confidence and financial backing of Birkbeck College, particularly Roger Johnson, Ken Thomas, and Guy Fitzgerald.

June 1997

Xiaohui Liu
Paul Cohen

Organization Committee

General Chair:
Xiaohui Liu
Department of Computer Science
Birkbeck College, Malet Street
London WC1E 7HX, UK
E-mail: hui@dcs.bbk.ac.uk

Technical Program:
Paul Cohen
Department of Computer Science
University of Massachusetts, Amherst
Amherst, MA 01003-4610, USA
E-mail: cohen@cs.umass.edu

Publicity&Publication:
Michael Berthold
Universität Karlsruhe, IRF
Am Zirkel 2, 76128 Karlsruhe, Germany
E-mail: berthold@ira.uka.de

Local Arrangements:
Trevor Fenner
Department of Computer Science
Birkbeck College, Malet Street
London WC1E 7HX, UK
E-mail: trevor@dcs.bbk.ac.uk

Finance:
Sylvie Jami
Department of Computer Science
Birkbeck College, Malet Street
London WC1E 7HX, UK
E-mail: s.jami@dcs.bbk.ac.uk

Sponsorship:
Mihaela Ulieru
Simon Fraser University
2357 Riverside Drive
North Vancouver, B.C.
Canada V7H 1V8
Email: ulieru@cs.sfu.ca

Exhibition:
Richard Weber
MIT GmbH, Promenade 9
52076 Aachen, Germany
E-mail: rw@mitgmbh.de

Program Committee

Table of Contents

Section III: Medical Applications

Section IV: Soft Computing

Section V: Knowledge Discovery and Data Mining

Section VI: Estimation, Clustering

Section VII: Data Quality

Section VIII: Qualitative Models

Intelligent Data Analysis: Issues and Opportunities

David J. Hand

Department of Statistics
The Open University

1   Introduction

Data analysis, as a scientific or technological discipline, is a new one. It has emerged as the product of the union of several disciplines: statistics, computer science, pattern recognition, artificial intelligence, machine learning, and others. The fact of its mixed parentage has created certain tensions within it, as well as certain benefits. Parallel streams of work, essentially on the same problem but from the different perspectives of different disciplines, have come together, the different perspectives providing a synergy which has led to even further development. One clear example of this is the work on supervised pattern recognition, or discriminant analysis, or classification [13, 15] over the past twenty or thirty years. Perhaps the oldest amongst the parents of data analysis is statistics. As a discipline, statistics has received its main motivation, at least until recently, from the different application areas to which it has been applied. Thus, for example, agricultural applications stimulated the development of experimental design, medical applications were a primary motivation underlying the development of survival analysis and graphical models, and psychological applications promoted the development of factor analysis, linear structural equation models, and item response theory. Of course, once developed, such methods rapidly spread to other disciplines. There is something of a parallel here, on a smaller scale, with the development of data analysis itself. The same method was sometimes developed within different disciplines, and accorded different names - and only later would recognition dawn that the methods were the same. An example is the development of multidimensional scaling or ordination techniques in psychometrics and ecology.

New application areas continue to provide stimulus and motivation for the development of new statistical and data analytic tools. One such area which I think will be important in the future is financial statistics [17]. Here the role of stochastic models in financial markets is already well-known but there are other sectors, such as consumer credit [20, 16], where the increasingly competitive financial environment is leading to new products and increasing need for sophisticated statistical models.

However, beyond the application areas, new stimuli have begun to have an impact. In particular, the computer has already revolutionised things. The opportunities provided by the computer have dramatically expanded what we mean

X. Liu, P. Cohen, M. Berthold (Eds.): "Advances in Intelligent Data Analysis" (IDA–97)
LNCS 1280, pp. 1–14, 1997. © Springer–Verlag Berlin Heidelberg 1997

by 'data', have radically altered the sorts of problems which can usefully be addressed, and have led to the development of entirely new classes of model. In the next three sections we consider each of these aspects in turn.

However, there is more than this. Just as the computer made possible the development of chaos theory, so it has also led to a deeper understanding of fundamental issues of model fitting and model building. One example of this is the new understanding of generalisability which has emerged from work on computational learning theory. The approach has major differences from the standard statistical approach.

This meeting is about 'intelligent data analysis'. In Section 5 we examine what this means - and in particular how the power of the computer can be harnessed to work in conjunction with the power of the human mind in effective, and intelligent, data analysis. I once wrote a book with the title Artificial intelligence and psychiatry [12]. I wanted to call it Artificial intelligence for psychiatrists, but the publishers wouldn't let me. They felt that this might be regarded as casting aspersions on the intellectual needs of psychiatrists. Something similar occurred while I was working on this paper. My thoughts kept returning to the adjective in the title. If the conference was to be about 'intelligent' data analysis, it implied that there was another sort - 'unintelligent' or perhaps even 'stupid' data analysis. What, I wondered, could this be? Presumably it didn't refer to the fact that we sometimes all made mistakes. Even the most able of us can do that. No, it had to refer to something deeper and more fundamental. Perhaps a school or style of data analysis which was, at its heart, fundamentally misconceived. At this point no doubt some eyes will light up, expecting me to start to criticise frequentist or Bayesian approaches to statistics. But that's not really what I'm after. Both camps have been occupied by extremely intelligent statisticians and data analysts, and it would be absurd to characterise either as unintelligent. In the end I decided that unintelligent data analysis was simply data analysis which went too far. Precisely what this means is the topic of Section 6.

2   New types of data

A traditional view of statistics might be that it deals with relatively small data sets, which are numerical, which are clean, and which permit straightforward enough answers. This view is understandable if one's perspective is gained from texts on the subject. By necessity, as a result of the limited space available in which to develop the techniques, these are characteristics of the data sets typically used to illustrate methods. But this is misleading. Modern statistics and, more generally, modern data analysis, must contend with data which depart from this ideal in many ways. Here are some examples.

Modern electronic data capture methods mean that vast databases are being accrued. These may contain tens of millions or even billions of records. Examples of this sort of thing are the transaction databases from modern industrial and commercial conglomerates. These may be on an international scale. Electronic point-of-sale data capture means that even the slightest sale goes recorded and is available for analysis. To take advantage of this rich source of data entirely novel techniques are required. They are slowly appearing - under the name of data mining, a subdiscipline of data analysis. Another example is the mass of information sent back from space probes and satellites. In both of these situations we have at our disposal vast bodies of data. But data, per se, is useless. What we want is information which is relevant to our objectives. We might even, in this context, define information as data which has been processed with some objective in mind. So, here is my definition of data analysis: data analysis is what we do when we turn data into information. And I can ally this to a definition of information, which is that which we extract from data when we attempt to answer some question. It is not always straightforward. Before extracting information which can shed light on a question, one must be clear about what that question is.

The number of records in some modern datasets is just one way in which one might measure their size. Another is the dimensionality (in statistical terms) or the number of fields (in computer terms) recorded for each record. Issues of high dimensionality open up entirely new classes of problem. Situations where high dimensionality arises can be divided into two types. One is where the variables are highly correlated. For example, spectral observations might be recorded at each of a large number of different wavelengths - and continuity considerations lead us to expect correlation between neighbouring wavelengths. The trick, again, is to process the large number of observations to extract that information which is relevant for the problem at hand - such as, for example, permitting one to distinguish between different classes of spectrum. The other type of high dimensional problem arises when the correlation structure is not evident a priori. Again, as an example, this could arise from a detailed study of staff satisfaction in a business, where hundreds of different questions might be asked and the answers recorded. The sorts of problems with which we have to contend when many variables have been measured are multicollinearity, singularity, and the simple problem of how to extract the relevant from the irrelevant. Problems of the first two kinds are obvious for certain classes of statistical technique (where, for example, a matrix may not be invertible) but they also apply to less formal, or 'model-free' data analytic methods.

The next section looks at the new kinds of problems which such vast data sets lead to. I should say and opportunities, since the data sets are collected because they are seen as potentially valuable sources of information. However, in many situations the aim will be to reduce the data sets to a manageable size, to summarise or describe them in some way. And summary data leads to its own issues. For example, one source of large bodies of data are governmental and official bodies. These data need to be summarised to give an accurate picture of the way the economy is functioning, the way lifestyles are changing, and so on. There are clearly merits in grouping numeric variables into qualitative categories (young, middle-aged, and old, for example). But then, if you happen to belong to the European Union, how do you merge such figures to yield an overall one when different definitions have been used by each member country? What if you are talking about unemployment statistics, where quite different definitions have been used? Perhaps even tougher, how do you combine different ways of measuring inaccuracy to yield a measure for the final overall figure? And, of course, even if you can develop ways to answer these questions, can you automate those ways?

The last illustration makes it particularly obvious, but it is a general truism that numbers are not data. Data are more than numbers. They are the numbers and the descriptive context in which those numbers arose, the units in which the numbers were measured, the meaning of those numbers. The force of this point is easy enough to illustrate by asking you to tell me the average (arithmetic mean) of the two numbers 10 and 350. Does your answer differ if I now tell you that they are angular measurements in degrees?

Table 1 shows another, more subtle, example. The body of the table shows counts of vehicles passing a particular point on a road, grouped by day (columns) and hour of the day (rows). Without this knowledge the table is just a set of numbers. With the contextual knowledge, not only do the numbers become data, but one can immediately spot (well, perhaps not immediately) that there is something wrong with the table and suggest how it might be corrected.

Such 'data about data' have been given the name of metadata [6, 8] and they

Hours	Day						
	Mon	Tue	Wed	Thu	Fri	Sat	Sun
0000-0100	41	10	12	5	15	19	35
0100-0200	29	4	5	5	3	8	10
0200-0300	17	3	3	0	4	6	9
0300-0400	7	5	5	3	5	3	4
0400-0500	2	6	9	6	7	6	6
0500-0600	10	24	20	20	20	26	18
0600-0700	20	60	63	62	70	68	50
0700-0800	42	219	232	220	225	222	89
0800-0900	91	429	425	473	447	466	207
0900-1000	163	270	274	272	285	281	314
1000-1100	222	201	206	190	210	227	391
1100-1200	276	224	227	233	235	236	395
1200-1300	280	261	240	262	281	272	398
1300-1400	193	234	231	251	232	293	404
1400-1500	202	240	299	282	317	313	353
1500-1600	226	315	316	299	322	402	320
1600-1700	203	286	309	293	339	381	301
1700-1800	220	405	377	393	386	362	281
1800-1900	206	303	295	270	302	299	213
1900-2000	193	257	238	242	254	242	219
2000-2100	123	128	170	153	177	156	122
2100-2200	99	106	111	109	102	98	79
2200-2300	74	66	95	85	82	86	66
2300-2400	54	39	54	50	59	86	80

Table 1. Road traffic data

illustrate the new kinds of issues of 'intelligent data analysis' that the computer is opening up for us. In particular, they raise the question of how such metadata should be analysed. That is, accompanying any data analysis is a metadata analysis, in which one uses the metadata to decide if certain data analytic operations are sensible (is averaging subjective ordinal preference scores meaningful?), what are the properties of the units which emerge from such an analysis (we can add tons of apples and tons of oranges, provided we label the result tons of fruit), and how to interpret the results.

Formalising metadata, in particular, formalising it to such an extent that it, like data, can be manipulated automatically by computer, is one of the major challenges of modern intelligent data analysis.

There seem to be two kinds of metadata. One is extrinsic metadata - that arising from the context in which the data are collected. This tells me that it is not sensible to add tons of fruit to tons of scrap metal, or to add someone's height to their weight, unless I am very careful about what I think the result measures. The other is intrinsic metadata - that arising from the internal properties of the measurements themselves. This tells me that it is not sensible to describe a Celsius temperature at time 1 as twice that at time 2 [14]. In general, statistical data can come in many shapes and forms - there is no neat all-encompassing taxonomy of data types which nicely summarises everything that we might need. Types of data which occur in different situations include categorical data, which may be nominal or ordinal, numerical data, which may be continuous or discrete, ordinal data, which may have numerical values assigned to it, and so on. Worse, or perhaps better if one appreciates the challenge, data are often structured. We have data collected on children, who are grouped into classes, within schools, within counties, and so on. We have the swirls and whorls which make up a fingerprint, and which cannot lie in arbitrary relationships to each other. We have the strokes of Chinese characters, where the meaning of the whole is defined not so much by the shape of each individual stroke as by its relationship to the others. We have the different objects within a scene. And we have spoken words, where their meaning is recognised partly by their own internal intensity and frequency content and partly by their relationship to the surrounding words. Structured data is of fundamental importance and represents a vital challenge to data analysts.

Time series and longitudinal data have been studied for time immemorial. But again the computer is leading to entirely different classes of problems. Electronic measurement means that continuous signals need to be processed - and in real time, at that. It is in such areas where one really appreciates the strength that comes from the merging of disciplines such as computer science and statistics. Indeed, ideas such as the Kalman filter and adaptive estimation lie right at the interface, benefiting from the ideas and developments of each discipline.

Such areas as forecasting, quality control charts, and process monitoring are already the familiar subject of data analysis. But new areas of interest are arising. For example, it has been recognised that populations change over time. A model built 'now' for predictive purposes may not be so good a year from now because the underlying population has altered. We shall return to this point below.

3   New types of problem

The different, and often novel, types of data described in the preceding section are associated with challenging new problems which confront the data analyst. The particular issues arising from huge data sets provide a nice illustration of how new problems have arisen. First there is the simple mechanical problem of how to handle huge data sets. Such issues are well outside the realm of conventional statistics. They include how best to store the data and how to estimate descriptors and parameters if not all of the data can be held in memory at any one time. Adaptive and sequential methods are necessary, and these have largely been developed outside the statistical community. Secondly, large data sets lead to fundamental questions about what a model is. With conventional, small, data sets, 'significance' means that an effect should be included in the model. (Non-significance does not mean a term should be excluded - other factors, such as the interpretability of the model also play a role.) But with very large data sets things are quite different. Here any effect you might care to test will show up as significant. In this kind of situation the distinction between statistical and substantive significance begins to play a more fundamental role. One will want to include in the model only the larger effects and only those aspects of model structure which are relevant to the questions in hand. Thirdly, new classes of software tools need to be developed to handle problems of large data sets. Consider, for example, anomaly detection in particle decay experiments or in credit card transaction records. Especially if the data is continuing to accumulate, one cannot simply rely on standard statistical summaries, designed for batch analysis of small samples. Instead one needs to develop software which continuously monitors the incoming data, searching for patterns of the specified kind. There are interesting technical problems here of how best to design such automatic data analytic tools so that they behave in as intelligent a manner as possible.

Sheer size is not the only novel issue which is generating new types of problem. Datasets, whether large or small, are being collected in more and more situations, with the hope that this quantitative information will lead to better decision making. But the number of data analysts being trained is not increasing concomitantly. This has the effect that people who are not expert are often called upon to undertake and interpret an analysis. The consequence, easy to imagine, is that data analysis can come to be regarded with suspicion. This type of effect has certainly contributed to the low popular regard in which statistics is held in some quarters. Clearly the computer is partly to blame for this. The computer facilitated the collection of the data in the first place and the computer has also permitted data analytic software to be placed on everyone's desk, whether or not they know how to use it properly. Fortunately, the computer can also come to our aid here - as outlined in Section 5.

Data analysis always occurs with an objective in mind - one doesn't set out simply to 'analyse' data (to do so would surely be a hallmark of unintelligent data analysis). One's aim might be a simple descriptive summary (i.e. what are the most important features of these data?), or it might be more specific (Does this pattern exist? Are there atypical observations?). It might be exploratory (Are there any 'interesting' patterns?) or confirmatory and highly specific (Does this variable increase as that one decreases?) Intelligent data analysis requires one to be as clear as possible about what that objective is.

Classically statistics is associated in most people's minds (most technical people, at least) with probability. But not everyone regards this as necessary. One can regard certain classes of tests (permutation tests, for example) as relating simply to the number of possible data configurations, rather than to a probabilistic interpretation. Moreover, descriptive statistics need not involve probabilistic interpretations at all: a mean is a summary regardless of how the data arose, an ordination can display relationships between the members of a population, not a sample, and dissection analysis can partition a population into commercially useful subgroups. These non- probabilistic interpretations are likely to sit well with data analysts from a non- statistical background. One might argue a case that they are more appropriate in many situations where probabilistic interpretations are applied as a matter of course. For example, clinical trials are typically analysed using sampling-theory based statistical methods, even though the population from which the 'sample' is drawn, given that it must also satisfy stringent inclusion/exclusion criteria, is hard to define. This is one reason why the numerical results of clinical trials do not generalise well to people outside the study group [1, 4]. Perhaps methods which simply interpreted things in terms of the proportion of alternative allocations which gave more extreme results would be more appropriate.

4   New types of model

The new types of data and new types of problem outlined in the previous sections have required new types of model to tackle them. The development of most of these lies in the computing camp as much as the statistical one. They really are data analytic tools in the most general sense. Examples include rule-based knowledge representations, as used in expert systems, hidden Markov models, neural networks, genetic algorithms, simulated annealing, generalised additive models, multivariate adaptive regression splines, Bayesian methods, computer intensive estimation methods such as Markov chain Monte Carlo methods, statistical databases which need operations beyond those found in standard relational databases (datacube, for example [9]), and so on.

Some comments about the use of the word 'model' are necessary here. On the one hand we can distinguish between structural and predictive models. And on the other hand, orthogonal to this distinction, we can distinguish descriptive and mechanistic models (physicists say phenomenological and theoretical, respectively). A structural model merely tries to summarise the structure in a set of data. One might, for example, want to know what are the strongest relationships. In contrast, a predictive model will be used to predict the 'value' of some variable from values of others. The word 'value' here is in inverted commas because one's prediction may be as simple as a statement that the score under condition A is likely to be larger than the score under condition B. A descriptive model assumes no knowledge of the underlying mechanism which generated the data. One will simply summarise the available data. Such models may be structural or predictive. Conversely, mechanistic models are based on some theory or model of the underlying processes. Again they can be structural or predictive. In a mechanistic model the parameters will typically have some meaningful interpretation, whereas in a descriptive model they will not. But this will not matter if the model is accurate - if the model accurately describes what goes on in the real world.

It is rather unfortunate that the term 'model' is used, within statistics at least, to describe all these different types of things. To many outside statistics, after all, a model is necessarily mechanistic - it is a simulacrum of the real thing.

Finally, a cautionary note about what I see as an overemphasis on modelling in data analysis in general. Our aim, when analysing data, is to answer questions. It is these questions, not the model per se, which must be paramount.

5   Intelligent data analysis: The human and the computer

Some of you may recall an important early book on artificial intelligence, Joseph Weizenbaum's Computer Power and Human Reason. That title neatly encapsulates the complementary 'abilities' of humans and computers. In a nutshell, computers are able to perform simple operations very rapidly, without error, without boredom, and time after time after time. In contrast, humans are able to abstract general impressions from complex data. This difference is particularly clearly manifest in strategies for playing the game of chess. Chess playing computers have now achieved world class standard. (Indeed, at the time of writing, the IBM computer Deeper Blue is making a very good showing against Garry Kasparov, the world champion. If a computer can play at this level, how does it compare to the chess ability of those of us at this meeting?) Computers are able to do this by adopting massive searches, examining the effects of all combinations of moves and countermoves that they have the time for. Their achievement is largely due to hardware and software advances permitting greater depth of search. In contrast, humans apply informal heuristics to rapidly review the general strength of positions and reject particular avenues as unlikely to be fruitful without exploring the ramifications of moves in detail. This means that the strategies adopted by computers are sometimes off-putting to their human opponents, since they are so different from the type of strategies they are used to. This, in itself, is indicative of something important. It means that we have a new style of approaches to solving problems. In turn this means we can solve tougher problems. An elegant example of this is the proof, obtained a few years ago, of the four colour map theorem. You will recall that (putting it informally) the human analytic approach succeeded in reducing the problem to some thousands of special cases, which exhausted the possible cases which could arise. Although there were many of these, there were not so many that they could not each be examined by computer. Thus the two types of problem solving strategy had worked in parallel yielding a synergy which allowed problems to be tackled which neither could handle alone. Note that both strategies are needed. A computer set simply to search through all possibilities may not be of great value if the set of such possibilities is infinite. Likewise, the powerful ability of the human to perceive patterns can go too far, even to the extent of perceiving pattern where there is none (think of the Martian canals, for example). In a sense, this has led us into a full circle. Finding patterns is what data analysis is all about.

Of course, I am simplifying. At a fundamental physiological level in the human, a huge amount of detailed computation goes on - some of it analogue, as in the ear. Also, in the computer, such approaches as rule-based architectures seek to apply informal high level strategies. And this is another important point. Computer hardware and software continues to develop at an awesome rate. It was scarcely twenty years ago that the first personal computers began to appear, and think how often one's desktop machine becomes laughably out of date. One of the drives is towards higher levels of processing. One can take an overview of these developments and define the objective of data analysis as being to integrate elementary observations, so that high level statements can be made about general structures. To do this, one begins with the very lowest level elements and combines them into higher level elements. These, in turn, are combined into higher level elements. And so on. A parallel to this is the way electrons, protons, and neutrons combine in various ways to form atoms, atoms combine to form molecules, molecules to form cells, and so on. An example is the way letters are combined to form words, words to form sentences, sentences to form paragraphs, and so on. A key aspect of this structure of levels is that there are emergent properties of the higher levels which do not exist in the lower levels. A sample of observations has a variance, a property not possessed by any individual element of that sample. Or, less prosaically, an atom cannot think, whereas a human can.

Perhaps I should interject a sobering thought here. Computers and software are continuing to develop rapidly. But humans are not. Researchers have been experimenting and developing software which parallels the high level integrative human approach for some time now. Progress has been slow but steady. I leave you to draw your own conclusion.

Coming back to data analysis, a nice illustration of the way that advantage can be gained by combining the complementary strategies of computers and humans is in graphics, especially interactive graphics. Not so long ago, the principle use of graphical tools in data analysis (apart from for communication) was as a quick and dirty mode of analysis - to avoid the tedious and lengthy hand calculations which were the only alternative. Nowadays the graphical probe has become an essential tool, the computer summarising possibly large masses of data into a simple display, to which the natural pattern detection abilities of the human eye can be applied. Thinking back to the comments on progress in hardware, mentioned above, this development is the direct result of the shift from line-printer output to live display screen. It is also worth commenting here that the process is not all one way. Graphical input as well as graphical output occurs, though it has not yet been developed to the same extent as output. Here, perhaps, is a ripe area for some exciting methodological data analytic research. Examples of graphical data input are in strategies for obtaining prior distributions for

Bayesian analysis and in the use of 'sliders' to determine good values for parameters. A more interesting notion here, is the idea of having the computer learn formal methods by watching a human's informal examinations of plots. To take a straightforward example, a machine could learn what to regard as a 'high' value of correlation by studying how a human treated a series of scatterplots. This sort of notion represents a generalisation of the ideas of supervised pattern recognition. It is important because research has shown that humans, even expert humans, are not always very good at articulating their expertise.

I have defined data analysis as the process of extracting information from data. Intelligent data analysis obviously refers to intelligent ways to do this. This suggests that we should examine closely the process of data analysis itself. Just as there are principles by which one chooses models (for example, that they should fit the data well, that they should not be unnecessarily complicated, and so on) so, presumably, there are principles which can be applied to the model fitting process. In fact, there have been some investigations of the process, often going under the term strategy. A strategy for data analysis describes the steps, decisions, and actions which are taken during the process of analysing data to build a model or answer a question. If strategies can be formalised - written down in some way so that they can be studied and compared - then one can identify good strategies. One might then even define a good strategy as being the hallmark of intelligent data analysis.

If strategy tells us how to put together the analytic bricks to build an analytic house, we find that almost all of the published work describing data analytic methodology is focused on low level bricks: how to fit a better regression, how to diagnose when a model is a good fit, what are the properties of a goodness-of-fit criterion, and so on. There is very little work on how to put those bricks together. Examples of published work in the area include [5, 11, 18, 3].

Progress in computer technology means that more and more people have access to powerful data analytic tools. Inevitably this means that a smaller proportion of those who do have access really understand what goes on within those tools. They are, to some extent (and a great extent in many cases) using those tools blind. This is not a criticism. We are all blind to some extent. How many readers of this paper, most of whom will regularly use a computer, can describe the structure and mechanism of the microchips within their machine? How many who drive a car can nowadays repair it? How many can describe the theory behind the particular type of mobile phone they happen to have? In general, a lack of understanding of how a tool works does not imply an inability to use that tool to excellent effect. That is, not understanding a tool does not mean it cannot be used as an effective part of an overall strategy. However, for data analytic tools effective use requires good training, or apprenticeship, or at least some other way of acquiring the proper skills. Or else the danger of misuse is great - just as great as the danger of misuse in driving a car, even though the consequences may be less obviously dramatic. I note parenthetically here that the consequences can be even more dramatic. The explosion of the boosters on the Challenger space shuttle is legitimately attributed to an incorrect data analysis, where observation points corresponding to very low temperatures were inappropriately omitted from the analysis. Less dramatic, but equally costly disasters can also occur at a more mundane level: if a data mining exercise applied to a supermarket's electronic point-of-sale data fails to detect an anomaly, feedback mechanisms may not be correctly applied, and loss may result. The same sort of phenomenon, at greater cost, may apply with governmental and official statistics.

From the above it follows that it is becoming increasingly important to develop some formalism for describing high level strategy. The dearth of research papers on this topic in the major journals is a serious cause for concern.

6   Unintelligent data analysis

In the above I have tried to indicate what I believe constitutes intelligent data analysis. Perhaps unintelligent data analysis can be defined by default as any data analysis which is not intelligent. This, of course, gives one plenty of scope since one can depart from an ideal in any number of directions. However, to make things more concrete, I want to draw attention to just one direction, because I think it nicely demonstrates why there is more to 'data analysis' than can be attributed merely to its statistical parent and because it is a topic of my current research.

It is fashionable, in some quarters, to describe Newton's theory of gravitation as wrong, and Einstein's as right. However, the fact is that neither is 'right' in the sense that it will predict precisely what occurs in the real world. There are always other influences, not allowed for in one's model, which interfere with the predictions of any model. One can refine one models ad infinitum, achieving an arbitrary degree of accuracy. Or, in the context of data analysis, one can use extremely sophisticated data analytic techniques in an attempt to squeeze more and more information out of a given sample of data, or to achieve even more predictive power. But - and this is one way in which intelligent data analysis differs from unintelligent data analysis - this should not be done out of context. One measure of genius is an ability to extract the important factors from the mass of material presented to us. That is, one defines a projection from the space spanned by all the quirks and foibles represented in the data to a space spanned by those factors which are 'most important' in describing the structure of the data - and a genius is someone who is good at this. This is easier for situations where there is an objective gold standard which has to be predicted - where one is trying to formulate a predictive rule - than it is for situations where one is simply trying to formulate effective descriptive/theoretical models. Examination of the developments of scientific theories provide beautiful (of course, retrospective) illustrations of this: look at Kepler's analysis of Tycho Brahe's data, for example, or, for a current case, look at the recent flurry of work on the conceptual basis of quantum mechanics. Achieving this projection is also our first objective in data analysis. Our second objective, to be carried out after the first, is to construct a model which describes the shape of the data in this reduced space. (The model may be very simple - perhaps as simple as saying that group

A has a larger mean than group B.) However, there is no point in overdoing the effort on making a good model in the reduced space if important aspects have been lost in the projection. This is one way in which intelligent data analysis differs from unintelligent data analysis: the solution must be directed towards solving the real problem, not towards solving a mathematical abstraction of the problem which ignores the fundamental uncertainties intrinsic to the real problem.

One kind of unintelligent data analysis, then, is over-refined data analysis: data analysis which squeezes the data to such an extent that the drops of information being extracted refer to an over-idealised problem which has lost contact with the real world. Note that I am not referring to overfitting issues. These have long been explored in the statistical literature and more recently, from a rather different perspective, in the computational learning theory literature. Overfitting describes the use of a model which is too flexible for the design data at hand (perhaps because it has too many terms, or because it is based on too many variables, for example). An overfitted model thus describes the available design data well, but fails to generalise well. And often it is generalisation ('inference') which is the objective. (Not always. Sometimes the aim is to describe data in a simple and convenient way, or to locate curious or anomalous data configurations in a body of data which is of itself of interest. These sorts of situations can often arise in official statistics. Overfitting cannot really arise in these sorts of situations, unless the term is taken to mean that the model is not simple enough for the purposes.) Discussions of overfitting take place in the context of well-defined underlying probability distributions - even though they will be unknown and, in some form or other, the object of the data analysis exercise. In contrast, our 'unintelligent data analysis' arises when a good model (neither under nor over-fitted) is made for a problem which has intrinsic variation or uncertainty. One's data analysis should only be pursued to the limits of one's certainty about what is the problem one is trying to solve. To go further is folly - unintelligent. One might argue that, while this kind of unintelligent data analysis is a waste of effort and resources, beyond that it does no harm. At best it gives the results to more significant figures than the data and problem statement can sustain. But, of course, that in itself is potentially harmful since it gives a false impression of accuracy and optimality. If decisions and actions are based on this, then who knows what the magnitude of the consequences could be. Ideally, such results should be supplemented by measures indicating the range of results which might have been obtained should the uncertainty in the problem have led to different statements of it. I am currently exploring such measures in joint work with Wojtek Krzanowski.

A concrete example might be helpful. It is well-known that objective statistical predictive models are more accurate than human judgmental models in a wide variety of situations. For example, for discussions of such issues in psychology, see [10], in medicine, see [12] (Chapter 5), and in financial credit scoring see [19] and [17]. However, this can only be determined if there is a crisply and objectively defined measure of performance. For example, if classification is the objective, one might determine that one method is superior to another in terms of overall misclassification rate. This is all very well but it hinges on the appropriateness of one's chosen measure of performance. If we replace misclassification rate by a more general loss measure, in which the different kinds of misclassification are weighted differently, then perhaps the order of superiority of the methods will change. If there is uncertainty about exactly what it is one is trying to do, then not only will there be uncertainty about which method is best, but this could have adverse consequences on, for example, those patients who are ill but are now classified as well. Such uncertainty should be reflected in and reported by one's data analysis, and not obscured by unintelligent analysis.

These sorts of dangers have become more important as time has passed and computer software has become more powerful. It is all too easy for researchers, no doubt extremely knowledgeable and skilful within their own discipline, to apply very sophisticated data analytic techniques in which they have no expertise whatsoever. The thrust in the development of data analytic software has, quite properly, been in this direction of making software easier to use. I once hoped to see a matching concern, in which software was written which contained not only data analytic expertise, but also problem specific expertise, so that users would be protected from rushing in and making silly mistakes. I now feel that it will be a while before we are able to do this. It might have to wait until we have developed true artificial intelligence - not the idiot savant intelligence of Deeper Blue sort of programs - which has an understanding of the world around us, so that it can make decisions about what sorts of questions it is sensible to ask.

7   Conclusions

I opened by saying that data analysis was a merger of disciplines, and that statistics was but one of them. This was deliberately provocative but is nevertheless true. There are many people working in different areas of data analysis who would not call themselves statisticians. For example, those who work on neural networks in engineering laboratories or on expert systems in AI laboratories for example. The key point is that modern data analysis requires skills which go beyond the classical boundaries. To analyse data effectively - intelligently - skills from a variety of disciplines are called upon. Methodology for data analysis is being developed in a wide variety of university departments and industrial research laboratories and this needs to be recognised and taken advantage of in order to gain full benefit from the powerful techniques which are now available. Data analysis is just as much a child of computer science as it is of statistics.

Having said this, one can learn from its parentage. One can learn, for example, that to develop useful data analytic tools it is necessary to get involved with real problems. It is from real problems that the important solutions emerge. It is by solving real problems that one changes the world. I think this is a fitting note on which to conclude this paper. I see too many researchers developing methods in the abstract - methods which might be ideal should data of the particular kind they have postulated ever arise. This is not only unnecessary, but it is also a waste - of their time, skills, expertise, and knowledge. There are many real problems out there requiring data analytic solution. If you want to develop new techniques you will soon find that there is need for this arising from the real problems. The alternative, of developing new techniques in a vacuum, unmotivated by real problems, may be fun, but is ultimately pointless.

References

1. Bailey K.R. (1994) Generalizing the results of randomized clinical trials. Controlled Clinical Trials, 15, 15-23.
2. Boden M.A. (1977) Artificial Intelligence and Natural Man. Harvester Press.
3. Brodley C.E. and Smyth P. (1997) Applying classification algorithms in practice. Statistics and Computing, 7, 45-56.
4. Cowan C.D. (1994) Intercept studies, clinical trials, and cluster experiments: to whom can we extrapolate? Controlled Clinical Trials, 15, 24-29.
5. Cox D.R. and Snell E.J. (1981) Applied Statistics: principles and examples. London: Chapman and Hall.
6. Darius P., Boucueau M., de Greef P., de Faber E., and Froeschl K. (1993) Modelling metadata. Statistical Journal of the United Nations - ECE, 10, 171-9.
7. Diday E. (1994) Towards a statistic of metadata for knowledge analysis. In AI and Computer Power: the Impact on Statistics. ed. D.J.Hand, Chapman and Hall, 23-35.
8. Froeschl K.A. (1996) A metadata approach to statistical query processing. Statistics and Computing, 6, 11-29.
9. Gray J. et al. (1997) Data cube: a relational aggregation operator generalizing group-by, cross-tab, and sub-totals. Data Mining&Knowledge Discovery, 1, 29-53.
10. Grove W.M. and Meehl P.E. (1996) Comparative efficiency of informal (subjective, impressionistic) and formal (mechanical, algorithmic) prediction procedures: the clinical-statistical controversy. Psychology, Public Policy, and Law, 2, 293-323.
11. Hand D.J. (1984) Patterns in statistical strategy. In Artificial Intelligence and Statistics, ed. W.A.Gale. Reading, Massachusetts: Addison-Wesley, 355-387.
12. Hand D.J. (1985) Artificial Intelligence and Psychiatry. Cambridge: Cambridge University Press.
13. Hand D.J., (1996) Classification and computers: shifting the focus. In COMPSTAT - Proceedings in Computational Statistics, Physica-Verlag. p77-88, 1996.
14. Hand D.J. (1996) Statistics and the theory of measurement (with discussion). Journal of the Royal Statistical Society, Series A, 159, 445-492.
15. Hand D.J. (1997) Construction and Assessment of Classification Rules. Wiley.
16. Hand D.J. and Henley W.E. (1997) Statistical classification methods in consumer credit scoring: a review. Journal of the Royal Statistical Society, Series A, 160.
17. Hand D.J. and Jacka S. (eds.) (1997) Statistics in Finance. Edward Arnold.
18. Pregibon D. (1984) A DIY guide to statistical strategy. In Artificial Intelligence and Statistics, ed. W.A.Gale. Reading, Massachusetts: Addison-Wesley, 389-399.
19. Rosenberg E. and Gleit A. (1994) Quantitative methods in credit management: a survey. Operations Research, 42, 589-613.
20. Thomas L.C., Crook J.N., and Edelman D.B. (eds) (1992) Credit scoring and credit control. Oxford: Clarendon Press.
21. Weizenbaum J. (1976) Computer Power and Human Reason. San Francisco: W.H.Freeman and Co.

Section I:

Exploratory Data Analysis, Preprocessing and Tools

Decomposition of Heterogeneous Classification Problems

Chidanand Apte, Se June Hong, Jonathan R. M. Hosking,
Jorge Lepre, Edwin P. D. Pednault, and Barry K. Rosen

IBM Research Division, T. J. Watson Research Center,
P.O. Box 218, Yorktown Heights, NY 10598, U.S.A.

Abstract. In some classification problems the feature space is heterogeneous in that the best features on which to base the classification are different in different parts of the feature space. In some other problems the classes can be divided into subsets such that distinguishing one subset of classes from another and classifying examples within such subsets require very different decision rules, involving different sets of features. In such heterogeneous problems, many modeling techniques (including decision trees, rules, and neural networks) evaluate the performance of alternative decision rules by averaging over the entire problem space, and are prone to generating a model that is suboptimal in any of the regions or subproblems. Better overall models can be obtained by splitting the problem appropriately and modeling each subproblem separately.
This paper presents a new measure to determine the degree of dissimilarity between the decision surfaces of two given problems, and suggests a way to search for a strategic splitting of the feature space that identifies regions with different characteristics. We illustrate the concept using a multiplexor problem.

1 Introduction

Many classification problems contain a mixture of rather dissimilar subproblems. This can occur when the features important for classification are different in distinct regions in the feature space, or when the decision rules that distinguish one group of classes from another are very different from those that separate the classes within each group. We use the terms *feature-space heterogeneity* and *class heterogeneity* to denote these two situations.

Feature-space heterogeneity is exhibited, for example, by some medical diagnosis problems: diagnosis may require quite different models for different sexes, for the relevant sets of symptoms can be very different. Here different regions in the feature space, split along the sex feature, exhibit distinct decision characteristics. As an instance of class heterogeneity, consider the problem of modeling severity levels of automobile insurance accounts. The payout amount is modeled as one of three severity classes: high, medium and low. However, the vast majority of examples have no claims (considered as having "no accident"). Assigning

X. Liu, P. Cohen, M. Berthold (Eds.): "Advances in Intelligent Data Analysis" (IDA–97)
LNCS 1280, pp. 17–28, 1997.   © Springer–Verlag Berlin Heidelberg 1997

these to a fourth class, "none," and generating one classification model for all four classes may be unwise. Whether a driver has an accident in a given time period may depend principally on the kind of driver and the driving area, while the severity given that an accident happened may depend more on the kind of automobile, its net value and the cost of its parts. Here one model would be appropriate for distinguishing "no accident" from "accident," but a quite different model would be required to classify the severity given that an accident occurred.

A logical first step towards solving a classification problem is to search for evidence of heterogeneity, and, if any is found, to try to decompose the problem into its constituent subproblems. This decomposition approach is potentially highly beneficial, because most widely used modeling techniques (including decision trees, rules, and neural networks) rely on measures that are computed over all the features and all the examples at hand, and are inevitably diffused by an averaging effect over the entire problem.

How then does one detect and separate out heterogeneity in a given classification problem? We answer this question in the two following sections. In Section 2 we present a new measure that reflects the degree of dissimilarity between the decision boundaries for two given subproblems. It is based on measures of feature merit, which we define in Section 3. In Section 4 we propose a tree strategy for identifying suitable regions in feature space. Once the problem is properly separated, each subproblem can be tackled by its own most appropriate model. This may even involve different model families: e.g., one subproblem may be modeled by a neural network and the other by a Bayesian classifier.

The well-known multiplexor function is an ideal example of feature-space heterogeneity. Under different control variable settings the output depends on entirely different signal input variables. From the classification point of view, the decision surfaces in each of the regions represented by the control variable settings are orthogonal to each other. Section 5 applies our methods to several classification problems based on the multiplexor function. For comparison, Section 6 illustrates the performance of two standard methods, C4.5 and CART, on those problems. Conclusions and some further discussion are given in Section 7.

2   A measure of dissimilarity

Given a classification problem and a proposed decomposition of it into two subproblems, we wish to find how dissimilar the decision surfaces are in the two subproblems. Differences in class probability distribution, or class probability profile, may give an indirect indication. We argue that a more direct indication comes from comparing the profiles of importance of the features of the two problems. When many features display widely different importance in the two subproblems the decision surfaces of the two subproblems must be quite distinct, although the converse may not hold in general.

To determine the degree to which the importance of features varies between two subproblems, we make use of the angle between the two vectors of the feature importance values in each subproblem. Let the importance measures of one subproblem be denoted by the vector of merits, $M_{a1}, M_{a2}, ..., M_{af}$, and of the other by $M_{b1}, M_{b2}, ..., M_{bf}$, where f is the number of features and M denotes a measure of feature importance (discussed further below). The angle formed by the two vectors in the f-dimensional space is

$$\frac{2}{\pi} \arccos \left(\frac{\sum_{i=1}^{f} M_{ai} M_{bi}}{\left(\sum_{i=1}^{f} M_{ai}^2\right)^{1/2} \left(\sum_{i=1}^{f} M_{bi}^2\right)^{1/2}} \right), \tag{1}$$

which we will call the Importance Profile Angle (IPA); strictly, (1) defines a normalized IPA that takes values between 0 and 1. Linear scaling of the vector does not change the angle, and hence the importance profile depends only on the relative magnitudes of the features' importance. We note also that as a result of splitting the universe along some features values, some features may become constant within each subproblem, yielding zero importance in any measure. The angle between two vectors that have zeroes in identical positions is the same as the angle between the two vectors with these zero elements deleted; thus the IPA is unaffected by deletion of features whose values are constant within each subproblem.

3   Feature merit measures

The importance or the merit of a feature is usually measured as the difference between an impurity measure of the classes and the resulting impurity of the classes given the fact that the feature value is known. Two of the most frequently used impurity measures are the Gini index, used in CART [1], and entropy, used in C4.5 [2]. We now define these impurity measures and feature merit measures.

We denote by $I(p_1, \ldots, p_m)$ the impurity of a set of probabilities $p_1, p_2, \ldots, p_m$, $p_1 + p_2 + \ldots + p_m = 1$. An impurity measure should satisfy $I(p_1, \ldots, p_m) = 0$ whenever $p_i = 1$ for some i, and should be maximized when $p_i = 1/m$ for all i. Reasonable forms for $I(p_1, \ldots, p_m)$ include the Gini measure

$$G(p_1, \ldots, p_m) = 1 - \sum_i p_i^2$$

and the entropy measure

$$H(p_1, \ldots, p_m) = -\sum_i p_i \log p_i.$$

In a categorical classification problem, there are N examples over which the dependent variable ("class") takes C distinct values with frequencies b_i, $i = 1, \ldots, C$. We consider an explanatory variable ("feature") that takes F distinct values with frequencies a_j, $j = 1, \ldots, F$. The conjunction of class i and feature value j occurs with frequency x_{ij}.

The impurity of the class variable is denoted by

$$I(C) = I(b_1/N, \ldots, b_C/N).$$

For those examples corresponding to a particular value of the feature, the impurity is denoted by

$$I(C|F = j) = I(x_{1j}/a_j, \ldots, x_{Cj}/a_j) \,.$$

When feature value j occurs with probability p_j the average impurity of the class variable, given the feature, is

$$I(C|F) = \sum_{j=1}^{F} p_j \, I(x_{1j}/a_j, \ldots, x_{Cj}/a_j) \tag{2}$$

For the set of N examples the proportion of occurrences of feature value j is a_j/n, and using this value as p_j in (2) yields

$$I(C|F) = \sum_{j=1}^{F} \frac{a_j}{N} \, I(x_{1j}/a_j, \ldots, x_{Cj}/a_j) \,. \tag{3}$$

This is the impurity that remains in the class variable after the information present in the feature variable has been used.

When using Gini's measure of impurity, (3) becomes

$$G(C|F) = \sum_{j=1}^{F} \frac{a_j}{N} \left\{ 1 - \sum_{i=1}^{C} \left(\frac{x_{ij}}{a_j} \right)^2 \right\}$$

$$= 1 - \frac{1}{N} \sum_{j=1}^{F} \sum_{i=1}^{C} \frac{x_{ij}^2}{a_j} \,. \tag{4}$$

The best feature is the one that achieves the lowest value of the Gini index (4), or, equivalently, the highest value of the "Gini gain" $G(C) - G(C|F)$.

When using the entropy measure of impurity, (3) becomes

$$H(C|F) = \sum_{j=1}^{F} \frac{a_j}{N} \sum_{i=1}^{C} -\frac{x_{ij}}{a_j} \log \frac{x_{ij}}{a_j}$$

$$= \frac{1}{N} \left(\sum_{j=1}^{F} a_j \log a_j - \sum_{j=1}^{F} \sum_{i=1}^{C} x_{ij} \log x_{ij} \right) \,. \tag{5}$$

The best feature is the one that achieves the lowest value of the entropy (5), or, equivalently, the highest value of the "Information gain" $H(C) - H(C|F)$.

Feature merit measures based on Gini and entropy measures are greedy, or myopic [3], in that they reflect the correlation between a feature and the class, disregarding other features. To overcome this problem, new merit measures for features have recently been developed. They take into account the presence of other features that may interact in imparting information about the class. They include the RELIEF measure developed by Kira and Rendell [4] and its follow-on RELIEFF developed by Kononenko et al. [3], and the "contextual merit" (CM) developed by Hong [5]. These new measures require more computation than the two myopic varieties, but are more robust in general.

We now describe CM in more detail. CM assigns merit to a feature taking into account the degree to which other features are capable of discriminating between the same examples as the given feature. As an extreme instance, if two examples in different classes differ in only one feature, then that feature is particularly valuable—if it were dropped from the set of features, there would be no way of distinguishing the examples—and is assigned additional merit.

To define contextual merit, first define the distance $d_{rs}^{(k)}$ between the values z_{kr} and z_{ks} taken by feature k for examples r and s. If the feature is symbolic, taking only a discrete set of values, define

$$d_{rs}^{(k)} = \begin{cases} 0 & \text{if } z_{kr} = z_{ks}, \\ 1 & \text{otherwise.} \end{cases}$$

If the feature is numeric, set a threshold t_k—Hong [5] recommends that it be one-half the range of values of feature k—and define

$$d_{rs}^{(k)} = \min(|z_{kr} - z_{ks}|/t_k, 1).$$

The distance between examples r and s is now defined to be

$$D_{rs} = \sum_{k=1}^{N_f} d_{rs}^{(k)},$$

N_f being the number of features. The merit of feature f is now defined as

$$M_f = \sum_{r=1}^{N} \sum_{s \in \bar{C}(r)} w_{rs}^{(f)} d_{rs}^{(f)},$$

where N is the number of examples, $\bar{C}(r)$ is the set of examples not in the same class as example r, and $w_{rs}^{(f)}$ is a weight function chosen so that examples that are close together, i.e. that differ in only a few of their features, are given greater influence in determining each feature's merit. Hong [5] used weights $w_{rs}^{(f)} = 1/D_{rs}^2$ if s is one of the k nearest neighbors to r, in terms of D_{rs}, in the set $\bar{C}(r)$, and $w_{rs}^{(f)} = 0$ otherwise; the number of nearest neighbors used by Hong was the logarithm (to base 2) of the number of examples in the set $\bar{C}(r)$.

An IPA can be defined for any measure of feature importance. Depending on the original merit function used, we speak of Gini gain IPA, information gain IPA, CM IPA, etc.

4   Strategic splitting of the feature space

Now that we have an effective means to tell whether two classification subproblems are similar, what remains is to generate candidate split regions in the feature space so that their dissimilarity can be measured. This is a difficult problem since one naturally wants to avoid exhaustive enumeration, and yet a reasonable set of candidates is desired. We suggest using the tree paradigm that splits along the values of a chosen "best" feature. We first consider cases in which all features are categorical and have binary values; we then discuss ways to generalize the idea to multiple-valued and numerical features.

To use the IPA statistic in practice, for each feature F one divides the feature space into two subregions corresponding to the different values of F and computes the IPA value using (1). If the largest of these values exceeds a suitable threshold, this is an indication of heterogeneity: the feature space, and hence the training set, is split according to the values of the feature that gives the largest IPA value and analysis proceeds separately on the two subproblems thereby defined.

The splitting process may proceed recursively until some stopping criterion is met. The computation at a node is analogous to a one-level look-ahead scheme in conventional tree building. We are still investigating what the practical stopping criterion should be. Since the goal is to generate subproblems to be modeled separately, this strategy tree would not be deep in general. In fact, one should guard against fragmentation which renders the distinct regions too small to generate a reliable model from: this is a perennial problem facing all tree-based modeling techniques. The stopping criterion should probably involve a threshold on the IPA value and a threshold on the number of examples in the subregions.

IPA is a useful measure so long as the problem can be decomposed into subproblems that are of lower effective dimension in that their decision surfaces involve fewer features than the original problem. At some point, reduction in effective dimension is no longer possible, and the objective changes to finding the best models within each lower-dimensional subspace. At this point the choice of which feature to split on should be based on other criteria such as entropy gain or contextual merit. The stopping criterion for the IPA method therefore needs to be sensitive to lack of reduction in effective dimension.

When a feature has multiple values, an IPA can be computed for each pair of the values, by considering splits between these two values and ignoring all examples in which the feature takes any other value. Values can then be merged recursively by grouping together the pair of values with the lowest IPA and re-generating the importance measure vector for the just-grouped values versus the rest, until the angles between each group are "acceptably" large. This approach can generate more than two groups—an added flexibility. The smallest of the angles between the final groups would be used as the IPA of the feature when deciding which feature to split on.

For numerical features, the split candidates can be generated from an initial discretization by any of the methods described in [6] or by the method of [5], based on contextual merit. However, this method may not be completely satisfactory: discretization methods use global analysis, which may not be a good approach when heterogeneity is present. Further study and experiments are needed before these ideas can be routinely used in practice.

5   Lessons from the multiplexor problem

As was mentioned above, the multiplexor is an ideal example of feature space heterogeneity. Our basic example is the 4-way multiplexor, which has two binary control inputs, X_1 and X_2, four signal inputs, Y_1 through Y_4, a number of irrelevant inputs—we use two irrelevant inputs, R_1 and R_2—and an output, or class, C, defined by

$$
C = \begin{cases}
Y_1 & \text{if } X_1 = 0 \ \& \ X_2 = 0, \\
Y_2 & \text{if } X_1 = 0 \ \& \ X_2 = 1, \\
Y_3 & \text{if } X_1 = 1 \ \& \ X_2 = 0, \\
Y_4 & \text{if } X_1 = 1 \ \& \ X_2 = 1.
\end{cases}
\tag{6}
$$

To be more realistic, we have devised the following variations of the 4-way multiplexor as classification problems (we have also carried out similar experiments on variations of 2-way and 8-way multiplexors and confirmed the similar behavior of IPA measures). They are progressively more "difficult" and approach more closely what some real problems might be.

Case 1: 4-way multiplexor with two random inputs and even distribution of feature values. Generate a 1000×8 random binary array, the values 0 and 1 being equally likely to occur. Generate the class variable according to (6).

Case 2: 4-way multiplexor with two random inputs, even distribution of feature values, and 5% noise. Take Case 1 and flip 50 randomly chosen class bits, thereby injecting 5% noise.

Case 3: 4-way multiplexor with two random inputs and uneven distribution of feature values. Generate a 1000×8 random binary array, the values 0 and 1 occurring with probabilities 2/3 and 1/3 respectively. Generate the class variable according to (6), except that the assignment $C = Y_2$ is replaced by $C = 1 - Y_2$ and $C = Y_4$ is replaced by $C = 1 - Y_4$; i.e., in these two cases the signal value is flipped when it is copied to the class.

The uneven distribution of 0 and 1 feature values is intended to simulate more realistic practical situations. Flipping some of the class values is done to achieve an even distribution of 0s and 1s in the class values; it does not change the multiplexor nature of the problem.

Case 4: 4-way multiplexor with two random inputs, uneven distribution of feature values, and 5% noise. Take Case 3 and flip 50 randomly chosen class bits, thereby injecting 5% noise.

The desirable decision tree for all of these cases should have exactly three levels of decision nodes, the first two levels consisting of the two control features in either order, and then the four signal features in the last level. There should be exactly seven decision nodes. These trees cannot be further pruned.

Tree generation should start by splitting on one of the control features. This splits the problem into two subproblems each of which is a 2-way multiplexor problem; thus at the second level of the tree the other control feature should be selected for splitting. Once the top two levels are constructed correctly, any

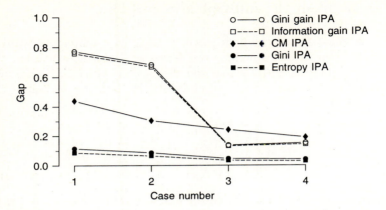

Fig. 1. Plots of the "Gap" values from Table 1 for each of the five IPA measures.

reasonable tree-generation method should be able to complete the third level nodes and stop.

In Table 1 we assess the ability of different feature importance measures and IPAs to identify which feature should be split on at the top-level node. Tabulated gain and IPA values are rounded to 3 decimal places.

In a multiplexor problem, the most effective split into subproblems is a split on a value of a control feature. A good measure of feature importance should therefore attach high importance to the control features and lower importance to signal and irrelevant features. Each "Gap" value in Table 1 is the difference between the smallest IPA value for either of the control features and the largest IPA value of any of the other features. A large "Gap" value means that the IPA measure performs well and indicates a clear preference for initially splitting on a control feature. The "Gap" values for each IPA are plotted in Figure 1.

Here are some observations from these experiments.

1. The myopic gain measures, Gini gain and information gain, are unable to identify appropriate variables to split on. In all four cases they choose a signal variable rather than one of the control variables for the first split.
2. The IPAs based on myopic gain measures, Gini gain IPA and information gain IPA, show good discrimination between control and signal features (a high "Gap" value) in Cases 1 and 2, but their performance is much poorer in the more difficult Cases 3 and 4.
3. The IPAs based on impurity measures, Gini IPA and Entropy IPA, are able to distinguish the control features. However, their resolution is much poorer than those of IPAs based on gain measures: the size of the "Gap" is rather small in each of Cases 1–4.
4. CM IPA shows the most consistent ability to discriminate between control and signal variable across all four Cases.

Table 1. Results for multiplexor classification problems.

Tabulated values are merit and IPA measures for each of the eight feature variables. Numbers of 0s and 1s in the 1000 examples are also given, both for the feature variables and for the class variable C. Starred values are the smallest of the IPA values for the control inputs and the largest of the IPA values for the signal and irrelevant inputs. "Gap" is the difference between the two starred values in the row.

		Feature								C	Gap
		X_1	X_2	Y_1	Y_2	Y_3	Y_4	R_1	R_2		
Case 1	Number of 0s	510	510	510	508	505	502	488	473	507	
	Number of 1s	490	490	490	492	495	498	512	527	493	
	Gini gain	.000	.000	.052	.029	.032	.027	.000	.000		
	Gini gain IPA	.986*	.993	.220*	.080	.125	.189	.071	.142		.766
	Gini IPA	.129*	.139	.013	.006	.012	.014*	.004	.008		.115
	Information gain	.000	.000	.076	.043	.045	.040	.000	.000		
	Information gain IPA	.986*	.993	.224*	.080	.128	.192	.072	.143		.762
	Entropy IPA	.096*	.103	.010	.005	.009	.011*	.003	.006		.085
	CM	1349	1406	1051	959	954	933	471	483		
	CM IPA	.496*	.509	.035	.057*	.037	.032	.045	.049		.439
Case 2	Number of 0s	510	510	510	508	505	502	488	473	511	
	Number of 1s	490	490	490	492	495	498	512	527	489	
	Gini gain	.001	.000	.046	.027	.024	.021	.000	.001		
	Gini gain IPA	.976*	.991	.299*	.205	.217	.274	.172	.244		.677
	Gini IPA	.102*	.115	.014	.011	.016*	.016	.008	.012		.086
	Information gain	.001	.000	.067	.038	.035	.030	.000	.000		
	Information gain IPA	.976*	.992	.303*	.206	.220	.277	.173	.245		.673
	Entropy IPA	.076*	.086	.011	.009	.013*	.012	.006	.009		.063
	CM	1306	1371	1103	1042	1005	959	602	607		
	CM IPA	.371*	.407	.047	.055	.019	.025	.067*	.031		.304
Case 3	Number of 0s	656	656	660	654	663	646	659	681	473	
	Number of 1s	344	344	340	346	337	354	341	319	527	
	Gini gain	.002	.057	.081	.024	.029	.011	.002	.002		
	Gini gain IPA	.947*	.997	.815*	.664	.649	.604	.150	.233		.132
	Gini IPA	.144*	.174	.098*	.076	.064	.057	.014	.020		.046
	Information gain	.000	.082	.110	.033	.040	.014	.000	.000		
	Information gain IPA	.954*	.998	.817*	.675	.653	.615	.150	.236		.137
	Entropy IPA	.113*	.139	.080*	.061	.050	.044	.010	.015		.033
	CM	1389	1685	1452	1017	874	633	439	435		
	CM IPA	.509*	.601	.267	.268*	.141	.179	.087	.060		.241
Case 4	Number of 0s	656	656	660	654	663	646	659	681	469	
	Number of 1s	344	344	340	346	337	354	341	319	531	
	Gini gain	.002	.045	.069	.016	.020	.010	.002	.002		
	Gini gain IPA	.966*	.992	.819*	.617	.598	.551	.136	.271		.147
	Gini IPA	.120*	.138	.079*	.059	.048	.041	.010	.018		.041
	Information gain	.000	.063	.099	.021	.026	.012	.000	.000		
	Information gain IPA	.969*	.993	.821*	.622	.599	.555	.137	.274		.148
	Entropy IPA	.092*	.106	.063*	.045	.036	.031	.007	.013		.029
	CM	1397	1619	1429	1052	934	716	585	580		
	CM IPA	.410*	.488	.218	.220*	.135	.115	.085	.053		.190

Table 2. Summary of the trees generated by C4.5 and CART for Cases 1–4.

Case	Method	Levels	Nodes	Top 3 levels of tree						
1	C4.5	5	43	Y_1	X_2	Y_3	Y_3	Y_4	X_1	X_2
2	C4.5	6	50	Y_1	Y_2	Y_3	X_1	Y_4	X_1	X_2
3	C4.5	5	35	Y_1	X_2	X_1	X_1	Y_2	X_2	Y_3
4	C4.5	6	35	Y_1	X_2	X_1	X_1	Y_2	Y_2	Y_3
1	CART	6	29	Y_1	Y_4	Y_3	Y_3	X_2	X_1	X_2
2	CART	6	25	Y_1	Y_2	Y_3	X_1	Y_4	X_1	X_2
3	CART	5	18	Y_1	X_1	X_2	X_2	Y_3	Y_2	X_1
4	CART	5	15	Y_1	X_1	X_2	Y_2	Y_3	Y_2	X_1
1–4	Best	3	7	X_1	X_2	X_2	Y_1	Y_2	Y_3	Y_4

5. The CM-based measures are the most robust in the presence of heterogeneity. In all four cases CM and CM IPA choose the control variable X_2 for the first split. This split separates the problem into two subproblems, each of which is a 2-way multiplexor with control variable X_1, and in each subproblem CM and CM IPA identify X_1 as having the highest merit. However, CM IPA performs better than CM: from Table 1 it can be seen that in all four Cases CM IPA gives both control variables higher merit than any of the other variables, whereas in Cases 3 and 4 CM does not do so.

These results are in agreement with theoretical considerations. We expect CM IPA to outperform Gini gain IPA and information gain IPA, because the vector of CMs provides an indication of the effective dimensionality of a problem (the number of features that contribute to classification) by estimating the importance of each variable to solving the problem. The information-gain and Gini-gain vectors, on the other hand, measure the one-step improvements in the degrees of fit to the data if splits are performed on the corresponding variables. One-step improvements in the degrees of fit are only indirectly related to the dimensionality of a problem. Moreover, because these measures are myopic, they may miss interactions between features that are important to efficient classification.

6   Comparison with C4.5 and CART

To illustrate the performance of standard decision tree algorithms in multiplexor problems, we applied C4.5 and CART, with default parameters. All pruned trees had either 5 or 6 levels with the number of decision nodes ranging from 15 to 50. None of the trees achieved the optimum misclassification rate (0 in Cases 1 and 3, 5% in Cases 2 and 4). Table 2 gives the number of levels, number of nodes, and the top three levels for the trees generated by C4.5 and CART, and also for the optimal ("Best") tree. The tree is shown by scanning each level left to right. It is noteworthy that neither C4.5 nor CART succeeds in identifying the control variables as being the best to split on: in each case the first split is on

Table 3. CART's measure of importance of features for Cases 1–4.

Case	X_1	X_2	Y_1	Y_2	Y_3	Y_4	R_1	R_2
1	100	75	32	79	36	53	2	2
2	100	92	33	45	46	70	7	4
3	65	100	57	57	70	39	0	1
4	58	100	56	33	61	26	0	1

Y_1, a signal variable, and only nine of the 32 paths to third-level nodes contain splits on both X_1 and X_2.

Curiously, the "measure of importance" values output by the CART program and defined in [1, p. 147] do consistently give the highest importance to one of the control features. These values are given in Table 3. Although CART has not generated optimal trees for these multiplexor problems, it has at least given some indication that its trees are not optimal.

7   Concluding remarks

Standard measures of the importance of features in classification problems can be misleading when the problem is really a mixture of distinct subproblems with different decision characteristics. To address this problem we define importance profile angles (IPAs), which directly measure the extent to which the relation between the class and feature variables is different in different parts of feature space. IPAs may be based on any of the conventional measures of the importance or merit of features. Though the details of the implementation of IPAs remain open, particularly for numerical and multiple-valued features, the initial results are promising. In several variants of the multiplexor problem, IPAs identified the features that led to the most efficient decomposition of the classification problem into subproblems, whereas measures of feature merit based on conventional impurity measures were unsuccessful. The IPA derived from contextual merit had the most consistent overall performance.

Another approach to decomposition is the twoing strategy of CART [1, pp. 104ff.]. In twoing, the decrease in impurity (originally, Gini gain) is computed for each feature and for each two-way grouping of the classes into (C_1^*, C_2^*) during the tree generation at each node. We quote from CART [1, p. 105]:

> The idea is then, at every node, to select that conglomeration of classes into two superclasses so that considered as a two-class problem, the greatest decrease in node impurity is realized.
>
> This approach to the problem has one significant advantage: It gives "strategic" splits and informs the user of the class similarities. At each node, it sorts the classes into those two groups which in some sense are most dissimilar and outputs to the user the optimal grouping C_1^*, C_2^* as well as the best split S^*.
>
> The word strategic is used in the sense that near the top of the tree, this criterion attempts to group together large numbers of classes that are similar in some characteristic.

Twoing can be effective when the groups of classes that it forms define subproblems that are more easily solved than the original problem. However, it does not address the question of whether different sets of features are needed to classify examples within the subgroups of classes, and it is this question that is the essence of class heterogeneity as we have defined it.

A measure of feature importance introduced by Fayyad and Irani [7] calculates the angle between the class probability profiles of the descendents of a two-way split of examples according to the candidate feature values. This measure is very much related to the twoing idea in that it attempts to identify the feature that best separates the data into two groups of predominately different classes. In this case, however, the emphasis is on separating the class statistics instead of the classes themselves. This subtle difference makes the Fayyad-Irani measure less costly to compute.

Like the IPA measure, the Fayyad-Irani measure is expressed as the angle between two vectors. However, the meanings of the vectors and, hence, the measures are quite different. In Fayyad and Irani's case, the class statistics are being separated and termination occurs when these statistics cannot be further refined. In the case of IPA, the importance of the features are being separated and termination occurs when the classification problem cannot be further decomposed.

In this paper we have sidestepped the issue of normalizing importance measures to compensate for the biases that particular measures have for certain features—for instance, those that take a large number of distinct values. The C4.5 method normalizes the information gain by the entropy of the feature. Hong et al. [8] have defined an alternative normalization scheme that also applies to the contextual merit. Feature merits should be properly normalized in practice. However, the ideas in this paper have been illustrated on examples in which all features have similar distributions of their values, so that the issue of normalization becomes moot.

References

1. L. Breiman, J. H. Friedman, R. A. Olshen, and C. J. Stone: Classification and regression trees. Monterey, Calif.: Wadsworth (1984).
2. J. R. Quinlan.: C4.5: programs for machine learning. Morgan Kaufmann (1993).
3. I. Kononenko, E. Simec, and M. Robnik: Overcoming the myopia of inductive learning algorithms with RELIEFF. Applied Intelligence, 7, 39–55 (1997).
4. K. Kira and L. Rendell: The feature selection problem: traditional methods and a new algorithm. Proceedings of AAAI–92 (1992), 129–134.
5. S. J. Hong: Use of contextual information for feature ranking and discretization. IEEE Trans. Knowl. Data Eng., 9, to appear (1997).
6. J. Dougherty, R. Kohavi, and M. Sahami: Supervised and unsupervised discretization of continuous features. Proceedings of ML–95 (1995).
7. U. Fayyad and K. Irani: The attribute selection problem in decision tree generation. Proceedings of AAAI–92 (1992), 104–110.
8. S. J. Hong, J. R. M. Hosking, and S. Winograd: Use of randomization to normalize feature merits. Information, Statistics and Induction in Science, eds. D. L. Dowe, K. B. Korb and J. J. Oliver. Singapore: World Scientific (1996), 10–19.

Managing Dialogue in a Statistical Expert Assistant with a Cluster-Based User Model*

Michael William Muller

Graduate School of Public and Development Management, University of the Witwatersrand, Johannesburg, P.O. Box 601, Wits 2050, South Africa.

Abstract. This paper is concerned with the management of dialogue between the user and an expert system that assists with the statistical analysis of preference data. The main focus of the paper is the development of a cluster-based user model that modifies the frequency, style and content of the messages produced by the expert system. Clusters in the population of potential users are identified by cluster analysis of the results of a preliminary survey dealing with knowledge of technical terms and style of interaction with a computer.

1 Introduction

1.1 Aim of the Research

As computer-based systems have become more sophisticated, the nature of the interaction between the user and the system has become increasingly complex. This complexity has led to problems that have seriously hampered the usefulness of these systems. One approach to achieving a more responsive system, which is the subject of this study, is to attempt to build an abstract model of the user which can be used by the system to structure and present information in a manner appropriate for each individual. The research described here is a case study in the use of a simple type of user modelling for a specific application. The aim of the study is to find out if a useful user-modelling system can be constructed for the given application using this simple approach.

Most of the attempts to build user models have been in the area of intelligent tutoring systems. There are critics of these systems such as Ridgway [13] who doubt the intrinsic feasibility of a user-modelling approach. It also appears that many projects in this area have resulted in systems which are costly to produce, difficult to modify and not very effective, although some limited success has been achieved [20]. In an attempt to deal with some of these problems Rich [12] suggests a simple model of stable user characteristics based on stereotypes. The term "stereotype" in every-day speech has a negative connotation that implies

---

* This research formed part of a research report submitted to the Faculty of Science of the University of the Witwatersrand, Johannesburg in partial fulfilment of the requirements for the degree of Master of Science.

X. Liu, P. Cohen, M. Berthold (Eds.): "Advances in Intelligent Data Analysis" (IDA–97)
LNCS 1280, pp. 29–40, 1997. © Springer–Verlag Berlin Heidelberg 1997

that an individual is being unfairly judged according to some supposed characteristic of a group. In user modelling a stereotype refers to a set of empirically based assertions about the way in which attributes co-occur in people. If there is a tendency for people to exhibit a predictable pattern of response with respect to a number of different characteristics then this information can be used to make a computer system more responsive to the user. This enables the system to modify its default behaviour with respect to several correlated attributes as soon as information is obtained about one of them.

The main focus of this paper is the use of a statistical clustering technique to find the stereotypes. Users of the system differ considerably in their technical knowledge and in their familiarity with personal computers. This makes it reasonable to consider a user-modelling approach. The selection of the attributes used to form the stereotypes is based on the literature on individual differences in cognitive style, and the dialogue structure of the system takes into account research on the nature of communication between humans and computers.

1.2   The Context: Multidimensional Unfolding Analysis

Multidimensional unfolding is a type of descriptive multivariate statistical analysis related to multidimensional scaling. Descriptions of multidimensional scaling can be found in many books on multivariate statistical analysis. The book [14] is devoted entirely to multidimensional scaling and unfolding and a short introduction to multidimensional unfolding alone may be found in [9].

Gale, Hand and Kelly in a review of statistical applications of AI [6] group the applications into several categories. The present system would fall into the category "Application of data analytic techniques". In the terminology of the review the system is a prototype rather than a commercial system. It follows the pattern of development shown in the systems reviewed, being a front-end for an existing statistical program, incorporating a rule-based expert system and providing commentary and graphical feedback to the user on the progress of the analysis. The authors note that pitching the interface at a level appropriate to the user's understanding is a key aspect of system effectiveness. The present system goes further than previous systems in this category by incorporating an explicit user model which can be used to modify the way that the system interacts with a user.

1.3   Research Procedure

Attributes that could be useful in identifying stereotypes were chosen as a result of observations made during preliminary trials and from results reported in the literature. A questionnaire based on a list of attributes compiled from both these sources was used to conduct a survey of potential users of the system. Cluster Analysis is used to identify these stereotypes empirically from the data obtained in the survey. Two separate sets of clusters are considered, one set relating to the technical knowledge needed to perform the preference analyses and a second set relating to the preferred style of interaction with a computer. A user-modelling system based on a two-way classification of users from these two sets of clusters is incorporated in a computer program for assisting users to perform preference analyses.

2   A Brief Review of some Approaches to Modelling Users

2.1   Modelling the User

User modelling is an idea which has gained wide currency in the last decade in the context of the Intelligent Tutoring System or ITS, where this approach is known as student modelling [20], [16]. A student model contains explicit representations of knowledge that can be used to make decisions or solve problems using reasoning processes. Advanced systems use knowledge about the student in conjunction with expertise about teaching methods to tailor the manner in which information is presented as well as the difficulty of the material.

Stereotype modelling is based on the idea that if attributes are observed to co-occur in people then a large number of plausible inferences can be made on the basis of a considerably smaller number of observations. Rich [12] defines a stereotype more formally as a knowledge structure consisting of a body, a set of triggers and an optional set of relations with other stereotypes. The body contains information that is typically true of users to whom the stereotype applies and the triggers are observable values that can activate the stereotype.

Rich emphasises adaptation in stereotype systems because of the difficulty of defining stereotypes beforehand. Values deduced for attributes that have not been directly observed are only guesses and can be changed. Mechanisms for resolving conflicts between assumed values and new systems because of the difficulty of defining accurate observations are an important and potentially complex part of this type of modelling system. This use of stereotypes is a form of default reasoning [10].

The Grundy system is an early system using stereotypes, which acts as a library consultant by building up profiles of people so that it can recommend books they may enjoy reading [11]. An initial profile is based on information supplied by the user in answer to questions about some personality traits. For each of these traits the system includes a parameter which records the user's assumed scale value and another parameter which estimates the confidence value for this choice.

Chin [3] uses a double stereotyping approach to model users' knowledge of UNIX. One set of stereotypes represents users' expertise and the other represents the difficulty of the information. Users are categorised as novice, beginner, intermediate and expert, while information is categorised as simple, mundane, complex and esoteric. The stereotypes are not represented as collections of attributes as in Grundy, but as categories within a multiple inheritance knowledge representation system.

2.2   Allowing for Individual Differences in Cognitive Style

Research into individual differences in cognitive style has brought to light some fundamental differences between people that could be very relevant to the way in which people interact with a computer system. Schmeck [15] gives a review of strategies and styles of learning which have been proposed as useful constructs. He suggests that all the cognitive styles proposed in the literature can be positioned on one broad dimension labelled "global versus analytic" following the terminology used by Kirby [8]. This construct is closely related to the extensively investigated psychological construct "field-dependence field-independence" or FD/FI. Witkin [21] gives more details of research on FD/FI.

Characteristics of people with an analytic style include focused attention, noticing and remembering details, sequential organisation of activities, consciously directed thought, separation of feelings and facts, and logical and critical thinking. Characteristics of people with a global style include scanning attention resulting in global impressions, random or multiple organisation of activities, intuitive thinking, entry of feelings into decisions and impulsive rather than analytic approach. From a developmental point of view Kirby [8], in a study of reading skill development finds that students begin with a global approach and that their skill improves as a more analytic approach is adopted. However, higher levels of reading comprehension require a blend of global and analytic styles.

3   Stereotype Construction

The approaches to modelling used in the applications mentioned above vary in the way in which the stereotypes are organised. In the present study it is found appropriate to use two sets of stereotypes which are treated in different ways as described below. One aspect in which the present study differs from all the research described above is that the problem of deriving the stereotype categories is explicitly addressed using cluster analysis. The data from the preliminary survey are cluster analysed to identify groups of users who could form the basis for the stereotype categories. These categories are thus derived empirically in a way that ensures that they meet as far as possible the basic requirement for a stereotype: that they should be characterised by attributes that co-occur within the stereotype groups.

3.1   The preliminary survey

A survey was done on 109 people, 71 of whom had work experience, in a Research Institute while 38 were students with no work experience. A questionnaire was drawn up incorporating biographical items, the items based on interaction style and items dealing with familiarity with technical terms relevant to the tasks to be performed.

The questionnaire on Interaction Style takes the form of thirty-one statements about ways of interacting with a computer; for example, "I often get frustrated with computers". Users were asked to describe how the statements applied to them by choosing one of the four options "definitely true", "more true than false", "more false than true" or "definitely false". The statements that proved useful in differentiating between clusters may be seen in summarised form in the next subsection.

Some Interaction Style items did not differentiate between stereotype clusters but received overwhelming endorsement from the entire sample. These items, for example "I like to understand why I am doing something" provide a guide to the way the interface should be constructed for this user group. The questionnaire on Knowledge of Technical Terms lists twenty-two technical terms. These items were constructed by first writing detailed explanations of the steps involved in the statistical analysis and then collecting all the technical terms used in these explanations. The options to be chosen for each item were "understand clearly", "understand vaguely", "do not understand" and "never seen the term". Examples of the terms used are given in the description of the clustering of terms in the next section.

3.2   Clustering Questionnaire Items

The way in which the items of the questionnaire group together is of interest, and grouping items together into similar item clusters based on the sample as a whole is a useful aid to the interpretation of the clusters of people. This is done using the hierarchical cluster analysis technique known as Ward's method [19] Stereotype clusters are formed separately for the items dealing with familiarity with technical terms and for the items based on interaction style.

The dendrogram for the cluster analysis for technical terms showed clear and well-defined clusters. These are listed with some examples of the terms in each cluster:

1. *Cluster One:* Computer-related terms. (PC, Menu, Expert System, Overflow)
2. *Cluster Two:* Basic statistical terms. (Ranked, Scatter, Outlier)
3. *Cluster Three:* Computational statistics terms. (Minimise, Degenerate, Convergence)
4. *Cluster Four:* Basic mathematical terms. (Dimension, coordinate, solve, exponential, absolute value)

The cluster analysis of Interaction Style items was, as could be expected, less clear-cut. There were two main groups that could loosely be described as "Information processing style" and "emotional factors". The first group could be split further into clusters relating to speed, thoroughness, and preference for an intuitive, graphic interaction style. The second group could be split into two clusters relating respectively to the important psychological dimension of primary-secondary functioning and to problems that could cause emotional distress.

3.3   Clustering Users Based on Interaction Style Items

Users were clustered using the k-means procedure (see, for example, [1]). The clustering procedure was used repeatedly to find a series of solutions with the number of clusters varying over a range of two to eight clusters. These solutions are not constrained to form a hierarchy and so do not yield a tree structures. Analyses were made for the raw data and for the data recoded to use only a dichotomous "true" or "false" coding. For all these potential solutions the profiles of the average scores on each item for each cluster were examined. The analyses based on the dichotomous data gave more interpretable solutions that also remained more consistent as the number of clusters varied.

Among the analyses for the dichotomous data the one giving the most meaningful clusters in terms of the items endorsed or rejected by the users within each cluster was a solution with five clusters. Each cluster can be characterised by listing the items for which the users in that cluster differ most in their percentage of endorsement from the average for the remainder of the sample. For example, in the case of Cluster One it was found that the item "I prefer to work at a slow but steady pace" was endorsed by 100 percent of the people assigned to this cluster. This can be contrasted with an endorsement level of 34 percent for the remainder of the sample.

A descriptive name was attached to each cluster to give some idea of the content of the stereotype. The clusters have distinct differences between them and the items on which each of them differ most strongly from the overall sample values give a coherent and interpretable picture of the cluster. These cluster descriptions are listed below:

Cluster One: Users in a hurry (n=27)

1. Do not like to work slowly but steadily.
2. Do not like detailed instructions beforehand.
3. Do not always read instructions carefully.
4. Fast readers.
5. Do not like a lot of encouragement.

Cluster Two: Attentive Users (n=28)

1. Do not like pictorial instructions.
2. Do not often get frustrated with computers.
3. Do not mind reading from a computer screen.
4. Do not get bored with long explanations.
5. Use intuition to guide thinking.
6. Do not like a lot of encouragement.

Cluster Three: Dependent users (n=33)

1. Like a lot of encouragement.
2. Like detailed instructions beforehand.
3. Impulsive.
4. Best to just accept problems that arise.
5. Like to work at a slow and steady pace.
6. Like pictorial explanations.

Cluster Four: Cautious users (n=14)

1. Do not like taking risks.
2. Work at a slow but steady pace .
3. Not fast readers.
4. Not good at remembering details.
5. Not impulsive.

Cluster Five: Frustrated users (n=7)

1. Not fast readers.
2. Sometimes get frustrated with computers.
3. Do not like detailed instructions beforehand.
4. Absent minded.
5. Like to act before worrying too much.
6. Not good at remembering details.

3.4   Clustering Users Based on Technical Knowledge Items

The 22 technical terms were cluster analysed using the k-means procedure described in the previous section. Analyses were made using only a dichotomous "true" or "false" coding corresponding to familiarity or unfamiliarity with the term, with the number of clusters varying from two to eight. All these potential solutions were examined by looking at the profile of items for each cluster relative to the profile of the sample as a whole. The four-cluster solution appeared to be the most useful. The clusters can be well characterised in terms of the groups of terms that were identified in the hierarchical cluster analysis of items mentioned above.

1. *Cluster One:* Familiar with all but the most technical computer terms but not experienced with using a mouse or a window-based system.
2. *Cluster Two:* Experienced with a mouse and a window-based system and familiar with all but the technical statistical terms.
3. *Cluster Three:* Experienced with a mouse and a windows-based system and familiar with most terms, but less familiar with the more complex statistical terms.
4. *Cluster Four:* Fairly familiar with basic mathematical terms but not familiar with any of the other terms and not experienced with a mouse or a window-based system.

4   Managing Dialogue in the Application System

4.1   Managing Human-Machine Dialogue

In this section some anthropological research on the nature of human-machine dialogue is considered. A theory of human-machine communication provides a framework for understanding how problems in the interaction between the user and the computer system can arise and this is central to the consideration of how the dialogue with a computer-based system should be structured.

Suchman [18] found two types of situation that could lead to a complete breakdown of communication. In the first case the user thinks that there is something wrong when in fact everything is proceeding correctly, while in the other there is a problem which has not been detected by either party. In the first case the instructions given by the system make no sense to the user, while the system has no way of detecting that there is a problem. In the second case the system does not detect any problem and so its responses may be interpreted as confirming an incorrect action. In summary, Suchman suggests that a system designer should be aware that the interaction between humans and machines requires the same interpretative effort needed in conversation between people, but that the means for interaction between a human and a computer are severely limited. Careful attention should be given to ways of increasing the amount of information available to the machine, making the limitations of the system clear to the user, and finding ways of compensating for these limitations as far as possible.

4.2   System Implementation

A design goal of the system was to provide sufficient information to the user so that the tasks to be performed would be unambiguous and the means for performing them would be clear, without overburdening the user with too many instructions. An important consideration, in the light of Suchman's research mentioned above, was to provide every opportunity for the user to obtain clarification if problems should arise. At all times the menu associated with the active window is available and the user can choose a "help" option. This can provide additional information, a summary of the course of the current analysis which indicates where the current task fits into the overall plan of the analysis, or an opportunity to comment on any aspect of the system. The unfolding assistant was written in Smalltalk/V286 [5] to run on an IBM-PC/AT style system and to interface with the existing FORTRAN program that performs the actual unfolding analyses.

4.3   The Domain Expert

The domain expert in this system is a rule-based expert system for unfolding analysis designed to provide specific information on how an analysis should proceed taking into consideration the progress made so far. The rules are specified as assertions about what conclusions will follow for the analysis if certain conditions are met. An example of a rule is the following: **if** the iterations have converged **and** the configuration is not degenerate **and** the stress is low **then** a good solution has been found.

The analysis procedure adds facts about the progress of the analysis to the rule base so that at any stage advice about the best way to continue the analysis can be given. The user can ask for advise on how to continue, or a predetermined "script" can ensure that at certain key points the analysis is interrupted automatically and users are invited to check their ideas for the continuation of the analysis against suggestions made by the expert system.

When the execution of the unfolding program is interrupted, control passes back to the unfolding assistant, which then passes the facts from the unfolding analysis to the inference engine. Based on these facts and the rules in the rulebase the inference engine then derives all the courses of action that can be recommended for the continuation of the analysis using forward chaining [2].

4.4 Generating Explanations for Different Clusters

The text of information screens displayed during a session needs to be built up taking into account information on both preferred interaction style and familiarity with technical terms from the user model. Each text component that can be displayed is stored together with a label, which indicates the nature of the component. The text for each information screen is generated afresh each time it is displayed using the current state of the user model to decide which of the alternative versions of each component should be displayed or whether that component should be displayed at all.

The knowledge level cluster to which the user is initially assigned specifies for each technical term whether the explanation should be included in the generated text or not. Once an explanation for a particular term has been displayed the explanation will no longer be included in generated text, although it can still be obtained by activating the hypertext button.

Information on the user's interaction style is not as easily obtained in the course of a session as information on the user's knowledge level and so these two aspects of the user model are dealt with in rather different ways. The initial style cluster is set after an initial set of questions and is not changed by the system after this. At any point the user can display the contents of the current user model and alter the style cluster if so desired.

The research discussed in Section 2.2, the results of the user survey and the observations made during the preliminary trials were used to suggest factors which could be varied to tailor the response of the system to suit the individual interaction style stereotype categories. The following factors were considered.

1. The level of detail given in initial task instructions and help information.
2. The amount of checking done to make sure that instructions have been read.
3. The use of encouraging remarks or reassurance after mistakes.
4. The extent to which the user is reminded of information which has been previously presented.
5. The layout of information on the screen.

Positions on the range of possibilities for these factors were postulated for each stereotype on the basis of the stereotype characteristics. For example, the following parameters were set for the "attentive users" in Cluster Two.

1. Can give quite long and detailed instructions.
2. Avoid unnecessary messages of encouragement or reassurance.
3. Do not need to provide too many opportunities to review previous decisions. (These have a cost as they slow down the analysis process.)
4. Can put more information on a screen than in the case of some other groups.

To answer the question of what the most appropriate system response characteristics should be for each style category from a psychological or pedagogical standpoint would require further research that is far beyond the scope of this study.

4.5   Experience with the System

The initial approach that was taken to defining the behaviour of the system for the knowledge-level stereotype categories was to display explanations for all terms that were assumed to be unfamiliar to the user. The confusion experienced by the users with a poor background in the area suggests that a better approach might be to make use of Goldstein and Carr's [7] concept of a frontier of learning. Explanations would then only be given for terms at the user's current learning frontier.

Stereotype modelling is based on the idea that if certain user attributes are correlated then knowledge of one or some of these attributes can be used to predict the others. The main emphasis in stereotype modelling is nevertheless on dividing the users themselves into explicit categories, as was done in this study using cluster analysis. In general one would expect each category to be defined by a subset of attributes which correlate relatively highly within that category. It might well be the case that two attributes, which correlate highly within one category, do not correlate at all within other categories. In this situation knowledge of one attribute is only useful in predicting another if the user's category membership is already known. For this reason the stereotype model includes trigger attributes for each category which must be known in order to assign a user to a specific category before attempting to predict the values of unobserved attributes from observed ones. This makes the successful operation of stereotype modelling highly dependent on finding effective trigger attributes. This may not be easy, especially when the stereotype category is of a complex and perhaps loosely defined nature, and this is often likely to be the case in areas such as cognitive functioning.

In some situations it is possible to adopt a simpler approach based directly on the correlation between attribute values. For example, in the case of knowledge level a cluster analysis of the terms showed that they could be grouped together into classes which could be consistently and meaningfully defined for the entire user group. The stereotype clusters can all be characterised by their knowledge level profile for each of the term clusters. In these circumstances it is unnecessary to look for trigger attributes for each cluster or even to consider to which stereotype cluster a user belongs in order to predict knowledge of a term once knowledge of any other term in the same class is established. The explicit stereotype clusters can still be used initially to set default values for all term classes given some limited preliminary information, but once information on the knowledge of terms in any class is available this can be used directly for predicting knowledge of other terms in the class.

This approach was implemented in the course of the final trials and worked without any problems. Not only is the process simpler than modelling by stereotypes but the prediction for each term class should be more accurate since it is operating in a more limited domain and makes use of all the relevant correlation information available from the entire user group.

5   Conclusions

5.1   Research Findings

Four distinct stereotype categories for levels of technical knowledge were found from the cluster analysis of the survey data. The technical terms used were of necessity very specific to the present application. Nevertheless the logical way in which they could be described; for example "familiar with all terms", "unfamiliar with any but basic terms" and "familiar with all but computer terms" suggests that similar results may well be found in other areas. This simple characterisation of the stereotypes also makes it particularly easy to make use of this information in a user model.

From the cluster analysis of the survey information on interaction style five stereotypes were identified which were labelled as "users in a hurry", "attentive users", "dependent users", "cautious users" and "frustrated users". These stereotypes were found purely as a consequence of the data analysis of the survey questionnaire and the names were only attached later for ease of reference. They appear to make good intuitive sense and to have definite implications for the style of instruction that should be used in each case. The approach of using cluster analysis based on a preliminary survey proved successful in constructing plausible stereotypes on which a user model could be based.

5.2   Suggestions for Further Research

The optimum use of the information on stereotype categories in the presentation of the text was beyond the scope of this study. A proper evaluation of the benefits of the user model would depend heavily on having carefully developed teaching material to rectify any problems of understanding which were detected, taking into account the preferred interaction style of the user. This is an obvious topic for further research, but the project described in this paper has been completed and the author is currently working on further development of the algorithm for the unfolding analysis and is not at present involved in research on the user interface.

References

1. Anderberg, M.R.: Cluster Analysis for Applications. Academic Press New York (1973)
2. Charniak, E. and McDermott, D.: Introduction to Artificial Intelligence. Addison-Wesley Reading, Massachusetts (1987)
3. Chin, D.N.: KNOME: Modelling what the user knows in UC. In Kobsa, A. and Wahlster, R. (Eds.) User models in dialog systems. Springer Berlin (1989)
4. Conklin, J.: Hypertext: a survey and introduction. Computer **20(9)** (1987) 17-41
5. Digitalk Inc.: Smalltalk/V 286 tutorial and programming handbook. Digitalk Inc. Los Angeles (1988)
6. Gale, W.A., Hand, D.J. and Kelly, A.E.: Statistical Applications of Artificial Inteligence. In Rao, C.R. (Ed.) Handbook of Statistics, Volume 9. North-Holland Amsterdam (1993)
7. Goldstein, I.P. and Carr, B.: The computer as coach: an athletic paradigm for intellectual education. Proceedings of the National ACM Conference (1977) 227-233
8. Kirby, J.R.: Style, strategy and skill in reading. In Schmeck, R.R. (Ed.) Learning strategies and learning styles. Plenum Press New York (1988)
9. Muller, M.W.: Unfolding. In Kotz, S. and Johnson, N.L. Encyclopedia of statistical sciences. Wiley New York (1988)
10. Reiter, R.: A logic for default reasoning. Artificial Intelligence **13** (1980) 81-132
11. Rich, E.: User modeling via stereotypes. Cognitive Science **3** (1979) 355-366
12. Rich, E. Stereotypes and user modeling. In Kobsa, A. and Wahlster, R. (Eds.) User models in dialog systems. Springer Berlin (1989)
13. Ridgway, J.: Of course ICAI is impossible ... worse though, it might be seditious. In Self, J. (Ed.). Artificial intelligence and human learning. Chapman and Hall London (1988)
14. Schiffman, S.S., Reynolds, M.L. and Young, F. W.: Introduction to multidimensional scaling. Academic Press New York (1981)
15. Schmeck, R.R.: Strategies and styles of learning: an integration of varied perspectives. In Schmeck, R.R. (Ed.) Learning strategies and learning styles. Plenum Press New York (1988)
16. Self, J.: Artificial intelligence and human learning. Chapman and Hall London (1988)
17. Sparck Jones, K.: Realism about user models. In Kobsa, A. and Wahlster, R. (Eds.) User models in dialog systems. Springer Berlin (1989)
18. Suchman, L.A.: Plans and situated actions: the problem of human-machine communication. Cambridge University Press Cambridge (1987)
19. Ward, J.H.: Hierarchical grouping to optimize an objective function. Journal of the American Statistical Association **58** (1963) 236-244
20. Wenger, E.: Artificial intelligence and tutoring systems. Morgan Kaufmann Los Altos, California (1987)
21. Witken, H.A. and Goodenough, D.R.: Cognitive styles: essence and origins - field dependence and field independence. International Universities Press New York (1981)

How to Find Big–Oh in Your Data Set (and How Not to)

C. C. McGeoch[1], D. Precup[2] and P. R. Cohen[2]

[1] Department of Mathematics and Computer Science, Amherst College
Amherst, MA 01002, USA. E-mail: ccm@cs.amherst.edu
[2] Department of Computer Science, University of Massachussetts
Amherst, MA 01003, USA. E-mail: dprecup,cohen@cs.umass.edu

Abstract. The *empirical curve bounding problem* is defined as follows. Suppose data vectors X, Y are presented such that $E(Y[i]) = \bar{f}(X[i])$ where $\bar{f}(x)$ is an unknown function. The problem is to analyze X, Y and obtain complexity bounds $O(g_u(x))$ and $\Omega(g_l(x))$ on the function $\bar{f}(x)$. As no algorithm for empirical curve bounding can be guaranteed correct, we consider heuristics. Five heuristic algorithms are presented here, together with analytical results guaranteeing correctness for certain families of functions. Experimental evaluations of the correctness and tightness of bounds obtained by the rules for several constructed functions $\bar{f}(x)$ and real datasets are described. A hybrid method is shown to have very good performance on some kinds of functions, suggesting a general, iterative refinement procedure in which diagnostic features of the results of applying particular methods can be used to select additional methods.

1 Introduction

Suppose the expected cost of algorithm A under some probabilistic model is described by an unknown exact function $\bar{f}(x)$ (where x denotes problem size). Function $\bar{f}(x)$ belongs to class $\Theta(\bar{g}(x))$ if there exist positive constants x_0, c_1 and c_2 such that $0 \leq c_1 \bar{g}(x) \leq \bar{f}(x) \leq c_2 \bar{g}(x), \forall x \geq x_0$ [8]. Function $\bar{g}(x)$ is an asymptotically tight bound for $\bar{f}(x)$. Asymptotic upper and lower bounds can be established separately for functions as well. A function $\bar{f}(x)$ has the asymptotic upper bound $O(\bar{g}_u(x))$ if and only if there exist positive constants x_0 and c_u such that $0 \leq \bar{f}(x) \leq c_u \bar{g}_u(x), \forall x \geq x_0$. An analogous definition can be formulated for the asymptotic lower bound, $\Omega(\bar{g}_l(x))$. $O(\bar{g}_u(x))$ and $\Omega(\bar{g}_l(x))$ characterize the efficiency of algorithm A.

Suppose that an experimental study of A produces a pair of vectors X, Y such that $E(Y[i]) = \bar{f}(X[i])$. The empirical curve-bounding problem, addressed in this paper, is: *Analyze (X, Y) and estimate complexity classes $O(g_u(x))$ and $\Omega(g_l(x))$ to which $\bar{f}(x)$ belongs*. (Functions f and g are *estimates* of functions $y = \bar{f}(x)$ and $\bar{g}$.) While a primary goal of traditional algorithm analysis is to identify complexity classes to which unknown functions belong, this empirical version of the problem appears to be new. We can find no techniques in the data analysis

X. Liu, P. Cohen, M. Berthold (Eds.): "Advances in Intelligent Data Analysis" (IDA–97)
LNCS 1280, pp. 41–52, 1997.

literature designed for finding bounds on data, although much is known about fitting curves to data (see sec. 5). Approaches to domain-independent function finding [13] might be adapted to curve bounding and some are considered here.

For any finite set of points X there are functions $\bar{f}(x)$ of arbitrarily high degree but indistinguishable from the constant c at those points. Therefore any heuristic producing an upper bound estimate can be fooled, and no curve-bounding method can be guaranteed correct. This paper presents five robust heuristics that can produce correct bound estimates (or clear indications of failure) for broad classes of functions and for functions that tend to arise in practice. We describe each rule R together with a justification that describes a class F_R: for any function $\bar{f} \in F_R$, the rule is guaranteed to find correct (sometimes exact) bounds when applied to data vectors $Y = \bar{f}(X)$. We also present empirical studies of the rules using constructed multi-parameter functions, and "typical" data sets from algorithm analysis. The experiments indicate the limitations of the rules and suggest an appropriate level of conservatism in their application. We also discovered that the rules can "diagnose" qualitative features of functions $\bar{f}$, and given these diagnoses, we can apply additional rules that are specific to functions with these features, and attain higher levels of performance.

The rules can be viewed as interactive tools or as offline algorithms. To accomodate both views, we present the algorithms with a small set of *oracle functions*. An oracle function establishes some basic property of a set of points drawn from some function; for example, an oracle can decide whether "residuals are concave upwards", or if the "function appears to be increasing". In interactive use, a human provides the oracle result; in offline use, a simple computation is used. However, the experiment in section 5 suggests that offline versions are far more efficient, and often more effective, than interactive versions.

2   Notation and the Heuristics

The vector X contains k distinct nonnegative values arranged in increasing order. Each heuristic takes a pair of vectors (X, Y) generated according to $Y = \bar{f}(X)$ or sometimes $E(Y) = \bar{f}(X)$. The heuristic reports a class estimator $g(x)$ together with a bound type, either *upper, lower,* or *close. Upper* signifies a claim that $\bar{f}(x) \in O(g(x))$, and *lower* signifies a claim that $\bar{f}(x) \in \Omega(g(x))$. A *boundtype = close* is returned when a data set does not meet the rule's criteria for upper or lower bound claims. An upper bound estimate $O(g(x))$ is *correct* if in fact $\bar{f}(x) \in O(g(x))$. A correct upper bound is *exact* if $g(x)$ labels the smallest correct class. For instance, for function $\bar{f}(x) = 0.5x^2$, $O(x^{1.9})$ is an incorrect upper bound, $O(x^{2.1})$ is correct but not exact, and $O(x^2)$ is both correct and exact. Analogous definitions hold for lower bound estimates. Some heuristics generate internal *guess functions* $f(x)$ before reporting the estimate $g(x)$. For convenience we assume that $g(x)$ and $\bar{g}(x)$ take the standard one-term form of complexity class labels (e.g., not $O(3x^2 + 5)$ but $O(x^2)$).

The following computations are performed by the oracle functions:

Trend(X, Y, c_r). Returns a value indicating whether Y appears to be increasing, decreasing, or neither. The function compares the correlation coefficient r (computed on X and Y) to a cutoff parameter c_r which is 0.1 by default. We have experimented with ways to implement this and the following oracles that are less sensitive to outliers in the data.

Concavity (X, Y, s). The function examines signs of smoothed residuals from a linear regression fit of X to Y. It returns "concave upward" if signs obey the regular expression $(+)^+(-)^+(+)^+$, "concave downward" if they obey $(-)^+(+)^+(-)^+$, and otherwise "neither." The default low setting on parameter s produces "less smooth" residuals and more frequent "neither" results.

DownUp(X, Y, s). The DownUp oracle checks whether successive differences in smoothed Y values obey the regular expression $(-)^+(+)^+$, returning True or False. The default low value of parameter s produces more frequent False results.

NextCoef$(f, direction, cstep)$ **and NextOrder**$(f, direction, estep)$. Some rules iterate over several guesses and require an oracle to supply the next guess. This implementation constructs functions $f(x) = ax^b$ for positive rationals a and b. $NextCoef$ changes a according to $direction$ (up or down) and the $cstep$ size. If a decrement of size $cstep$ would give a negative coefficient, then $cstep$ is reset to $cstep/10$ before decrementing. $NextOrder$ changes the exponent b according to the $estep$ size. Default $estep$ is .001 and initial $cstep$ is .01.

2.1   Guess Ratio

The first heuristic is called the Guess Ratio (GR) rule. To justify GR, let F_{GR} contain $\bar{f}(x) = a_1 x^{b_1} + a_2 x^{b_2} + \cdots + a_t x^{b_t}$, with rationals a_i positive, and b_i such that $b_1 > 0$, $b_i \geq 0$, and $b_i > b_{i+1}$. Let the guess function be of the form $f(x) = x^b$. Then the ratio $\bar{f}(x)/f(x)$ has the following properties: (1) When $\bar{f}(x) \in O(f(x))$, the ratio decreases to a nonnegative constant as x increases; (2) When $\bar{f}(x) \notin O(f(x))$ the ratio eventually increases and has a unique minimum point at some location x_r. If $x_r > 0$, then the ratio shows an initial decrease followed by an eventual increase. These properties are established by an application of Descartes' Rule of Signs [17], which bounds the number of sign changes in the derivative of the ratio.

If a plot of a finite sample of the ratio $(X \text{ vs } Y/f(X))$ shows an eventual increasing trend, then (2) must hold. If only a decrease is observed, then cases (1) and (2) cannot be distinguished. The Guess Ratio rule begins with a constant guess function and increments b until the ratios $Y/f(X)$ do not appear to eventually increase. The largest guess for which an eventual increase is observed is reported as a "greatest lower bound" found. When $\bar{f}(x) \in F_{GR}$ and $k \geq 2$, the correctness of GR can be guaranteed simply by defining "eventual increase" as $Y[k-1] < Y[k]$ (recall, k is the number of design points or X values). However our implementation uses the Trend oracle for this test because of possible random noise in Y. For any data set (X, Y) and our Trend oracle, the rule must eventually terminate.

2.2　Guess Difference

The Guess Difference (GD) rule evaluates differences $f(X) - Y$ to produce an upper bound estimate. This rule is effective for the class F_{GD} which contains functions $\bar{f}(x) = cx^d + e$ where c, d and e are positive rationals. Let the guess have the form $f(x) = ax^b$. Consider the *difference curve* $f(x) - \bar{f}(x)$. When $f(x) \notin O(\bar{f}(x))$ this curve eventually increases and has a unique minimum at some location x_d. Also, x_d is inversely related to the coefficient a: for large a the difference curve increases everywhere ($x_d = 0$), but for small a there might be an initial decrease. In the latter case we say the curve has the DownUp property.

The GD rule starts with an upper bound guess $f(x) = ax^b$ and searches for a difference curve with the DownUp property by adjusting the coefficient a. If a DownUp curve is found, the rule concludes that $f(x)$ overestimates the order of $\bar{f}(x)$, so it decrements b and tries adjusting a again. The lowest b for which the rule finds a DownUp curve is reported as a "least upper bound" found. Using an analysis similar to that for GR, we can show that when $\bar{f}(x) \in F_{GD}$ and $k \geq 4$, and X is fixed, there exists an a such that $f(X) - \bar{f}(Y)$ will have the DownUp property. If the rule is able to find a DownUp curve in its finite sample, then the upper bound it returns must be correct. We can also show that the DownUp property cannot guarantee correctness for functions from F_{GR}. In our implementation, if the rule is unable to find an initial DownUp curve within preset limits, it stops and reports the original guess provided by the user.

2.3　Power Rule

The Power Rule (PW) modifies a standard method for curve-fitting (see [12]). Suppose that F_P contains functions $\bar{f}(x) = cx^d$ for positive c and d. Let $y = \bar{f}(x)$. Transforming $x' = \ln(x)$ and $y' = \ln(y)$, we obtain $y' = dx' + c$. The Power Rule applies this log-log transformation to X and Y and then reports d, the slope of a linear regression fit on the new scale. The Concavity oracle, applied to residuals from the regression, determines whether an upper or lower bound (or neither) is claimed. If $Y = \bar{f}(X)$ and $\bar{f}(X) \in F_P$ then the Power rule finds d exactly. If $Y = \bar{f}(X) + \epsilon$ and the random noise component ϵ obeys standard assumptions of independence and lognormality, then confidence intervals on the estimate of d can be derived.

High-End Power Rule (PW3). When $\bar{f}(x)$ has low-order terms (such as $ax^b + e$), the transformed points do not lie on a straight line, and regression using only the j highest design points might give a better asymptotic bound than one using all k design points. The PW3 variation tested in this paper applies the Power rule to the data points for $X[k - 2]$, $X[k - 1]$, $X[k]$.

Power Rule with Differences (PWD). The *differencing* variation on the power rule attempts to straighten out plots under log-log transformation by removing constant and logarithmic terms. This variation is applicable when the X are chosen such that $X[i] = \Delta \cdot X[i-1]$ for a positive constant Δ. The variation applies the Power rule to *successive differences* in adjacent Y values.

To justify this rule, suppose F_{PWD} contains $\bar{f}(x) = cx^d + e$ where c, d and e are positive constants, and let $Y = \bar{f}(X)$. Set $Y'[i] = Y[i+1] - Y[i]$ and $X'[1..k-1] = X[1..k-1]$. Then $Y' = c'X'^d$ (with a new coefficient and with e gone), to which the basic power rule can be applied. When $\bar{f}(x) \in F_{PWD}$, $Y = \bar{f}(X)$ and $k > 2$, the PWD rule finds d exactly. Differencing affects other kinds of terms: for example, taking differences twice will remove logarithms.

2.4 The BoxCox rule

A general approach to curve-fitting is to find transformations on Y or on X, or both, that produce a straight line in the transformed scale. For example, if $Y = X^2$, then a plot of X vs $\sqrt{Y}$ would produce a straight line, as would a plot of X^2 vs Y.

The Box-Cox ([1], [5]) transformation on Y is parameterized by λ. This transformation is applied together with a "straightness" statistic that permits comparisons across different parameter levels. The transformation is as follows:

$$
Y^{(\lambda)} = \begin{cases} \frac{Y^{\lambda}-1}{\lambda \bar{Y}^{\lambda-1}} & \text{if } \lambda \neq 0 \\[2ex] \bar{Y}\ln(Y) & \text{if } \lambda = 0 \end{cases}
$$

where $\bar{Y}$ is the geometric mean of Y, equal to $\exp(\text{mean}(\ln{(Y)}))$. The "best" transformation in this family minimizes the Residual Sum of Squares (RSS) statistic which is calculated from X and Y^{λ}.

Our BC rule iterates over a range of guesses $f(x) = x^b$, evaluating $Y^{(\lambda)}$ with $\lambda = 1/b$. The Concavity of residuals from the best transformation found determines the type of bound claimed. When $\bar{f}(x) = F_{PW}$, $Y = \bar{f}(X)$, $k > 2$, and the NextGuess oracle includes $\bar{f}(x)$, this rule finds the function exactly. With standard normality assumptions about an added random error term, it is possible to calculate confidence intervals for the estimate b; see [1] or [5] for details.

2.5 The Difference Rule

The **Difference** heuristic extends Newton's divided difference method for polynomial interpolation (see [15] for an introduction) to be defined when Y contains random noise and nonpolynomial terms. The method iterates numerical differentiation on X and Y until the data appears non-increasing, according to the Trend oracle. The number of iterations d required to obtain this condition provides an upper bound guess x^d. When $\bar{f}(x)$ is a positive increasing polynomial of degree d, $k > d$, and $Y = \bar{f}(X)$ then this method is guaranteed correct. Much is known about numerical robustness, best choice of design points, and (non)convergence when $k \leq d$.

3 Experimental Results

The rules have been implemented in the S language [2], designed for statistical and graphical computations. The experiments were carried out on a Sun SPARC-station ELC, using functions running within the Splus statistical/graphics package; some supporting experiments were conducted using the CLASP statistical/graphics package, and the method labelled HY in Table 1 was implemented in C. Timing statistics would be misleading in this context and are not reported in detail. Roughly, the Power rules required a few microseconds, and the iterative rules usually took no more than a few seconds per trial. The Guess Difference rule required a coarser *estep* value in the NextOrder oracle (.01 instead of .001) to produce comparable running times.

3.1 Parameterized Functions

The first experiment studies the sensitivity of the rules to second order terms, using functions $\bar{f}(x) = ax^b + cx^d + e$ (with no random term). Very roughly, the particular constants for this test were chosen after several months of exploration to highlight the boundary between functions that are "easy for all rules" and "hard for all rules." Vector X takes powers of two between 8 and 128. In Table 1, the notations **l, u, c**, indicate the type of bound claimed. An underline marks an incorrect bound, and an X marks a case where the heuristic failed to return a meaningful result. We will defer discussing the results in column HY until section 4.

No	Function	GR	GD	PW	PW3	PWD	BC	DF	HY
1	$3x^{.2}+1$	.171l	(2.26).24u	.171l	.174l	.2u	.178l	1u	$.03x^{.60}+4x^{0.15}$
2	$3x^{.2}+10^2$	.011l	(2.26).24u	.011l	.012l	.2l	.012l	1u	$.12x^{.56}+103x^{.01}$
3	$3x^{.2}+10^4$	.0001l	(2.27).24u	.0001l	.0004l	.2l	X	1u	$.12x^{.55}+10003$
4	$3x^{.8}+10^4$	.004l	(1.0)1u*	.004l	.006l	.8l	X	1u	$1.55x^{.91}+10003$
5	$3x^{.8}+x^{.2}$	.775l	(1.0)1u*	.774l	.784l	.793l	.792l	1u	$.22x^{1.1}+4x^{.67}$
6	$3x^{.8}-x^{.2}$	.825l	(1.0)1u*	.829u	.817u	.807u	.809u	1u	$-.34x^{1.26}+2x^{1.03}$
7	$3x^{.8}+10^4x^{.2}$	.201l	(1.0)1u*	.202l	.202l	.206l	.203l	1u	$x^{.97}+10003x^{.2}$
8	$3x^{.8}+x^{.6}$	.771l	(1.0)1u*	.771l	.775l	.778l	.778l	1u	$.03x^{1.2}+4x^{.75}$
9	$3x^{.8}-x^{.6}$	.838l	(1.8).88u	.841u	.834u	.829u	.819l	1u	$-.05x^{1.26}+2x^{.89}$
10	$3x^{.8}+10^4x^{.6}$	.600l	(1.0)1u*	.600l	.600l	.600l	.600l	1u	$.12x^{1.15}+10003x^{.6}$
11	$3x^{.8}-10^4x^{.6}+10^6$	-.01l	(1.0)1u*	-.059u	-.086u	X	X	0u	$-3361x^{.77}+990003x^{-.01}$
12	$3x^{1.2}+10^4$	0.035l	(2.8)1.22u	.032l	.056l	1.2l	X	2u	$2.4x^{1.25}+10003$
13	$3x^{1.2}+x^{.2}$	1.187l	(2.8)1.22u	1.187l	1.194l	1.198l	1.2u	2u	$.48x^{1.4}+4x^{1.01}$
14	$3x^{1.2}+10^4x^{.2}$	0.213l	X	0.212l	0.220l	0.263l	0.231l	1u	$2.03x^{1.27}+10003x^{.2}$
15	$3x^{1.2}+x^{.8}$	1.169l	(3.1)1.21u	1.168l	1.175l	1.178l	1.183u	2u	$.11x^{1.54}+4x^{1.11}$
16	$3x^{1.2}-x^{.8}$	1.235l	(2.2)1.26u	1.238u	1.227u	1.223u	1.218l	2u	$-.18x^{1.65}+2x^{1.36}$
17	$3x^{1.2}+10^4x^{.8}$	0.800l	(1.0)2u*	0.800l	0.801l	0.801l	0.801u	1u	$.48x^{1.45}+10003x^{.8}$

Table 1. Parameterized nonrandom functions

The functions tend to track large positive second terms. For example, for the function $\bar{f}(x) = 3x^{.8} + x^{.2}$, most of the methods estimate b to be in the range

0.77 to 0.79, which are correct and close lower bounds on the true value of .8. But for the function $\bar{f}(x) = 3x^{.8} + 10^4 x^{.2}$, these same methods estimate b to be .2, tracking the exponent of the larger second term. Negated second terms can present problems, particularly for the GR method. GD does remarkably well at estimating the coefficient of the first term, although it is an iterative algorithm and its performance is sensitive to the choice of initial guess and step size. The starred entries mark cases where the rule failed to find a DownUp curve and returned the user-supplied guess which was either $1x^1$ (functions 1 through 11) or $1x^2$ (functions 12 through 17). Both PW3 and PWD give tighter bounds than PW; not only does PWD successfully eliminate constants, but it is slightly better than PW and PW3 when the second term is non–constant. The BC rule provides very competitive bounds when it works, but it goes into an infinite loop on functions with a large-magnitude constant as a second term; the failure of BC on these functions is an intrinsic property of the λ transformation. Like PWD, the differencing operation of DF makes it insensitive to large constant terms. Because DF returns an integer exponent its bound is never tight on this test set. Function 11 is disastrous for all the rules because the negated second term causes Y to be decreasing within its range.

Larger Problem Size An obvious remedy to the problem of a dominant second-order term is to use larger problem sizes. A second experiment uses the same functions as above, except X takes values at powers of two in the range $8 \ldots 256$ rather than $8 \ldots 128$. That is, the largest problem size is doubled. This had very little effect on the bounds returned by Guess Ratio and the three Power Rules. The *change* in estimate is generally only in the third decimal place, and incorrect bounds remain incorrect. We can argue that GR would probably be least affected by larger problem sizes, but one might expect greater responsiveness of PW3 because the new point should have greater leverage. The greatest improvement is found in the Guess Difference rule on functions 4 through 9 (excepting 7). In the previous experiment the rule failed to find an initial DownUp curve at all—now the rule finds upper bounds within .05 of the true exponent. BC also shows some very slight improvement; in two cases the rule produces *close* bound claims (which are hard to evaluate) where previously it had been incorrect.

Adding Random Noise. We added a random term to three easy functions (1, 5, and 13) to learn how rule performance degrades with increased variance. We let $Y = \bar{f}(X) + \epsilon_i$ with and $i = 1, 2, 3$. The ϵ_i are drawn independently from normal distributions with means 0 and standard deviations set to 1, 10 and the function means $\bar{f}(X[j])$, for $i = 1, 2, 3$ respectively. We ran two independent trials for each i.

The quality of results returned by all rules degrades as variance increases and the replication of tests in each category demonstrates that many correct bounds are spurious. Conversely, rule performance improves when variance decreases. We suspect that some of the decrement in performance is due to the oracles, some of which are not particularly robust. For example, a robust linear fit would probably make more sense than linear regression as an oracle for slope. Encouragingly, it is usually possible to reduce variance in data by increasing the number of trials or by applying variance reduction techniques [10].

With greater variance in Y the Power and the BoxCox rules more frequently return claims of *close*, which are hard to evaluate. Large variance has less impact when the *change* in Y is large. Our implementations of the BC and PWD rules encounter difficulties with negative values and negative differences in case ϵ_3; the former can be remedied by adding a large positive constant to the data, but this introduces new inaccuracies.

3.2 Algorithmic Data Sets

This experiment applied the rules to eight data sets drawn from previous computational experiments by the first author. The data sets were not originally intended for this purpose and may give more realistic indications of performance. Data sets 1 and 2 are the expected costs of Quicksort and Insertion Sort, for which formulas are known exactly [9]. Sets 3 through 6 are from experiments on the FFD and FF rules for bin packing [3], [4]. Sets 7 and 8 are from experiments on distances in random graphs having uniform edge weights [11]. The X vectors have various ranges and intervals; except for the first two cases, the Ys represent means of several independent trials.

Results appear in Table 2. The left column presents the best analytical bounds known for each. The entries NA for PWD mark cases where this rule was not applied because design points were not in required format. Results in column HY are discussed in the following section.

	Known	GR	GD	PW	PW3	PWD	BC	DF	HY
1	$y=(x+1)(2H_{x+1}-2)$	<u>1.2l</u>	1.24u	1.221u	1.181u	NA	1.181c	2u	$2.1x^{1.25}+11.23x^{.64}$
2	$y=(x^2-x)/4$	2.0l	2.03u	3.003u	3.001u	NA	2.0l	2u	$.08x^{2.71}-.01x^{3.11}$
3	$E(y)=x/2+O(1/x^2)$	.99l	1u*	0.996l	.999u	1.0002c	1.203c	2u	$1.5x^{1.58}-.01x^{2.12}$
4	$E(y)\in\Theta(x^{.5})$	<u>.52l</u>	1u*	0.555c	.5716u	.7785c	0.999c	1u	$.2x^{.57}-.01x^{.8}$
5	$E(y)\in O(x^{2/3}(\log x)^{1/2})$ $E(y)\in\Omega(x^{2/3})$	<u>.68l</u>	.72u	0.689c	.695u	.692c	.687c	1u	$.22x^{.71}-0.00x^{1.07}$
6	$E(y)\le.68x$	.90l	1u	0.893l	.954l	<u>1.269l</u>	.976c	1u	$0.00x^{1.13}+.68x^{.47}$
7	$x-1\le y\le 13.5x\log_e x$	<u>1.13l</u>	1.18u	1.142u	<u>1.125l</u>	NA	1.109c	2u	$1x^{1.21}-0.00x^{1.81}$
8	$x\log_e x<y<1.2x^2$	1.30l	1.47u	1.318u	1.201l	NA	1.203c	2u	$.15x^{2.01}-0.01x^{2.31}$

Table 2. Tests on Algorithmic Data

Contrary to experience with the constructed functions, GR obtains a correct and tight bound when a negated second term is present (case 2), but in four cases GR produces results violating known bounds. GD and the Power Rules rarely violate known bounds, although without tighter analyses it is impossible to tell whether the rules are always correct. BC nearly always returns a "close" bound which is difficult to evaluate. Interestingly, every incorrect bound produced by the rules is a lower bound.

The most interesting results are in cases 6,7 and 8, which have gaps in the known asymptotic bounds. In 6, the rules provide consensus support for a conjecture that $\bar{f}(x)$ is closer to linear than, say, to $\sqrt{x}$. In 7, there is some very slim support for super-linear growth in the data set, but the rules are not really powerful enough to make such fine discriminations. In 8 the results give consensus support for a conjecture of sub-quadratic growth.

4   Iterative Refinement, Combining Methods

One known pathology of the heuristics presented so far is their sensitivity to low-order terms with large coefficients. For this type of function, the heuristics tend to track the low-order term.

This problem can be overcome by an iterative diagnosis and repair technique that combines the existing heuristics to produce improved models. The technique is designed to find upper bounds for functions of the form $ax^b + cx^d$ with rational exponents $b > d \geq 0$ and real coefficients $a << c$. This method represents a departure from our approach up to now: The earlier methods were intended to be general, but this one is specific to functions with relatively large coefficients on low order terms. This suggests a new role for the methods we have discussed so far: Instead of using them to guess at the order of a function, they can provide diagnostic information about the function (e.g., whether $a << c$), and then more specific, purpose-built methods, designed for particular kinds of functions, can estimate parameters.

To illustrate this new approach, we developed a three-step hybrid method for functions of the form $\bar{f}(x) = ax^b + cx^d$;

1. Apply a discrete derivative (the Difference rule) to the datasets, in order to find the integer interval of the exponent b.
2. Refine the guess for the exponent using the Guess Ratio rule. We start with the known upper and lower bound for the exponent, u and l. At each step we consider the model $x^{(u+l)/2}$ by plotting x against $y/x^{(u+l)/2}$. If the plotted points appear to be decreasing, then $(u+l)/2$ is overestimating the exponent, and we replace u by $(u + l)/2$. If the points are increasing, then l will be replaced. The estimates are refined until u and l get within a desired distance ϵ of each other. At this point, if the dataset $y/x^{(u+l)/2}$ has a DownUp feature, then we know that function $\bar{f}$ must have a relatively high coefficient c on a low order term. This diagnosis invokes the next step.
3. If, as we suspect, the current result is tracking a low-order term with a high coefficient, then this term will dominate $\bar{f}$ for small values of x. Thus we can approximate the upper bound for small x's to be cx^d. Let (x_1, y_1) and (x_2, y_2) be two points from the beginning part of the curve. If we consider that $y_1 \approx cx_1^d$ and $y_2 \approx cx_2^d$, then d can be approximated by $\frac{\log y_1 - \log y_2}{\log x_1 - \log x_2}$, and c is $\frac{y_1}{x_1^d}$. Now we can correct the model using these estimates, in order

to make the high-order term appear. For all points (x, y), we transform y into $\frac{y}{x^d} - c$. Now we can apply the same procedure as above to find the a and b parameters, assuming that $y \approx ax^b$. In this case, though, we use for our estimates two points that have high values of x, as the influence of the high-order term is stronger for these points.

This technique illustrates a way in which models can be improved by generating data and comparing it against the real values to obtain diagnostic information (step 2), which suggests a method specific to the diagnosis—in this case, a method specific to functions with large coefficients on low order terms. (We envision similar diagnostics and methods for functions with negative coefficients, but we haven't designed them, yet.)

The results of this method are found in the columns labelled HY in Tables 1 and 2. The results are tight upper bounds when $\bar{f}$ does in fact contain a low order term with a large coefficient (functions 7, 10, 11, 14, 17 in Fig. 1). In fact, these bounds are tighter than those returned by the other methods, and, remarkably, this hybrid method estimates coefficients and low order exponents very well. When the functions do not contain low order terms with large coefficients, the bounds returned by this method remain correct but they are looser than those given by other methods. Interestingly, this situation is often indicated by very low estimated coefficients on the high order terms; for example, in funtion 1 (Fig. 1), the coefficient of the first term is .03. The only cases when the technique fails are those in which negative coefficients appear in the low-order terms. The failure is probably due to the sensitivity of the Guess Ratio heuristic to such circumstances. This new method was also tested on noisy datasets but the noise had negligible effects. The new method used different oracles and different implementations of oracles from the previous methods, which might account for the relatively robust performance. Or, the small effects of noise might be due to a different method for sampling data from the given functions. Clearly, the effects of noise on these methods are still poorly understood.

5   Remarks

In our informal explorations and designed experiments with little or no random noise in the data, the rules generally get within a $\sqrt{x}$ factor of the exact bound. On data from algorithms, the rules can get within a factor of x and sometimes within $\sqrt{x}$. The rules are not reliable in discerning lower-order and logarithmic factors (this holds even when logarithms are added to the NextOrder oracle), and it doesn't seem likely that taking larger problem sizes would help.

Most rules do not respond much to larger problem sizes. However the quality of bound obtained is very responsive to variance in the data. This is good news for algorithm analyzers when Y is correlated with runtime, since variance can be reduced by taking more random trials, and trials are easier to get when Y grows slowly.

Can Humans Do Better? In one experiment, the third author was given the 25 data sets presented here, without any information about their provenance, and was allowed to use any data analysis tools to bound the function. He was more frequently incorrect than any of the implemented rules, and the human/machine interactions took considerably more time to accomplish. A second experiment involved strict application of the heuristics, but with a human oracle (the first co-author) who was familiar with the eight algorithmic data sets. Again, interactive trials require much more time to perform. Very preliminary results indicate that: GR produces worse (less close) bounds with a human Trend oracle; the human Concavity oracle tends to agree with the implemented one in the Power rules (no improvement); an interactive GD is more successful at finding DownUp curves (more frequent success, but not tighter bounds); an interactive BoxCox can be more successful by providing bounds that bracket the estimate rather than optimizing the transformation.

Removing Constant Terms. In many applications it may be possible to remove a constant from Y before analysis, either by testing with $x = 0$ or by subtracting an estimated constant. Our preliminary results suggest that subtraction of a known constant uniformly improves all the rules, but subtracting an estimated constant gives mixed results.

Some Negative Results. A basic requirement is that a heuristic be internally consistent. That is, should not be possible to reach the contradictory conclusions "Y is growing faster than X^2" and "Y is growing more slowly than X^2" on the same data set. Surprisingly, two plausible approaches turn out to have exactly this failure. The first, which is perhaps the most obvious approach to the bounding problem, is to use general regression to fit a function $f(x)$ and to read its leading term, using regression analysis to determine an upper/lower bound claim. In preliminary tests with this approach it quickly became clear that the results were primarily artifacts of the regression technique: contradictory bound claims, such as $\Omega(x^{2.2})$ and $O(x^{1.8})$ were easy to obtain by small changes in the regression method. This approach was abandoned early in this research. The second is based on Tukey's [16] "ladder of transformations." This approach also gives contradictory results depending on whether the transformation is applied to Y or X.

References

1. A. C. Atkinson (1987) *Plots, Transformations and Regression: an Introduction to Graphical Methods of Diagnostic Regression Analysis,* Oxford Science.
2. R. A. Becker, J. A. Chambers, and A. R. Wilks (1988) *The New S Language: A Programming Enviornment for Data Analysis and Graphics,* Wadsworth & Brooks/Cole.
3. J. L. Bentley, D. S. Johnson, F. T. Leighton, and C. C. McGeoch (1983) "An experimental study of bin packing," *Proceedings of the 21st Allerton Conference on Communication, Control, and Computing,* University of Illinois, Urbana-Champaign. pp 51–60.

4. J. L. Bentley, D. S. Johnson, C. C. McGeoch and L. A. McGeoch (1984). "Some unexpected expected behavior results for bin packing," *Proceedings of the 16th Symposium on Theory of Computing*, ACM, NY. pp 279–298.

5. G. P. Box, W. G. Hunter, and J. S. Hunter (1978) *Statistics for Experimenters*, Wiley & Sons.

6. J. M. Chambers et al. (1983) *Graphical Methods for Data Analysis*, Duxbury Press.

7. P. R. Cohen (1995) *Empirical Methods for Artificial Intelligence*, the MIT Press.

8. T. Cormen, C. Leiserson and R. Rivest (1990) *Introduction to Algorithms*, the MIT Press.

9. D. E. Knuth (1981), *The Art of Computer Programming: Vol. 3 Sorting and Searching*, Addison Wesley.

10. C. C. McGeoch (1992), "Analyzing algorithms by simulation: Variance reduction techniques and simulation speedups," *ACM Computing Surveys*. (245)2, pp. 195–212.

11. C. C. McGeoch (1995) "All pairs shortest paths and the essential subgraph," *Algorithmica* (13), pp. 426–441.

12. J. O. Rawlings (1988) *Applied Regression Analysis: A Research Tool*, Wadsworth & Brooks/Cole.

13. C. Schaffer (1990) *Domain-Independent Scientific Function Finding*, Ph.D. Thesis, Technical Report LCSR-TR-149, Department of Computer Science, Rutgers University.

14. R. Sedgewick (1975), *Quicksort*. Ph. D. Thesis, Stanford University.

15. J. Soer and R. Bulirsch (1993) *Introduction to Numerical Analysis*, Springer-Verlag.

16. J. W. Tukey (1977) *Exploratory Data Analysis*, Addison-Wesley.

17. L. Weisner (1938) *Introduction to the Theory of Equations.*, Macmillan.

Data Classification Using a W.I.S.E. Toolbox

Ian Berry and Paul Gough

Space Science Centre,University Of Sussex, Brighton UK

Abstract. This paper describes a toolbox for manipulating huge databases that are created in many scientific domains. The Whole Information System Expert (W.I.S.E.) was originally developed to fulfill the need for this type of toolbox in Space Science. Its effectiveness is shown using an example from the Remote Sensing field. We discuss the methods of converting image data into a useable form for presentation to Artificial Neural Networks and weigh up the pros and cons of each technique.

1 Introduction

Within many fields of scientific research, vast quantities of data are being continually produced. Much of this data is stored in huge databases that are rarely accessed by researchers who do not have the time to fully exploit terabytes of data. This situation is particularly prevalent in Space Science where it is not uncommon for the researcher who has created the instrument to be the only person able to completely understand and hence fully exploit the data. In addition, these researchers will often only concentrate on specific events in the data. Although it is possible to train people to perform the (often prohibitive) time consuming analysis, it would be difficult for them to recognise any new features in the data even after long periods of familiarisation. If these databases are left untouched, however, important scientific results may be ignored and eventually lost.

W.I.S.E. is a system that greatly reduces the time required analysing data sets. As most data analysis requires looking repetitively for patterns that correspond to known and unknown information, it is logical to look to computers to tackle the problem. In particular the application of Artificial Neural Networks (ANNs) is ideal for this kind of activity as they can be trained on sample data and then applied to other sections of the data with relative ease. Once the ANN has been trained, the expert user labels the resulting classes, and any computer literate user can operate the system. If the ANN fails to classify a feature, it can be taken to the expert for labelling and added to the ANN training set. This relieves the main burden of analysis from the experts who can then spend more of their time examining new phenomena rather than studying what is already known.

Similarly, inexpert users are often employed to generate a list of known features in the data, i.e. a morphology of events. This can now be performed au-

X. Liu, P. Cohen, M. Berthold (Eds.): "Advances in Intelligent Data Analysis" (IDA–97)
LNCS 1280, pp. 53–64, 1997.   © Springer–Verlag Berlin Heidelberg 1997

tomatically by W.I.S.E. and thus the expert users can concentrate on detailed analysis of individual events or new phenomena identified by W.I.S.E.

This tool, called the *Whole Information System Expert* (W.I.S.E.) [5] [6] [1], provides a toolbox for manipulating and classifying data. The classification is performed by unsupervised ANNs. A suite of pre-processing techniques has been included to manipulate the data before its application to the ANN by the chosen presentation method. The method of presentation of the data to the ANN is discussed in Section 3. At present, the system has been applied to several different data set types including upper atmosphere satellite data and Earth remote sensing data, but other data sets could be easily included.

2   The System

An overview of the W.I.S.E. architecture can be seen in Figure 1. The databases are generally held locally, but with the increasing use of the internet, it is anticipated that access to remote study data through this medium would be a useful addition. Once a representative section of these databases has been loaded into the system, the data manipulation can be performed as required.

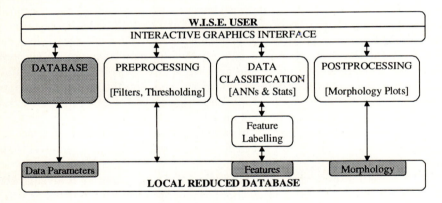

Fig. 1. An Overview of the W.I.S.E. architecture

The operation of the W.I.S.E. system is shown in Figure 2. As can be seen, the system still relies on Expert user input, but the amount of input required is greatly reduced.

The algorithms available for manipulating the data sets are very general, to cope with the variety of data sets that can be processed. Some data sets require more data–specific techniques that have limited scope for other data sets. One example is how the data is to be presented to the ANN. This is discussed in Section 3.

The expert user is only required to choose the pre-processing techniques necessary for the particular data set. The task of controlling the classification

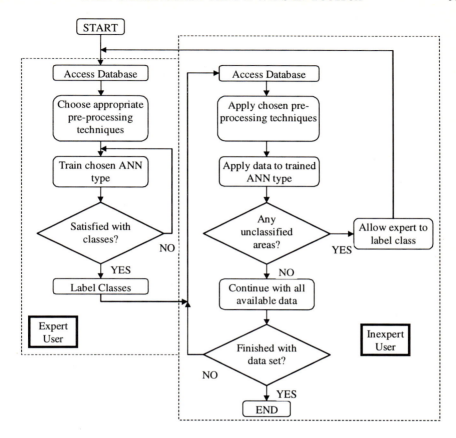

Fig. 2. Flow Chart of W.I.S.E. Operation

of the remainder of the data is left to an "inexpert" user - someone who has knowledge of the computer environment but not necessarily any understanding of the data set. In studying the data using the W.I.S.E. system, this inexpert user could become knowledgeable using the class labels that were defined by the Expert User.

It is envisaged that a collection of both ANNs and statistical methods will be available for the classification of the data. Currently, however, W.I.S.E. is limited to three ANN types: the Adaptive Resonance Theory (ART), created by Carpenter and Grossberg [4], the Self-Organising Map created by Kohonen [8] and the Associated List Memory (ALM) created by Gough [7]. ALM is a new ANN specifically developed for large input sizes and for easy access to the discovered knowledge.

Each of these three ANN architectures supports unsupervised learning, considered essential for this type of data study. In using unsupervised methods, we are making no assumptions about the content of the data and this makes it easier to identify new features.

The post-processing that is available for the classified data will depend largely on the data set. For example, when analysing space plasma wave-particle interaction data, plotting the location of each feature in relation to satellite orbit position (morphology plots) can provide an insight into the data. Another example, in the Remote Sensing field, would be where the results of classification are often compared with ground truth data to create a confusion matrix. These matrices are often associated with ANN studies to show the effectiveness of the classification process. All these post-processing techniques enable the user to estimate the accuracy of the system and get an overall picture of how the data set has been classified.

3   Preparation of data for Classification

Source data is generally an array of intensity levels that are converted into image form for the convenience of viewing (e.g. frequency – time spectrograms). The question is what parameter(s) do we extract from the image to present as inputs to the ANN? Several possible methods are available, all of which have arguments both for and against. These methods are:

3.1   Shape Extraction

This involves the analysis of the image and the extraction of shapes from the data. The image must be first separated into its component parts by a sequence of thresholding and edge detection algorithms, and the shape features found. The parameters that define the shape of features are extracted and used to train the classifier. Figure 3 shows features extracted from a section of an original image; the data comes from the CRRES (Combined Release and Radiation Effects Satellite) Swept Frequency Receiver Instrument. Shape parameters that could be used include edge entropy, degree of region spread and major region angle [3].

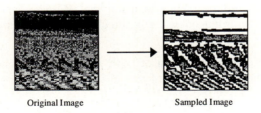

Original Image   Sampled Image

Fig. 3. Shape extraction from CRRES wave spectra time series

3.2   Pixel Intensity

This data is searched for areas of peak pixel intensity, and a snapshot of samples is taken from around that area. These samples are presented as inputs to the classifier. For example, Figure 4 shows the application of a simple peak search algorithm.

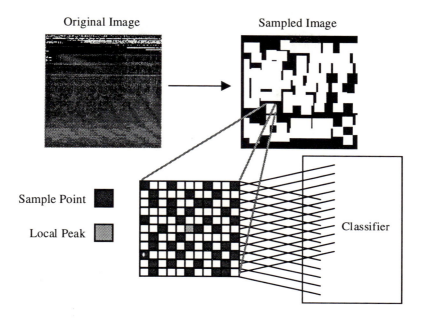

Fig. 4. Peak searching, sampling and application to the ANN

3.3   Data Driven Methods

Some data set formats have more intrinsic types of presentation to the ANN. Remote sensing data provides a good example of this. These data sets generally recorded simultaneously in specific electromagnetic bands, hence producing several overlapping data sets. For example, the Thematic Mapper Instrument on the Landsat 5 satellite contains data from seven electromagnetic bands (described in Table 1) for each location. Here it is logical to take pixel intensity from each band as a parallel input to the classifier with little or no pre-processing. Extensions to this procedure involve using composite images of the band intensities as additional inputs. These composites often give clearer images than a single band. Standard band combinations for this instrument are: 3,2,1 (visible colour composite); 4,3,2; and 7,4,3.

These data sets could be looked at using the other methods described above but it would depend largely on the type of processing required. For example, with the remote sensing data, if the study was the field shape over a certain area then you could choose to sample shape. However, in general land use surveys we concentrate on relative band contributions to each pixel (as shown by the example in the Results)

Band	Wavelength Interval (μm)	Spectral Response	Resolution (m)
1	$0.45 - 0.52$	Visible Red	30
2	$0.52 - 0.60$	Visible Green	30
3	$0.63 - 0.69$	Visible Blue	30
4	$0.76 - 0.90$	Near InfraRed	30
5	$1.55 - 1.75$	Middle InfraRed	30
6	$10.4 - 12.5$	Thermal InfraRed	120
7	$2.08 - 2.35$	Middle InfraRed	30

Table 1. Landsat Thematic Mapper Band Description

3.4   Pre-processing in General

A comparison of pre-processing techniques (Table 2) reveals a trade-off between the computer processing time used in performing the pre-processing and that used in performing the ANN processing. For example in the data-driven remote sensing, pre-processing is minimal but there is an input presentation to the ANN for each pixel. Shape extraction conversely involves intensive processing while the ANN is limited to one presentation per shape corresponding to a large group of pixels.

However, in general, for a maximum discovery, pre-processing should be minimised so that the analysis is not biased by pre-existing knowledge. Remote sensing data is ideal with its small input size to the ANN. Other data sets, such as space plasma physics data would require ANNs capable of very large input size ($> 10^3$ bits) if no significant pre-processing were used.

When choosing the pre-processing, it is the aim of the study which effectively defines the choice. The ANN is only as good as the information that it has been taught. For example, for a land use study, shape information might not necessarily add anything useful to the ANN learning.

4   Results

The results presented here are for the classification of a set of remote sensing images of London, England taken by the Landsat remote sensing satellite on 18th August 1984 and is in the form of seven 512x512 pixel images. The image "ground truth" (an interpretation of the actual ground classes), Figure 5, was created using local maps, photographic evidence and 'local' knowledge. The resultant classifications by the three ANNs currently available in the W.I.S.E. system can be seen in Figures 6, 7 and 8. The confusion matrices, from the comparison of the classified images with the ground truth, can be seen in Tables 3, 4 and 5 [2]. Confusion matrices are tables that compare the ANN classes with the actual ground classes on a pixel by pixel basis to provide some indication of the effectiveness of the classification procedure. In theory, a good classification would result in a diagonal matrix.

Shape Extraction	
Advantages	Disadvantages
– Limits noise used in training since most will be filtered in shape data extraction – Good classification results for particular image regions – Concentrates on most obvious data features	– Preconceived concept of phenomena types expected - little discovery – Loses any relation with the original raw data – Limited response with feature dense data – Processor intensive
Pixel Intensity	
Advantages	Disadvantages
– More related to raw data – No expectation of phenomena shapes – Simple extraction algorithm	– Affected by noise – Limited discovery – Concentrates on phenomena which have distinct pixel intensity peaks above background
Data Driven Methods	
Advantages	Disadvantages
– ANN input preparation is simple – Composite images can be included – ANN input size is smaller – Less pre-processing – More discovery	– Affected by noise – Shape information not included – More ANN processing – Larger number of input presentations to ANN

Table 2. Summary of ANN presentation techniques

In each case, the number of ANN classes is greater than the number of ground truth classes. Therefore, some ANN classes are combined to reduce their number to the number of ground classes, hence producing a square matrix. When a row is empty (all zeros), it means that the ANN has not created a class which corresponds to the ground truth class. For a given ground class, the "classification accuracy" is defined as the fraction of pixels that are classified into the associated ANN class. For a given ANN class, the "classification confidence" is the fraction of that class that belongs to the associated ground truth class. Thus, the accuracy and confidence percentages are calculated by dividing the 'diagonal' value by the total for that column or row. The overall classification accuracy is calculated by dividing the sum of all the greyed elements by the total number of image pixels (512x512).

The classified images are represented using grey scale, with each grey level representing one output class from the ANN.

These results show that W.I.S.E. is producing satisfactory results: above

70% classification accuracy. This has been mirrored in tests carried out on other data sets and compared with other techniques [2] and show these results, using unsupervised methods, compare favourably with both supervised ANNs and other, more traditional, statistical techniques.

5   Conclusion

The W.I.S.E. system will allow a user to apply a wide range of techniques to the analysis of scientific data. Using Artificial Neural Networks as part of this toolbox permits the simple classification of images and the labelling of classes. This allows an inexpert user to be left to complete testing of a large data set while the expert is free to examine only those important parts of the data that may contain new features. In doing this, the inexpert user is becoming acquainted with the data set, hence opening the data to a wider scientific audience.

6   Acknowledgements

The authors would like to thank David H. Brown, Curator, The Observatorium (http://observe.ivv.nasa.gov/), Jon W. Robinson, and Nicholas M. Short Sr. of Goddard Space Flight Center for help with the remote sensing data. We would also like to thank Lars Liden, Department of Cognitive and Neural Systems, Boston University for the ART Gallery source code and Roger Anderson at the University Of Iowa, USA for kindly supplying the CRRES data set

References

1. Berry, I.M., Reeder, B.M., Gough, M.P.: The Use Of Artificial Neural Networks For Pattern Recognition in Satellite Data. EUFIT '96 Proceedings, Verlag Mainz, Aachen **3** (1996) 1524–1527
2. Berry, I.M., Gough, M.P.: A comparison of ART and SOM Artificial Neural Networks for unsupervised classification of Remote Sensing data. submitted for publication 1997.
3. Brückner, J.R.: Automatic Pattern Recognition and Learning for Information Systems. Thesis, University Of Sussex 1995.
4. Carpenter, G.A., Grossberg, S.: The ART of Adaptive Pattern Recognition. IEEE Computer **21** 3 (1988) 77–88
5. Gough, M.P.: Automatic Feature Recognition and Morphology Generation. 'Earth and Space Science Information Systems', ed. A.Zygeielbaum, American Institute Of Physics. (1993) 653–658
6. Gough, M.P. and Brückner, J.R.: Geophysical Phenomena Classification by Artificial Neural Networks. 'Visualisation Techniques in Space and Atmospheric Sciences', Ed E.Szuszczewicz and J.Bredekamp, NASA SP-519. (1995) 243–251
7. Gough, M.P.: Associative List Memory. Neural Networks, awaiting publication (1997)
8. Kohonen, T.: Self Organization and Associative Memory. Springer-Verlag (1984)

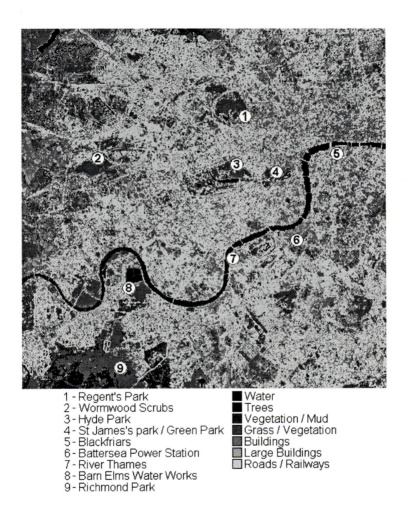

1 - Regent's Park
2 - Wormwood Scrubs
3 - Hyde Park
4 - St James's park / Green Park
5 - Blackfriars
6 - Battersea Power Station
7 - River Thames
8 - Barn Elms Water Works
9 - Richmond Park

■ Water
■ Trees
■ Vegetation / Mud
■ Grass / Vegetation
■ Buildings
■ Large Buildings
□ Roads / Railways

Fig. 5. The Ground Truth for the London Remote Sensing Data

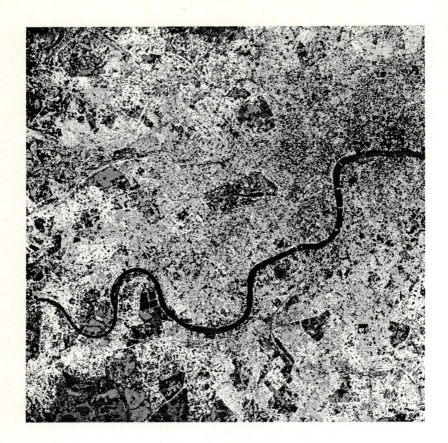

Fig. 6. ART Classification of London Remote Sensing Data

		Actual classes							Total	Conf
		1	2	3	4	5	6	7		
A	1	4864	34	407	0	20	32	298	5655	86.0%
N	2	2	3665	7	256	40	21	76	4067	90.1%
N	3	58	0	218	0	7	46	142	471	46.3%
	4	19	1153	123	33398	1630	49	1353	37725	88.5%
C	5	1	4214	23	766	39687	87	6169	50947	77.9%
L	6	31	0	187	138	1387	8766	3197	13706	64.0%
A	7	247	1399	299	175	41124	2446	103883	149573	69.5%
S	Tot	5222	10465	1264	34733	83895	11447	115118		
S	Acc	93.1%	35.0%	17.2%	96.2%	47.3%	76.6%	90.2%		

Table 3. Confusion Matrix for ART Network on London Data. Total Class Numbers = 220. Overall Classification Accuracy = 74.2%. Operation Time = 23 minutes 14 seconds.

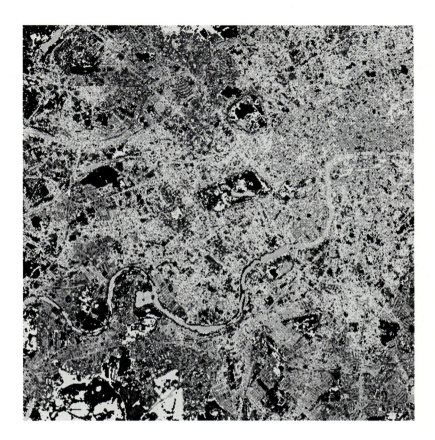

Fig. 7. SOM Classification of London Remote Sensing Data

		\multicolumn{7}{c}{Actual classes}		
		1	2	3	4	5	6	7	Total	Conf
A	1	2988	25	382	439	79	52	383	4348	68.7%
N	2	0	633	0	0	316	0	1	950	66.6%
N	3	0	0	1	0	0	0	0	1	100.0%
	4	0	4575	21	32141	1586	169	1263	39755	80.8%
C	5	2045	1439	442	1474	59592	4747	6393	76132	78.3%
L	6	12	1	151	250	576	4855	431	6276	77.4%
A	7	177	3792	267	429	21746	1624	106647	134682	79.2%
S	Tot	5222	10465	1264	34733	83895	11447	115118		
S	Acc	57.2%	6.0%	0.1%	92.5%	71.0%	42.4%	92.6%		

Table 4. Confusion Matrix for SOM Network on London Data. Total Class Numbers = 119. Overall Classification Accuracy = 78.9%. Operation Time = 1 hour 54 minutes 7 seconds.

Fig. 8. ALM Classification of London Remote Sensing Data

		Actual classes							Total	Conf
		1	2	3	4	5	6	7	Total	Conf
A	1	4822	34	387	0	24	23	265	5546	87.0%
N	2	1	2660	5	338	937	0	304	4245	62.7%
N	3	161	5	349	0	12	79	155	761	45.9%
	4	0	3536	21	33648	1255	31	1204	39695	84.8%
C	5	3	3297	40	625	53273	66	14643	71947	74.0%
L	6	38	531	147	10	681	8846	4386	14639	60.4%
A	7	197	402	324	112	27713	2406	94161	125315	75.1%
S	Tot	5222	10465	1264	34733	83895	11447	115118		
S	Acc	92.3%	25.4%	27.6%	96.9%	63.6%	77.8%	81.8%		

Table 5. Confusion Matrix for ALM Network on London Data. Total Class Numbers = 214. Overall Classification Accuracy = 75.4%. Operation Time = 1 hour 24 minutes 28 seconds.

Mill's Methods for Complete Intelligent Data Analysis

Tremaine A O Cornish

Institute of Child Health, University College London
&
Maxillofacial & Dental Department
Great Ormond Street Hospital for Children, London, WC1N 3JH
T.Cornish@ich.ucl.ac.uk

Abstract. If we are to implement Intelligent Data Analysis into computer systems we must instantiate the essential aspects of scientific method. In the quest to systematically unearth hidden information and bring forth knowledge from complex, noisy and incomplete data, we must be sure that such systems are first able to determine all possible patterns, which are suggestive of such information and knowledge from all possibly relevant, clean and complete data sets.

It will be shown that John Stuart Mill's methods as extrapolated in his 'System of Logic' provide such a set of algorithms. The value of all possible relevant data sets is determined prior to the coverage of each of Mill's methods. It is then shown that they are distinct from each other and given any particular data set, only one of the 'methods' is applicable, that is that they are disjunct. It is then shown that the methods cover all relevant data sets.

Attention is then directed to some Intelligent Data Analysis systems to determine which of Mill's methods they implement. None of them implement all of them or are aware of the applicability of the methods to the complex task of Intelligent Data Analysis.

1 Introduction

In this so called 'Information Age', we are constantly being told that we accumulate vast stores of data which cry out to be mined for nuggets of information and knowledge. If it is the case that there is useful information in these data warehouses, then we should endeavour to systematically extract all that we may. While considerable work has been undertaken to extract any and all possible information these systems are unable to indicate all possible patterns in clean and complete data sets. This being the case what hope is there of finding hidden information in noisy and incomplete data sets? It is thus necessary to determine a set of algorithms which are applicable to all possible data sets.

Mill's methods (Mill, 1843/1973), as formulated in his 'system of logic', are introduced as an ideal candidate and shown, with a slight extension, to cover

X. Liu, P. Cohen, M. Berthold (Eds.): "Advances in Intelligent Data Analysis" (IDA–97)
LNCS 1280, pp. 65–76, 1997.   © Springer–Verlag Berlin Heidelberg 1997

all possible relevant data sets (Cornish, 1995 & 1996). It will be shown that the methods are disjunct and that accordingly any data set can only be interrogated by one of the methods. It is then shown that Mill's methods are co-extensive with all relevant data sets. It is asserted that researchers are generally unaware of the contribution that Mill has to make to Intelligent Data Analysis (IDA) and that, because of the complete coverage of the methods, no one is working outside of them.

2 All Relevant Data sets

Before determining the coverage of any particular method, it is necessary to determine the extent of all possible relevant data sets. Let the number of variables in an investigation be v and the number of determinants d. From the number of variables we can determine the maximum possible number of distinct scenarios is $s = 2^v$. This is the maximum number of possible distinct scenarios in a particular experiment. The maximum number of possible distinct scenarios which exhibit the phenomenon is $p = 2^v - d$. Given that no pattern or rule of causality may be determined from just one trial, and that only non-empty sets are of utility, the smallest relevant data set, that is those which are of possible use, is one with at least two scenarios, at least one of which displays the phenomenon. The number of such relevant data sets is the set of observations and is given as:

$$|Rel| = (2^p - 1)(2^{s-p} - 1) + (2^p - p - 1) \tag{1}$$

Where the first term represents those data sets which consist of one or more scenarios displaying the phenomenon and one or more counter examples. The last term represents all those data sets containing two or more distinct scenarios all displaying the phenomenon. This simplifies to:

$$|Rel| = 2^s - 2^{(s-p)} - p \tag{2}$$

Having evaluated the extent of all relevant data sets, it is necessary to turn to Mill's methods as a set to cover all these relevant data sets.

3 Mill and his Methods

In 1843, John Stuart Mill (1806 - 1873) published his 'System of Logic' which was in part a formulation of Francis Bacon's '*Novum Organon*' (Bacon, 1620/1960). Mill stated that the purpose of his work was to "embody and systematise" the thoughts of writers and accurate practitioners of scientific inquiries (Mill, 1843/1973).

Mill asserted that his methods were the only possible modes of experimental inquiry - of direct induction *a posteriori*. These he asserted, are the limits of the mind for ascertaining the laws of the succession of phenomena. The different antecedents and consequence being, so far as the case requires, ascertained and discriminated from one another we are to inquire which is connected with which (ibid.).

These methods are extrapolated and the number of data sets to which each method applies is quantified. It will be shown that the methods, with a slight extension, partition the set of all possible relevant data sets. I.e. for any particular set of variables, there is but one which can be employed to explore that particular set of data.

3.1   Four Methods in Five Canons

Mill defined induction as the operation of discovering and proving general propositions. He formulated four methods expressed in five canons, those being:

1. The canon of agreement
2. The canon of difference
3. The joint method of agreement and difference or the method of indirect difference
4. The canon of residues
5. The canon of concomitant variation

The methods of agreement, difference and the joint method each deal with previously unsorted data and may be thought of as the principal methods. The other two methods have quite a different quality. The canon of residues relies on previously analysed data, and the canon of concomitant variation deals with variances, while they apply to binary or categorical data. Each of the principal methods are addressed in turn and their coverage evaluated.

The Canon of Agreement   The general formulation of this is as follows: *If two or more instances of the phenomenon under investigation have only one circumstance in common, the circumstance in which alone all the instances agree is the cause (or effect) of the given phenomenon.* This may be expressed as:

circumstances phenomenon

a b c   p

a d e   p

result therefore a $\Rightarrow$ p

The implication arrow '$\Rightarrow$' here is not meant to be read in the strict manner which is generally the case in mathematical logic, a looser interpretation is more appropriate. It is something of the order: "accordingly, it follows, from the evidence so presented, that 'a' appears to be the cause or effect or the probable cause or probable effect of the phenomenon 'p' or an important part thereof." A common example of this method occurs in determining the cause of an incidence of food poisoning, where using this method for such a scenario, one will determine which food the persons so affected have eaten in common. Since this method is only applicable to data sets where all the scenarios display the phenomenon, the extent of the coverage of the method of agreement is the number of such sets, i.e.:

$$|Agr| = (2^p - p - 1) \qquad (3)$$

This is determined from the number of all possible combinations of p, less that set which is a singleton and the empty set. However extreme caution must be exercised in the application of this method, for false conclusion may easily follow from the presented facts. This is readily demonstrated from the case of the scientific drinker.

Scientific Drinker: On being pressed by his colleagues that he should do something about his excessive drinking habits, he resolved to conduct a series of experiments and act on the results (table 1).

Circumstances			Phenomenon
Monday	Scotch	Soda	Drunk
Tuesday	Bourbon	Soda	Drunk
Wednesday	Brandy	Soda	Drunk
Thursday	Rum	Soda	Drunk
Friday	Gin	Soda	Drunk

Table 1. Outcome: The scientist vows never to drink soda again!

This effectively demonstrates that having come to a conclusion with the method of agreement, it is always necessary and appropriate to endeavour to undermine the new theory by seeking cases which refute it. It shows the importance of looking at instances where the phenomenon is not present. This brings one to a consideration of the method of difference.

The Canon of Difference   In contrast to the method of agreement, the method of difference exploits our knowledge of cases where the incident does not occur, by looking for differences in the surrounding circumstances. The formulation of this method is: *If an instance in which the phenomenon under investigation occurs, and an instance in which it does not occur, have every circumstance in common save one, that one occurring only in the former; the circumstance in which alone the two instances differ is the effect, or the cause, or an indispensable part of the cause, of the phenomenon.* The representation of this is:

circumstances phenomenon

$$b \; x \; y \qquad\qquad q$$
$$\neg b \; x \; y \qquad\qquad \neg q$$
result therefore b $\Rightarrow$q

Here we find that in a comparison of two cases, one exhibiting the phenomenon 'q ' and one not exhibiting it, that for the instance exhibiting 'q' there is a corresponding evidence that 'b' is present, which is absent where the phenomenon 'q' is absent. Accordingly, from the available evidence, it is asserted that 'b' is probably causally related to the phenomenon 'q'. It is important to note that this may also be interpreted as: 'q' may always be present except where 'b ' is absent. That is to say that the absence of 'b' results in the absence of 'q'. This is a slight modification on that presented by Mill, but the essence is retained. This method is applicable to all data sets with exactly two scenarios, one of which displays the phenomenon and one counter example. There are p possible scenarios exhibiting the phenomenon and s - p which do not, the number of possible data sets is given by:

$$|Dif| = p(s - p) \tag{4}$$

Having looked at the method of difference, attention is turned to the joint method.

The Joint Method of Agreement and Difference or The Method of Indirect Difference   In general it is not possible to secure examples which agree and differ in precisely the one circumstance associated with the phenomenon as is required for the methods of agreement and difference. Thus the methods of agreement and difference, even used in sequence cannot always produce a clear indication of the determinant circumstances. The joint method is an extension to these methods and is a very powerful technique which calls into evidence both those instances which exhibit the phenomenon and those which do not. The general formulation of this method is: *If two or more instances in which the phenomenon occurs have only one circumstance in common, while two or more instances in which it does not occur have nothing in common save the absence of that circumstance, the circumstance in which alone the two sets of instances differ is the effect, or the cause, or an indispensable part of the cause, of the phenomenon.* Which in turn may be represented thus:

1st application of the method of agreement

circumstances	phenomenon
f b c	r
f d e	r

result therefore f $\Rightarrow$ r by agreement

2nd application of the method of agreement

circumstances	phenomenon
b c	¬ r
d e	¬ r

result therefore ¬f $\Rightarrow$ ¬r by agreement

Application of the method of difference

1st result	$f \Rightarrow r$	by agreement
2nd result	$\neg f \Rightarrow \neg r$	by agreement
overall result	f $\Rightarrow$ r	by difference

At least two cases exhibiting the phenomenon of interest and two cases not exhibiting it are considered. With each group the method of agreement is applied, and the results compared through an application of the method of difference. So, to take the above illustration, after a search through all cases exhibiting the phenomenon of interest 'r', followed by a similar search through all those not exhibiting the phenomenon, it is determined that the common factor is 'f'. This double application of the method of agreement is highly effective in determining patterns suggestive of cause and effect which are not tractable with either of the aforementioned methods. The extent of the coverage of this method is $(2^p - p - 1)$ $(eqn.3)$ being all scenarios displaying the phenomenon, with all counter examples $(2^{s-p} - s + p - 1)$ and is determined thus:

$$|Jnt| = (2^p - p - 1)(2^{s-p} - s + p - 1) \tag{5}$$

which in turn simplifies to:

$$|Jnt| = 2^s - 2ps + 2^p p - 2^p - p2^{s-p} + ps - p^2 - 2^{s-p} + s + 1 \tag{6}$$

Having considered each of the principal methods, before determining the extent of the coverage of them as a set of sets, it is necessary to introduce an additional method.

4   An Extension to Mill's Methods

On considering the form of the data sets which may be addressed by the principal methods it becomes apparent that there is a sub-group of possible data sets which Mill overlooked. Having considered sets: where all exhibit the phenomenon; where there are only two scenarios, one each exhibiting and not exhibiting the phenomenon; and the set where a sub- set similar to that of agreement are considered alongside a similar sub-set which do not exhibit the phenomenon of interest, and each of these sub-sets contain at least two scenarios. A possible group of scenarios which was not conceived of by Bacon or Mill is that where all but one of the scenarios exhibits the phenomenon of interest or, all but one does not exhibit the phenomenon of interest. This shall be termed the method of agreement plus one.

4.1   The Method of Agreement Plus One and Its Coverage

To determine the coverage of this extension to the methods one may take all data sets with two or more scenarios displaying the phenomenon and exactly one not exhibiting the phenomenon, thus:

$$(2^p - p - 1)(s - p) \tag{7}$$

This strategy may then be inverted to look at counter examples for agreement using a single scenario displaying the phenomenon for confirmation. This, for its part, may be determined thus:

$$(2^{s-p} - s + p - 1)\, p \tag{8}$$

Hence the sum of equations 7 and 8 gives us the following:

$$|Ind| = (2^p - p - 1)(s - p) + (2^{s-p} - s + p - 1)\, p \tag{9}$$

which simplifies to:

$$|Ind| = 2^p s - 2^p p - 2ps + 2p^2 - s + p2^{s-p} \tag{10}$$

Having addressed the matter of all possible relevant data sets for each of the methods, it will be established that the extended set of principal methods are complete as a set and disjunct from each other.

5   The Methods are Complete and Conjunct

Having determined the coverage for each method in turn, it is necessary to determine how this relates to the set of all relevant data sets. In particular:

   * are there any relevant data sets which may not be explored by any method? and

   * are there any relevant data sets which can be explored by more than one method?

If it could be shown that the answer to the first question is no, (i. e. the sets Agr, Dif, Ind and Jnt are disjunct) then the answer to the second question may be established simply be looking at the size of the sets. To establish that the methods apply to disjoint sets, consideration is given in turn to the sets Agr, Dif, and Ind. It is determined that each method has no elements in common with any other method.

Since agreement applies only to data sets with no counter examples and all other methods require at least one counter example, it can be concluded that that:

$$Agr \cap (Dif \cup Ind \cup Jnt) = \emptyset \tag{11}$$

Further, since difference applies to only pairs of results and the remaining two methods require sets of at least three observations, it can be concluded that:

$$Dif \cap (Ind \cup Jnt) = \emptyset \tag{12}$$

Finally, it is noted that for all data sets where agreement plus one applies, there is either a single instance of the phenomenon or a single counter example, but the joint method requires multiple instances of the phenomenon and multiple counter examples. Thus:

$$Ind \cap Jnt = \emptyset \tag{13}$$

From these results, it follows that:

$$Dif \cap Ind = \emptyset \qquad \wedge \qquad Jnt \cap Dif = \emptyset \tag{14}$$

and that:

$$Agr \cap Dif = \emptyset \quad \wedge \quad Agr \cap Ind = \emptyset \quad \wedge \quad Agr \cap Jnt = \emptyset \tag{15}$$

Thus it has been established that there are no data sets which may be explored using more than one of the extended principal methods.

Regarding the second question, if the total number of cases covered by all of the methods is the same as the number of relevant data sets, then there is no relevant data set which may not be explored and each data set may only be explored by one of the methods. That is, the set of methods is complete. It is necessary to evaluate:

$$|Agr| + |Dif| + |Ind| + |Jnt| \tag{16}$$

Substituting equations 3, 4, 6, and 10, we get:

$$(2^p - p - 1)(p(s - p)) + (2^p s - 2^p p - 2ps + 2p^2 - s + p2^{(s-p)}) +$$

$$(2^s - 2ps + 2^p p - 2^p - p2^{(s-p)} + ps - p^2 - 2^{(s-p)} + s + 1) \tag{17}$$

This reduces to:

$$2^s - 2^{(s-p)} - p \tag{18}$$

From equation 2, it can be seen that this is the number of all relevant data sets. I.e.

$$|Agr| + |Dif| + |Ind| + |Jnt| = |Rel| \tag{19}$$

Thus, it is concluded that the coverage of the extended principal methods is complete.

6   Implementation of the Methods

Given that Mill's methods are the only set of algorithms shown to be complete in their coverage, it is appropriate the measure all other systems against them, to use them as the benchmark. Consideration is given to a number of IDA systems these being: BACON, DENDRAL, ID3, the General Unary Hypothesis Automaton (GUHA) and the RX Project, to illustrate the limited scope of most systems.

6.1   BACON

Langley et al. (1986 & 1987) noted that scientific discovery is a complex enterprise involving many components. This understanding lead them to construct four Artificial Intelligence systems; being: BACON, GLAUBER, STAHL and DALTON. Langley (1981) noted that Francis Bacon believed that if one gathered enough data that regularities would leap out at the observer, he then proceeds to state that the BACON program discovers empirical laws in just the same way. He makes no other reference to Bacon and none to Mill. Nonetheless, Langley does state that the program is named after the great man. It is interesting to note that while the developers of the BACON system were aware of the Mill's work, they did not incorporate all of Mill's methods into BACON.

Differences are discarded or ignored rather than using them as an important and valuable tool in the discovery process, and is thus an application of the method of agreement and possibly difference (table 2). It is noted that BACON is presented with the evidence, and only the evidence, that could lead it to the correct answer while it is the case that after considering a wide set of variables we eliminate some and include others in our final hypothesis. It is problematic as to whether the joint method is implemented, as only a small set of independent terms can be manipulated at a time (table 2).

6.2 DENDRAL

DENDRAL has three fundamental parts, those being: PLAN, GENERATE and TEST. Regarding any determination of which if any of Mill's methods are implemented by the programme, PLAN is the important part. The automatic inference of constraints and applying them to the available data to determine all the consistent combinations of a set of atoms. The constraints are listed in two parts: molecular fragments (clusters of atoms) that must be in the final molecular structure and fragments that must not appear in the final structure (Buchanan & Feigenbaum, 1969).

PLAN is essentially executed in the first of five sub-programmes, being the Preliminary Inference Maker. The others being the Data Adjuster; the Structure Generator; the Predictor and the Evaluation Function. The programme runs up to three tests for each structure.

1. Is the empirical formula of the structure compatible with the empirical formula of the molecule? If not get the next structure.
2. Is any necessary condition falsified by the spectrum? If so put this structure on the BADLIST and get the next structure.
3. Are all sufficient conditions satisfied by the data? If so put this structure on the GOODLIST and get the next structure. (Buchanan & Feigenbaum, 1969).

So we find that on being presented with a spectra, the programme compares it with a series formula and seeks to determine whether the spectra presented is consistent with them each in turn. This may be seen as consecutive applications of the method of agreement. The cases where the condition is falsified may for their part be seen as an application of the method of difference. Overall the effect is that the joint method is implemented (table 2). Regarding the method of residues, while it is the case that the knowledge base may be extended it does not 'learn ' from experience thus it does not implement the method of residues. As for the method of concomitant variation the programme is not looking for regularities of differences over a series of events, rather it seeks common categorical values and as such it is not implementing the method (table 2).

6.3 ID3

ID3 or 'Iterative Dichotomizer Three' implements a hill climbing algorithm of decision trees (Quinlan, 1986). It assumes that the simplest decision tree which covers all of the training set, has the greatest likelihood of accurately classifying an unseen population.

ID3 was based upon Hunt's Concept Learning System (CLS) (Hunt et al., 1966). Both CLS and ID3 are given a set of situations (or scenarios) each described in terms of features, attributes or values, and a class value, induces a situation classification rule. In ID3 the algorithm is extended to enable it to grow a small example set from an exhaustive file of pre-classified examples, allowing it to induce rules incrementally. It did this by adding a selection of counter examples to the working set, re-inducing an improved rule on each iteration, thus the principal methods are covered (table 2). Given that the systems 'knowledge' can be expanded through iterations it also implements, at least in part, the method of residues (table 2). Quinlan did not refer to the work of Bacon or Mill while dealing with ID3, whereas Shapiro (1987) who built upon that with Interactive ID3 does note that Bacon set out some rules for inductive inference. It is however the case that he did not identify these rules and that he was apparently unaware of Mill's formulation of them.

6.4 General Unary Hypotheses Automaton

The General Unary Hypotheses Automaton (GUHA) (Hájek et al., 1966) was presented as an application of 'mathematical' logic to research problems of the concrete sciences. It was proposed as an application to all problems where it is required to obtain unknown laws, relations or causal connections. It was claimed that the basis of mathematical logic makes it possible to describe all assertions which might be hypotheses. The aim was to determine generally valid relations such as dependencies and connections between properties, as such this approach is seen to have broadly similar aims to those of Mill (1843/1973).

The GUHA reduced all data to a binary format. In the second stage of the process, properties of objects are identified, and their presence or absence indicated accordingly thus the methods of agreement, difference are implemented as is the joint method. It also exhibits some properties of the method of residues as it filters subsequent clauses in the explanation on the basis of what has been discovered during the previous iterations of the algorithm (table 2). However, this filtering behaviour is limited to the results attained in the current execution of the program. In some respects it is a crude example of the method of concomitant variation (table 2).

6.5 The RX Project

Blum & Wiederhold, in the RX Project (Blum & Wiederhold, 1978; Blum, 1982), proposed an implementation to automate the process of hypothesis generation and the exploratory analysis of data in large databases. It embodies the method of agreement within the method of concomitant variation which calculates Spearman's rank correlation coefficients for each patient and for each observable time delay between all pairs of variables. This attempts to reduce the work load by exploiting a knowledge base of known cause effect relationships, as in the GUHA this can be seen as an application of the method of residues to filter out known effects which might confound the discovery process (table 2). The system displays significant insights into the implementation of the methods of residues and concomitant variation.

6.6   Coverage and Awareness of Mill's methods

Consideration has been given to the coverage of Mill's methods in other systems (table 2). There are a number of aspects which deserve to be drawn out from this table but first it is noted that only one of the authors of these systems (Langley et al., 1987) referred directly to the work of Mill. Accordingly it is reasonable to assume that the remainder were unaware of the applicability of Mill's methods to the discovery process.

Package Name	Authors	Year	Method of Agreem.	Method of Difference	Joint Method	Method of Residues	Method concomitant variation
BACON	Langley et al.	87	yes	possibly	no	partial	partial
DENDRAL	Lederberg, Feigenbaum, & Buchanan	64/5	yes	yes	yes	no	no
GUHA	Hájek et al.	66	yes	no	yes	yes	partial
ID3	Quinlan	86	yes	yes	partial	partial	no
RX Project	Blum & Wiederhold	82	yes	yes	no	partial	partial

Table 2. Coverage of Mill's methods by others

It is noted that the system which has been all but overlooked by all in the knowledge discovery domain, the GUHA, is the one system with the greatest coverage of the methods, and is the only one to really incorporate the joint method of agreement and difference. It is also very interesting to observe that the work of Blum and Wiederhold (Blum and Wiederhold, 1978: Blum, 1982) covers the most complex of the methods, concomitant variation. It is further noted that none of these systems proposed any methods which clearly fall outside of those formulated by Mill.

7   Summary

It has been shown that Mill's methods cover all possible relevant data sets and that developers are unaware of the valuable contribution that they have to make to IDA in general. It has further been shown that none of the systems considered here implement all of the methods. In Machine Learning through to the current work in IDA, Knowledge Discovery in Databases (KDD) and Data Mining, there is an unstated belief that any system which is capable determining large numbers of possible causal patterns, is finding all the patterns or at least we are finding more patterns than we can deal with already. Certainly systems do highlight a large number of possible causal patterns and there are considerable problems regarding the noisy and incomplete data, but this does not negate the high probability of not identifying all possible causal relationships. Over successive generations in Machine Learning only passing comment has been made to Francis Bacon's work on scientific method and no detailed reference is made to Mill's formulation of the same. It is suggested that the general absence of awareness of even the existence of Mill's work is because it is a work of philosophy, rather than computer science or engineering. This is unfortunate, given that the impetus behind Artificial Intelligence came from both engineering and philosophy.

Bibliography

Bacon, F., (1620/1960) The New Organon, [Edited by Anderson, F.H.], 1960 The Bobbs Merrill Co. Inc., Indianapolis, New York.

Blum, R.L., (1982) Discovery and Representation of Causal Relationships from a Large Time-Oriented Clinical Database: The RX Project, Ph.D. Thesis, Department of Computer Science, Stanford University, January 1982, Report No. STAN-CS-82-900.

Blum, R.L., & Wiederhold, G., (1978) Inferring knowledge from clinical data banks utilizing techniques from artificial intelligence, In the proceedings of the 2nd Annual Symposium on Computer Applications in Medical Care, Washington, D.C., IEEE, November, pp. 303-307.

Buchanan, B.G., & Feigenbaum, E.A. (1969) Heuristic DENDRALL A Program for Generating Explanatory Hypotheses in Organic Chemistry, in Meltzer & Michie, (1969).

Cornish, T.A.O., (1995) It's not always bunk, honest!, New Scientist, Forum, Dec, 9.

Cornish, T.A.O., (1996) Historical Perspectives on Information Science, International Journal of Systems Research and Information Science, Vol. 7, No. 2, pp. 105-116.

Hájek, P., Havel, I., & Chytil, M., (1966) The GUHA Method of Automatic Hypotheses Determination, Computing, Vol. 1, pp. 293-308.

Hunt, E.B., Marin, J., & Stone, P.T., (1966) Experiments in Induction, Academic Press, New York.

Langley, P., (1981) Data-Driven Discovery of Physical Laws, Cognitive Science, Vol. 5, pp. 31-54.

Langley, P., Zytkow, J.M., Simon, H.A., & Bradshaw, G L. (1986), The Search for Regularity: Four Aspects of Scientific Discovery, in Michalski et al. (1986).

Langley, P., Simon, H.A., Bradshaw, G.L. & Zytkow, J.M., (1987) Scientific Discovery: Computational explorations of the creative process, MIT Press, Cambridge MA.

Mill, J.S., (1843/1973) Collected Works, Volume VII, A System of Logic, Ratiocinative and Inductive: Being a Connected View of the Principles of Evidence and the Methods of Scientific Investigation, Books I-III (Ed. Robson, J.M.) University of Toronto Press, RKP.

Meltzer, B., & Michie, D., (1969) Machine Intelligence 4, Edinburgh University Press.

Michalski, R.S.; Carbonell, J.G.; & Mitchell, T.M., (1986) Machine Learning: An Artificial Intelligence Approach, Volume II, Morgan Kaufman, Los Altos.

Quinlan, J.R., (1986) Induction of Decision Trees, Machine Learning, Vol. 1, No. 1, pp. 81-106.

Shapiro, A. D., (1987) Structured Induction in Expert Systems, Addison Wesley, Wokingham.

Integrating Many Techniques for Discovering Structure in Data

Dawn E. Gregory and Paul R. Cohen

Experimental Knowledge Systems Laboratory
Computer Science Department, LGRC
University of Massachusetts
Box 34610, Amherst, MA 01003-4610

Phone: (413) 545-3616
Fax: (413) 545-1249
E-mail: {gregory,cohen}@cs.umass.edu

Abstract. This paper describes a formal representation of the discovery process that integrates of any number of data analysis strategies, regardless of their differences. We have implemented a system based on this formalization, called the *Scientist's Empirical Assistant* (SEA). SEA employs several analysis strategies from the discovery literature, including techniques for function finding, causal modeling, and Bayesian conditioning. It uses high-level knowledge about the discovery process, the strategies, and the domain of study to coordinate the selection and application of analyses. It relies on the skills and initiatives of an expert user to guide its search for structure. Finally, it designs and runs experiments with a simulator to verify its findings. SEA's primary sources of power are its abstraction of the discovery process and its numerous analysis strategies.

1   Motivation

Data analysis is not simply a matter of applying a formula to a set of data: it is a complex process that involves selecting an appropriate representation, designing an analysis strategy, gathering data into an appropriate format, applying the formula(e), and explaining the results. Thus, if we want data analysis programs that behave intelligently, they must be designed to accomodate each of these tasks.

Recent research in Machine Learning and Scientific Discovery indicates that systems with a variety of representations are more proficient at uncovering structure in data. There are several reasons to suspect this is true. First, in having several techniques to choose from, a system can select the one most suited to the analysis problem at hand (e.g. [1]). Second, integrated systems have the advantage of supplementing one kind of result with others, information which often constitutes an *explanation* of previous findings (e.g. [6]). Finally, interesting

X. Liu, P. Cohen, M. Berthold (Eds.): "Advances in Intelligent Data Analysis" (IDA–97)
LNCS 1280, pp. 77–88, 1997.   © Springer–Verlag Berlin Heidelberg 1997

structure is often discovered in the process of shifting from one representation to another [9]. Thus, there is a significant advantage to integrated discovery systems.

This paper describes our approach to the integration of analysis techniques and a system, called the *Scientist's Empirical Assistant* (SEA), that implements this view. Section 2 presents an example of complex data analysis that employs a variety of strategies and representations. We then make several high-level observations about the analysis process to motivate our design, which is described in section 4. Finally, we describe the key contributions and identify areas of future work.

2   Example

We begin with a detailed example of data analysis to show how different strategies and representations might be employed to discover structure within data. This analysis was originally performed manually, as described in [4], and has subsequently been replicated by SEA under the guidance of a human user.

The example focuses on a real-world problem from the domain of parallel computing. Parallel architectures employ several processing units that work in conjunction to solve problems more quickly than is possible on a single processor. In theory, P processors should reduce the overall running time by a factor of P; for example, N unit-time tasks should be executed in time $T = N/P$. In practice, connectivity among the processors and interdepencies among the tasks have a significant impact on performance, because some processors may be forced to remain idle for long periods of time. To deal with this problem, designers are highly concerned with policies for *balancing* the load among processors, to ensure that idle periods are a brief as possible.

In this problem, data analysis is used to discover the properties of two load-balancing policies for a particular parallel architecture on a specific kind of computing task.[1] Simulations of the architecture and stochastically-generated input problems generate the data used in analysis. The simulation is configured by three parameters: the number of processors, P, the load-balancing policy, Π, and an input parameter, α. Among the values output by the simulator are the net running time, T, and the number of unit-sized tasks, N. Simulation of both policies on architectures of different sizes and various input settings yielded a preliminary dataset of 360 data points containing the variables Π, P, α, N, and T.

The purpose of the analysis is to determine whether optimal performance is achieved; that is, we want to determine if $T = N/P$. Given the preliminary dataset, the first strategy is to try a *t-test* on the paired samples of T and N/P. The test indicates that T and N/P are not significantly different, which lends support to the hypothesis but cannot be considered conclusive evidence (i.e.

---

[1] Detailed descriptions of the architecture, the task, and the policies are irrelevant to the ensuing discussion, so we refer the interested reader to [4] for more information.

the test cannot *accept* the null hypothesis $T = N/P$). Thus we must consider additional strategies for testing the hypothesis.

Now we consider the possibility that the error term, $\epsilon = T - N/P$, could be different than zero under specific conditions. To identify the factors that influence ϵ, it is necessary to bring in a new representation for the relationship between ϵ and other variables. In this case, we are interested in *dependencies* among variables; for example, whether the value of ϵ depends on P. Heuristic rules generate hypotheses about potential dependencies: ϵ might depend on any of P, Π, α, or N.[2]

The new representation indicates a different set of analysis strategies which are applied in turn. P, Π, and α are independent variables and can thus be considered discrete-valued for analysis, so one-factor analysis of variance (ANOVA) is employed. N is strictly numeric so its influence is tested with linear regression. From these analyses, we find that P, Π, and N all have significant effects on ϵ, but α does not.

At this point, there are several possibilities for further analysis. We can use two-factor ANOVA to determine whether there is an interaction between P and Π. We can explore the separate effects of P or Π on the relationship between N and ϵ. Or, we might skip both of these tasks and immediately start exploring the combined effects of P and Π on the relationship between N and ϵ.

Regardless of the approach chosen, we eventually arrive at several interesting results: the slope of the regression of ϵ on N increases as P increases; the slope is consistently larger for one of the policies given each value of P; the slope is non-zero for some combinations of P and Π, but not all. These results can be explained by external knowledge about the load-balancing problem. As P increases, it is more likely that some processors will become idle during the computation, regardless of the load-balancing policy employed. One policy does a better job at balancing the load, and this policy apparently behaves optimally (i.e. the slope is approximately 0) for some values of P. These results are quite useful to the designer, who now has empirical evidence that a certain configuration will utilize its resources effectively.

3   Observations

The example described above has several important features that are common to data analysis. First, it shows that analysis is not simply a matter of applying a formula to obtain a result — it is an *iterative* process that relies heavily on context and previous results to identify which formula is appropriate and how the results should be interpreted. Second, the goal of data analysis is to develop a model that accurately predicts the values of variables, and this goal is attained through *variance reduction* techniques. Finally, the model should correspond to the true structure of the world, so its predictions must be reconciled with higher-level knowledge of the domain. Often, one or more of these features is overlooked by the designers of data analysis systems.

---

[2] By definition ϵ depends on T, so this hypothesis is not generated.

3.1   A Complex, Knowledge-Intensive Process

Data analysis is an iterative process consisting of several stages. First, the question to be addressed by analysis is formalized as a *hypothesis*; for example, the question of whether an architecture performs optimally is initially represented by the hypothesis $T = N/P$. The form of the hypothesis produces *strategies* for analysis: the "=" in $T = N/P$ indicates that we might try a t-test. Next, *observations* of real behavior are collected and arranged in the required format. The analysis strategy supplies a formula that is applied to the observations, yielding a *result*. Finally, the result must be explained by considering the *reasons* it may occur; for example, $T = N/P$ can be explained by $\epsilon = 0$.[3] The task of explanation often generates new hypotheses, and the whole process repeats.

Every data analysis system must consider each of these stages, but none has automated all of them. Statistical packages provide facilities for hypothesis testing, but must rely on the user to formulate hypotheses, select the appropriate analysis, prepare the data, and explain the results. Intelligent discovery systems, such as Bacon [5], Tetrad [3], and C4.5 [7], formulate hypotheses of a specific form and then perform analysis in this context, but rarely have facilities for gathering and preparing data or explaining the results of analysis. Other discovery systems have addressed these latter concerns in the context of certain domains; for example, Fahrenheit [10] deals with issues of experiment design within the domain of chemistry. Unfortunately, none of these systems supports all the stages in a flexible, domain-independent manner.

There are good reasons why intelligent data analysis has been restricted to certain domains or specific types of hypothesis. First, each stage of the process may produce several potential courses of action, and selecting among these alternatives is a nontrivial task. By focusing on a specific domain or hypothesis test, it is possible to minimize the number of choices and to formalize policies for decision-making. Second, experiment design and data collection are difficult to automate, because they rely heavily on domain knowledge and the ability to interface with the physical world. Again, restricting the context makes automation possible. Finally, explaining results requires a deep semantic knowledge of the domain, the source of the data, the underlying assumptions, and valid interpretations of the analysis. In sum, it is the complexity of the analysis process and the need for high-level knowledge that restricts the cabilities of data analysis systems.

3.2   Reduction of Variance

The desired outcome of analysis is a formal model of behavior that yields accurate predictions about the values of variables. Prediction accuracy is measured by *variance*, approximating the difference between predictions and actual observations. Data analysis is a mechanism for explaining variance, to identify

---

[3] The reader may question whether $\epsilon = 0$ is truly an *explanation* of the finding $T = N/P$. However, it is not the hypothesis $\epsilon = 0$ itself, but the subsequent findings surrounding this hypothesis that constitutes an explanation.

which factors contribute to it. When successful, analysis indicates that a model will continue making accurate predictions; on failure, it means that the model should be modified to account for more of the variance. Thus, variance reduction is a major principle behind intelligent data analysis.

Let us consider how this variance reduction principle takes form in some more common analysis strategies. Analysis of variance is a statistical technique that decides whether a categorical variable X influences the value of numeric variable Y. This decision is directly based on the variance reduction principle: if the variance of Y is significantly reduced by accounting for X, then X influences Y. The t-test decides whether two values can be considered equal given the background variance; significant results are more likely when variance is reduced. Linear regression determines if a linear relationship exists between two numeric variables; again, smaller variance makes a significant result more likely. All of these techniques are based on the *generalized linear model*, which decomposes net variance into a sum of effects for each factor, interactions between factors, and underlying background variance.

Other techniques that do not explicitly incorporate statistical variance still make use of this principle. Bayesian conditioning, for example, compares the *prior* probability of an event A, given only background knowledge, with its *posterior* probability given a specific condition B. When the posterior probability is large compared to the prior, we conclude that B is useful for predicting A, thereby reducing the prediction error, or variance.

3.3   Identifying True Structure

Models that make accurate predictions are desirable because, in principle, the most accurate model is one that captures the true structure of the world. In practice, this is not always the case: background variance, missing or censored data, measurement error, and faulty conclusions may lead to an accurate model that has little to do with the mechanisms that generate behavior. Thus, it is important to distinguish real effects from those that appear (or do not appear) in the data.

Unfortunately, the problem of deciding whether an effect is real or not may never be resolved. This is due to several limitations imposed by the discovery process itself. First, analysis is limited to the observations that are gathered and evaluated. Since it is not feasible to record all possible observations, the range of behaviors captured in any dataset is restricted. Of course, through random sampling we can *assume* that interesting behaviors will be reflected in the data, but we cannot *guarantee* this is so.

Second is the notorious *latent variable problem*: analysis focuses on a specific set of variables, and we cannot be sure that all relevant variables have been included in the set. Thus, it is desirable to be able to incorporate new variables as analysis proceeds, an event that often leads to deeper understanding of behavior. For example, consider the scientific advances following the discovery of the "atom" in chemistry or the "tectonic plates" in geology.

Finally, the models themselves impose a specific semantic perspective from which to view the data. For example, the model $T = N/P$ depicts equivalence between the values of T and N/P, but it does not consider any underlying causal relationships or the precedence among variables. On the other hand, a causal model describes direct, causal influences, but does not expose the mechanism responsible for these effects. Thus, any type of model is restricted in the reality it can represent.

These problems lead to the observation that no level of data analysis can identify the true structure in data. That being the case, how is it that data analysis has become a widely accepted approach to scientific modeling? It is because analysis is only part of the process: the real power lies in the correspondence between analysis and theory, the mapping between the semantics of the model and the mechanisms underlying behavior. In establishing and justifying such relationships, data analysis leads to a sound theory grounded in the real world.

4   The Scientist's Empirical Assistant

We have incorporated these principles into an intelligent discovery agent, the Scientist's Empirical Assistant (SEA). To circumvent some of the trickier problems with the discovery process, we make two simplifying assumptions. First, SEA works in conjunction with a human user who is an expert in the domain of study and can supply external domain knowledge, make strategic decisions, and establish exploration goals. Because human scientists rarely work in complete isolation, it is reasonable that our fledgling scientist should rely on the experience and knowledge of an expert. Second, SEA assumes that empirical observations are generated by a simulator, so it can design and run its own experiments on-line during analysis. Again, this is a reasonable assumption because simulations are intended to reflect the underlying processes and relevant properties of their real-world counterparts.

Our goal with SEA is to provide a formalization of the discovery process that works with any data analysis strategy. We have developed an open architecture that allows users to design their own strategies for data collection and analysis.[4] We have implemented several such strategies, such as analysis of variance and contingency tables, causal modeling, line and function fitting, and computer-intensive hypothesis tests.

It is important to note that the specific analysis being performed is independent of the analysis process. A valid analysis always follows the same abstract procedure: formulate the problem, design an experiment, gather observations, explain the results. SEA exploits this notion by decomposing analysis strategies into separate rules for each of these steps. This provides the generality we are shooting for and also supports efficient, intelligent control using current planning technology. At this point, all planning decisions are made by the user; our

---

[4] Methods for automating this process are still under development.

immediate interest in this work revolves around the problem of automating such decisions.

4.1   Model Representation

SEA constructs a model based on user inputs and data collected from a simulation program. The model consists of all that is known (to SEA) about the simulation domain, including rules for running the simulator, the parameters and outputs of the simulation (the experiment variables), known and expected behaviors. In this latter set are the "models" of scientific discovery: statements of causal influence, dependence, functional relation, and so on. In SEA, we call these *statements* and reserve the term "model" for the entire collection of knowledge.

Statements are the focal data structure in a discovery agent. For our purposes, statements are always predictive in nature. Because they are predictive, it is possible to measure both their accuracy and coverage. For example, a statement that represents an assumption within the domain, such as $T \geq N/P$, is automatically assigned accuracy and coverage of 1.0. Others, such as $\epsilon = 0$, are evaluated against experiment data. The tradeoff between accuracy and coverage can be used to direct exploration with a simple rule: if a statement has low accuracy, then try limiting its context - in reducing the coverage we should expect increased accuracy.

Statements are syntactic forms, composed from variables and parameters, that are manipulated by the analysis procedures. The variables are associated with the simulation, linking the predictions to real data. The parameters specialize the statement to a particular value; for example we might propose that $\epsilon = N/P$ when $P > 8$; in this case the "8" is a parameter. The procedures give *meaning* to the form by identifying underlying *mechanisms*; for example, $T = N/P$ suggests that T is generated by a mechanism that divides N items into P sets.

The mechanisms associated with a statement define a statement class. The various classes of statements form a hierarchy; it is both an object-class hierarchy and a search tree. As an object-class hierarchy, it depicts inheritance relationships among classes of statements. For example, the "depends" class inherits the notion of association from the root class "predicts", and adds the notion of inseparability: not only is the relationship predictive, it is direct. The "equals" class infers a different semantics: the *value* of one variable is predicted by the value of another. Each descendant class supplies additional mechanisms, and the mechanisms drive analysis.

The class hierarchy also depicts an approach for iteratively increasing the complexity of a statement, traversing the tree as a search hierarchy. The search rule is: to deepen the understanding of a hypothesis, generate new hypotheses for each child class. By nature, the child classes are semantically more complex, thus any child hypothesis is likely to be more informative.

4.2   High-level tasks of Scientific Discovery

Each class of statements is associated with procedures for performing each of the four major tasks of data analysis: formulation, design, observation, and explanation. SEA selects and applies these procedures according to user directives and a basic understanding of the discovery process. Here we see a major advantage of a class hierarchy: statement classes can inherit analysis procedures from their ancestors. For example, a set of observations for $T > N/P$ is not significantly different than those for $T = N/P$, so this class inherits its observation procedures from the parent class.

Formulation. Formulation is the problem of identifying hypotheses. Initially, formulation involves a parameterization of the field of study: what are the important variables? What is already known? Is there experiment data available? What interesting behaviors should be explored? When initially defining a model, SEA solicits much of this information directly from the user.

Formulation is not simply a problem of definition. The major challenge here is *hypothesis generation.* Heuristic rules for generating hypotheses fall into this category, because the first stage of hypothesis testing is always to form a hypothesis — whether or not the hypothesis survives subsequent stages will yet be determined. What this means for SEA is that each statement class must provide weak, heuristic rules for generating new statements, based on various conditions that may occur during execution. For example, when all else fails, SEA will generate all possible "predicts" statements (for all pairs of variables) and begin evaluating them.

Design. Design is the problem of deciding how to test a hypothesis; that is, what values will be used to determine its validity. This task obviously includes the design of data collection protocols and the selection of hypothesis tests; it also includes the critical problem of deciding *whether or not the hypothesis can be tested at all.* Often, this latter problem is implicitly included in the formulation stage; SEA exploits the distinction so that every hypothesis can be used in reasoning procedures regardless of its testability. Often an untestable hypothesis motivates a testable one that would not otherwise have been discovered.

Design procedures take a hypothesis as input and generate an experiment protocol for use by the observation procedures. In order to test a hypothesis, it is necessary to identify counter-hypotheses and criteria for rejecting each counter-hypothesis. For example, "equals" proposes the counter-hypothesis that T and N/P are different, motivating the t-test of section 2. Given the criteria, a set of requirements and constraints for data collection are created. These requirements, along with the analysis procedures, are passed along in the experiment protocol.

Observation. Observation procedures take the data collection protocol generated in stage 2 and generate results based on these instructions. This is an abstract notion of observation, which includes simulation experiments, statistical calculations, hypothesis testing, resampling procedures, and subjective conclusions. All values used in analysis are derived during the observation stage. Thus, what is commonly considered "analysis" is considered observation by SEA; selection and computation are explicitly placed in separate stages.

SEA's experiment protocol consists of a set of requests for various computations. Some will be requests for simulation experiments, some for computations, and some for conclusions. In most cases, the requests in the set will be inter-dependent: a statistical test relies on a statistical computation which relies on an experiment dataset. Part of the design problem is to determine a valid ordering for the various requests, and to avoid requesting experiment data unless necessary. The observation procedures simply carry out the instructions of the protocol and return result values.

Explanation. The final and perhaps most important stage of data analysis is to explain the results and conclusions and determine their context and implications. The purpose of the explanation stage is *accountability*: all conclusions are held accountable to prior and future knowledge. The truest test of a theory is its applicability: good theories are used over and over, while bad ones sit on shelves collecting dust. Effectively, explanation is for deciding whether new results are consistent with other possible perspectives on the problem. SEA's explanation procedures are based on the notion of *views*. Views are collections of statements that generate consistent predictions. The explanation procedures are rules for integrating statements into views and for creating new views when inconsistencies arise. At any point, SEA may be working with multiple views that represent competing theories of operation. Comparison of competing theories is a useful tactic for generating new hypotheses.

4.3   Control Architecture

SEA accomodates the iterative, multi-stage data analysis through partial–hierarchical planning (PHP) [2]. *Planning* is the problem of selecting a sequence of actions that move the system from its current state to some desired, or *goal*, state. SEA has the high-level goal of deriving a model, and this goal is decomposed into the four major tasks described above. Because the goals of data analysis can change so frequently, the planner is *reactive*, which means that it can modify its course of action to accomodate changing goals.

Actions in PHP are structured control constructs called partial plans, or simply *plans*. Plans may post new goals, which in turn trigger some set of matching plans. This results in a hierarchy of goals which induces a hierarchy of potential actions. Plans may also contain iteration, conditionals, executable statements, variable bindings, and subgoals. The planning language is essentially a high-level programming language.

SEA's planner was developed by St. Amant [8] and applied to the problem of exploratory data analysis. In addition to the basic PHP mechanisms mentioned above, the planner provides a facility for managing the decision points in planning through a *meta-planner*. Whenever a goal can be satisfied by several plans, the planner creates a *focus point* to keep track of the alternatives. When a focus point is created or otherwise encountered during execution, the meta-planner is invoked to redefine the ordering of alternatives and possibly shift effort to another focus point. Thus the planner manages planning decisions, while the meta-planner manages focusing decisions.

The user's strategic expertise is also incorporated through the focusing mechanism. Whenever a focus point is encountered, the absence of relevant meta-plans indicates that the user should decide which option to pursue. Thus, the user is not only the expert but also the "default meta-planner".

4.4   The Example Revisited

Here we return to the example of section 2 and discuss the procedures SEA uses to run it. All of these procedures have been implemented, but the actual course of exploration is highly dependent on the user, who makes all control decisions.

SEA begins in the formulation stage and posts a goal to formulate a new hypothesis. Because there is no working model, a variety of preliminary definitions are required: load and name the simulator, define variables and the "hooks" into the simulation, identify givens and initial hypotheses, load existing datasets, and so on. Once all the preliminary information has been provided, the existence of an initial hypothesis ($T = N/P$) signals that the formulation goal has been satisfied. Now SEA posts a new goal: design an experiment to test this hypothesis.

In the design stage, SEA chooses a t-test to determine if $T \neq N/P$. Although this is a weak test (it cannot conclude $T = N/P$) it can reject the hypothesis outright if it is successful. In the design stage, the fact that $T \neq N/P$ is a counter-hypothesis motivates the choice of a t-test. Given that a valid test was generated, SEA has satisfied the design goal and switches to the observation stage.

In order to observe the value of the t-test, an experiment dataset must be collected. SEA has a default rule to run a "pilot experiment" when no data is available, as well as rules to refine and extend existing experiments. These rules use the CLIP program to run the experiment and return experiment data. Given a pilot dataset, SEA computes the t-statistic and evaluates the t-test: there is no difference between T and N/P. Now all the observations have been computed and SEA switches to the final stage.

In this case, the results are as expected and easily explained: we did not want to see a difference in T and N/P, and indeed we didn't. Yet this is a weak test and really says little about the relationship between T and N/P. SEA shifts back to the formulation stage in search of deeper understanding.

This time through SEA proposes the variable $\epsilon = T - N/P$, based on the rule that "equality" can be extended by considering the difference between the two sides. A weak heuristic (that independent variables predict other variables) generates the hypotheses $\epsilon \mid P$ and $\epsilon \mid \Pi$. Each of these hypotheses is run through the remaining stages in turn. When both are found to be valid, other heuristics combine the two and evaluate their interaction.

5   Conclusions and Future Work

SEA integrates a variety of analysis strategies for different representations. Integrating representations and their various strategies is an important problem in intelligent data analysis.

SEA incorporates much-needed external knowledge, both in terms of domain facts and analysis strategies. Techniques for incorporating domain knowledge are encoded in the plans, inducing a tradeoff between generality and expressive power.

SEA provides an opportunity for studying the data analysis strategies employed by expert scientists. These can be addressed formally, by symbolically encoding these strategies as meta-plans, or informally, by observing the behavior of expert scientists using the system. An important area of future work is to incorporate unsupervised and/or supervised learning methods to automatically develop high-level analysis strategies through experience with an expert user.

SEA needs to be formally evaluated. First we need to evaluate its generality by trying a variety of analysis problems. Then we need to determine its proficiency as a scientific assistant by considering performance with and without the guidance of a human user.

SEA is a testbed for the study of analysis and discovery methods. It provides the opportunity to compare and contrast alternative strategies and/or representations. The application of these methods to real-world problems helps determine their strengths and weaknesses. It also provides the opportunity to develop new discovery heuristics, including those based on domain knowledge and tactical considerations. This is due to its formalization of the data analysis process.

Acknowledgements. This work is supported by DARPA/Rome Laboratory under contract number F30602-93-C-0076, and by a National Science Foundation Graduate Research Fellowship. The U.S. Government is authorized to reproduce and distribute reprints for governmental purposes not withstanding any copyright notation hereon. The views and conclusions contained herein are those of the authors and should not be interpreted as necessarily representing the official policies or endorsements, either expressed or implied, of the Defense Advanced Research Projects Agency, Rome Laboratory, the National Science Foundation, or the U.S. Government.

References

1. Carla E. Brodley. Addressing the selective superiority problem: Automatic algorithm/model class selection. In *Proceedings of the Tenth International Machine Learning Conference*, pages 17–24, 1993.
2. Michael P. Georgeff and Amy L. Lansky. Procedural knowledge. *Proceedings of the IEEE Special Issue on Knowledge Representation*, 74(10):1383–1398, 1986.
3. Clark Glymour, Richard Scheines, Peter Spirtes, and Kevin Kelly. *Discovering Causal Structure: Artificial Intelligence, Philosophy of Science, and Statistical Modeling*. Academic Press, Orlando, FL, 1987.

4. Dawn Gregory, Lixin Gao, Arnold L. Rosenberg, and Paul R. Cohen. An empirical study of dynamic scheduling on rings of processors. In *Eighth IEEE Symposium on Parallel and Distributed Processing*, 1996. To appear.

5. Pat Langley, Herbert A. Simon, Gary L. Bradshaw, and Jan M. Zytkow. *Scientific Discovery: Computational Explorations of the Creative Processes*. MIT Press, Cambridge, MA, 1987.

6. Bernd Nordhausen and Pat Langley. An integrated approach to empirical discovery. In Jeff Shrager and Pat Langley, editors, *Computational Models of Scientific Discovery and Theory Formation*. Morgan Kaufmann, 1990.

7. J. Ross Quinlan. *C4.5: Programs for Machine Learning*. Morgan Kaufmann, 1993.

8. Robert St. Amant and Paul R. Cohen. A planner for exploratory data analysis. In *Proceedings of the Third International Conference on Artificial Intelligence Planning Systems*, pages 205–212. AAAI Press, 1996.

9. Raul E. Valdes-Perez. Some recent human/computer discoveries in science and what accounts for them. *AI Magazine*, 16(3):37–44, 1995.

10. Jan M. Zytkow, Jieming Zhu, and Abul Hussam. Automated discovery in a chemistry laboratory. In *Proceedings of the 8th National Conference on Artificial Intelligence (AAAI-90)*, pages 889–894, 1990.

Meta–Reasoning for Data Analysis Tool Allocation

Robert Levinson and Jeff Wilkinson

Department of Computer Science, University of California
Santa Cruz, CA 95060
E-mail: levinson@cse.ucsc.edu

Abstract. It is desirable that data analysis tools become more autonomous in managing their computational resources, minimizing risk
and cost, assessing their errors, developing new representations, and integrating with other data analysis tools. To aid in this development we
introduce a design for using Meta-Reasoning in a Data Analysis Tool
Allocation system that employs analogical and structural reasoning to
learn from and across domains and its experience. We give a formal definition of a "Data Analysis Game" that allows the implementation of
a hierarchical learning framework suitable for describing and exploring
real-world analysis problems. The framework may also be used to analyze the performance of the Tool Allocation system itself, allowing it to
self-optimize. If the integration of tools is performed correctly, it should
allow a cost-efficient level of performance not obtainable with a single
tool alone or with unsystematic use of a group of tools.

1 Motivations and Objectives

In recent years both automated data analysis and the research community's understanding have improved. However, there are a number of fundamental problems to be addressed before the technology may be considered fully mature:

- Current data analysis systems typically do not manage their own computational resources. Resources are either allocated and coordinated by hand, or
  the problem is ignored.
- The economics of computational resources and information as utility, risk,
  and cost is not appreciated and hence has been under-exploited.
- Potential synergy among multiple statistical modules needs to be considered.
  This is especially pertinent to distributed computing systems.
- Knowledge recoding and compression is not considered, or is implemented
  by hand rather than as a fundamental aspect of the design. Given representations remain fixed even if they later become unwieldy. Within a domain,
  progress beyond a particular representation scheme is limited.
- Representational rigidity also prevents automated generalization from previously represented domains to new but structurally related domains. New
  representations must be generated for each new domain encountered.

X. Liu, P. Cohen, M. Berthold (Eds.): "Advances in Intelligent Data Analysis" (IDA–97)
LNCS 1280, pp. 89–100, 1997.   © Springer–Verlag Berlin Heidelberg 1997

  – The full expressive power of graphs, structural relations, and their use in analogical reasoning is not employed (due to presumed intractability). This cripples the mathematical power of data analysis.

But, clearly *all* of these problems are inextricably related and fundamental for realizing practical applications. Taking this perspective, to allocate resources successfully, data analysis methods must categorize or approximate the worth and reliability as they are encountered. Not only must they learn the characteristics of individual resources, but also what it is reliably possible to accomplish when using them in concert. They should also examine and predict the combined efficacy of resources which are at their disposal, and choose an analysis strategy based on this prediction and the acceptable level of risk. Externally provided or internally developed analysis tools should be selectively employed according to their demonstrated effectiveness in recognized contexts, and supplied with appropriate levels of computational resources. This is superior to using any single tool alone, using all tools in all cases, or using the right tool but using too much or too little of the available resources to meet the desired level of accuracy.

2   Meta-reasoning

Meta-reasoning may be described literally as "reasoning about reasoning"[15]. When meta-reasoning is applied to an underlying analytical procedure, the behavior of the procedure may be modified according to outcome of the meta-reasoning process. The objective when performing meta-reasoning is to improve the efficiency with which the underlying analytical procedure makes use of computational resources to achieve a set goal. Ideally, the additional resources invested in meta-reasoning will be outweighed by the resources which are saved in execution of the underlying procedure. If the same goal is achieved at a lower cost (or a better goal achieved at the same cost), the meta-reasoning may be called successful. Meta-reasoning may even occur recursively, when reasoning occurs about whether or not to perform meta-reasoning.

    Our system, the Meta-Reasoning Data Analysis Tool Allocator (MRDATA), relies on mathematical and statistical techniques for discovering the inherent structure of domains and modifying itself to exploit them. MRDATA may be viewed as a hierarchical reinforcement learner [16] but goes beyond: due its autonomy, use of meta-reasoning, and synergy with other tools. MRDATA is a generalization of the Adaptive-Predictive Search model[5] employed in our previous and current research, in which powerful pattern-matching and state-monitoring facilities are exploited.

    To fully exploit meta-reasoning, the analytical optimization of MRDATA's performance in the domain is cast as a new data analysis problem. This allows MRDATA to characterize and adjust its own learning performance. Thus, the system learns how to optimize itself not only for analysis but also for learning how to analyze, that is, learning how to learn.

3 Defining a Data Analysis Game

In order to describe MRDATA in both a rigorously defined and intuitively under-standable setting, we describe problem domains as games. Games have precisely defined rules and yet supply a point of natural access for human thinking. Further, game rules provide a player with clear feedback about their performance within the game, across games, and in comparison to other players. We begin by describing a simple representation of a data domain, and then describe an interactive analysis game which may be played on the domain.

3.1 Foundational Definitions

The domain is represented as a "spreadsheet" table composed of cells, which contain data. The cells are grouped into rows and columns. Each cell belongs to exactly one row and exactly one column. A cell's location in the table is uniquely and fully specified by its row and column. For now, attention is restricted to tables in which all the cells in a given column may only hold a single type of data (e.g. if one cell in a column holds a natural number, then all cells in that column may only hold natural numbers). As a conceptual aid, the columns may each be viewed as a variable. Under appropriate conditions, the relationships between or among column variables may be viewed as functions. If the contents of a column do not change during an analysis, the column is called "static", else it is called "dynamic". A table is called "static" if all of its columns are static. If even one column of a table is "dynamic", the entire table is called "dynamic".

An analysis game is defined on a dynamic table when one column variable is defined as a record of the actions of a player at each timestep of the game. The cells are filled in over time as the player performs the actions. A game with instant feedback also has a column variable defined to record some measure of the performance of the player at each timestep. For now, attention is restricted to games where the player's possible actions are drawn from a finite discrete set, while the performance measure is drawn from a continuous set.

Finally, a player must possess a "reward" function which maps from the values in the performance column to a single value (assumed to be of the same continuous type as the performance measure) which represents overall performance. It must be possible to evaluate this function at each timestep, as well as at the end of the game. Thus, the player not only receives the incremental feedback of the performance measure, but also the changing overall feedback of the reward function. The goal of the player is to maximize the final value of the reward function (a continuous value) by choosing actions (from a discrete set) at each timestep, based on the information already stored in the table.

3.2 Example 1: A Baseball Player

A simple example illustrates the main features of the framework. Consider a baseball player who must choose at the start of each season whether to bat left or right handed. This decision may not be changed until the next season begins. At the end of each season, the player's batting average is computed. Play continues for 10 seasons. The player wishes to maximize career average. [1]

Current Game State		Options Presented to Player				New Game State	
HAND	**AVERAGE**	**HAND**	**AVERAGE**	**HAND**	**AVERAGE**	**HAND**	**AVERAGE**
Right	0.213	Right	0.213	Right	0.213	Right	0.213
Left	0.343	Left	0.343	Left	0.343	Left	0.343
		Left		Right		Left	0.183
.	.	.	.	.	.	.	.
.	.	.	.	.	.	.	.
.	.	.	.	.	.	.	.

Fig. 1. The boxed set of tables (with two members: Left and Right), derived from the current game state and representing action options, is presented to the player. The player chooses one action. The monitor finishes filling in the new game state.

3.3   Example 2: The Personal Physician

As a more complex example, which is closer in spirit to actual problems to which MRDATA could be applied, consider a doctor examining the medical records of a series of patients at risk for heart trouble, and determining a course of treatment. The treatment options are limited to:

- Performing an n-tuple-bypass operation (risky, expensive, but effective).
- Prescribing a regimen of medication (reliable and relatively inexpensive).
- Suggesting a low-fat, low-sodium diet, and an occasional glass of red wine (no cost).
- Delaying treatment selection and instead performing an additional test.

If the doctor chooses to perform a test, the patient must be seen again when the test results come back. At this time, the doctor may choose a treatment with the help of the new information, or elect to perform an additional test. If no tests are left to perform, the doctor must choose a treatment.

---

[1] The translation of this problem into the game format is straightforward.

1. The choice of hitting left or right handed is an action drawn from a discrete set.
2. The seasonal batting average is the performance variable.
3. The timesteps are the seasons, so the table has 10 rows.
4. The overall reward function at any time is the average of all seasonal performance values so far.

Further, the doctor has a record of all previous patients, which includes their responses to a (free) questionnaire about their health history, lifestyle, and current symptoms. The records also include the order and results of any additional (not free) diagnostic tests which were performed, and the treatment option eventually selected. The records also include a measure of effectiveness: the quantity of quality-adjusted years of life ("quals") the patient enjoyed after the treatment. The quality-adjustment takes into account the total cost of all the tests and the treatment performed.

Finally, note that the doctor's practice is *very* exclusive, treating only the current ruling monarch of a dynasty with a history of heart trouble. Thus only one patient is in treatment at a time, and evaluation of a new patient only begins once the previous patient has died. [2]

This final game rule is included to ease the explanation of the translation into the data analysis game format, and is absolutely non-essential to the general methods of MRDATA. For example, the next patient to be seen after a test or treatment is selected might be drawn from a pool of waiting patients. When a test is selected for a given patient, the patient would be returned to the pool, and return for further consultation (with the new test results) later. When a treatment is selected, the patient would be removed from the pool, and a new one might be added.

3.4   Game Mechanics

At each timestep, the game monitor examines the game state (the current table) and determines which actions are available to the player. For each of these actions, the monitor makes a copy of the table, records the particular action in the next empty cell of the action column. The set of these hypothetical tables is then presented to the player. The player selects one option, informing the monitor. The monitor replaces the old game state with the selection, the new values of the dependent column variables (including the performance measure) are calculated, and the game advances to the next timestep.

---

[2] This complex game translates into the same format as that of the baseball player:

1. The combined lists of tests and treatments form a finite discrete set of actions.
2. The ratio of the monarch's quality-adjusted lifespan to the national average is the performance variable.
3. The timesteps are the sequence of selections between tests or treatments. After a test selection, the same patient returns in the next timestep, with the test results added to their record. The same patient returns again and again until a treatment option is selected. Examination of the next patient begins in the timestep following the treatment selection. There is no limit to the number of rows.
4. There is also a column for the names of the patients, a column for each question on the questionnaire, and a column for each test (default value: "not taken").
5. As only final outcomes matter, the overall reward function is the average of the performance values from only the rows in which a treatment option was selected.

4   Defining a Data Analysis Tool

As a most general definition, a data analysis tool takes a set of tables as input and produces a set of tables as output, consuming some preset quantity of computational resources in the process. Some tools may also accept parameters which affect their behavior. For a data analysis tool to be considered useful, the contents of the output tables should be some function (hopefully entropy reducing) of the contents of the input tables.

HAND	AVERAGE
Right	0.213
Left	0.343
Left	0.183
.	.
.	.
Right	0.280

1 →

Left	0.343
Left	0.183
.	.
Left	0.078

Right	0.213
Right	0.259
.	.
Right	0.280

→ 2

HAND	MEAN	DEVIATION	DEV. OF MEAN
Right	0.271	0.053	0.019
Left	0.156	0.135	0.042

Fig. 2. Tool 1 separates out Left and Right rows into different tables. Tool 2 reports statistical information about the contents of the tables. Tool 3 results in a promising choice of HAND. This chain of transforms could be built up from simple tools and used to implement a more advanced tool.

Any computational or statistical package which accepts tabular input and produces tabular output can be considered a data analysis tool under this definition. This definition is perhaps more inclusive than most non-rigorous conceptions of a "data analysis tool". The most important types of tools for use in a data analysis game are ones with some specific input-output characteristics:

Evaluator: Takes one input table with the game column structure, and produces an output table with a single cell containing an estimate of the desirability of the state represented by the table.

Advisor: Takes a set of input tables with the game column structure, and produces an output table with two columns and as many rows as there are input set members. The first column holds copies of the input tables. The second column estimates the desirabilities of the associated tables.

Sage Advisor: Takes the same input as a normal Advisor, but the output table has a third column which gives ratings of the errors expected in each estimate. Both types of Advisor could be implemented using a pool of Evaluators or Actors (see below) whose outputs are collated (perhaps weighted by their historically reliability) to form a single Advisor format output.

Executive: Takes one or more input tables in Advisor output format. The single-cell output table contains the element in the first column of a row elected from one input table. This output is suitable for presentation to the game monitor.

Actor: Takes an input table in game format, and produces a single cell output table containing a selected action, suitable for presentation to the game monitor. Such a tool might be composed of an Executive and a pool of Advisors, but need not be. As long as input and outputs to the tool satisfy the conditions it can reasonably be called an Actor.

We are most interested in tools which are able to function with varying levels of computational resources. For example, an averaging tool could use less resources by calculating fewer significant figures. It is desirable for an advanced data analysis tool to be able to take the resources which it has been allocated and (possibly after spending some resources on meta-reasoning) delegate appropriate resources to its various sub-tools. Learning how to perform tool allocation optimally in an arbitrary domain is simply another game of data analysis. This view is one of the driving concepts in the theory behind MRDATA.

5   Design and Implementation of MRDATA

For concreteness, the basic design of the Meta-Reasoning Data Analysis Tool Allocator will be illustrated using the format of the above defined games. As the same data analysis tool is being applied to each segment of each task, generalization to any domain describable within the framework is natural.

If a player is capable of detecting statistical regularities in the data, and can identify predictive relationships between the variables (including the performance measure and reward function), actions can be selected which predict better results. In particular, identification of these regularities and relationships may be achieved by the application of data analysis tools.

The detection of regularities and relationships allows the transformation of the data into higher level representations. Once simple transformations have occurred, higher level regularities and relationships may then be recognized. The output of each level of analysis is used as the input to the next. The collection of tools which performs an analysis process may be viewed as a tool itself. This allows new tools to be built up from compositions of user provided analysis tools (if any) and the generic methods of MRDATA itself.

5.1   Overview of Learning Methods Employed

MRDATA has been designed with three fundamental methods in mind: K-Nearest Neighbor (KNN) tables, regression trees (RTrees) [2], and an extension of Temporal Difference (TD) learning. Initial exploration of a domain calls for a direct storage and lookup system such as KNN. In many domains, KNN tables have been shown to outperform neural networks while using less resources [3, 13]. Once data has begun to accumulate, it is redescribed into more explicit representations, in this case, RTrees. Finally, TD learning assigns credit and reward based on experience [16].

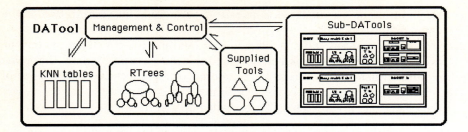

Fig. 3. Meta-reasoning is carried out by Management and Control, which allocates resources to other components, and gives them feedback. The other components function as Advisor types while Management and Control functions as an Executive. Sub-tools have the same structure as both tools and sub-sub-tools.

These methods are employed in an expandable hierarchy of data analysis tools. Tools use meta-reasoning to allocate resources to their component parts. These parts are implemented as pools of KNN tables, RTrees, sub-tools, and any domain specific tools which may be supplied by the user for integration. TD performance feedback is given by a tool to its components, including the sub-tools which then provide feedback to their own components.

5.2   K-Nearest Neighbor Tables: Simple Starting Structure

The data analysis tool uses KNN tables to record the desirability of various states it has encountered in its experience. The desirability of a state is the expected value of the overall reward function, if the state is selected. In theory, a KNN table is composed of a list of game states associated with a previously observed return values. The format of the the game state tables matches this precisely. Even if the KNN table was erased after every selection, the monitor provides a full game history (indeed, one copy for each option) with each timestep.

To process an option, all empty rows are first deleted. The partially completed row representing the possible move is then separated from the rest of the table. To rate the desirability of the move, comparisons of the move are made to the remaining rows in the table, using a learned metric to determine the closeness of possible matches. The metric may be as simple as the number of column variables which match (for discrete values) or are "close" to matching (for continuous values). The metric may also weigh different variables using an estimate of the sensitivity of the outcome to their values, or be some other more complicated learned measure. The K closest matches are retained, and the final value is formed for the state from a combination function of the K values. A KNN object is described by the functions used to process the input table into Evaluator output format. A pool of KNN tables is a collection of such KNN objects.

During play, the Management and Control presents the KNN pool with the possible action states. The pool calls each KNN object once for each of the actions. For each action, the ratings of the KNN objects are combined, using a learned weighted measure, to produce the final rating used in the pool's Advisor output format. The resource consumption of each function used in this process is tallied, and the net usage must be less than the amount allocated by Management and Control. In order to meet this requirement, the accuracy to which each function calculates its results may be reduced, for example.

5.3   Regression Trees: Redescription of Representations

A newly created MRDATA module uses only a KNN pool as an Advisor. In complex domains, the KNN tables will eventually grow large. Unless cleverness is employed [1], every element of the table must be examined for the closeness of its match to the current state. As data analysis methods must cope with limited time and space, large tables with large access times are unacceptable. At some point, the investment of time in the conversion of the information into a different, simpler representation will become cost effective.

There are well known algorithms [6] for converting from tabular representations (e.g. NN tables) to hierarchical representations (e.g. RTrees). Regression trees[2] are related to decision trees in that at each branch in the tree, one or more tests are performed, with direction of branching depending on the outcome(s). However, they are more general in that a continuous-type value is returned upon arrival at a leaf, rather than a discrete class assignment.

The relatively compact structure and rapid access time of RTrees make them an attractive format for re-representing the learned utility data. Also, useful features and relationships may be easier to recognize in tree format, assisting the data analysis tool in generating more powerful representations.

The functioning of a pool of RTrees is analogous to that of a KNN pool. Each RTree Evaluates each option once, and the results are collated. Resource constraints may be met by reducing the accuracy of combination functions, or by checking the assertions of the branching tests less rigorously.

5.4   Sub-tools: Higher Level Control

Sub-tools are passed state information, and use their time and space allotment to consult their own KNN tables, RTrees, and sub-sub-methods [3], to select the action option they feel is best. The pool keeps track of the accuracy of the sub-tools, and combines their Actor outputs (again using a learned combining function) into a single Advisor output.

As a group of Advisors can be made into an Actor by attachment to an Executive, and groups of Actors can be made into an Advisor by organization into a pool, a hierarchy of Meta-Reasoning Data Analysis Tool Allocators may be constructed. During decision making, information about the current state spreads down the hierarchy, and analysis percolates back up. After the decision is made, the essential feedback required for adaption must be propagated back down. This

---

[3] Sub-sub-methods perform the same activities.

weight changing mechanism resembles connectionist algorithms, such as back-propagation[14] but the nodes are much more complex than simple neurons–they are entire data analysis modules or communities of modules.

5.5   Temporal Difference Learning: Hierarchy Diffusion Variation

Temporal Difference learning is appropriate in domains such as medicine (e.g. the Personal Physician) or checkers in which a series of action choices are often made before significant feedback is obtained.[17]

TD learning alters the rated values of each of the states in the sequence that led to the payoff. The feedback value is used as the "reinforcement value" of the final state in the sequence. The old rating, R_{old}, of the state is combined with the reinforcement value, R_{TD}, using an equation such as $R_{new} = ALPHAR_{TD} +(1- ALPHA)R_{old}$. MRDATA's learning rate, $ALPHA$, is parameterized by the received payoff. Exceptionally good or bad payoffs are taken more seriously (i.e. a higher $ALPHA$). After the new rating of a state is calculated and stored, it is used recursively as the reinforcement value for the state which preceded it.

In our extension for hierarchical data analysis, we employ standard TD learning for games such as the Personal Physician. We *also* use a structural version of TD, even in games where feedback occurs immediately, such as the Baseball Player. As an analysis percolates up the hierarchy to the top level player, a sequence of decisions are made at each level by the tools, with no feedback until the final decision is evaluated by the monitor. Although only one action was taken in the top-level game, the sequence of decisions made is treated analogously to the temporal sequence in standard TD. Since the reinforcement learning spreads through the structure of sub-tools, we refer to this as Hierarchy Diffusion.

At each timestep, each module receives a feedback value and a payoff. In the case of the top-level game player, the feedback is the newly calculated performance measure, and the payoff is the recalculated value of the reward function. At other levels of the hierarchy the payoff and feedback values need not be simple. When a tool receives a point payoff and reinforcement value it alters its KNN and RTrees pools, and sends feedback and payoffs to its sub-tools.

5.6   Global Meta Learning

All statistical modules in the hierarchy face the same problem: maximizing payoff given time and space allotments. They must make decisions about how heavily to invest resources in the search for new domain representations (basic research), as well as decisions about when to convert KNN tables to regression trees (a capital investment of time) and when to form new sub-tools (division of labor). Lessons learned at any level of the hierarchy may be useful at any other. Although the domain problems faced by data analysis methods vary from level to level in the hierarchy, the uniformity of the design across levels allows lessons to be learned that transcend hierarchy location. Knowledge gained from the decision experiences of each data analysis tool in the hierarchy is stored in a globally available KNN table. This table is subject to the same decision making and knowledge compression process as the rest of the data structure.

6 Implementation and Results to Date

Our original research [8] provides a sound basis for our current work on implementing MRDATA. Morph learners use "patterns" to evaluate states and make decisions about them. Highly optimized methods for storing and retrieving patterns give Morph powerful facilities for monitoring and playing abstract analytical games. The features of Morph used by MRDATA include:

- **A domain-independent data analysis and learning mechanism based on understanding a Game of Abstract Mathematical Relations.**
- **Analogical pattern creation and matching.** The original Morph system, was applied to chess. Its graph pattern language, although more expressive than most, was limited[12]. Morph's graph patterns have been subsumed by relations in Morph II[7]. At the same time we have restricted the pattern language to a small set of mathematical primitives from which all complex patterns can be derived. These primitives include cardinality, and, or, not, join, modulus, factorability, permutation, periodicity and cycle. [4] Using its pattern learning methods, MorphII learns to play small domains including Nim, Tic-Tac-Toe, and Hexpawn nearly optimally.
- **Automated derivation of combining function for weights.** In the original Morph, the function that combined the weights of patterns (for state evaluation) was human-engineered. MorphII uses linear regression, gradient descent, nearest neighbor and other function learning methods to derive its own combining functions at all level of the hierarchy.
- **Optimized incremental pattern-matching.** Maintaining a large set of predicates is a critical operation. Our current pattern-matcher is now 10 to 100 times faster than our previous implementation, when matching similar sets of patterns. Further, storage requirements have been reduced 25-fold. These improvements are achieved by a procedure that compiles the declaration of a given domain into a RETE-like network[4, 11].
- **Bit-level processing.** The storage and retrieval scheme above is described in [9]. Since the time of writing [9], we have taken things one step deeper. By compiling the relational tables into bit-arrays, we are exploiting the logic operations of the CPU. Thus, we are now able to process individual domains at speeds comparable to domain-engineered programs.

7 Conclusions and Perspective

We hope that the Meta-Reasoning Data Analysis Tool Allocator will be one of a new generation of tools that operate synergisticly with other tools to provide cost-efficient and reliable analysis to support real-world decision making.

---

[4] Some researchers are beginning to consider the redescription of their representations into MorphII's pattern language [10]

It is our intention that the ability of MRDATA to self-optimize make it a useful and robust tool for researchers and organizations. The Data Analysis Game
format allows the implementation of a hierarchical learning framework suitable
for describing and exploring real-world analysis problems. As MRDATA devel-
ops, we hope to obtain an even better understanding of the effectiveness of
meta-reasoning, coupled with analogical and structural representation, as an
organizing principle for already powerful statistical methods.

References

1. J. Allen, E. Hamilton, and R. Levinson, "New advances in adaptive pattern-
   oriented chess," in *Proceedings of 8th Annual Conference on Advances in Computer
   Chess*, (Maastricht, Neatherlands), June 1996.
2. L. Brieman, *Classification and regression trees.* The Wadsworth statis-
   tics/probability series, Wadsworth International Group, 1984.
3. S. Cost and S. Salzberg, "A weighted nearest neighbor algorithm for learning with
   symbolic features," *Machine Learning*, vol. 10, pp. 57–67, 1993.
4. C. Forgy, "Rete: A fast algorithm for the many pattern/many object patern match
   problem," *Artificial Intelligence*, vol. 19, no. 1, pp. 17–37, 1982.
5. J. Gould and R. Levinson, "Experience-based adaptive search," in *Machine Learn-
   ing:A Multi-Strategy Approach* (R. Michalski and G. Tecuci, eds.), vol. 4, pp. 579–
   604, Morgan Kauffman, 1994.
6. P. Langley, *Elements of Machine Learning.* Morgan Kaufmann Publishers, Inc.,
   1996.
7. R. Levinson, "General game-playing and reinforcement learning," *Computational
   Intelligence*, vol. 12, pp. 155–176, February 1996.
8. R. Levinson and R. Snyder, "Adaptive pattern oriented chess," in *Proceedings of
   AAAI-91*, pp. 601–605, Morgan-Kaufman, 1991.
9. R. A. Levinson, "Uds: A universal data structure," in *Proc. 2ND International
   Conference on Conceptual Structures*, (College Park, Maryland USA), pp. 230–
   250, 1991.
10. S. Markovitch and Y. Sella, "Learning of resource allocation strategies for game
    playing," *Computational Intelligence*, vol. 12, pp. 88–105, February 1996.
11. D. P. Miranker, "Treat: A better match algorithm for ai production systems," in
    Proceedings of AAAI-87, pp. 42–47, 1987.
12. E. Morales, "Learning patterns for playing strategies," *International Computer
    Chess Association Journal*, vol. 17, pp. 15–26, March 1994.
13. S. Omohundro, "Efficient algorithms with neural network behavior," Tech. Rep.
    UIUCDCS-R-87-1331, University of Illinois, April 1987.
14. E. D. Rumelhart and J. L. McClelland, *Parallel Distributed Processing*, vol. 1–2.
    MIT Press, 1986.
15. S. Russell and E. Wefald, *Do The Right Thing: studies in limited rationality.* Cam-
    bridge, Massachusetts: MIT Press, 1991.
16. R. Sutton, "Special issue on reinforcement learning," *Machine Learning*, 1991.
17. R. S. Sutton, "Learning to predict by the methods of temporal differences," *Ma-
    chine Learning*, vol. 3, pp. 9–44, August 1988.

Navigation for Data Analysis Systems

Robert St. Amant

Department of Computer Science
North Carolina State University
Box 8206
Raleigh, NC 27695-8206
stamant@csc.ncsu.edu

Abstract. Statistical strategies are formal descriptions of the actions and decisions involved in applying statistical tools to a problem. One difficult problem, beyond strategy design issues, is ensuring an effective interaction between the the user and the implementation of the strategy. In this paper we review some existing techniques that have been effective in mediating the interaction between users and expert statistical systems for data analysis. We explore the relationship between *navigation* and statistical decision-making and consider features common to both areas. Our analysis leads to the identification of navigation functions that we believe could improve almost any system for statistical data analysis.

1 Introduction

Statistical strategies are formal descriptions of the actions and decisions involved in applying statistical tools to a problem [7]. Strategies have been designed and implemented for simple and multiple linear regression [4], MANOVA [7], collinearity diagnosis [11], variable selection in specific contexts [3], and other statistical procedures [5]. The potential benefits of general-purpose, automated strategies are enormous, greatly reducing the work load of statisticians and providing non-statisticians with easily accessible expert advice.

Unfortunately, the promise of automated, strategic statistical reasoning has not been fulfilled. In 1986, John Tukey optimistically wrote [18], "By 1995 or so, the largest single driving force in guiding general work on data analysis and statistics [will be] to understand and improve data-analytic expert systems..." Today there are no commercial statistical systems that perform sophisticated strategic reasoning, and many research projects have turned out to be single efforts with little or no follow-on work [5].

Perhaps the most difficult problem faced by the developer of a statistical strategy is dealing with *contextual knowledge* [11, 10]. Contextual knowledge, or subject-matter knowledge, can be essential in selecting an initial course of analysis, making appropriate decisions, and interpreting the importance of results. In analyzing a medical dataset, for example, we would probably interpret age observations of "9999" not as evidence of extraordinary longevity, or even as

X. Liu, P. Cohen, M. Berthold (Eds.): "Advances in Intelligent Data Analysis" (IDA–97)
LNCS 1280, pp. 101–109, 1997.   © Springer–Verlag Berlin Heidelberg 1997

simple outliers, but rather as a data input convention for unknown values. Also, knowing that `age` has a theoretical lower bound of 0 and a practical upper bound of 100 or so influences the way we might model the variable. This kind of judgment, or meta-data analysis, requires experience, common sense, and intuition, properties we cannot expect to find in automated statistical strategies.

Though there is some pessimism about the possibility of general-purpose strategies that can handle contextual knowledge, there nevertheless remains strong interest systems that implement restricted forms of automated strategic reasoning. Rather than acting as a complete, autonomous statistical consultant, an expert statistical system might limit its decision-making to procedures that are relatively independent of context; it might only make suggestions, rather than taking unilateral action; in cases where it does act, it might provide several possible analysis scenarios and defer to human judgment for its selection between them [12, 10, 2, 14].

For these approaches to be effective, we must shift our attention away from the strategies themselves, toward user interaction issues. By dividing decision-making responsibility between the user and an automated system, we cast the system as a collaborator, rather than a passive tool. Collaborative systems must address issues such as reasoning about shared evolving goals, planning and coordination, implicit and explicit communication, evaluation of shared progress, among many others [17]. These are difficult issues, with only partial general solutions; they have barely been touched upon in the statistical strategy literature.

In this paper we review some existing techniques that have been effective in mediating the interaction between users and expert statistical systems. We concentrate mainly on systems for data analysis and techniques that make the process of the analysis explicitly accessible to the user. These techniques can be best understood in the framework of *navigation*. We explore the relationship between navigation and statistical decision-making and consider features common to both areas. Our analysis leads to the identification of navigation functions generally useful for intelligent, strategic data analysis.

2   Access to the data analysis process

Data analysis is an inherently constructive process. As Peter Huber observes [8, p. 69], "Data analysis is different [from word processing and batch programming]: the correctness of the end product cannot be checked without inspecting the path leading to it." This view is part of any approach to statistical strategy, and most statistical expert systems have made this process-oriented view explicit to the user. There are several ways in which this can be done, displaying relationships between data, statistical operations, elements of strategic knowledge, and interactions between knowledge and the data.

- The system can show the relationships between data elements and their aggregates. For example, in TESS, the Tree-based Environment for Statistical

Strategy [10], for example, one might see a graphical display of the relationship (x, y), its component variables x and y, transformations of the relationship or its individual variables, and so forth. In this display, nodes represent data objects, arcs the relationships between them. The graph or network expands as new operations generate new transformations, compositions, and reductions of the data.

— A system can show relationships between the data and the models generated to describe or explain the data. In the data network described above, additional nodes might represent regression or cluster models. Models give rise to other data and model nodes, such as residual structures or cluster subsets.

— A system can show the statistical operations by which models and descriptions are generated. In some cases there is little difference between showing an operation and showing its result; however, for complex sequences of operations leading to a single result, or iterative procedures, the distinction can be valuable. In the display under discussion, these nodes would be intermediates between data nodes and model nodes. In an active display these can be used to modify the system's behavior.

— A system can display discrete elements of statistical knowledge, and the relationships between them. There are close ties between research in expert systems and statistical strategies; after all, strategies are a kind of expert knowledge. Just as expert systems tend to rely heavily on objects and rules, so do implementations of statistical strategies. Objects provide an explicit representation of statistical procedures; the relationships between objects correspond to the potential ways these procedures can interact with one another. Augmenting these objects with rules is one way of specializing a strategy's behavior for specific data. For example, in REX, the Regression Expert system, a strategy is encoded as a hierarchy of frames. A set of rules is associated with each frame that guide the system from one frame to the next, giving in effect a static decision tree [4, 13]. The system can display the object hierarchy as well as the rules to give the user a better understanding of its potential behavior. This can be especially effective if the system can act as a knowledge-based help facility, as with KENS, a statistical knowledge enhancement system [6].

— A system can display its representation of statistical knowledge as it applies to a specific dataset. That is, as with an expert system, there is a difference between the static rules in the knowledge base and the set of instantiated rules that apply in a specific situation. Analogously, a statistical knowledge base may contain a large number of individual elements, only some of which may applicable to any given data. There may also be a number of different ways in which these elements combine to result in an appropriate sequence of decisions. We can think of a particular set of elements, combined in a specific way, as an instantiation of a strategy. A system can display this instantiation to help the user understand its behavior given the current data.

Of these, the last type of display is the most comprehensive. The instantiated knowledge structures contain references to models and the data structures from which they are derived, as well as the static elements, such as rules, in the knowledge base. Through appropriate filtering and linking, the other types of displays can easily be generated.

3   Navigation and the data analysis process

It will be convenient to discuss interaction techniques in the framework of a specific system. Our discussion will center on AIDE, an assistant for intelligent data exploration, which we have developed over the past several years [15, 14]. AIDE is a knowledge-based system that incrementally explores a dataset, guided by user directives and its own evaluation of indications in the data.

AIDE's knowledge base implements simple strategies for exploratory data analysis (EDA). These strategies are represented as plans, a refinement of the objects/rules representation described earlier. In AIDE's representation, objects of different types correspond to goals, statistical operations, and abstract statistical plans. AIDE's library contains about a hundred plans, at different levels of detail. These plans are intended to capture elements of common EDA practice, such as the examination of residuals after fitting a function to a relationship, the search for refinements and predictive factors when observing clustering, the reduction of complex patterns to simpler ones, and so forth.

AIDE explores as follows. When a dataset or relationship is presented to the system, a goal is established for its exploration. The planner searches through its library for an appropriate plan and expands it, that is, establishes a set of new subgoals to be satisfied. These subgoals are satisfied in turn by plans from the library. Goals can also be satisfied directly by primitive actions, which execute code directly rather than establishing new subgoals. Often several plans in the library can satisfy a single goal, and there may be an unlimited number of ways to bind a given plan's internal variables to different values. The planner relies on specialized control rules to make these decisions; these rules provide explicit justification for the system's actions. As the exploration proceeds, a network of decisions is generated, connecting statistical operations with one another and tying data to results. The process continues until the goal at the top level has been satisfied.

AIDE casts exploration as navigation through this network of decisions, data, and results. A set of navigational operations supports close interaction with the user. Our further discussion is based mainly on experience with these operations.

Most users of computer systems are familiar with the metaphor of navigation. One navigates through a sequence of menu selections, dialog boxes, pages on the World Wide Web, topics in an automated help system, and so forth. The navigation metaphor is based on a relationship between the target concept (e.g., menu selection or Web page retrieval) and movement, sometimes called wayfaring, over a physical landscape [9]. The effectiveness of the metaphor depends on specific properties of the target domain, which is usually some kind of information space. First, the state space of the domain can be tied to a spatial representation, in the simplest case a two-dimensional representation. Second, the current state of the system can be clearly defined, and can be associated with a spatial location occupied by the user. In the WWW domain, this current state is given by the page displayed by a Web browser. Third, the user can move through the state space. This movement may be discrete or continuous, depending on the granularity of the space. In the WWW domain, this means causing different pages to be displayed. Continuous movement is more common in domains with a definite spatial interpretation, such as monitoring of physical processes, or domains with very high connectivity, such as general information retrieval [1].

The navigation metaphor can be very effective when these properties are present. Systems commonly support these navigation operations:

Go: The user specifies a state to visit. In some situations the new state is a successor to the current state. In other cases it can put the user in a new, otherwise unexplored part of the state space.

Back: The system records the sequence of decisions the user makes in moving through the space to be navigated. This operation returns the user to the state immediately preceding the current state.

Forward: This operation provides a reversal of a *Back* operation; having gone back from state B to state A, the user can move forward to revisit state B again.

Any: In general, we can view the user's activities as expanding the state space on the fringe of a directed graph, with the starting state as the root node. This operation lets the user revisit any state previously visited.

History: This operation lets the user revisit those states that lie in a sequence between the starting (root) state and the current state.

Bookmark: This operation stores the current state on a list. The user can later revisit any of the states on this list, without having to traverse the state space explicitly.

These are only a selection of possibilities, which may also include aids such as guided tours, landmarks, zooming, node comparisons, and different types of overviews [9]. The operations described above, however, are a sufficient set for navigation.

The graph or network representation of a space of decisions, along with navigation operations, provides a useful framework for data analysis. The representation acts as an explicit record of the user's actions, can support semi-autonomous action on the part of the system, and can supply a means of communication between system and user. We consider each of these areas in turn.

3.1   User actions

Each of the navigational operations discussed above causes a new result to be generated or revisited, which results in some graphical or textual display. Because the structure of the state space for data analysis is not fixed in advance, as with many other domains, some additional operations become useful.

Go/Step: The user specifies a data analysis operation: a transformation, a regression computation, selection of variables, or some other atomic operation. The result is displayed as the new current state.

Back: The user returns to a previous decision to modify it: selection of a different set of variables, a different setting for a model parameter, a different type of model or description.

Forward, Any, History: Each of these operations brings the user to a result already seen, generated by a decision made earlier.

Copy/Paste: Sometimes an analyst may carry out a set of procedures on one relationship, and then find that the same procedures are appropriate for another. This can happen, for example, when the analysis of different versions of the same data is called for. In such a case, the user can select and copy the relevant sequence of decisions applies to one relationship and then paste them onto the other relationship. The new procedures are carried out, in appropriately modified form, on the new data.

Copy and paste operations are a short cut, but one that must be taken with care. Because procedures may depend on specific features of the data to which they are applied, it may not always be clear how to interpret these features in another dataset. For example, we may remove an observation from some sequence because it had a value of "9999", or because it was more than three standard deviations from the mean, or because it occurred as part of a known invalid observation, or some other reason.

Delete: Data analysis is a constructive process. Because results are generated unevenly, with deadends commonly encountered, a much greater space can be examined than is eventually needed. To more easily manage the set of decisions to consider, the user can delete sets of decisions that have become irrelevant.

Replay. Sometimes the easiest way to understand a complex process is to watch it as it happens. It can sometimes be useful to have the system repeat an analysis procedure automatically, pausing at each step to show intermediate results, rather than have the user step through the process manually.

3.2   System actions

One of the goals of implementing a statistical strategy is to provide a system with some of the autonomy necessary to carry out an analysis. Given general instructions, the system can then make its own decisions about how to carry them out. This behavior requires a refinement of the navigation operations described above.

Go/Jump: In addition to carrying out an operation requested by the user (via the *Go/Step* operation), the system can sometimes follow up with a plausible sequence of follow-on activity. For example, if the user has run a regression, examined the residuals, identified outliers, and removed them, the natural next step is to rerun the regression. A statistical strategy can give the system a good notion of where a course of action will lead, and can let it jump directly to the destination.

Look: Go/Step and *Go/Jump* follow a single line of analysis; *Look* tests many possible lines of analysis, in parallel, and presents intermediate findings. The goal is to facilitate a more informed decision on the part of the user. For example, if the user selects a specific clustering criterion for a relationship, this can eventually lead to the identification of an interaction with another variable or relationship and opportunities for further exploration. The system can explore the relevant possibilities, searching for strong patterns, and present its results to the user, annotated by the information it has gained.

Refine: A strategy may follow a chain of decisions that is almost but not quite correct. For example, given an approximately linear bivariate relationship, the system might observe points that could be considered outliers, take action to remove or downweight the points, fit a line, examine the residuals, and stop. The user, stepping in at this point, might then observe that patterns in the relationship would be more apparent if the data were slightly transformed, say by $y = x^{1.25}$, rather than relying on the linear relationship. The user can insert the transformation in the appropriate place in the path, and have the system repeat its earlier activities.

These operations highlight the process-oriented side of navigation. In domains such as hypertext, navigation is a way of getting from one information-rich node to the next. In data analysis, a great deal of necessary information is contained in the relationships *between* decision nodes. Navigation is appropriate for both types of domain.

3.3 Communication

The most important benefit of maintaining a graph of decisions is that it allows the user to communicate with the system through an explicit representation of the data analysis process. A display of the graph shows user actions and system actions (both already executed and planned for the future.) Many operations are possible to tailor the representation to the needs of the analysis:

Overview type: In addition to displaying decision nodes, the system can display nodes representing raw data, results, or primitive operations.

Justify: The system can display the knowledge structures (rules) relevant to a specific decision it has made. Interestingly, the system can also give a justification for decisions the user has made, providing a possible opportunity for learning, or strategy acquisition.

History: The system can limit its display to the sequence of decisions between the starting state and the current decision.

Zoom: The system can limit its display to a small neighborhood surrounding the current decision.

Magnify: The system can magnify the area around the current decision. In contrast to zooming, in which the surrounding decisions are removed from the display, magnification may be continuous, even nonlinear, giving a fisheye view of the decision graph.

Filter: The system can display only those nodes that meet criteria set by the user. For example, the user might wish to check for outliers in single variables by filtering out all other variables.

Landmark: The system can highlight a specific set of nodes to help the user maintain a sense of context while browsing through other nodes. The root node, for example, is a natural landmark.

Paths: The system can highlight sequences of nodes (paths) through the graph that meet criteria set by the user. This might be a specific derivation, or a set of related results, or some other possibility.

4   Conclusion

In an empirical evaluation of AIDE [16], evidence suggested that navigation is a significant factor in user performance. AIDE maintains an explicit representation of the data analysis process, including relationships between operations, decisions, justifications, and other relevant information. In the AIDE environment, the user's ability to move from one decision to another, to gain an overall view of the space of decisions, to modify and extend the current set of decisions—in short, the ability to navigate—proved to be important to the success of the analysis. The navigation facilities included most of the user and system actions described above, but only the first few of the communication features.

We believe that all of the navigational functions we have discussed could be put to good use in conventional data analysis systems. Further, they provide a useful framework for strategic reasoning, especially in the difficult task of communication between the user and the semi-autonomous statistical system.

References

1. Matthew Chalmers. Visualization of complex information. In *Third European Workshop on HCI.* 1993.
2. Julian J. Faraway. On the cost of data analysis. *Journal of Computational and Graphical Statistics,* 1(3):213–229, 1992.

3. Edward B. Fowlkes, Ramanathan Gnanadesikan, and Jon R. Kettenring. Variable selection in clustering and other contexts. In C. L. Mallows, editor, *Design, Data, and Analysis: by some friends of Cuthbert Daniel*. John Wiley & Sons, Inc., 1987.

4. W. A. Gale. REX review. In W. A. Gale, editor, *Artificial Intelligence and Statistics I*. Addison-Wesley Publishing Company, 1986.

5. William A. Gale, David J. Hand, and Anthony E. Kelly. Statistical applications of artificial intelligence. In C. R. Rao, editor, *Handbook of Statistics*, volume 9, chapter 16, pages 535–576. Elsevier Science, 1993.

6. D. J. Hand. A statistical knowledge enhancement system. *Journal of the Royal Statistical Society Serial A*, 150:334–345, 1987.

7. D.J. Hand. Patterns in statistical strategy. In W.A. Gale, editor, *Artificial Intelligence and Statistics I*, pages 355–387. Addison-Wesley Publishing Company, 1986.

8. Peter J. Huber. Languages for statistics and data analysis. In Peter Dirschedl and Ruediger Ostermann, editors, *Computational Statistics*. Springer-Verlag, 1994.

9. Hanhwe Kim and Stephen C. Hirtle. Spatial metaphors and disorientation in hypertext browsing. *Behaviour and Information Technology*, 14(4):239–250, 1995.

10. David Lubinsky and Daryl Pregibon. Data analysis as search. *Journal of Econometrics*, 38:247–268, 1988.

11. R. Wayne Oldford and Stephen C. Peters. Implementation and study of statistical strategy. In W.A. Gale, editor, *Artificial Intelligence and Statistics I*, pages 335–349. Addison-Wesley Publishing Company, 1986.

12. R. Wayne Oldford and Stephen C. Peters. DINDE: towards more sophisticated software environments for statistics. *SIAM Journal of Scientific and Statistical Computing*, 9(1):191–211, 1988.

13. Daryl Pregibon. Incorporating statistical expertise into data analysis software. In *The Future of Statistical Software*, pages 51–62. National Research Council, National Academy Press, 1991.

14. Robert St. Amant and Paul R. Cohen. Control representation in an EDA assistant. In Douglas Fisher and Hans Lenz, editors, *Learning from Data: AI and Statistics V*, pages 353–362. Springer-Verlag, 1996.

15. Robert St. Amant and Paul R. Cohen. A planner for exploratory data analysis. In *Proceedings of the Third International Conference on Artificial Intelligence Planning Systems*, pages 205–212. AAAI Press, 1996.

16. Robert St. Amant and Paul R. Cohen. Interaction with a mixed-initiative system for exploratory data analysis. In *Proceedings of the Third International Conference on Intelligent User Interfaces*, 1997.

17. L. G. Terveen. Intelligent systems as cooperative systems. *International Journal of Intelligent Systems*, 3(2–4):217–250, 1993.

18. John Tukey. An alphabet for statisticians' expert systems. In W.A. Gale, editor, *Artificial Intelligence and Statistics I*, pages 401–409. Addison-Wesley Publishing Company, 1986.

An Annotated Data Collection System to Support Intelligent Analysis of Intensive Care Unit Data

Christine L. Tsien, SM[1,2] and James C. Fackler, MD[3]

[1] Harvard Medical School, Boston, MA 02115, USA
[2] M.I.T. Laboratory for Computer Science, Cambridge, MA 02139, USA
[3] Johns Hopkins School of Medicine, Baltimore, MD 21287, USA

Abstract. Without a proper system for collecting data, the extent to which intelligent analysis can be performed on that data is limited. This is especially true in the study of time-ordered data, in which the gold standard for analysis techniques often comes from knowledge of discrete event occurrences amongst continuous data streams. An illustrative example is the study of patient monitor data from the Intensive Care Unit (ICU), which holds the promise of improving patient care in a setting where data overload and false alarms currently make that goal difficult at best. Here, monitor data is practically useless without corresponding knowledge of what clinical events were taking place to produce the observed data. A method of collecting both monitor data and clinical event annotations, and subsequently being able to correlate the two, has been developed and is being used at Children's Hospital in Boston. Preliminary results indicate that this type of data collection system is a viable tool for facilitating the intelligent analysis of temporal data, such as that from the ICU.

1 Introduction

Anyone who has worked in an Intensive Care Unit (ICU) understands the dual feelings of fortune and frustration evoked by modern technology in the form of alarmable patient monitors. Caregivers on the one hand, can be alerted to medical emergencies, but on the other hand, can be needlessly stressed as they waste valuble time and energy chasing after false alarms[1]. In this setting, teeming with bedside monitoring devices that generate enormous amounts of data, there are insufficient medical providers to continuously attend to each patient; alarm soundings thus become crucial indicators of a patient's deteriorating condition or need for assistance. Unfortunately, most of these alarms are actually false alarms[2, 3]. Various efforts are being attempted to alleviate this problem, such as developing specialized algorithms or intelligent monitoring systems aimed at producing more accurate alarms[4, 5, 6, 7, 8, 9, 10, 11]. A more informative alarm system could improve medical staff response time and thus improve

X. Liu, P. Cohen, M. Berthold (Eds.): "Advances in Intelligent Data Analysis" (IDA–97)
LNCS 1280, pp. 111–121, 1997.   © Springer–Verlag Berlin Heidelberg 1997

patient care[12]. Intelligent data analysis of physiologic signals clearly has great potential for making feasible such systems.

Development of intelligent data analysis techniques, however, depends integrally on having suitable data, and thus integrally on being able to properly collect such data. In terms of time-ordered data, suitable data consists of not only the temporal data streams, but also corresponding knowledge of the discrete event occurrences which may have influenced those data streams. These events, moreover, should be recorded with resolution as close as possible to that of the data streams. For example, the study of minute-to-minute changes in a company's stock prices would not be helped by knowledge only of the year in which a corporate takeover occurred, nor of just the month. Knowledge of the day the takeover occurred would be more useful, while knowledge of the time at which the news story was publicly announced would be most helpful for correlating this event with its immediate effects on the company's stock prices.

In the classification of non-temporal data, the gold standard is often well-understood. For example, consider the prediction of whether or not a patient with a given set of chest pain attributes is having a myocardial infarction (MI); the gold standard in this case is the actual presence or absence of MI. In contrast, the gold standard for analysis of temporal data is less clear-cut. Proper recording and synchronization of discrete event occurrences amongst temporal data streams is an approach to better understanding what comprises the gold standard of time-ordered data analysis. In terms of the ICU, proper recording requires collection of not only physiologic value streams for a given patient, but also annotations of the corresponding time-stamped, clinical events that were taking place. In this way, various data analysis techniques can then be explored and compared through retrospective study of the patient monitor data in conjunction with the annotations. A similar observation regarding suitable data has been made in Neurophysiology and Neurotraumatology monitoring[13].

This paper presents one such system for collecting bedside monitor data in conjunction with annotations of relevant clinical events, or "annotated data." The goal is for intelligent analysis of such annotated data to be able to aid in the development of "smarter" alarm algorithms. This would enable bedside monitors to issue more accurate alerts, thereby streamlining the work done by medical personnel and thus improving patient care in the ICU. The basic components of an annotated data collection system for temporal data will first be presented, followed by a detailed description of the system designed and implemented for the Multidisciplinary ICU (MICU) at Children's Hospital in Boston. Finally, an evaluation of this system, based on actual collection of and preliminary use of annotated data, will be discussed.

2   Collection System Components

There are three essential components of a general annotated data collection system for settings in which temporal data is the focus, such as the ICU. These include: a way to collect the data itself (the "primary data"), a way to collect the annotations for that data (the "secondary data"), and a way to unify the two. Typically, the primary data, in this case consisting mainly of dense, time-ordered physiologic signal values, are readily available (e.g., from a bedside monitoring device) and thus can be automatically stored into a database or other file. The secondary data, in this case consisting of notes describing the occurrence of relevant clinical events, are manually created during primary data collection time and are by nature sporadic. Secondary data need to be generated in an easy, accurate, and consistent manner, as well as need to be stored with an accurate time-stamp.

The final step in producing annotated data–the proper unification of primary data with secondary data–is crucial. In the ICU, where clinical events can last anywhere from a couple seconds to many minutes, poor unification makes that much harder, if not impossible, the already difficult challenge of gleaning understanding from plain time-ordered data. It is feasible to imagine, for example, that temporal misalignment of annotations with their data can cause false correlations to be drawn via intelligent data analysis techniques.

3   Children's Hospital ICU Collection System

The Multidisciplinary ICU at Children's Hospital is a 16-bed unit which admits each month approximately 120 patients (not including children with primary cardiac disease, who are admitted to a separate ICU). The annotated data collection system created for and tested in the MICU is like the general system described, except that collection of secondary data has two parts. It consists of both a computer-based collection program, and a trained observer to record annotations into that program. As in the general system, there is also a way to collect the primary data, and a way to unify this data with the annotations. Each of these system components will be described below.

3.1   Automated Collection of Bedside Monitor Data

Each bedside in the MICU is equipped with a SpaceLabs monitor (SpaceLabs Medical, Redmond, WA) that can display values from, for example, the electrocardiogram (ECG), pulse oximeter, and arterial line. Medical staff can indicate to each SpaceLabs monitor the upper and lower threshold values for a given signal being monitored, which determines when "threshold alarms" should sound.

The primary data in the MICU consists of the various time-stamped physiologic signal values. Every five to six seconds, the available values (i.e., those being monitored at the time of data transfer) can be obtained from the SpaceLabs monitor via a serial line connected between the monitor and a laptop computer placed at the bedside. These values are written to a file as shown in Figure 1.

Occasionally, a new set of heading labels (e.g., "TIME," "ECG HR," etc.) are presented by the monitor on the serial line; these are also inserted into the stored file, one for every signal that is available at that time. These headers occur whenever a "page" of data (approximately 100 rows of text and spaces) has

TIME	ECG HR	LEAD	RESP RATE	ART mmHg SYS/DIA	MEAN	SPO2 %
********	********	****	****	************		****
8:50:35	127	I	10	67/ 40	52	98
8:50:40	132	I	11	68/ 39	52	98
8:50:46	128	I	0	67/ 39	52	98

Fig. 1. Unprocessed Monitor Data–Typical Format

been written, or whenever the set of monitored signals has changed, whichever is sooner. Figure 2 illustrates a change from monitoring the ECG heart rate (HR), ECG lead number, respiratory rate, and arterial oxygen saturation (SpO2), to monitoring those same four signals with the addition of arterial line systolic, diastolic, and mean blood pressures. The relative placement of the different signal columns remains unchanged (e.g., SpO2 if present always follows the arterial blood pressures, which if present always follow respiratory rate).

TIME	ECG HR	LEAD	RESP RATE	SPO2 %
********	********	****	****	
10:34:37	131	I	9	100
10:34:42	134	I	9	100
10:34:47	134	I	9	100

SPACELABS MEDICAL PC PATIENT DATA LOGGER
PATIENT NAME   Bed # BS16   DATE 22AUG96

TIME	ECG HR	LEAD	RESP RATE	ART mmHg SYS/DIA	MEAN	SPO2 %
********	********	****	****	************		****
10:34:51	133	I	8	0/ 0	0	100
10:34:56	134	I	8	132/ 29	31	100

Fig. 2. Unprocessed Monitor Data–Change in Set of Monitored Signals

The laptop computer, under human control, specifies when to start or stop primary data collection from the SpaceLabs monitor. A new data file is automatically created whenever data collection begins for a new patient or a new date.

3.2 Computer-based Collection of Data Annotations

The secondary data in the ICU consists of annotations detailing the relevant clinical events occurring in the monitored bedspace, as well as annotations detailing the various threshold alarm values set on the monitoring devices. Examples of events include (but are not limited to): equipment malfunction, disconnection, or re-connection; interventions performed by medical staff; patient movements; and alarm soundings, which can be due to any of the other events or which can be for another reason. The duration of an event can vary from minutes (e.g., suctioning performed by nurse) to seconds or less (e.g., false alarm due to ECG wire movement). As a result, accurate recording in the annotation of the time of the event is extremely important. Another consideration is that events can occur very closely together, even simultaneously, such that the method of recording an annotation must be a quick one.

To meet these needs, specifically-tailored "forms" on the laptop computer at the bedside, running the Access database program (Microsoft, Redmond, WA), have been developed. Two types of computer forms are used: one ("Limit Settings") for recording alarm limit settings for a particular day, bedspace, and monitoring period, and a second ("Events Log") for recording clinical events that occur. Figures 3 and 4 show examples of these forms, respectively. The Limit Settings form stores the upper and lower limits of each monitoring device in use, as well as the begin and end time of each data collection period.

The Events Log form makes possible the recording of events quickly and easily by checking off a few boxes and by choosing from a list of likely event descriptions. (This list was compiled over the course of several days by manually recording, for each alarm observed in the ICU, the type of alarm and its cause.) Events are also categorized into alarm silencing episodes (during which audible alarms are disabled), non-alarm events (events that do not cause an alarm to sound), or alarm events. Alarm events are further divided into five types: true alarms with clinical relevance, true alarms without clinical relevance, false alarms, false negatives (lack of alarm when one was expected), and true negatives (correct lack of alarm because none was expected). This information is recorded to aid in retrospective analysis and correlation between the monitor data and the occurrence of alarms.

Another useful feature of the annotation collection program is that the laptop's system clock time can be automatically entered into a new annotation to promote data consistency and accuracy.

Finally, the annotation collection program maintains controls for starting and stopping collection of primary data from the SpaceLabs monitor. This facilitates collection of annotated ICU data by only storing primary data when event annotations are being simultaneously recorded. Furthermore, the laptop computer time is automatically stored to file with the primary data at the beginning of a data collection session. The importance of this feature is described in Section 3.4, Post-collection Data Unification.

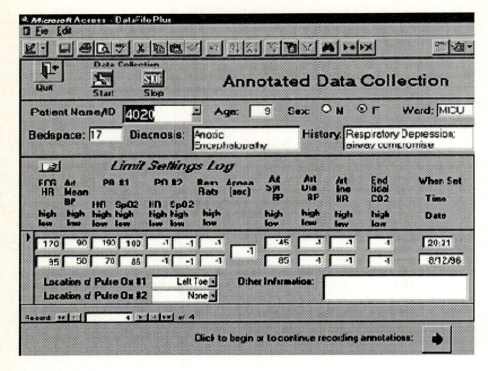

Fig. 3. Collection Program, Alarm Limit Settings Screen

3.3 Role of Trained Observer

Until this point, no details have been given about who is performing the work of annotating the ICU data. Clearly, the work is labor-intensive. To further burden a nurse or clinician with this task is infeasible. But it would be equally unacceptable to depend fully on a medically-ignorant person for recording medically-related information. The solution at the Children's Hospital MICU has been to use a trained observer as a link between the nurse and the annotation collection program. Three factors make this solution possible: First, the annotation collection program has been created to shoulder as much of the need for medical knowledge burden as possible. Second, the role of the trained observer has been filled by affordable yet quick-learning and meticulous college students who are taught how and when to make event annotations. And finally, the nurse at the bedside verbally (thus quickly and easily) verifies the medically-relevant information for the trained observer to record when an event occurs.

3.4 Post-collection Data Unification

The last step in the process of collecting useful annotated ICU data is the step of unifying the primary data (physiologic values) with the secondary data (clinical

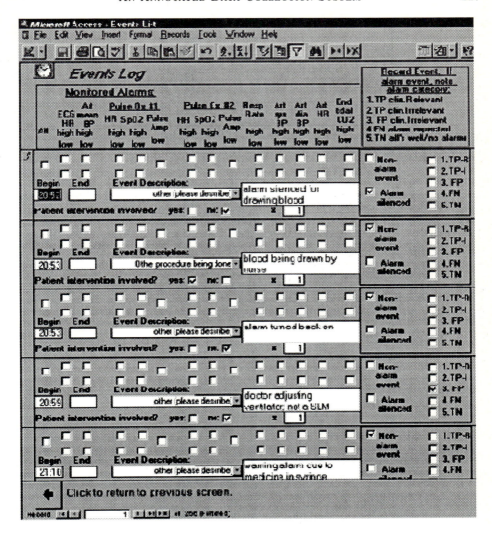

Fig. 4. Collection Program, Annotations Screen

events). This is achieved through processing with a series of computer programs (written in the C language), referred to as post-collection processing programs.

Of foremost importance is correcting the data time-stamps to adjust for the difference between the laptop computer time (thus the annotation time-stamps) and the SpaceLabs monitor time (thus the physiologic value time-stamps). Recall that upon starting the annotated data collection, the laptop computer time is recorded into the primary data file. For the data file from which the example in Figure 2 was taken, the laptop time was recorded at the beginning of the file as 08:37:05, and then was immediately followed by the first data stream of monitor values, having the time 08:42:11. The Transposition post-collection processing program not only calculates the time offset (00:04:06 in this example) and adjusts the data appropriately, but also transposes the data format from that shown in Figures 1 and 2 to that shown in Figure 5. Each line of transposed data, one per single physiologic value, contains the bedspace number, a numerical code to specify its type (e.g., code 3: ECG HR; code 64: SpO2), the value itself, the corrected time-stamp, and the date. This format is much easier for use in subsequent data analysis programs because the dynamically changing data format of the original data file (due to addition or deletion of one or more monitored signals) has been eliminated.

```
16 3 131 10:29:31 08-22-96
16 4 I 10:29:31 08-22-96
16 81 9 10:29:31 08-22-96
16 64 100 10:29:31 08-22-96
16 3 134 10:29:36 08-22-96
16 4 I 10:29:36 08-22-96
16 81 9 10:29:36 08-22-96
16 64 100 10:29:36 08-22-96
16 3 134 10:29:41 08-22-96
16 4 I 10:29:41 08-22-96
16 81 9 10:29:41 08-22-96
16 64 100 10:29:41 08-22-96
16 3 133 10:29:45 08-22-96
16 4 I 10:29:45 08-22-96
16 81 8 10:29:45 08-22-96
16 14 0 10:29:45 08-22-96
16 16 0 10:29:45 08-22-96
16 12 0 10:29:45 08-22-96
16 64 100 10:29:45 08-22-96
16 3 134 10:29:50 08-22-96
16 4 I 10:29:50 08-22-96
16 81 8 10:29:50 08-22-96
16 14 132 10:29:50 08-22-96
16 16 29 10:29:50 08-22-96
16 12 31 10:29:50 08-22-96
16 64 100 10:29:50 08-22-96
```

Fig. 5. Transposed Monitor Data With Adjusted Times

The transposed primary data can then be used in conjunction with textual equivalents of the event annotations. These textual equivalents are created by simply exporting the event annotation data from the Access database program into text files. This makes available for each annotated data collection period: the date plus begin and end times of the collection period, the signals monitored, the corresponding alarm threshold settings, and the alarm and non-alarm events that occurred during that period

The remaining post-collection processing programs are used to facilitate and/or perform various data analyses on the annotated ICU data. One, called the Standard Threshold Filter, for a given date and monitoring period reads the relevant alarm threshold settings and then the relevant transposed data file to "find" and tally threshold alarms. Another program, the Median Threshold Filter, does the same except that it first performs a moving median of N values (where N can be specified at runtime) and then "finds" and tallies where the median values surpass the alarm thresholds.

Count Alarms, a different type of processing program, works in conjunction with the filter programs. It first reads the alarm annotations, and then searches for each amongst the alarms found by the filter programs. These alarms are tallied by signal type (e.g., ECG heart rate, SpO2, etc.) as well as by categorization into clinically-relevant true alarms, false alarms, etc., so that the effects of different filters on eliminating false alarms without eliminating true ones can be compared.

4   Results and Discussion

The described data collection system has been used to collect patient monitor values along with annotations of clinical events for more than 800 hours. The collection has gone smoothly, and preliminary analysis efforts using the annotated data indicate that this is a beneficial system for aiding in the development of more accurate ICU alarm algorithms.

For example, a trial of the post-collection processing programs on 223 data files (a total of 35118 minutes of monitoring) has shown the ease of comparing the effects of taking a median of two, three, four, or any number of values on different categories of alarms. For the ECG heart rate high threshold alarm type, Count Alarms with Standard Threshold Filter found 31 clinically-relevant true alarms, 39 clinically-irrelevant true alarms, and 67 false alarms. With Median Threshold Filter of 2 values, it found 30 (97%), 31 (79%), and 38 (57%) of those same alarms, respectively. Using a median of 12 values, Count Alarms found 22 (71%), 12 (31%), and 7 (10%) of those alarms.

Such filter comparisons are made possible by having accurate and synchronized, time-stamped annotations of the alarms that had sounded. Had these not been collected, it is difficult to imagine how this sort of automated detection of alarms by type, and thus comparison of filters, could be accomplished. Imagine, for example, a different approach to studying the effects of clinical events that depends on retrospectively gleaning the event information from nurses' charts. This method would certainly lack accurate synchronization of events with their underlying data streams at least for those events that are very brief, if not for all events. It is also improbable that nurses would record each and every alarm sounding in their charts, thus this information would be unavailable at data analysis time.

While using a median filter has been described as an example of how the annotated data can help in filter algorithm development, it is in no way implied as a solution for intelligent alarming. In fact, Makivirta *et al.* concluded that a single median filter seems insufficient to improve the alarm problem[14]. The example has been presented, rather, to illustrate the ease with which different filter algorithms can be compared.

One disadvantage of the annotation collection system described is that it exists separate from the SpaceLabs monitor, which in practice means that primary data values can only be obtained once every five to six seconds. More sophisticated analysis techniques might work better on data collected at frequencies of once per second or greater, as in the digital data acquisition and analysis system developed by Amlab[15].

Others have also described a project aimed at developing an ICU data collection system[16]. It is not clear, however, that their system would be able to store the type of clinical annotations in the time-stamped and synchronizable manner that has been described here. Storage of such information is felt to be crucial for providing the gold standard used in development and evaluation of intelligent ICU alarms. In any case, there are few examples of intelligent systems that have been successfully put into practice despite the number of efforts made to develop such systems[5], thus continued efforts are still needed.

Acknowledgments

The authors would like to thank Isaac Kohane, MD PhD for providing invaluable advice; Peter Szolovits, PhD for encouraging independent investigation; Banny Wong and Scott Pauker for carefully collecting alarm annotations; Jonathan Tien for assistance with the SpaceLabs monitor communications; and the MICU nurses for their clinical judgment, patience, and cooperation. This work was supported in part by SpaceLabs Medical and the National Library of Medicine.

References

1. Topf M, Dillon E: Noise-induced stress as a predictor of burnout in critical care nurses. Heart Lung. **17** (1988) 567–574
2. Lawless, S: Crying wolf: False alarms in a pediatric intensive care unit. Critical Care Medicine. **22** (1994) 981–985
3. Tsien C, Fackler J: Poor prognosis for existing monitors in the intensive care unit. Critical Care Medicine. **25** (1997) 614–619
4. Crew A, Stoodley K, Lu R, Old S, Ward M: Preliminary clinical trials of a computer-based cardiac arrest alarm. Intensive Care Medicine. **17** (1991) 359–364
5. Uckun, S: Intelligent systems in patient monitoring and therapy management. A survey of research projects. Int J Clin Monit Comput. **11** (1994) 241–253
6. Bosnjak A, Bevilacqua G, Passariello G, Mora F, Sanso B, Carrault G: An approach to intelligent ischaemia monitoring. Medical & Biological Engineering & Computing. (1995) 749–756

7. Aukburg S, Ketikidis P, Kitz D, Mavrides T, Matschinsky B: Automation of physiological data presentation and alarms in the post anesthesia care unit. Symposium on Computer Applications in Medical Care. (1989) 580–582

8. Dumas C, Wahr J, Tremper K: Clinical evaluation of a prototype motion artifact resistant pulse oximeter in the recovery room. Anesth Analg. **83** (1996) 269–272

9. Visram A, Jones R, Irwin M, Bacon-Shone J: Use of two oximeters to investigate a method of movement artefact rejection using photoplethysmographic signals. British Journal of Anaesthesia. **72** (1994) 388–392

10. Sukuvaara T, Koski E, Makivirta A, Kari A: A knowledge-based alarm system for monitoring cardiac operated patients – technical construction and evaluation. Int J Clin Monit Comput. **10** (1993) 117–126

11. Adams J, Inman I, Abreu S, Zabaleta I, Sackner M: A computer algorithm for differentiating valid from distorted pulse oximeter waveforms in neonates. Pediatric Pulmonology. **19** (1995) 307–311

12. Westenskow D, Orr J, Simon F, Bender H, Frankenberger H: Intelligent Alarms Reduce Anesthesiologist's Response Time to Critical Faults. Anesthesiology. **77** (1992) 1074–1079

13. Guedes de Oliveira P, Cunha J, Martins da Silva A: The role of computer based techniques in patient monitoring: technical note. Acta Neurochir. [**Suppl**] 55 (1992) 18–20

14. Makivirta A, Koski E, Kari A, Sukuvaara T: The median filter as a preprocessor for a patient monitor limit alarm system in intensive care. Computer Methods and Programs in Biomedicine. **34** (1991) 139–144

15. Hall G, Colditz P: Continuous physiological monitoring: an integrated system for use in neonatal intensive care. Australasian Physical & Engineering Sciences in Medicine. **18** (1995) 139–142

16. Metnitz G, Laback P, Popow C, Laback O, Lenz K, Hiesmayr M: Computer assisted data analysis in intensive care: the ICDEV project – development of a scientific database system for intensive care. Int J Clin Monit Comput. **12** (1995) 147–159

A Combined Approach to Uncertain Data Analysis

Hairong Yu and Arthur Ramer

School of Computer Science and Engineering
The University of New South Wales, Sydney 2052, Australia
{hairong,ramer}@cse.unsw.edu.au

Abstract. In the real world, various kinds of uncertain information are very popular in decision making. It is highly desirable to represent them in a database form. We develop a new *uncertain relational database* model that includes using possibility distributions as values of attributes along with probabilities associated with tuples. This new model can represent all kinds of uncertainty by providing rich structure to support realistic uncertain data manipulation more completely. Relational algebra based on the new uncertain database model is introduced and a query processing example is presented.

1 Introduction

During the course of daily life much of human reasoning is approximate in nature. Sometimes we would rather say Mary is young than rigidly state that Mary is twenty-one years ten months seven days six hours old. The former called an uncertain data and the latter a perfect data.

There are many kinds of uncertain data. Using a classic student database, we make an example to show their many aspects:

Incomplete/Probable data:
It is highly probable that Alex will graduate this session.

Imprecise/non-specific/Vague data:
Alex logs in to the school system two to six times a week.

Fuzzy data:
Alex is a good student.

Ambiguous data:
Alex is doing either Master or PhD from this session.

Inconsistent data:
Alex submitted assign. 2 last week, and he will submit assign. 2 next week.

Ignorant data:
Do not know / Not applicable.

Erroneous/Irrelevant data.
Caused by unreliable operation.

X. Liu, P. Cohen, M. Berthold (Eds.): "Advances in Intelligent Data Analysis" (IDA–97)
LNCS 1280, pp. 123–134, 1997.   © Springer–Verlag Berlin Heidelberg 1997

There is a need for expressing uncertainty in data modelling directly and also coping with uncertain schemas and queries more efficiently [Pet95]. To represent and manipulate uncertain information in database and information systems is not a trivial problem. Since the initial effort of Codd [Cod79], management of uncertain information in relational databases has been an active research area. An ideal information system should take all these forms of uncertain information into account. The existing work generally considers only one form of uncertainty in any given framework except a few ones are except, paper like [RY96], but still without comprehensive theoretical basis.

Although we outlined several forms of uncertainty, and quite a few more have been considered, most can be expressed quantitatively using one of two approaches. One expresses a degree of imperfection either as probability or possibility. In both cases we assign a value in the range [0,1] to express uncertainty, but use different rules of computation and combination. Thus the first approach could be suited for the 'probable' or 'ambiguous' data, while the second would fit 'imprecise' or 'fuzzy' data. Handling data subject to imperfections stemming from different causes require a model which can manipulate all these kinds of uncertain information simultaneously.

This paper proposes an extension of the relational model which enables us to deal with fuzzy values and imprecise facts, probable information and also ignorant attributes. Moreover, queries comprising vague predicates can also be posed and answered.

In next section, we give the definition of uncertain database with assumption. The semantics of uncertain relations is described in Section 3. Then the generalised relational algebra is given in Section 4. Section 5 is devoted to associated query processing by an example. The last section presents conclusions to emphasise the differences and similarities with other approaches.

2   Uncertain Database Model

The uncertain database model is presented as a set of definitions describing how real world data is conceptually represented in a computerised form. The available operations are also described to access and update the information.

We define uncertain relation as generalisation of Codd's relation model [Ull83] by associating possibility with attributes to express possibilistic data and an additional attribute $p(\mathbf{t})$ to indicate the probability that tuple $\mathbf{t}$ belongs to a uncertain relation $\mathcal{R}$.

A scheme of a relation is a finite collection of attributes $A_1, \ldots, A_m$ (which may be considered a part of the universal scheme [UW97]). Each attribute has its associated crisp domain $dom(A_i)$, To model various kinds of uncertainty we require a notion of a fuzzy domain of A_i. We would like to consider a set of all possibility assignments $Poss(A_i) = \{\pi : dom(A_i) \rightarrow [0,1]\}$ as a potential *fuzzy domain*.

We would also like to have a special symbols to denote concepts like 'unknown' or 'unavailable'. A full set of possibility assignments is uncountable infinite, hence not suitable for a typical database application. We require that our domain is a finite subset $U_i \subset \{\pi \in Poss(A_i)\}$ and that it is augmented with three designated elements. They are *void* a special value interpreted as *not yet defined*, (meaning that we intend to fill in the value), *none* interpreted as no possibility for the value of a object could exist, and *unknown* a possible possibility for any value of a object. We have

Definition 1. *Fuzzy domain U_i of an attribute A_i*

$$fdom(A_i) = U_i \subset \{\pi : dom(A_i) \rightarrow [0,1]\},$$

$$\{void, none, unknown\} \subset U_i, \ card(U_i) < \infty.$$

Definition 2. An uncertain database $\mathcal{D}$ is defined as a set of uncertain relation $\mathcal{R}_i$, where $i = 1, 2, \ldots, n$, i.e. $\mathcal{D} = \{\mathcal{R}_1, \mathcal{R}_2, \ldots, \mathcal{R}_n\}$.

Definition 3. Let $R(A_1, A_2, \ldots, A_m)$ be a relation schema where $fdom(A_i)$ is fuzzy domain of attribute A_i, $i = 1, 2, \ldots, m$, which is defined as above. An *uncertain relation* $\mathcal{R}$ of scheme R is:

$$\mathcal{R} \subseteq \{(\mathbf{t}, p(\mathbf{t})) \mid (\mathbf{t}, p(\mathbf{t})) = \langle u_1, u_2, \ldots, u_m, p(\mathbf{t}) \rangle,$$

$$(u_1, u_2, \ldots, u_m) \in U_1 \times U_2 \times \cdots \times U_m, 0 \leq p(\mathbf{t}) \leq 1\}$$

Here $p(\mathbf{t}) \in [0,1]$ is a probability, the symbols $\subseteq$, $\times$ and $\in$ and the like denote, depending on context, either ordinary or fuzzy set-theoretic operations.

A complete tuple $(\mathbf{t}, p(\mathbf{t}))$ in uncertain relation $\mathcal{R}$ includes also a probability $p(\mathbf{t})$ is associated with fact $\mathbf{t}$. Conversely, the probability of $\mathbf{t}$ which does not belong to $\mathcal{D}$ is $1 - p(\mathbf{t})$. This approach agrees with the closed world assumption [Rei78], expressing that if the tuple $\mathbf{t}$ is fully related to the relation $\mathcal{R}$, then its probability of belonging to $\mathcal{R}$ becomes 1. If $\mathbf{t}$ does not appear in the relation $\mathcal{R}$, its probability of belonging to $\mathcal{R}$ should be 0.

Under this interpretation $p(\mathbf{t})$ is the probability that $\mathbf{t} \in \mathcal{R}$. This treats $\mathcal{R}$ as a form of a random variable. Another closely related interpretation would be to consider $p(\mathbf{t})$ as the (probabilistic) truth value of the assertion $\langle \mathbf{t}$ belongs to $\mathcal{R} \rangle$.

In our uncertain databases we assume stochastic independence of the distinct tuples, which implies that

$$p(\mathbf{t}_k \in \mathcal{R}_i) \times p(\mathbf{t}_l \in \mathcal{R}_j) = p((\mathbf{t}_k \in \mathcal{R}_i) \wedge (\mathbf{t}_l \in \mathcal{R}_j))$$

for any $\mathbf{t}_k, \mathbf{t}_l, \mathcal{R}_i, \mathcal{R}_j$ except $\mathbf{t}_k = \mathbf{t}_l$ and $\mathcal{R}_i = \mathcal{R}_j$.

From here we can see clearly that our *uncertain model* is an extension of the traditional relational model. The latter could be viewed as the special case of our model with value p taking only values 1 or 0, i.e. for a tuple $(\mathbf{t}, p(\mathbf{t})), p(\mathbf{t}) = 1$, if $\mathbf{t} \in \mathcal{R}$ or $p(\mathbf{t}) = 0$, if $\mathbf{t} \notin \mathcal{R}$, the relation become a traditional one.

Our uncertain relation is represented as a table where p is placed in an extra column. For instance

Students

Name	Age	Course	p
Daniel	about 30	*none*	0.9
Anna	29	PhD	0.2
Stefan	*void*	PhD or Master	1
Karen	20 or 21	*void*	0.8
John	*unknown*	Undergraduate	0.5

The meaning of this relation is that Daniel is about 30, where we use possibility distribution to express 'about 30' $= \{0.8/29, 1/30, 0.9/31\}$. But if he does not enrol in any course in the university, *none* signifies that there is no possibility the value of $A(x)$ be present in the universe of discourse U. In other words, $\pi_{Age}(u) = 0$, for all u in U. The value of p shows that the probability of Daniel's $\langle Age, Course \rangle$ occurring in this relation is 0.9. Next, for Anna, like in a classical database, all facts have crisp values. She is 29 years old and a PhD candidate. And the likelihood of the complete tuple is 0.2 in our database. Let us now discuss the tuple corresponding to John. He is an Undergraduate student, but we are not sure of his age. *Unknown* is perceived as the possibility that $A(x)$ could be any value in U, and as we do not have any information to decide about it, we put $\pi_{A(x)}(u) = 1$ for all u in U. This fact has probability 0.5 in our example. Back to Stefan, we can use possibility distribution to spell out 'PhD or Master' $= \{1/PhD, 1/Master\}$. For *Age:void*, we do not know whether it is defined or we do not care and it is neither *none* nor *unknown*. The probability of this happening is assumed to be one hundred percent. And now it is easy to interpret Karen's tuple. Karen's $Age = \{1/20, 1/21\}$, without any idea about her course, and the probability of this tuple is 0.8.

3   Semantics of Uncertain Relation

To each uncertain relation scheme $\mathcal{R}$ we attach a set of pairs $< \mathcal{T}, p >$ where $\mathcal{T}$ is a traditional relation, p is a real number at interval $[0, 1]$. This set of pairs represents all kinds of cases happening under $\mathcal{R}$, where p is incorporated with each case. These pairs are named the *possible worlds* of $\mathcal{R}$. A *possible world* is a consistent and complete description of how things might have been or might be in actuality [Gen89]. It may or may not be true in real world, but it is true in the miniworld of the database. The *actual world* is a possible world that is true – the complete description of how things are.

So far $\mathcal{R} = \{(\mathbf{t}_1, p(\mathbf{t}_1)), (\mathbf{t}_2, p(\mathbf{t}_2)), \ldots, (\mathbf{t}_m, p(\mathbf{t}_m))\}$ represents at most 2^m traditional relations, i.e. the power set of $\mathbf{t}_1, \mathbf{t}_2, \ldots, \mathbf{t}_m$. We define a mapping $\mathcal{PW}$ that computes all the possible cases of $\mathcal{R}$ and each case (in possible world) is attached by a probability which is grater than 0. We are only interested in listing the possible ones whose probabilities are greater than zero and not the impossible ones whose probabilities are equal to zero.

Definition 4. If $\mathcal{R} = \{(\mathbf{t}_1, p(\mathbf{t}_1)), (\mathbf{t}_2, p(\mathbf{t}_2)), \ldots, (\mathbf{t}_m, p(\mathbf{t}_m))\}$ is an uncertain relation, then the possible worlds $\mathcal{PW}(\mathcal{R})$ of $\mathcal{R}$ is a mapping such that

$$\mathcal{PW}(\mathcal{R}) = \{(\mathcal{T},p) \mid \mathcal{T} = \mathbf{t}_{i_1}, \mathbf{t}_{i_2}, \ldots, \mathbf{t}_{i_j},$$

$$p = p(\mathbf{t}_{i_1}) \times p(\mathbf{t}_{i_2}) \times \cdots \times p(\mathbf{t}_{i_j}) \times (1 - p(\mathbf{t}_{i_{j+1}})) \times (1 - p(\mathbf{t}_{i_{j+2}})) \times \cdots \times (1 - p(\mathbf{t}_{i_m})) > 0\}$$

where $\mathbf{t}_{i_1}, \mathbf{t}_{i_2}, \ldots, \mathbf{t}_{i_j}$ is a combination out of $\mathbf{t}_1, \mathbf{t}_2, \ldots, \mathbf{t}_m$ and $\exists j (0 \leq j \leq m)$.

Theorem 5. *For every possible worlds $\mathcal{PW}(\mathcal{R})$ of $\mathcal{R}$, its members are* $(\mathcal{T}_1, p_1), (\mathcal{T}_2, p_2), \ldots, (\mathcal{T}_{2^m}, p_{2^m})$, $p_1 + p_2 + \cdots + p_{2^m} = 1.0$.

Proof.

Since from uncertain relation $\mathcal{R} = \{(\mathbf{t}_1, p(\mathbf{t}_1)), (\mathbf{t}_2, p(\mathbf{t}_2)), \ldots, (\mathbf{t}_m, p(\mathbf{t}_m))\}$, only two cases, with or without $\mathbf{t}_i$, are concerned as the member of $\mathcal{T}$, where $0 \leq i \leq m$. We have at most 2^m traditional relations, i.e. $\mathcal{T}_1, \mathcal{T}_2, \ldots, \mathcal{T}_{2^m}$. And each of them associates one p as $p_1, p_2, \ldots, p_{2^m}$ respectively.

For $m = 1$: result is noticeable and trivial.

For $m = 2$: $\mathcal{R}_2 = \{(\mathbf{t}_1, p(\mathbf{t}_1)), (\mathbf{t}_2, p(\mathbf{t}_2))\}$, and

$$\mathcal{T}_1 = \phi, \mathcal{T}_2 = \{\mathbf{t}_1\}, \mathcal{T}_3 = \{\mathbf{t}_2\}, \mathcal{T}_4 = \{\mathbf{t}_1, \mathbf{t}_2\}$$
$$p_1 = (1 - p(\mathbf{t}_1)) \times (1 - p(\mathbf{t}_2)), p_2 = p(\mathbf{t}_1) \times (1 - p(\mathbf{t}_2))$$
$$p_3 = p(\mathbf{t}_2) \times (1 - p(\mathbf{t}_1)), p_4 = p(\mathbf{t}_1) \times p(\mathbf{t}_2)$$

So,
$$p = p_1 + p_2 + p_3 + p_4$$
$$= (1 - p(\mathbf{t}_2))(1 - p(\mathbf{t}_1) + p(\mathbf{t}_1)) + p(\mathbf{t}_2)(1 - p(\mathbf{t}_1) + p(\mathbf{t}_1))$$
$$= 1 - p(\mathbf{t}_2) + p(\mathbf{t}_2) = 1.$$

Suppose now by induction when $m = n$ the proposition holds.

For $m = n+1$: $\mathcal{R}_{n+1} = \{(\mathbf{t}_1, p(\mathbf{t}_1)), (\mathbf{t}_2, p(\mathbf{t}_2)), \ldots, (\mathbf{t}_n, p(\mathbf{t}_n)), (\mathbf{t}_{n+1}, p(\mathbf{t}_{n+1}))\}$
$$= \mathcal{R}_n \cup \{(\mathbf{t}_{n+1}, p(\mathbf{t}_{n+1}))\}$$

From the above formula, we could see, besides the 2^n traditional relations which are the same as they are when $m = n$ and without $\mathbf{t}_{n+1}$, there are other 2^n traditional relations and each with tuple $\mathbf{t}_{n+1}$. Total number is $2^n + 2^n = 2^{n+1}$.

The former with probabilities called $(\mathcal{T}_1, p_1), (\mathcal{T}_2, p_2), \ldots, (\mathcal{T}_{2^n}, p_{2^n})$, and latter with probabilities are $(\mathcal{T}_{2^n+1}, p_{2^n+1}), (\mathcal{T}_{2^n+2}, p_{2^n+2}), \ldots, (\mathcal{T}_{2^n+2^n}, p_{2^n+2^n})$.

To calculate p, since the only distinction between p_i and p_{2^n+i} could be the former has factor $1 - p(\mathbf{t}_{n+1})$ while the latter has $p(\mathbf{t}_{n+1})$ and all the other factors are same, where $1 \leq i \leq 2^n$. We could add both together to have partial probability value of p^{n+1} as $p_i^{n+1}[1 - p(\mathbf{t}_{n+1}) + p(\mathbf{t}_{n+1})]$, where that power of p is $n + 1$ means that we are at stage $m = n + 1$.

$$p^{n+1} = p_1^{n+1} + p_2^{n+1} + \cdots + p_{2^n}^{n+1} + p_{2^n+1}^{n+1} + p_{2^n+2}^{n+1} + \cdots + p_{2^n+2^n}^{n+1}$$
$$= (p_1^{n+1} + p_{2^n+1}^{n+1}) + (p_2^{n+1} + p_{2^n+2}^{n+1}) + \cdots + (p_{2^n}^{n+1} + p_{2^n+2^n}^{n+1})$$
$$= [1 - p(\mathbf{t}_{n+1}) + p(\mathbf{t}_{n+1})] \sum_{i=1}^{2^n} (p_i^n)$$

according to assumption: when $m = n$, $\sum_{i=1}^{2^n} (p_i^n) = 1$. Then
$$p^{n+1} = 1 - p(\mathbf{t}_{n+1}) + p(\mathbf{t}_{n+1}) = 1. \qquad \square$$

We remark that there is another shorter but less direct argument. Denoting $q_i = 1 - p_i$, we have

$$(p_1 + q_1) \cdots (p_n + q_n) = 1$$

expanding the left side as a sum of products $p_{i_1}, \ldots, p_{i_k}, q_{i_{k+1}}, \ldots, p_{i_n}$. We find each term p of Definition 4 exactly once, which proves the sum of the probabilities of all the possible worlds is 1. To illustrate it we take the previous example regarding each tuple as t_i, where $i = 1, \ldots, 5$. We only list the worlds of positive probability; we consider probability value 0 as indicating the potential world as impossible. In our case each world must include t_3, hence there are $2^4 = 16$ possible cases.

Possible worlds of uncertain relation **Students**:

5 tuple worlds					
$\{t_1, t_2, t_3, t_4, t_5\}$	0.072				
4 tuple worlds					
$\{t_2, t_3, t_4, t_5\}$	0.008	$\{t_1, t_3, t_4, t_5\}$	0.288	$\{t_1, t_2, t_3, t_5\}$	0.018
$\{t_1, t_2, t_3, t_4\}$	0.072				
3 tuple worlds					
$\{t_3, t_4, t_5\}$	0.032	$\{t_2, t_3, t_5\}$	0.002	$\{t_2, t_3, t_4\}$	0.008
$\{t_1, t_3, t_5\}$	0.072	$\{t_1, t_3, t_4\}$	0.288	$\{t_1, t_2, t_3\}$	0.018
2 tuple worlds					
$\{t_1, t_3\}$	0.072	$\{t_2, t_3\}$	0.002	$\{t_3, t_4\}$	0.032
$\{t_3, t_5\}$	0.008				
1 tuple world					
$\{t_3\}$	0.008				

4   Uncertain Relational Algebra

Having established a proposed structure of uncertain database, we need to formalise its relational algebra.

The relational algebra operations are usually divided into two groups [Cod72, EN94]: (a) set theoretical operations union, intersection, difference and Cartesian product, (b) and the operations specific for relational databases select, project and join. We must broaden them to the uncertain relations.

Consider two n-ary uncertain relations $\mathcal{R}$ and $\mathcal{S}$ in a universe of discourse $U = \{U_1 \times U_2 \times \cdots \times U_n, p\}$.

Union $\mathcal{R} \cup \mathcal{S} = \{max(\pi_{\mathcal{R}}(u_i), \pi_{\mathcal{S}}(u_i))/u_i, (p_{\mathcal{R}} + p_{\mathcal{S}} - p_{\mathcal{R}} \times p_{\mathcal{S}}) : u_i \in U_i\}$, where $i = 1, 2, \ldots, n$. [1]

Intersection $\mathcal{R} \cap \mathcal{S} = \{min(\pi_{\mathcal{R}}(u_i), \pi_{\mathcal{S}}(u_i))/u_i, p_{\mathcal{R}} \times p_{\mathcal{S}} : u_i \in U_i\}$.

Duality between union and intersection is preserved as $1 - p_{\mathcal{R}} p_{\mathcal{S}} = (1 - p_{\mathcal{R}}) + (1 - p_{\mathcal{S}}) - (1 - p_{\mathcal{R}})(1 - p_{\mathcal{S}})$.

---

[1] Since we assume that the union event expression is a Boolean combination of independent events, we can apply sieve formula to compute the probability [Ros93].

Difference $\mathcal{R} - \mathcal{S} = \{max((\pi_{\mathcal{R}}(u_i) - \pi_{\mathcal{S}}(u_i)), 0)/u_i, (1 - p_{\mathcal{R}} + p_{\mathcal{R}}p_{\mathcal{S}}) : u_i \in U_i\}$, where $-$ on the left denotes set subtraction.

Cartesian product We define the Cartesian product of a m-ary uncertain relation $\mathcal{R} \in U_1 \times U_2 \times \cdots \times U_m$ and a n-ary relation $\mathcal{S} \in V_1 \times V_2 \times \cdots \times V_n$ as follows:
$\mathcal{R} \times \mathcal{S} = \{t = (a_1, \ldots, a_m, b_1, \ldots, b_n), p_{\mathcal{R}} \times p_{\mathcal{S}} : a_i \in U_i, b_j \in V_j, 0 \leq i \leq m, 0 \leq j \leq n\}$.

Selection The selection constructs a subset of tuples within named relation that satisfy some conditions. Let $\mathcal{R}$ be as above and $\mathcal{C}$ a k-place predicate from $U_{i_1}, U_{i_2}, \ldots, U_{i_k}$ and one threshold value θ. Then the selection from an uncertain relation is stated as:
$\sigma_{C(A_{i_1}, A_{i_2}, \ldots, A_{i_k}, \theta)}(\mathcal{R}) = \{t, p(t) \mid max[\pi_{U_{i_j}}(u)\mu_{A_{i_j}}(u)] \geq \theta : u \in U, p(t) \in \mathcal{R}, 1 \geq \theta \geq 0, 1 \leq j \leq k\}$, where μ is membership function of fuzzy set A_{i_j}.

Projection The projection is an operation which constructs a new relation by selecting the specified attributes and discarding others. Let $\mathcal{R}$ be as above, whose attributes are $A_1, A_2, \ldots, A_m$, and let the new relation $\mathcal{Q}$ be a projection of $\mathcal{R}$ onto the attributes $A_{i_1}, A_{i_2}, \ldots, A_{i_k}$. It is defined as: $\pi_{A_{i_1}, A_{i_2}, \ldots, A_{i_k}}(\mathcal{R}) = \{\mathcal{Q}(A_{i_1}, A_{i_2}, \ldots, A_{i_k}, p_{\mathcal{Q}}(t)) : p_{\mathcal{Q}}(t) = p_{\mathcal{R}}(t), \mathcal{Q} \in \mathcal{R}\}$ where more than one matching k-tuple is combined together and its probability value is the maximum of them.

Theorem 6. *The probability of projection resulting relation $p_{\mathcal{Q}}(t)$ is never less than the probability of original relation $p_{\mathcal{R}}(t)$.*

Proof.
From the semantics of relational database $R(A_1, A_2, \ldots, A_m)$, any A_i is independent from A_j, where $i, j = 1, 2, \ldots, m$ and $i \neq j$. We have $p_{A_1}(t) \times p_{A_2}(t) \times \cdots \times p_{A_m}(t) = p(t)$ according to axiom in [Ros93]: two events E and F are independent if $P(EF) = P(E)P(F)$. This

$$p_{A_{i_1}}(t) \times p_{A_{i_2}}(t) \times \cdots \times p_{A_{i_k}}(t) \times p_{\{A_1, A_2, \ldots, A_m\} - \{A_{i_1}, A_{i_2}, \ldots, A_{i_k}\}}(t) = p_{\mathcal{R}}(t)$$

stands as well. Seeing that $max(p_{\{A_1, A_2, \ldots, A_m\} - \{A_{i_1}, A_{i_2}, \ldots, A_{i_k}\}}(t)) = 1$, clearly gives $p_{\mathcal{Q}}(t) = p_{A_{i_1}}(t) \times p_{A_{i_2}}(t) \times \cdots \times p_{A_{i_k}}(t) \geq p_{\mathcal{R}}(t)$. $\quad\square$

Natural Join The join operation is used to combine related tuples from two relations into single tuples over all the combined attributes. Let m-ary $\mathcal{R}$ and n-ary $\mathcal{S}$ join over attributes $A_{i_1}, A_{i_2}, \ldots, A_{i_k}$ in $\mathcal{R}$ and $B_{i_1}, B_{i_2}, \ldots, B_{i_k}$ in $\mathcal{S}$ and one threshold value th, the natural join operation which produces $(m+n-k)$-ary relation is defined as:
$\mathcal{R} \bowtie_{A_{i_1}, \ldots, A_{i_k}, B_{i_1}, \ldots, B_{i_k}} \mathcal{S} = \{t = (a_1, \ldots, a_m, b_1, \ldots, b_n) - (a_{i_1}, \ldots, a_{i_k}), p_{\mathcal{R}}(t) \times p_{\mathcal{S}}(t) : max[min(\pi_{A_{i_l}}(u), \pi_{B_{i_l}}(u))] \geq \theta, 1 \leq i \leq m, 1 \leq j \leq n, 1 \leq l \leq k \leq m$ or $n, A_{i_k} \in U_i, B_{i_k} \in V_j, U_i = V_j, u \in U\}$.

5 Query Processing

Because of the limited space we cannot present a full explanation. Generally we can transform our uncertain queries into algebra expressions and use relational operations to evaluate the results.

A simple example will illustrate retrieval processing in the uncertain relational database. We have Students relation in section 2, and Sports relation below:

Sports

Age	Sports	p
around 20	soccer playing	0.6
close to 30	swimming,badminton playing	0.3
40	playing bridge	0.5
mid 50	playing golf	0.7

Here 'around 20' = {0.9/19,1/20,0.9/21}, 'close to 30' = {0.8/29,0.9/30} and 'mid 50' = {0.8/54,1/55,0.7/56 }.

Query: List young students participating in sports,
where 'young'={0.3/15,0.6/16,0.8/17,1/18,1/19,1/20,1/21,1/22,1/23,
                1/24,1/25,0.9/26,0.9/27,0.8/28,0.7/29,0.6/30,0.6/31}
Relational algebra expression for the query is

$$(Students \bowtie_{(Age,0.5)} Sports)_{where} C(Age = 'young', 0.65)[Name, Sports]$$

We divide its processing into three steps.

Step 1: $\mathcal{R}_1 = (Students \bowtie_{(Age,0.5)} Sports)$

For Daniel, we compare 'about 30' and 'close 30' for attribute Age from the join definition $max[min(\pi_{C_i}(u), \pi_{D_j}(u))] = max\{min(0.8, 0.8)/29, min(1, 0.9)/30,$ $min(0.9, 0)/31\} = max\{0.8, 0.9\} = 0.9 \geq 0.5$ and $p_{Students}(\mathbf{t}) \times p_{Sports}(\mathbf{t}) = 0.9 \times 0.3 = 0.27$. Anna, $max\{min(0.8, 1)/29\} = 0.8 \geq 0.5$, $p = 0.2 \times 0.3 = 0.06$. As Stefan's Age is $void$, we could not put it into the joined relation. Karen's case is straightforward, just calculate between {1/20,1/21} and 'around 20':$max\{min(0, 0.9)/19, min(1, 1)/20, min(0.9, 1)/21\} = max(1, 0.9) = 1 \geq 0.5$, then $p = 0.8 \times 0.6 = 0.48$. Next for John, as our join is based on common Age attribute and John's Age is $unknown$, John's $Sports$ could be also $unknown$ and similarly the probability value is missing represented by $*$.

$$\mathcal{R}_1 = (Students \bowtie_{(Age,0.5)} Sports)$$

Name	Age	Course	Sports	p
Daniel	about 30	*none*	*swimming,badminton playing*	0.27
Anna	29	PhD	*swimming,badminton playing*	0.06
Karen	20 or 21	*void*	*soccer playing*	0.46
John	*unknown*	Undergraduate	*unknown*	*

After this we obtain an intermediate relation, to which we apply the following step

Step 2: $\mathcal{R}_2 = \mathcal{R}_{1\,where}\mathcal{C}(Age = \text{'young'}, 0.65)$

First is Daniel, from selection definition $max[\pi_{U_{i_j}}(u)\mu_{A_{i_j}}(u)] = max\{(0.8 \times 0.7)/29,$
$(1 \times 0.6)/30, (0.9 \times 0.6)/31\} = 0.6 \ngeq 0.65$. So he is not that young, out of $\mathcal{R}_2$. Next, for Anna, $max[\pi_{U_{i_j}}(u)\mu_{A_{i_j}}(u)] = max\{(1 \times 0.7)/29\} = 0.7 \geq 0.65$. Of course for Karen $max[\pi_{U_{i_j}}(u)\mu_{A_{i_j}}(u)] = max\{(1 \times 1)/20, (1 \times 1)/21\} = 1 \geq 0.65$. Last one, John is *unknown*, which means it could be anything, which is defined as having possibility 1.

$$\mathcal{R}_2 = \mathcal{R}_{1\,where}\mathcal{C}(Age = \text{'young'}, 0.65)$$

Name	Age	Course	Sports	p
Anna	29	PhD	*swimming, badminton playing*	0.06
Karen	20 or 21	*void*	*soccer playing*	0.46
John	*unknown*	Undergraduate	*unknown*	*

The final calculation is

Step 3: $\mathcal{R}_3 = \mathcal{R}_2[Name, Sports]$

Simply we keep columns *Name, Sports*, and according to Theorem 5, p as well.
$$\mathcal{R}_3 = \mathcal{R}_2[Name, Sports]$$

Name	Sports	p
Anna	*swimming, badminton playing*	0.06
Karen	*soccer playing*	0.46
John	*unknown*	*

6 Conclusion

Our model is motivated by many earlier works on uncertain databases [Uma82, PT84, ZK85, AR86, BP86, GH86, Lee92, BGMP92, Pit94, FR95, CVGS95]. Despite that, our model is rather unique by managing the probabilistic and possibilistic data simultaneously.

Umano proposed the possibility distribution relational model of fuzzy database both on attribute and tuple level [Uma83]. Buckles and Petry defined a fuzzy database whose attributes are subject to a similarity relation [BP82]. M. Anvari and G.F. Rose applied 'distinguishability' to measure the user's view of the difference between the values for that particular query [AR86]. Prade and Testamale proposed a retrieval method by using a possibility measure and a necessity measure to satisfy the condition [PT86]. Zemankova and Kandel modified those measures [ZK84].

In fact, there are mainly two approaches dealing with the uncertain data, namely fuzzy (or possibilistic) models and probabilistic models. Both ideas of possibility and probability are meanings of representing and manipulating uncertainty. But they are expressing different aspects of uncertainty. You could not use possibility to give the likelihood of the event happens.

Probabilities have been used in many database models. Barbará et al. showed a model which included probabilities associated with the values of attributes [BGMP89]. Lee accommodated probabilities with attributes and tuples [Lee92]. Pittarelli defined a classical relation in terms of a relational system and then extended this definition to a probabilistic system by adding a supplementary column in probability values which are summed to 1 within the relation [Pit90]. N. Fuhr and T. Rölleke in [FR95] generalised non-first-normal-form (*NF2*) and weighted tuples both by probability. E. Zimányi outlined a *trace relation* to handle a particular type of probabilistic information [Zim92].

The strength of our model is in providing the concepts to represent uniformly and manage all kinds of uncertainty. The uncertain database has much more expressive power than the traditional database, being able to cover wide range of imperfect data.

So far we have not addressed the matter of a key attribute whose value can be used to identify each entity uniquely. Description of a key of the relation is critical in database systems and it brings us to investigate uncertain functional dependency in normalisation for relational databases consequently. Work in these areas will be pursued by authors. Another topic for further research is to examine the problem of lossless join decomposition of uncertain relations for a given set of uncertain functional dependencies. This can be expected to help rationalising relational database schema design.

References

[AR86]   M. Anvari and G.F. Rose. Fuzzy relational databases. In J.C. Bezdek, editor, *Artificial Intelligence And Decision Systems: Analysis of Fuzzy Information*, volume III, pages 203–212. CRC Press, Boca Raton, FL, 1986.

[BGMP89] D. Barbará, M. Garcia-Molina, and D. Porter. A probabilistic relational data model. Technical Report CS-TR-216-89, Department of Computer Science, Princeton University, 1989.

[BGMP92] D. Barbará, H. Garcia-Molina, and D. Porter. The management of probabilistic data. *IEEE Tran. Knowledge and Data Engineering*, 4(5):487–502, 1992.

[BP82]   B.P. Buckles and F.E. Petry. A fuzzy representation of data for relational databases. *Fuzzy Sets and Systems*, 17(3):213–226, 1982.

[BP86]   B.P. Buckles and F.E. Petry. Generalized database and information systems. In J.C. Bezdek, editor, *Artificial Intelligence and Decision Systems: Analysis of Fuzzy Information*, volume III, pages 177–201. CRC Press, Boca Raton, FL, 1986.

[Cod72]   E.F. Codd.   Relational completeness of data base sublanguages.   In
           R. Rustin, editor, *Data Base System*, pages 65–98. Prentice-Hall, Engle-
           wood Cliffs, NJ, 1972. (Courant Computer Science Symposium 6, May
           24-25, 1971).

[Cod79]   E.F. Codd. Extending the database relatonal model to capture more mean-
           ing. *ACM Trans. Database System*, 4(4):397–434, 1979.

[CVGS95]   R. De Caluwe, R. Vandenberghe, N. Van Gyseghem, and A. Schooten. In-
           tegrating fuzziness in database models. In P. Bosc and J. Kacprzyk, editors,
           Fuzziness in Database Management Systems, pages 71–113. Physica-Verlag,
           Heidelberg, 1995.

[EN94]   R. Elmasri and S.B. Navathe. *Fundamentals of Database System*. Ben-
           jamin/Cummings, Redwood City, CA, 2nd edition, 1994.

[FR95]   N. Fuhr and T. Rölleke.   A probabilistic NF2 relational algebra for
           imprecision in databases.   on-line, 1995.   http://charly.informatik.uni-
           dortmund.de/projects/fermi/papers.html 'Fuhr-Rolleke-94a.ps'.

[Gen89]   H.J. Gensler. *Logic: analyzing and appraising arguments*. Prentice Hall,
           Englewood Cliffs, 1989.

[GH86]   E. Gelenber and G. Hebrail. A probability model of uncertainty in data-
           bases. In *Proc. Int. Conf. Data Engineering*, pages 328–333, 1986.

[Lee92]   S.K. Lee. Imprecise and uncertain information in databases: An evidential
           approach. In *Proc. IEEE Int. Conf. Data Engineering*, pages 614–621, 1992.

[Pet95]   F.E. Petry.   Information systems for the twenty-first century:'a fuzzy-
           oriented manifesto'? In *Proc. VI IFSA World Congress*, volume I, pages
           3–6, Sao Paulo, 1995.

[Pit90]   M. Pittarelli. Probabilistic databases for decision analysis. *Int. J. Intelli-
           gent Systems*, 5:209–236, 1990.

[Pit94]   M. Pittarelli. An algebra for probabilistic databases. *IEEE Trans. Knowl-
           edge and Data Engineering*, 6(2):193–303, 1994.

[PT84]   H. Prade and C. Testamale. Generalizing database relational algebra for
           the treatment of incomplete or uncertain information and vague queries.
           Information Sciences, 34:115–143, 1984.

[PT86]   H. Prade and C. Testemale. Representation of soft constraints and fuzzy
           attribute values by means of possibility distributions in databases. In J.C.
           Bezdek, editor, *Artificial Intelligence and Decision Systems: Analysis of
           Fuzzy Information*, volume III, pages 213–229. CRC Press, Boca Raton,
           FL, 1986.

[Rei78]   R. Reiter. On closed world data bases. In H. Gallaire and J. Minker, edi-
           tors, *Logic and Data Bases*. Plenum Press, New York, 1978.

[Ros93]   S.M. Ross. *Introduction to probability models*. Academic Press, Boston, 5th
           edition, 1993.

[RY96]   A. Ramer and H. Yu. Similarity, probability and database organisation. In
           Proc. 1996 Asian Fuzzy System Symposium, Kenting, Taiwan, pages 272–
           277, 1996.

[Ull83]   J. Ullman.   *Principles of Database Systems*.   Computer Science Press,
           Rockville, MD, 1983.

[Uma82]   M. Umano.   Freedom-0: A fuzzy database system.   In M. Gupta and
           E. Sanchez, editors, *Fuzzy Information and Decision Processes*, pages 339–
           347. North-Holland, New York, 1982.

[Uma83]   M. Umano. Retrieval from fuzzy database by fuzzy relational algebra. In
           Proc. IFAC Conf. on Fuzzy Information, Knowledge Representation, and

Decision Processes, Marseille, pages 1–6, 1983.

[UW97]   J.D. Ullman and J. Widom. *A First Course in Database Systems*. Prentice Hall, Englewood Cliffs, 1997.

[Zim92]   E. Zimányi. Probabilistic relational databases. Technical Report RR-92-02 of INFODOC, Université Libre de Bruxelles, 1992.

[ZK84]   M. Zemankova and A. Kandel. *Fuzzy Relational Data Bases - A Key to Expert Systems*. Verlag TUV Reinland, Köln, Germany, 1984.

[ZK85]   M. Zemankova and A. Kandel. Implementing imprecision in information systems. *Information Sciences*, 37:107–141, 1985.

Section II:

Classification and Feature Selection

A Connectionist Approach to the Distance–Based Analysis of Relational Data

Kristina Schädler and Fritz Wysotzki

Technical University of Berlin
FR 5-8, Franklinstr. 28/29, D-10587 Berlin, Germany
Phone 0049-30-314 25491, e-mail: schaedle@cs.tu-berlin.de

Abstract. Objects with higher structural complexity often cannot be described by feature vectors without losing important structural information. Several types of structured objects can be represented adequately by labeled graphs. The similarity of such descriptions is difficult to define and to compute. However, many algorithms in machine learning, knowledge discovery, pattern recognition and classification are based on the estimation of the similarity between the analysed objects. In order to make similarity based algorithms like nearest neighbor classifiers, clustering, or generalised prototype learning accessible for the analysis of relational data, a connectionist approach for the determination of the similarity of arbitrary labeled graphs is introduced.

Using an example from organic chemistry, it is shown that classifiers based on the connectionist approach to structural similarity to be considered in this paper perform very satisfactorily in comparison with recent logical and feature vector approaches. Moreover, being able to handle relational data in a natural way without any loss of structural information, the algorithms need only a subset of the given features of the objects for classification.

1 Introduction

By now, an immense number of algorithms for the solution of different tasks in the analysis of datasets and the extraction of knowledge from collections of data is available. Many methods in intelligent data analysis require the estimation of the similarity between the entities of the data base and the query objects. Clustering algorithms divide the dataset into clusters with high intra-class and low inter-class similarity. Distance based classifiers use the similarity between classified objects and the query object for determining its class. In machine learning and knowledge discovery [26, 27], often the aim is to find common characteristic features of objects having similar properties or belonging to the same class. If the objects are represented by feature vectors, some kind of Euclidian or generalised Minkowski metric might be employed.

Objects having a complex structure often cannot be described as fixed length feature vectors without losing important structural information (see [13, 14, 15]).

X. Liu, P. Cohen, M. Berthold (Eds.): "Advances in Intelligent Data Analysis" (IDA–97)
LNCS 1280, pp. 137–148, 1997.   © Springer–Verlag Berlin Heidelberg 1997

In machine learning, these objects often are represented in a logical framework (see [1, 23] for an overview). Inductive Logic Programming is used to tackle the problems of learning and classification of structured objects. In many real world applications, it is more natural to describe complex objects or other structures by labeled graphs, for instance chemical structures by structural formulas or computer programs by trees or flow charts, respectively. In this paper, the estimation of the similarity between graphs is discussed and a connectionist approach for computing the similarity of two graphs is introduced. It is a neural net approach to the graph theoretic problem of subgraph isomorphism. The net is able to find an approximate solution of the NP-complete graph matching problem efficiently including domain knowledge about the similarity of objects. For more details about the algorithm and its application to case-based reasoning, see [29, 28].

Based on these results, quite a lot of similarity or distance based algorithms can be made accessible for the treatment of relational data. Two groups of classification algorithms and their performance on a specific problem using the neural approach to similarity estimation described in Section 5 are considered in detail in this paper. The first group consists of two generalised prototype classifiers, learned from examples by a new *similarity-based inductive graphtheoretic* learning algorithm called SIG-Learning and a prototype learning algorithm from [37] which reduce the set of given instances to a smaller set of generalised prototypes used for classification. The second group of algorithms performs a weighted nearest neighbor classification. The results of the algorithms are compared with those produced by some recent logic, graphtheoretic and feature vector based algorithms, applied to the discovery of cause-effect relationships of some organic compounds (mutagenesis data), a typical data mining problem.

2   Generalised Prototypes: Learning and Classification

2.1   Relational Learning Algorithms

Learning tasks for structured objects often aim at the discovery of certain substructures of a set of objects, for instance the part of the structure that causes a specific common property of the objects. One main approach to relational learning models objects and relations in a logical framework, for instance in the field of Inductive Logic Programming (ILP). In this paper, elementary objects that constitute a structured object and the (binary) relations between them are described by the nodes and edges of graphs representing the structured objects, respectively[1]. Colors (labels) of nodes denote one-place relations, i.e. properties of elementary objects, and colors of edges represent the names of the binary relations between objects. A graph is a more general description than another graph if it has less nodes or relations, i.e. it is a part of the latter. A data mining or classification learning task is to find a general description of objects showing or not showing a common property or behavior. Using a graph representation

---

[1] For a comparison of graph theoretic methods with ILP see [21], Chapter 6. [32] and [31] discuss some more aspects of the graph vs. logic generalisation.

of structured objects, this generalisation is a set of common subgraphs that are believed to cause or to prevent the property or behavior in question. An example is a certain substructure in a chemical structural formula the presence or absence of which causes some biological activity.

Common subgraphs provide an injective mapping between the nodes and relations of two structures to be generalised, i.e. one node of one graph is mapped exactly onto one node of the other (see Section 5, Eq.(6)). As well as in ILP, the selection of the appropriate subgraph is the key problem in every step of generalisation. Usually a largest common subgraph between the current hypothesis and the new example is selected as a new hypothesis. Another approach is to choose a subgraph which is the best with respect to the aim of the learning, for instance the Minimum Description Length criterion used in [7] which may result in very large search spaces.

2.2   The Similarity-Based Prototype Algorithm

In generalisation based learning the aim is to produce some general descriptions of objects of the same class. A new object is believed to belong to class K if a generalisation of objects of class K exists which covers the new object. In similarity based learning, an object is assigned class K if it is more similar to some stored instances of class K than to instances of another class. Generalised Prototype algorithms [37] can be considered as a kind of mixture between generalisation and similarity based classification. Prototypes are descriptions of generalised objects which can be interpreted as abstract, typical objects of some class. Prototypes are constructed by using some similarity-based notion of generalisation whereas the classification procedure can use similarity or generalisation alone or a combination of both. The algorithms presented in this paper produce general descriptions of objects in the form of common subgraphs which can be interpreted as typical partial structures of objects of a class. The computation of the common subgraph of two descriptions gives *simultaneously* a similarity measure (see section 5). Although the fast neural net algorithm described in section 5 is used, the computation of the appropriate best mapping between two graphs is still of high complexity. Thus the learning algorithm *is combined with some similarity based heuristics* during the generalisation.

The following method for determining a set of prototypes for feature vector representations of objects has been introduced in [37]:

(a) Take the set of all examples S
(b) Choose an example x from S, remove it from S and let the initial set of prototypes be **P**={x}
(c) WHILE examples in S exist
    i. Take the next example x from S, remove it from S
    ii. Find the Prototype P from **P** most similar to x
    iii. IF P belongs to the same class as x THEN replace P by the generalisation of P and x ELSE set **P**=**P** ∪ {x}

Using the results of Section 4 and 5, this algorithm can be used to produce a set of prototypes not only for feature vectors, but for labeled graphs, too. During the construction process, the prototypes are built by generalising similar objects of the same class. So every example is generalised by exactly one of the prototypes. Thus, the algorithm divides the set of examples into disjunct subclasses of similar objects of the same class. This results in a heuristic where the most specific generalisation is chosen, similar for instance to the minimum inductive-leap heuristic in [21] in ILP. So the generalisation is combined with the method of agglomerative single linkage clustering of the instances by subsequently adding the most similar instances to a cluster described by the current prototype and containing all structures generalised by the prototype including the seed and the instances added in the generalisation step. This is similar to the method of conceptual clustering (see for example [11] and [16]). A similar procedure can be used to find clusters of structured objects with some common structural pattern in a database by unsupervised learning, i.e. where no classification is given a priori.

The second algorithm, called SIG-Learner, learns only one class at a time. It generalises a set of examples of the same class by constructing prototypes P as follows:

1. Take a set of examples S' of class K,
   mark all examples in S' as not_processed
2. WHILE examples with mark not_processed in S' exist
   (a) S:=S', choose an example y marked not_processed from S, remove y from S, P:=y
   (b) WHILE S not empty
       i. choose the x from S most similar to P, remove x from S
       ii. IF the generalisation of P and x does not cover any example from another class than K THEN P := the generalisation of P and x ,mark x as processed FI
   (c) save the new prototype P

In this algorithm, some examples can be generalised by more than one prototype. The algorithms finds prototypes which are generalised descriptions of disjunct as well as overlapping disjunctive subclasses.

As a result, the learner reduced the set of examples to some sets of class prototypes which are used for the distance-based classification of new examples. The results shown in Table (1) are obtained by assigning an unknown object the class of the nearest prototype according to Eq. (5) in Section 4.

The second method of generalisation requires some more computational effort than the first. So the similarity of all remaining examples to the current prototype has to be computed which is reduced in the actual computation by dividing S into two subsets which are processed in sequence. The first subset contains all examples not yet covered by a prototype. The second subset of S consists of all other examples, so in the second step the overlapping of subclasses is detected.

The overgeneralisation test in step (2(b)ii) is the most time-consuming part of this algorithm. In order to detect overgeneralisation as fast as possible, a heuristic is introduced in this step. Since the overgeneralisation test provides a measure of similarity between the current prototype and the tested counter-example, the counter-examples are sorted in every step by their similarity to the current prototype. So another property of the generalisation by subgraph detection can be utilized to reduce the number of matches in the overgeneralisation test. The prototype is reduced in size in every generalisation step. So, if we are looking for a counter-example that is covered by the current prototype, it is useful to test the counter-examples most similar to the old, more specific prototype first and to omit all counter-examples where the old match contains less nodes and edges of the old prototype than the new prototype consists of.

In the first algorithm, no overgeneralisation is checked. Every example is generalised by exactly one prototype. As a result, this algorithm performs much faster than the second one. But, using the heuristics and pruning conditions explained in this section, the number of graph matches in the second algorithm can be reduced to a manageable extent. As it can be seen in Table (1), the additional effort of in the second algorithm does not result in better classification accuracy for the given dataset. However, it reduces the number of prototypes generated and used in the classification of the test examples from the bigger dataset from about 60 for the first algorithm to about 26 for the second one.

3   The Weighted Nearest Neighbor Classifiers

Instance-based classifiers classify unknown objects by comparing them with a set of stored examples whose classes are known. The most common instance-based classifier is the k-nearest neighbor (k-NN) classifier, where the class of a new example is determined by considering the classes of its k nearest neighbors. The success of a k-nearest neighbor classifier depends heavily on the choice of an appropriate distance measure. In order to check the usefulness of the chosen similarity concept as well as the goodness of the approximation of this similarity function by the neural net, three nearest neighbor classifier have been chosen for testing the connectionist similarity estimator from Section 5.

The first classifier ("Simple k-NN" in Table (1)) has been the simple k-NN algorithm where the class of the unknown object is determined by a voting of the k nearest neighbors. The second classifier ("2-NN predictor" in Table (1)) takes advantage of the fact that the classes in the given datasets are produced by the discretisation of a continuous range of values for a certain property p of the structures. The classifier tries to predict this value p_x using the values of this property measured for the two nearest neighbors y, z by the simple formula

$$p_x = \frac{(1 - sim(x,y)) * p_y + (1 - sim(x,z)) * p_z}{2 - sim(x,y) - sim(x,z)}, \tag{1}$$

which works surprisingly well for the given datasets.

Finally a weighted nearest neighbor classification, the variable-kernel similarity metric (see [34]) and David G. Lowe's extension described in [25] has been implemented. This algorithm takes into account the distribution of the distances $d(x,y) = 1 - sim(x,y)$ in the set of given instances. The influence of any of the k neighbors on the classification is weighted by a number proportional to a Gaussian function of its distance to the instance to be classified. The parameters of the Gaussian are estimated using the distances of the M (M<k) nearest

neighbors. The evidence of belonging to class c for an instance x with the set of k nearest neighbors $\{x_1, x_2, ..., x_k\}$ is given by the formulae:

$$e_c(x) = \frac{\sum_{j=1, x_j \in c}^k w_j}{\sum_{j=1}^k w_j}, w_j = exp\left(-\frac{1}{2}\frac{d(x, x_j)^2}{\sigma^2}\right) \qquad (2)$$

If the evidence for a class c exceeds a given threshold, for instance 0.5, x is assumed to belong to class c.

As it is proposed in [25], the memory requirements and the classification effort was reduced by removing instances which are not important for classification. This is done in [25] by deleting all instances whose k nearest neighbors all belong to the same class when classified using Eq.(2). In our tests, this rule has been changed slightly, such that only those instances are deleted whose k neighbors are assigned the same class like the instance in question. Lowe eliminates instances which are surrounded by examples of another class because he assumes them to be outliers or exceptions. However, the datasets considered in this paper are comparatively small and the apparent outliers can be the representatives of subclasses as well. The algorithm reduces the number of instances in the dataset to about 60% of the original number of examples which cuts down the number of necessary matches for classification considerably.

The results presented in Section 6 show that the algorithms of this and the previous section perform well. The prerequisites of this success and even the availability of the algorithms from the last two sections are the existence of an appropriate similarity measure and a method to its efficient computation. The next two section describe how these requirements can be provided.

4   Structural Similarity of Relational Descriptions

In contrast to objects represented by feature vectors, no appropriate mapping of graph representations into the Euclidian vector space of real numbers exists. Thus the similarity of graphs cannot be determined using a metric in the Euclidian space. Caused by the growing interest in relational descriptions, different measures of similarity of relational descriptions have been proposed in the last years. Only a few of them have metric properties. Subgraphs are used in the similarity detection of cases, for instance in [35], where the largest isomorphic subgraph of two graphs is called the structural similarity of the graphs but no quantitative measure of similarity is given. Shapiro and Haralick defined in [6, 33] a structural difference of relational descriptions with metric properties for graphs of the same size. The well-known Dice and the Tanimoto-Coefficient for feature vector representations can also be adapted for graph representations. In [8], Emde and Wettschereck propose a distance-based learning algorithm using a recursive similarity measure for relational structures described by predicate logic.

The similarity measure used in this paper is based on these concepts and the ZELINKA-Metric [40] and its derivates [20], where the common parts of the structures are related to the number of nodes and edges of the larger of the two graphs. In the following sections colored graphs are described by $G(N, V, l, e, L, E)$, where N and $V = N \times N$ [2] are the nodes and edges of the graph, respectively, L and E some arbitrary sets of colors or labels and $l : N \to L$ and $e : V \to E$ the coloring functions for the nodes and edges of the graph. Depending on the task, a similarity measure based on the following definition is chosen:

$$sim_a(x, y) = \frac{|(n_i, n_j) \in V_x : e_x(n_i, n_j) = e_y(\varphi(n_i), \varphi(n_j))|}{|N_x| * (|N_x| - 1)} \qquad (3)$$

where φ is the chosen mapping between nodes of the graphs $x = G(N_x, V_x, l_x, e_x, L_x, E_x)$ and $y = G(N_y, V_y, l_y, e_y, L_y = L_x, E_y = E_x)$. It is assumed that φ is a graph morphism that maps a node only onto a node with the same label and provides a bijective partial mapping between the nodes of the graphs. So the measure relates the number of corresponding relations (matching edges) between corresponding nodes of the two graphs to the number of these relations when graph x is mapped onto itself, or in short: The value of $sim_a(x, y)$ measures the fraction of edges of x contained in y. In general, φ is chosen in such a manner that it gives the best mapping between y and x with respect to the similarity in Eq.(3). As it can be seen, the measure is an asymmetrical one. $sim(x, y)$ can be defined as $sim_a(x, y)$ or $sim_a(y, x)$ depending on the task and the properties of the graphs involved. In nearest neighbor classification as described in Section 3 the symmetric definition

$$sim(x, y) = \left\{ \begin{matrix} sim_a(x, y) \text{ if } max(|N_x|, |N_y|) = |N_x| \\ sim_a(y, x) \text{ otherwise} \end{matrix} \right\} \text{ turned out to be the best}$$

choice. This measure is a metric in the space of labeled complete graphs. The similarity measure used in the prototype classification in Section 2 reflects the asymmetry of the generalisation/specialisation relation between the generalised description P of a set of examples and the single example x. It is given by the formula

$$sim_l(P, x) = \frac{|(n_i, n_j) \in V_x : e_x(n_i, n_j) = e_P(\varphi(n_i), \varphi(n_j))|}{|N_x| * (|N_x| - 1)} \in [0, 1] \qquad (4)$$

during learning where the instance covered best by the current prototype has to be determined and

$$sim_t(P, x) = \frac{|(n_i, n_j) \in V_x : e_x(n_i, n_j) = e_P(\varphi(n_i), \varphi(n_j))|}{|N_P| * (|N_P| - 1)} \in [0, 1] \qquad (5)$$

in overgeneralisation testing and classification where the question is asked whether the instance x belongs to the set of instances containing P as a subgraph or not. The mapping φ which has to be determined for computing the similarity in Eq.(4) is used to compute the generalisation of P and x as well. This generalisation contains all nodes and edges which are mapped by φ from P to x.

5   A Neural Net for Similarity Estimation

In general the computation of the similarity of relational objects requires a lot of effort because every measure is based on the (NP-complete) search of some best

---

[2] That means all graphs are considered complete graphs.

mapping between the objects (see [4, 3]). If the similarity of graphs with respect to Eq. (3) is to be determined, a mapping φ must be found that maximizes this similarity between the graphs. Replacing φ by a corresponding relation $\rho \subseteq N_x \times N_y$ between nodes of the graphs with the same or similar labels and allowing ambiguous mappings between the graphs, the graph matching problem can be transformed into an optimization task where matches between pairs of nodes $(n_i^x, n_j^y) \in \rho$ with $e_x(n_i^x, n_j^x) = e_y(n_k^y, n_l^y)$ for related nodes $(n_i^x, n_k^y) \in \rho, (n_j^x, n_l^y) \in \rho$ are rewarded by a positive weight w in order to maximize the numerator in Eq. (3) where the denominator remains constant. On the other hand a penalty $-w_I$ is introduced to mappings of the same node of one graph to more than one node of the other graph. The graph matching problem can then be reformulated as follows:

$$\rho^*(x,y) = max_{\rho \subseteq N_x \times N_y} \{w * c1_\rho - w_I * c2_\rho\}, w, w_I > 0 \tag{6}$$

$$c1_\rho = |\{(\rho(n_i^x, n_k^y), \rho(n_j^x, n_l^y)) | e_x(n_i^x, n_j^x) = e_y(n_k^y, n_l^y)\}| \tag{7}$$

$$c2_\rho = |\{(\rho(n_i^x, n_k^y), \rho(n_j^x, n_l^y)) | (n_i^x = n_j^x \wedge n_k^y \neq n_l^y) \vee (n_i^x \neq n_j^x \wedge n_k^y = n_l^y)\}| \tag{8}$$

The relation ρ can be represented by a twodimensional array $o_{[N_x, N_y]}$ where $o_{ik} = 1$ iff $\rho(n_i^x, n_k^y)$ and $o_{ik} = 0$ otherwise. So, the objective function of the optimization tasks is

$$f(o^*) = max_{o \in \{0,1\}^{|N_x|} \times \{0,1\}^{|N_y|}} \sum_{i,j=1}^{N_x} \sum_{k,l=1}^{N_y} w_{ij,kl} o_{ik} o_{jl} \tag{9}$$

where $w_{ij,kl} = \left\{ \begin{array}{c} w \\ -w_I \\ 0 \end{array} \right\}$ according to Eq.(6). It has been shown ([18, 17, 2, 19, 24, 5, 12, 36]), that such quadratic optimization tasks can be solved by Hopfield-like Artificial Neural Nets, i.e. bi-directional associative memories. In contrast to the most other implementations which use Hopfield Nets with binary or continuous sigmoidal output function, in this work an approach described in [9, 10, 38] is used where the output of the neurons is restricted by the non-differentiable ramp function $r(x) = \left\{ \begin{array}{l} 0 \text{ if } x < 0 \\ x \text{ if } 0 \leq x \leq 1 \\ 1 \text{ if } x > 1 \end{array} \right\}$. In addition the units of the net receive a part of their own output as an input, setting the diagonal of the connection matrix to a weight $1 > w_d > 0$. As usually in the domain of artificial neural nets, the net's results depend strongly on the chosen parameters like w, w_I and w_d. In the previous implementations, these parameters have been chosen ad hoc or experimentally. As a result, the algorithms turned out to be less reliable and could not be transferred to some more problems. In [30], for the first time an extensive analysis of the approach of Feldman [9, 10] and Wysotzki [39, 38] with regard to the graph matching problem is given that describes the algorithm and provides its theoretical foundations, delivering the parameter settings which guarantee that the net reaches a stable state that represents a good solution of the problem. This work provided the foundation of the incorporation of this method in other algorithms which need reliable results of the graph matching algorithm in a reasonable time. The ramp function used as the output function of the neurons which raises some difficulties in the theoretical analysis of the behavior of the net came out to force the convergence of the net to a stable state. Experiments have shown that the algorithm approximates the optimum solution within $O((|N_x| * |N_y|)^2)$, i.e. polynomial time which can be accelerated considerably using parallel hardware.

6   Results and Conclusions

The algorithms above have been applied to two datasets of chemical compounds provided by Dr. R. King. The sets contain nitro aromatic compounds and their mutagenicity. The aim is to produce a classifier that predicts the mutagenicity of such compounds. The first set contains 188 compounds which could be classified successfully using regression while the second set of 42 compounds caused some more difficulties. The compounds are described by graphs of atoms and bonds, providing information about the chemical elements, the kind of bonds between them, the information about the mutagenicity (a real number or "active" / "not active") and some additional data which were not used. The algorithms have been tested using tenfold crossvalidation for the first dataset and leave-one-out for the smaller second one. The data on the left of Table (1) is taken from [14] and [22] where the dataset is described in more detail.

In the prototype algorithms as well as in the k-NN classifiers less information than in the former algorithms displayed on the left side of Table (1) has been used. The authors assumed that the structure of the compounds alone causes all other chemical and physical properties of the compounds except the variances caused by stereo-chemical effects. Thus, the classifier should be able to predict the mutagenicity of the compounds *using only the structural information.* So in contrast to the other algorithms the classifiers described in this paper did not use additional data like atom charges or chemical or physical properties of the compound as a whole. The assumption turned out to be true. In PROGOL-S2 [22] and INDIGO [14] structural information was included in the features of the atoms by using external information (PROGOL-S2) or context information generated by the algorithm (INDIGO). In both cases this results in classification errors near to the ones obtained by the algorithms described in this paper, so the differences are not significant for the given dataset where about $\frac{2}{3}$ of the examples belong to one class and the rest to the other.

The SIG-Learning reduced the bigger dataset to sets of ca. 26 prototypes and achieved the classification rates shown in the Table (1). The algorithm from [37] showed the same classification error with about 60 prototypes. The authors suppose that the reason for the smaller number of prototypes produced and used for classification in the SIG-Learner is the overgeneralisation testing which leaves only prototypes characteristic for exactly one of the classes. The prototype classifiers have been outperformed by the variabel kernel k-nearest neighbor classifier introduced in this paper and some of the former classifiers but they show the advantage of producing very *few prototypes* which in addition contain *new knowledge about substructures causing the mutagenetic activity of the substances.* In addition, the simplicity of the algorithms offers several possibilities to improve their performance as it is mentioned below.

The nearest neighbor classifiers all show an outstanding classification accuracy. The only exception is the simple 3-NN algorithm for the small dataset, supposedly due to the very sparse data. Increasing k to 5 results in 83% accuracy for this set.

To achieve these good results the nearest neighbor classifiers need the whole training set or, in the case of Lowe's algorithm, a considerably larger sample of examples (of at least 85 examples in case of the bigger dataset) for the classification of new objects, as compared to the prototype classifiers.

	188	42			188	42
Linear Regression	0.85	0.67	SIG-NN		0.80	0.81
Neural Net (Backprop)	0.86	0.64	Generalised Prototypes from [37]		0.80	0.81
CART	0.83	0.83	Simple k-NN	k=3	0.88	0.79
Progol	0.81	0.86	2-NN predictor		**0.89**	**0.86**
Progol-S2	0.88	0.83	Variable Kernel k-NN	k=3	**0.91**	**0.83**
INDIGO	0.86	0.89	Variable Kernel k-NN (reduced	k=6	0.88	0.86
			set of instances)	k=10	**0.88**	**0.88**

Table 1. Results for the mutagenesis data

The results in Table (1) show, that the chosen similarity measure as well as the connectionist algorithm for the approximation of the graph similarity are feasible for processing data represented by labeled graphs. Various, new and well-known similarity-based algorithms can be applied to relational data using the similarity estimator from section 5. The intrinsic complexity of graph algorithms could be reduced far enough to be able to process a data base like the mutagenesis dataset. The authors are optimistic about its use for larger databases because the algorithm bears a lot of possibilities to further reduce the complexity, for instance by using parallel hardware, reducing the number of processed instances or some preprocessing of the data, including the use of domain similarity knowledge and transforming the instances into smaller graphs by producing more abstract descriptions of them. All these improvements will be implemented in the future.

References

1. D.W. Aha. *Inductive Logic Programming*, chapter Relating Relational Learning Algorithms. Academic Press, London, 1992.
2. Y. Akiyama et al.. Combinatorial optimization with gaussian machines. In *IEEE Int. Conf. on Neural Networks*, vol. I, pp. 533–540. 1989.
3. J.E. Ash, W.A. Warr, and P. Willett. *Chemical Structure Systems. Computational Techniques for Representation, Searching and Processing of Structural Information.* Ellis Horwood, 1991.
4. J.M. Barnard. Substructure Searching Methods: Old and new. *J. Chem. Inf. Comp. Sci.*, 33:532–538, 1993.

5. L.I. Burke and J.P. Ignizio. Neural networks and operations research: An overview. *Computers Ops.Res.*, 19(3/4):179–189, 1992.
6. L. Cinque et al.. An improved algorithm for relational distance graph matching. *Pattern Recognition*, 29(2):349–359, feb 1996.
7. D.J. Cook and L.B. Holder. Substructure Discovery Using Minimum Description Length and Background Knowledge. *Journal of Artificial Intelligence Research*, 1:231–255, 1994.
8. W. Emde and D. Wettschereck. Relational instance-based learning. In L. Saitta, editor, *Proc. of the 13th Int. Conf. on Machine Learning*, pages 122–130. Morgan Kaufmann, 1996.
9. J. A. Feldman and D. H. Ballard. Computing with Connections. TR 72, University of Rochester, April 1981.
10. J. A. Feldman, M. A. Fanty, N. Goddard, and K. Lynne. Computing with Structured Connectionist Networks. TR 213, University of Rochester, April 1987.
11. D. H. Fisher. Knowledge acquisition via incremental conceptual clustering. *Machine Learning*, 2(2):139–172, 1987.
12. N. Funabiki, Y. Takefuji, and K.-C. Lee. A neural network model for finding a near-maximal clique. *Journal of Parallel and Distributed Computing*, 14(3):340–344, March 1992.
13. P. Geibel, K. Schädler, and F. Wysotzki. Begriffslernen für strukturierte Objekte (Concept Learning for Relational Structures). In *Proc. of FGML-95, Dortmund, Germany*, 1995.
14. P. Geibel and F. Wysotzki. Learning relational concepts with decision trees. In L. Saitta, ed., *Machine Learning: Proc. of the 13th Int. Conf.*, pages 166–174. Morgan Kaufmann Publishers, San Fransisco, CA, 1996.
15. P. Geibel and F. Wysotzki. Relational learning with decision trees. In W. Wahlster, editor, *Proc. of the 12th European Conf. on Artificial Intelligence*. John Wiley and Sons, Ltd., 1996.
16. J. H. Gennari, P. Langley, and D. Fisher. Models of Incremental Concept Formation. *Artificial Intelligence*, 40:11 – 61, 1989.
17. J. Hopfield and D. Tank. Neural computations of decisions in optimization problems. *Biological Cybernetics*, 52:141–152, 1986.
18. J.J. Hopfield. Neurons with graded response have collective computational properties like those of two-state neurons. In *Proceedings of the National Academy of Sciences USA 81*, pages 3088–3092. 1984.
19. A. Jagota. Efficiently approximating MAX-CLIQUE in a hopfield-style network. In *Proc. of Int. Joint Conf. on Neural Networks '92 Vol. II*, pages 248–253, 1992.
20. F. Kaden. Graphmetriken und Distanzgraphen. In *Beiträge zur angewandten Graphentheorie*, ZKI-Informationen. Berlin, Juni 1982.
21. J.-U. Kietz. *Induktive Analyse Relationaler Daten*. PhD thesis, TU Berlin, FB 13, 1996.
22. R. D. King, M. J. E. Sternberg, A. Srinivasan, and S. H. Muggleton. Knowledge Discovery in a Database of Mutagenic Chemicals. In *Proceedings of the Workshop "Statistics, Machine Learning and Discovery in Databases" at the ECML-95*, 1995.
23. N. Lavrac and S. Dzeroski. *Inductive Logic Programming: Techniques and Applications*. Ellis Horwood, New York, 1994.
24. C. Looi. Neural network methods in combinatorial optimization. *Computers and Operations Research*, 19(3/4):191–208, 1992.
25. D. G. Lowe. Similarity metric learning for a variable-kernel classifier. UBC-TR-93-43, University of British Columbia, Vancouver, November 1993.

26. H. Mannila. Aspects of data mining. In *Proceedings of the Workshop "Statistics, Machine Learning and Discovery in Databases" at the ECML-95*, 1995.

27. M. Moulet and Y. Kodratoff. From machine learning towards knowledge discovery in databases. In *Proceedings of the Workshop "Statistics, Machine Learning and Discovery in Databases" at the ECML-95*, 1995.

28. K. Schädler, U. Schmid, B. Machenschalk, and H. Lübben. A neural net for determining structural similarity of recursive programs. In R. Bergmann and W. Wilke, eds., *Proc. of the German Workshop of Case-Based Reasoning*, pages 199–206, Technical Report Univ. of Kaiserslautern, LSA-97-01E, Kaiserslautern, 1997.

29. K. Schädler and F. Wysotzki. Klassifizierungslernen mit Hilfe spezieller Hopfield-Netze. In W. Dilger, M. Schlosser, J. Zeidler, and A. Ittner, eds., *Beiträge zum 9.Fachgruppentreffen "Maschinelles Lernen"*, number CSR-96-06 in Chemnitzer Informatik-Berichte, pages 96–100. TU Chemnitz-Zwickau, August 1996.

30. K. Schädler and F. Wysotzki. Theoretical foundations of a special neural net approach for graphmatching. Technical Report 96-26, TU Berlin, CS Dept., 1996.

31. K. Schädler and F. Wysotzki. A connectionist approach to structural similarity determination as a basis of clustering, classification and feature detection. In *Proc. of the 1st European Symposium on the Principles of Data Mining and Knowledge Discovery*, LNAI. Springer, to appear 1997.

32. T. Scheffer, R. Herbrich, and F. Wysotzki. Efficient θ-subsumption based on graph algorithms. In W. Dilger, M. Schlosser, J. Zeidler, and A. Ittner, eds., *Beiträge zum 9.Fachgruppentreffen "Maschinelles Lernen"*, number CSR-96-06 in Chemnitzer Informatik-Berichte, pages 96–100. TU Chemnitz-Zwickau, august 1996.

33. L.G. Shapiro and R.M. Haralick. A metric for comparing relational descriptions. *IEEE Trans.Pattern Anal. Mach.Intell.*, 7(1):90–94, 1985.

34. B. W. Silverman. *Density Estimation for Statistics and Data Analysis*. Chapman and Hall, London New York, 1986.

35. A. Voß. Similarity concepts and retrieval methods. FABEL Report 13, GMD, Sankt Augustin, 1994.

36. Jun Wang. *Progress in Neural Networks*, volume 3, chapter 11: Deterministic Neural Networks for Combinatorial Optimization, pages 319–340. Ablex Publishing Corporation, Norwood, New Jersey, 1995.

37. Ch. Wisotzki and F. Wysotzki. Prototype, nearest neighbor and hybrid algorithms for time series classification. In N. Lavrac and S.Wrobel, editors, *Machine Learning: ECML-95*, number 912 in LNAI, pages 364–367. Springer, 1995.

38. F. Wysotzki. Artificial Intelligence and Artificial Neural Nets. In *Proc. 1st Workshop on AI*, Shanghai, September 1990. TU Berlin and Jiao Tong Univ. Shanghai.

39. F. Wysotzki. Artificial intelligence and artificial neural nets. In L. Budach, editor, *Neural Informatics.*, number 12/1989 in Informatik Informationen Reporte, pages 43–51, Berlin, 1989. Akademie der Wissenschaften der DDR.

40. B. Zelinka. On a certain distance between isomorphism classes of graphs. *Časopis pro pěstování matematiky*, 100:371–373, 1975.

Efficient GA Based Techniques for Automating the Design of Classification Models

Robin Glover and Peter Sharpe

Intelligent Computer Systems Centre
Faculty of Computer Science and Mathematics
University of the West of England
Bristol BS16 1QY.
{rpg,pks}@ics.uwe.ac.uk +44 (0)117-9656261

Abstract. As genetic algorithms are able to perform extensive global search they have the potential to find the optimal set of model parameters for a classification algorithm. However if test set error is used to calculate fitness, the computational costs can be high and there is a danger that over-fitting to the test set can occur. This paper empirically examines the over-fitting problem in a feature selection context and then proposes techniques for modifying the fitness function to improve speed and accuracy. It is shown that test set sampling can dramatically speed up the evaluation function and hence enable the GA approach to be feasibly applied to large data sets. A technique is then proposed which combines the use of Occam's razor with statistical confidence tests to determine the number of samples utilized by the evaluation function.

1 Introduction

Whilst classification algorithms such as neural networks, decision trees and local learning systems have some ability to uncover the properties of a data set, the quality of the results produced is often limited by a pre-determined choice of design parameters. These parameters may include choice of features to use, the size and shape of the architecture and the learning rate. Such parameters are defined here as 'meta-level' parameters, they define the form of the classification algorithm and the environment in which it operates and consequently cannot be determined by the classification algorithm itself.

In practice the task of finding suitable 'meta-level' parameters for a specific data set often takes place manually and involves a combination of domain knowledge, past experience and experimentation. This process can be time consuming, the search space of parameter combinations may be immense and there is a danger that the optimal choice will be overlooked. It is desirable therefore to automate the search for 'meta-level' parameters and hence reduce the amount of human expertise needed to find a near optimal classification model for a given data set.

X. Liu, P. Cohen, M. Berthold (Eds.): "Advances in Intelligent Data Analysis" (IDA–97)
LNCS 1280, pp. 149–160, 1997.   © Springer–Verlag Berlin Heidelberg 1997

A Genetic Algorithm (GA) [9] can act as such a search engine and several recent papers have described their use in this manner. Examples include the use of GAs in finding the architectures of neural networks (reviewed in [1]), and in determining the features to be used by a classifier [2]. GAs typically perform a global search, examining a large number of areas of search space in parallel and consequently they have the potential to be more robust than local search techniques such as hill climbing methods. A brief description of GAs and their use in the determination of meta-level parameters is given in Sect. 2.

A common approach to evaluating competing models in a GA context is via accuracy on a test set or on cross-validation sets. However this can be computationally costly and the benefits of performing a wide search are compromised by the fact that the estimates of the generalization abilities of competing models are subject to noise. This paper discusses these problems and provides suggestions for how the evaluation function can be modified to improve efficiency and generalization and how it can self-adapt to suit the data and different phases of the search process. One problem which occurs when searching for suitable meta-level parameters, regardless of whether it is done manually or is automated, is that when two or more models are evaluated via test set score and the best one is chosen, the test set score of the best is a biased indicator of generalization abilities [3]. Generally the scale of the bias can be expected to increase as the number of models evaluated is increased and hence over-fitting to the test set can occur. In a GA context the over-fitting problem has the potential to be particularly troublesome as typically several hundred or thousand strings are evaluated. This problem is discussed in more detail in Sect. 3 and examined empirically in a feature selection context.

The problem of over-fitting is reduced as test set size increases but in some contexts a large test set is not feasible due to the increased computational costs in model evaluation. Section 4 however shows that there are certain properties of a GA which can be exploited to dramatically speed up the evaluation function and allow the GA approach to become both feasible and effective when applied to large data sets. The section demonstrates empirically that each model only needs to be tested on a small random sample taken from the test set for solutions to evolve with high classification accuracy on the whole test set. Investigation are made into how few samples are needed per evaluation and how accuracy is affected by sample size, population size and number of evaluations.

Section 5 describes initial work into producing a more 'intelligent' evaluation function which adapts to suit the quality of the competing models and the amount of data available. A method is proposed which uses statistical confidence tests to determine the quantity of data used to evaluate competing models. The principle of Occam's razor (i.e. select the simplest) is employed when no clear winner emerges in an attempt to reduce over-fitting and improve generalization. Initial results suggest that this approach is promising.

2   GAs and Their Use in Model Determination

In the GA approach considered in this paper a population of strings (or chromosomes) is created with each string defining a competing set of whatever meta-level parameters are to be determined by the search process. Co-evolutionary approaches where different strings are used to solve different parts of the problem are not considered here. The 'traditional' GA uses strings which are of fixed length and in binary form although techniques for handling variable length strings and non - binary coding schemes have also been developed.

Associated with each string is a fitness score which is evaluated via an evaluation function. Generally when finding meta-level parameters for classification algorithms, the evaluation function implements the classification algorithm using the parameters defined in the string and derives the fitness score from generalization performance on a set of test data.

New strings are formed by 'mating' members of the population to produce a child containing a random mix of genetic material from the two parents, a process known as cross-over or recombination. Parents are selected for cross-over using a selection strategy where the probability of a string being chosen as a parent depends on its relative fitness. To increase diversity in the population each offspring has a small probability of being mutated before being added to the population. The population size is kept constant by removing certain strings via some deletion strategy which usually takes account of string age or fitness.

The initial population is generally constructed using random strings. As the process of reproduction and deletion continues, similarities between strings in the population will tend to increase often eventually leading to the population being dominated by a single solution, a process known as convergence.

For simplicity, in the experiments reported below the meta-level parameters encoded in the GA strings were feature subsets and the classification algorithm was 1 -nearest neighbour. Nevertheless the techniques for improving the evaluation function developed in this paper could be easily adapted for use in a number of more complex classification models potentially leading to greater classification accuracy and more concise models. The choice of prototypes to be used could be encoded within the GA as in [4], as could conflict resolution parameters (e.g. k in k nearest neighbour.) Likewise feature weights rather than choice of features could be encoded within the GA [5] or local feature weighting / selection could be implemented with different features / feature weights for different prototypes.

The speed of the evaluation function could also be improved by employing a clustering algorithm (supervised or unsupervised) to reduce the number of prototypes prior to the feature selection stage. Generalization may also be improved after the termination of the GA search by utilizing the best string in a more complex classifier as in [2] or by replacing the 1 nearest neighbour conflict resolution method with a set of 'Radial Basis Function' weights.

The main limitation to the type of model which can be constructed using the techniques described in Sect(s). 4 and 5 is that when implementing a classifier for evaluation the main computational cost must be in the calculation of test set accuracy rather than in the training phase. As the evaluation function will typically be implemented at least several hundred times it is unfeasible to implement a classification algorithm which must perform extensive additional search during each evaluation. Such additional search however can be avoided when constructing many local learning systems by either setting the relevant parameters in advance or encoding them within the GA.

Success also depends on the effectiveness of the cross-over operators utilized for the particular coding scheme used This has proved to be especially true when using GA's to design neural network architectures as discussed in [1].

3 Over-Fitting and Sample Error

In the discussion below it is assumed that the classification algorithms used are deterministic in the sense that when a fixed test set is used a particular model will always classify the same set of patterns correctly. Without this constraint the level of noise in fitness scores may be considerably higher. It is also assumed that any data utilized was sampled from the environment without bias.

Let the true accuracy of a particular classification model, (measured by the percentage correct as sample size approaches infinity / 100) be $atrue$. Let S be the number correct on an independent test set of size n. For reasonably large test sets the distribution of S/n is approximately Gaussian with mean $atrue$ and a variance of $atrue(1 - atrue)/n$. The difference between S/n and atrue will be referred to here as the 'sample error' in the evaluation score. Each fitness score can therefore be viewed as consisting of two components '$atrue$ + sample error'.

When a GA is used to find a set of parameters which minimize the error for some fixed test set this sample error places an upper bound on the accuracy of the evaluation scores and is valid for the evaluations in the initial random population. For a given test set and GA search space there may exist a number of points in the search space which have a much higher sample error potentially producing highly overrated fitness scores. There is a danger therefore that the GA will direct search towards strings with a high positive sample error at the expense of those with a high $atrue$ value. The 'noise' in the evaluation function caused by sample error must therefore limit the benefits of performing extensive search.

The effectiveness of the GA search is partly dependent on the size of the deviations in the $atrue$ scores of the current population in relation to the size of the sample errors. As the relative size of the sample errors increases, the probability that the apparent superiority of one string over another is genuine, decreases. If the relative size of the sample error is too high, strings with a positive sample error are more likely to be selected for recombination than those with a higher true score. In such circumstances the search direction is being governed by the wrong criteria and at best the search process is inefficient and at worst it is counterproductive.

As the GA search progresses from initial random strings towards convergence the competing models in the population will tend to increase in similarity and consequently will tend to have more similar fitness values. It is likely therefore that the deviations in the *atrue* scores for a population decrease over time. As the GA search will be directed towards points which have a high positive sample error as well as those with high *atrue* scores, the average sample error of the evaluation scores is also likely to increase over time. Consequently as the GA search progresses it would appear that sample error has a greater influence on search direction and the benefits of further search are reduced.

3.1 Initial Feature Selection Experiments

To assess the practical consequences of the above problem, experiments were performed to compare the quality of feature subsets found using a GA search with those found using the well known incremental hill climbing method, Forward Sequential Selection (FSS) [6]. For both search types, on each run 500 patterns were picked at random for use in the evaluation function and the rest were used to validate the quality of the 'best' string found. Prior to each run, 250 patterns were picked from the 500 to act as prototypes for a nearest neighbour classifier. The evaluation score was obtained via classification accuracy on all 500 data patterns. If when a test pattern was being evaluated, a corresponding prototype existed, that prototype was temporarily covered (i.e. 'leave-one-out cross-validation' was employed.)

The experiments utilised an incremental GA where strings were created and deleted sequentially (as opposed to a generational GA where a whole population is created to replace the current population). The population size was set at 100 and a mutation rate of $(1/(stringlength * 2)$ was used. Parents were selected using tournament selection which involved picking two strings at random and choosing the parent with the highest fitness 85% of the time. Strings were selected for deletion via a 4-way inverse tournament with the worst string always being deleted. A conventional coding scheme was utilised where each bit in the GA string corresponds to a different potential feature and contains a '1' if the feature is to be used, '0' otherwise. .

Table 1. Description of data sets used

Name	Num. patterns	Num. features
Australian credit card (AUS)	690	14
Vehicle recognition (VEH)	846	18
DNA identification (DNA)	3186	60
Satellite image rec. (SAT)	4435	36
Image segmentation (SEG)	2310	18

The five data sets shown in Table 1, taken from the Statlog library [8], were utilized. Each experiment was repeated 10 times with different random splits of the data. After each run of FSS and the GA, the string with the best evaluation score was evaluated on an independent validation set. The best string in the initial population and at 1000 evaluations was also tested to monitor the benefits of increasing search time. The number of offspring to be created was set to 3000 for all data sets except for the DNA data set (10,000). As the evaluation function was deterministic a look-up table of past evaluation scores was implemented which considerably reduced the number of full evaluations needed.

The results in Table 2 show the average highest evaluation score found and the average corresponding score on the validation data set. The first row shows the results obtained using all the available features and hence the table clearly shows that with the exception of the SAT data, in addition to creating simpler models, accuracy can be much improved by using a feature selection algorithm. The GA on average found higher evaluation scores than FSS for all data sets. On the independent validation sets, the GA has higher scores on four occasions but the results are only significant for VEH and SAT. The results also show that the estimates of accuracy obtained via the evaluation function are optimistically biased with the scale of the bias tending to increase as the number of evaluations increases. Several recent feature selection papers report results without using independent validation sets and the results demonstrate that this can be highly misleading.

Extreme over-fitting occurs on the two data sets with the most features and hence the widest search space. On three of the data sets, generalization scores are actually reduced as search continues with the validation score obtained after 1000 evaluations being higher than that obtained after 3000 evaluations. These results are consistent with the ideas put forward in Sect. 3, the improvements to fitness found between 1000 and 3000 evaluations appearing to be due to solutions being found with higher sample errors rather than higher true scores.

Kohavi [7] suggests that the over-fitting problem may not matter for tests of size over 250 and hints that on average the feature set with the highest evaluation score may be the one which generalizes best. The results above must question this.

Table 2. Average accuracy over 10 runs on test set and validation set

	AUS test / validation	VEH test / validation	DNA test / validation	SAT test / validation	SEG test / validation
All features	79.8/ NA	64.7/NA	73.3/ NA	84.5/ NA	91.1/ NA
FSS	86.3 / 85.3	68.7 / 65.9	90.2 / 87.5	85.8 / 82.4	94.0 / 92.4
GA: Initial population	85.2 / 83.3	69.1 / 67.3	79.5 / 77.2	85.9 / 84.1	93.5 / 92.1
GA 1000 evaluations [1]3000 for DNA	86.9 / 85.8	71.3 / 69.2	87.5/82.1[1]	87.1 / 84.5	94.2 / 92.2
GA 3000 evaluations [2]10000 for DNA	87.1 / 85.7	72.2 / 68.9	90.3/83.1[2]	88.7 / 84.3	94.4 / 92.5

Although the above GA approach is seen to be generally superior to FSS in terms of function optimization and generalization, the number of models evaluated by the GA is considerably greater and hence the process is much slower. The one data set which the GA generalizes more poorly than FSS on is the DNA data set. Section 5 however shows how the principle of Occam's razor which is arguably implicitly incorporated within FSS can be utilised to significantly improve the results on the DNA data.

In the GA approach the vast majority of processing time is taken up executing evaluation functions. The rest of the paper examines methods for improving the speed of the GA approach as well as its generalization abilities by using evaluation functions other than full test set score.

4   Test Set Sampling

GA theory suggests that with a sufficiently large population, a GA may still be able to find the optimal value for a function in the presence of noise. The argument [10, 11] is derived from Holland's schema theorem [9] which states that in addition to individual strings, selection acts on hyperplanes representing sub-partitions of search space. If random Gaussian noise is added to the evaluation function the average net effect of the noise on the hyperplanes fitness is zero. As the size of the population increases, search direction is less influenced by noise in individual fitness scores and hence the ability to handle noise would appear to increase.

It appears that this theory can be put to practical use when GA's are used to construct classification models. As mentioned in Sect. 3, when an unbiased sample of previously unseen data patterns is used to estimate the classification abilities of a model, variations in fitness score are distributed in approximately Gaussian fashion around the true mean with variance depending on sample size. This suggests that if an unlimited supply of test data was available, each evaluation may only need to utilize a small random sample from this test set for an optimal solution to evolve.

In practice the amount of test data available is not infinite and hence when choosing a sample of data for use by the evaluation function one must sample from a sample (i.e. the full test set) and hence performance is still limited by the size of the full test set and over-fitting may still occur. Nevertheless, the above suggests that if each evaluation only uses a small proportion of the test set, solutions may evolve with optimal performance on the whole test set. As the size of the overall test set increases the problem of over-fitting decreases. By using test set sampling the speed of the evaluation is no longer related to test set size and consequently an increase in the size of the test set may improve the quality of the solutions without increasing search time. (To avoid confusion below, the term 'sample size' will exclusively refer to size of sample sampled from test set.)

There are however problems which suggest that there are limits to how small the sample size may be. For example the noise caused by the sample error approximates a Gaussian as sample size increases and consequently the effectiveness of the approach when using very small sample sizes must be questioned. For example, selection pressure, the extent to which strings with high fitness are favoured as parents, can decrease as sample size decreases: If tournament selection is used and the string with the highest fitness is chosen for recombination then the probability that noise in the fitness function causes the inferior string to win increases as sample size decreases. If selection pressure is too low convergence may never be achieved.

Whilst reducing sample size increases speed, larger populations and more evaluations may be required as compensation. Fitzpatrick and Grefensette [10] demonstrated this empirically on an image registration problem but empirical investigations in a classification context seem rare. Brill [2] employed a type of test set sampling where the samples were used to determine prototypes as well as test patterns but the range of experiments performed was limited. An alternative method for increasing the speed of evaluation functions is that of 'fitness inheritance' as proposed by Smith [14] where evaluation functions are only performed on certain offspring and the rest inherit fitness from their parents. The reported experiments however were performed on function optimization problems rather than classification problems.

To examine the effects of test set sampling, the experiments in Sect. 3 were modified. Instead of implementing a deletion strategy which utilizes fitness, the oldest string in the population was always deleted. The reasoning behind this is that individual fitness scores are unreliable - genetic material is proven to be robust if it survives for many generations and hence has been evaluated several times. The mutation rate was also lowered to reduce the amount of unfamiliar material produced.

At the end of the GA search, three methods were tried for selecting the best string: The first method was to evaluate the final population using the whole test set and pick the highest scoring. Whilst this method ensures that a good string is picked as the final solution the process of fully evaluating the entire population may be computationally costly for large population sizes. A second method was to pick the string with the highest current fitness (obtained using test set sampling) which is far quicker but may be unreliable particularly for small sample sizes. The final method attempted was to form a composite string by examining each bit position in turn and setting it to a '1' if the majority of the strings in the population contained a '1' in the corresponding position. and 0 otherwise.

In the first experiment, a fixed population of 100 was utilized and the size of the sample taken from the test set was varied between 1 and 500. 10,000 evaluations were performed. Table 3 shows the average score on unseen data over 10 runs with a full evaluation being used on the final population in order to find the 'best' string.

The results obtained using the composite string were generally very similar and hence in a feature selection context could be used for increased speed. Picking the string with the highest sample based fitness score was, as expected, generally

Table 3. Results on unseen data using test set sampling averaged over 10 runs

Sample size	AUS	VEH	DNA	SAT	SEG
1	82.5	92.4	67.2	83.3	77.5
5	84.7	92.2	67.9	83.9	78.2
10	83.2	93.0	68.6	84.5	79.4
20	83.9	92.5	67.7	84.1	79.7
50	83.7	92.9	68.9	84.0	81.5
75	86.1	92.7	69.2	84.3	82.1
100	85.2	92.4	68.0	83.7	82.0
150	85.4	92.9	68.1	84.2	81.8
200	85.9	92.5	69.1	84.0	82.2
500	85.7	92.5	68.9	84.3	83.1

inferior particular for very small sample sizes.

The results demonstrate that test set sampling can greatly increase the speed of the evaluation function with little loss of accuracy. For example the evaluation with 10 samples is 50 times quicker than that utilizing all samples yet on two of the data sets the classification results are actually higher.

The results are poor for the DNA data set. As it operates with a much larger search space it is envisaged that larger populations and extra evaluations may be required. Further experiments were therefore set up for the DNA data in which the sample size was kept small (10, 20 or 50) and various population sizes were tried. The composite method for selecting the final string was utilized in order that the computational costs of an increased population size are minimal and to overcome the problem that the larger the population, the greater the likelihood that it may contain 'good' strings purely by chance. The total number of evaluations was increased to 40000 and the process was repeated 10 times with different splits of the data.

Table 4. Average score on 2686 unseen DNA patterns

Sample size	Pop size 100	Pop size 200	Pop size 300	Pop size 500
10	79.4	82.5	82.6	83.9
20	81.9	82.4	83.5	84.0
50	83.5	83.9	83.9	82.4

The results show that for the DNA data, an increase in population size can improve the quality of the models produced. The scale of improvement appears to be greater for smaller sample sizes. With a sample size of 50 and population size of 500 the average evaluation score obtained using the whole test set was actually the highest (90%+) but this over-fitting causes accuracy on unseen data to decrease.

Whilst test set sampling in itself doesn't significantly improve classification accuracy, it appears to be far more efficient than using the full test set. This increase in efficiency may allow larger tests to be utilized and hence permit solutions to evolve which are likely to generalize better on independent data sets. Such techniques may be particularly useful in data mining scenarios where data is available in abundance but using the full test set for each evaluation would be infeasible.

5   The Utilization of Confidence Tests and Occam's Razor

This section briefly describes initial attempts to improve the evaluation function to enable it to become more adaptable to the current state of the search and the data available. The technique involves the integration of the principle of Occam's Razor with the idea of using statistical significance tests to determine how many samples are to be used.

Variations on the concept of 'Hoeffding races' [12, 13] have been used in situations where a number of competing models have been created. All the models are evaluated in parallel with the number of samples given to each model gradually increasing. Statistical significance tests are performed and any model which is significantly worse than the best is removed and the process continues until a winner emerges. No examples of this technique being applied in a GA context have been found.

In the experiments utilized below a variation to the technique of Hoeffding races was utilized in a GA during tournament selection. Two strings were picked at random from the population and samples were added until either a maximum number of samples was reached or a clear winner emerged. In the situations with no clear winner the principle of Occam's razor was applied and the simplest string was chosen (i.e. the one containing the fewest features). The idea behind this is that the evaluation function can adapt the number of samples required to suit the strength of the competing strings whilst the utilization of Occam's razor is intended to reduce over-fitting and encourage minimal solutions. As well as arguably generalizing better, minimal solutions can convey the essential properties of the data to the user.

The specifics of the experiments were similar to those described previously. The population size was kept at 100 and the number of evaluations set at 5000. The size of the full test set remained at 500 but the maximum number of samples taken was set at 120.

When performing a tournament, initially 30 random samples were given to both competitors, a significance test was performed and the process was repeated with 30 more samples being added until a significant result emerged or the limit (120) was reached. A winner emerges if $Z > Zmax$ or $Z < -Zmax$ with Z defined as:

$$Z = \frac{(S1 - S2)/n}{\left(\sqrt{\hat{p}(1-\hat{p})}\right)\left(\sqrt{\frac{2}{n}}\right)} \tag{1}$$

Where n is the sample size, $S1$ and $S2$ are the respective number of sample patterns correct for the two competitors and $\hat{p} = (S1 + S2)/2n$.

Whilst different values of $Zmax$ were experimented with, there are reasons why lower values than those used for usual tests of significance are appropriate.

The reasoning here is twofold: Firstly, to encourage diversity in the population and prevent pre-mature convergence the inferior string should on occasions be allowed to 'win' and hence the statistical test should occasionally select the wrong model. Secondly if $Zmax$ is too high then a winner may only rarely emerge and the bias towards small models may limit the quality of solutions obtainable as suggested by the results in Table 5.

The results in Table 5 show that the above approach is promising. On three of the data sets (AUS, SEG and DNA), average generalization scores were obtained which were higher than any of those found using the previous evaluation functions. The scores obtained on the DNA data were particularly high in comparison, with a bias towards small feature sets reducing the amount of overfitting which occurs and improving generalization. However the optimal value of $Zmax$ appears to be problem dependent and as suggested below overcoming this is a goal for future research.

Table 5. Average accuracy on unseen data when using confidence testing + Occam's razor

Zmax	AUS	SEG	VEH	SAT	DNA
0.6	87.5	92.5	68.2	84.1	81.9
0.8	85.6	93.5	67.9	83.1	84.3
1.0	84.0	92.9	67.6	82.7	88.0
1.96	84.7	92.8	62.2	82.4	86.3

6 Conclusions and Further Work

The above has shown that the utilization of test set sampling, statistical tests and the principle of Occam's razor in the fitness function has the potential to increase the efficiency of the GA approach to model determination and hence make it more amenable to large data sets. Further work is necessary to examine the mechanics of these approaches in more detail. Ideally the form of the evaluation function would automatically change to suit the properties of the data and the current population. Whilst Sect. 5 describes a method where sample size is adjusted to suit the quality of competing models, other parameters in the evaluation function could perhaps also be automatically adjusted to suit the data and current population. The optimal size of test set sample in Sect. 4 and maximum sample size and the value of $Zmax$ in Sect. 5 probably vary as search progresses from the initial population towards convergence. It seems likely that sample size should increase over time to take account of the fact that fitness scores tend to become more similar as search progresses and hence need more accurate assessments of fitness but further study on this topic is necessary.

In the context of finding meta-level parameters, the GA process can be viewed as that of adapting the classification algorithm to suit the data. By increasing the range of parameters encoded in the GA, flexibility and classification accuracy can be potentially increased. Whilst an increase in the number of parameters may increase the scope for over-fitting, the techniques described in this paper suggest ways for overcoming this.

References

1. I Kuscu, C Thornton: "Design of Artificial Neural Networks using Genetic Algorithms: review and prospect", Technical Report, Cognitive and Computing Sciences, University of Sussex (1994)
2. F Brill, D Brown et al.: "Fast genetic selection of features for neural network classifiers", IEEE Transactions on Neural Networks, Vol.3 No.2 (1992) 324-328.
3. S Salzberg: "A Critique of Current Research and Methods", Technical Report JHU-95/06, John Hopkins University, Department of Computer Science. (1995).
4. R Forsyth: "IOGA: An Instance-Oriented Genetic algorithm", Parallel Problem Solving from Nature 4, (1996) 482-493.
5. J Kelly, L Davis: "A hybrid genetic algorithm for classification.", Proceedings of the Twelfth International Joint conference on Artificial Intelligence, Morgan Kaufmann (1991) 1022-1029.
6. A Miller: "Subset Selection in Regression", Chapman and Hall (1990).
7. R Kohavi, D Sommerfield: "Feature subset selection using the wrapper model: Overfitting and dynamic search space topology", First International conference on Knowledge Discovery and Data mining. (1995) 192-197
8. Statlog data and documentation at ftp.ncc.up.pt/pub/statlog
9. J Holland: "Adaption in natural and artificial systems", University of Michigan Press(1975).
10. J Fitzpatrick, T Grefenstette: "Genetic Algorithms in noisy environments", Machine Learning Vol. 3. No. 2/3 (1985) 101-120.
11. S Rana, D Whitley et al.: "Searching in the presence of noise", Parallel Problem Solving from Nature 4, (1996) 198-207.
12. O Maron, A Moore: "Hoeffding Races: Accelerating Model Selection Search for Classification and Function Approximation", Advances in Neural Information Processing Systems 6. Morgan Kaufmann (1994).
13. A Moore, M Lee: "Efficient algorithms for minimizing cross validation error", Machine Learning: Proceedings of the Eleventh International Conference, Morgan Kaufmann (1994)
14. R Smith, E Dike et al.: "Inheritance in Genetic Algorithms", Proceedings of the ACM 1995 Symposium on Applied Computing, ACM Press (1994).

Data Representations and Machine Learning Techniques

C. P. Lam and G. A. W. West and T. M. Caelli

Department of Computer Science,
Curtin University of Technology,
GPO Box U1987, Perth 6001, Western Australia,
{penglam, geoff, tmc}@cs.curtin.edu.au

Abstract. In recent years there has been some interest in using machine learning techniques as part of pattern recognition systems. However, little attention is typically given to the validity of the features and types of rules generated by these systems and how well they perform across a variety of features and patterns. A comparison of the classification performance of two different types of decision tree techniques and their associated feature types is presented.

1 Introduction

There are now available numerous techniques for classifying multi-variate data into a number of classes. Such techniques include linear classifiers, linear descriminants, decision trees and neural networks. These techniques essentially determine how to partition feature attribute spaces by generating and refining rules to maximally evidence different classes from the training data. They all assume that the features presented have characteristics that make this process viable i.e. that, for each class, the feature values cluster in n-dimensional space such that each class is ideally represented by a set of coherent clusters. Most current models (specifically, decision trees and neural networks) do not assume any specific spatial density model, such as Gaussians, but they do assume that such regions evidence different classess.

A number of issues such as pattern decomposition, rule generation and evaluation are important in the design and application of machine learning techniques in high level vision. In the area of pattern decomposition, it is our contention that it is important to compare the different methods of representing patterns (e.g. global *vs* local feature representation) for classification of known and novel patterns because the data representation is crucial and highly coupled to the learning techniques used.

In fact, little research has been performed on what are the best features to use for a particular learning task. It is very much the case that intuition and some knowledge of the different classes are used to decide the features. For example, simple features, such as area and perimeter, are very good when

X. Liu, P. Cohen, M. Berthold (Eds.): "Advances in Intelligent Data Analysis" (IDA–97)
LNCS 1280, pp. 161–172, 1997.170, 1997. © Springer–Verlag Berlin Heidelberg 1997

dealing with discrimination between simple shapes such as circles, squares and triangles. However their justification for more complex shapes such as spanners and pliers, which have concavities, fine and coarse detail and variable shape, is less convincing. The choice of features is simpler when features have a direct shape interpretation (such as polygonal approximations and Fourier descriptors) either in terms of encapsulating parts of objects or more general descriptions at different scales.

In this paper we investigate these issues using two problem domains that involve two different decision tree classifiers namely C4.5 [7] and Conditional Rule Generation (CRG) [2]. C4.5 induces rules over attributes without considering data indexing - for example, part labels, whereas CRG uses parts and relational attributes over labelled parts to generate a tree of conditional feature spaces (see Section 3.2). In this study we have investigated the use of integral 'global' shape features compared with part descriptions of shape. For example, given a polygonal description of the boundary of an object, C4.5 would use attributes for all lines (length, orientation) whereas CRG would use these attributes as well as relationships between adjacent lines (angle, for example). That is, CRG uses '*Recognition by Parts*' (RBP). In this paper, the features used with C4.5 are Fourier descriptors which describe the shape of a closed curve. For CRG, a polygonal approximation is used with unaries and binaries as described above.

Of specific interest here are part-indexed representations more robust than simple attribute-indexed systems in problems involving complete as well as partial data (for example, when objects/pattern are partially occluded).

In the area of rule generation and application, an issue that past research [2, 4, 6] has not fully addressed concerns the representative power of the training set. The most common requirement is that the training data be representative of various classes. However, this may not be guaranteed and there are many situations where it is necessary to generate samples by perturbing the raw input images or the extracted feature values. Such a procedure raises issues about the validity of models for the processes of sample and rule generation. Equally, even with observed data, it is quite possible to generate descriptions or rules which are simply not valid if only specific features or attributes are used.

2   Feature Extraction and Perturbation of Feature Data

We have explored two methods for representing the shape of 20 objects — local and global. One global method is to parameterise the boundary using Fourier descriptors. Unlike global descriptors such as area and perimeter, Fourier descriptors carry shape information and thus are useful in the reconstructing the original shape of objects.

Features can be divided broadly into *information non-preserving* and *information preserving* groups. Information preserving features include Fourier descriptors, chain codes and polygonal descriptions. As such these vector features allow either good or perfect reconstruction of the input pattern and each feature has meaning e.g. increasing Fourier descriptors describe how the boundary of the pattern is modified at increasingly fine detail in a physically realisable way. Because of this, it is much easier to choose the right features for a recognition task as the descriptive and discriminative power are more easily understandable. In a similar way, part and relational feature attributes can be used to encode shapes in many ways which allow physically meaningful interpretations - if not unique.

Of interest, is whether the physical or meaningful interpretation of features/attributes adds to their discriminatory power. To study such issues we have used three classes of handtools, namely pliers, screwdrivers and spanners (see Figure 1) for the reasons that:

- All shapes are not convex and possess a high degree of fine detail.
- The objects in each class of handtools are similar in shape but vary in size (as for spanners).
- The handtools are suitable for generating polygonal approximations as well as Fourier descriptors — high and low frequency. In other words, the objects produce rich global as well as local feature descriptions. The detailed and complex shapes result in many Fourier descriptors and many lines describing the boundary.

2.1   Global Representation

Fourier descriptors are used here since they describe the shape in terms of the complete boundary of the object. They are computed by taking the Fourier transform of the object's boundary in the following manner. A point moving along the boundary generates the following complex function:

$$u(n) \cong x(n) + jy(n), \qquad n = 0, 1, 2, \cdot, N - 1. \tag{1}$$

The discrete Fourier Transform representation is:

$$u(n) \cong \frac{1}{N} \sum_{k=0}^{N-1} a(k) \exp\left(\frac{j2\pi kn}{N}\right), \qquad 0 \leq n \leq N - 1, \tag{2}$$

and the corresponding inverse is:

$$a(k) \cong \sum_{n=0}^{N-1} u(n) \exp\left(\frac{-j2\pi kn}{N}\right), \qquad 0 \leq k \leq N - 1. \tag{3}$$

The DC component is the centroid of the object and the fundamental frequency (real and imaginary components) defines an ellipsoidal approximation to the shape. Other frequencies are similar to epicycles that will deform this fundamental ellipsoidal shape to produce the original shape. Thus, the high frequency components of the Fourier descriptors account for fine details while the low frequency accounts for global shape. Thus, a few Fourier descriptors can be used to capture the gross essence of a boundary (see Figures 2(a) and 2(b)).

In these cases, the shape of the object was approximated using the lowest ten Fourier descriptors. Ellipses were fitted using the Fourier descriptors and the associated parameters such as the major and minor axes and the orientation were extracted. These parameters were used as attributes in C4.5. This is used because Fourier descriptors by themselves convey no physical interpretation. Figure 2(c) shows an example of the reconstructed ellipse for a spanner using the first (lowest non DC) Fourier descriptor.

2.2　Part Extraction

To segment a curve or boundary into features of a single type the complete pixel list is approximated by the required feature. The list is then split into two at the point of maximum deviation between the approximation and the pixel data. This process of approximation and splitting is repeated recursively on each of the two lists and halts when the pixel lists are too small to properly fit the representation (Lowe's method [5]. An important well known point is that all existing methods for polygonising boundaries are flawed in some way as they do not represent the data accurately. This is brought about from the method used to choose constraints such as number of lines to represent the boundary, and the definition of vertices. Note for a polygonal approximation to a curve, vertices will not correspond to any in the data. Various definitions of vertices can be used such as position of maximum deviation, maximum curvature etc. All the above methods are ultimately flawed as how the boundary of an object should be polygonally represented depends, to a large extent, on the context and scale of processing.

Given this multi-scale part-based representation, the unary and binary attributes associated with parts of a handtool are obtained in the following manner. A series of lines were extracted from the boundary of the handtool. Based on the analysis shown in [8], we have chosen the following unary and binary attributes. Unary attributes associated with each of the lines are the lengths and the orientation of the line. Binary attributes were attributes associated with pairs of lines: (1) angles between lines, (2) whether the pairs of lines are adjacent or parallel and (3) distance between the centroids of each of the two adjacent lines.

2.3　Perturbation of data

The training data used to generate the classifiers typically contains only a small number of examples of each class of object. In order to generate more examples, the existing training examples can be perturbed in a valid way e.g. statistically. The data in this paper are perturbed in the following way:

1. Generate a series of lines from the given boundary of the object.
2. Perturb the end points of the lines using a Gaussian model with a particular value of σ.
3. Generate the pixel list representing the perturbed boundary from step 2.

The pixel list obtained in step (3) is subsequently used to generate the Fourier descriptors and the unary and binary attributes associated with parts of the object. This procedure is used instead of simply perturbing the positions of each edge pixel to simplify processing. Perturbing every pixel's coordinates would mean many checks would have to be made for self-intersection leading to reduced frequency of the production of a legal perturbation.

From initial experiments, the values of σ used in step 2 above were chosen to be 1.0, 2.0 and 2.5 pixels. There is no particular significance in using these values of σ but values of $\sigma \geq 2.5$ have not been used because the resulting perturbed objects are often not legal and the process of testing for legality in the objects can be very time consuming. Examples of illegal instances of the screwdriver are shown in Figures 3(a) and 3(b).

2.4   Partial Data

The partial objects are obtained by editing the polygonal representation of the complete object by removing lines. Hence partial data represent occluded shapes and missing features. This process is valid for generating the partial data as the error between the polygon and shape that occurred here would be *much* smaller than the errors in the gross shape caused by missing data. Figure 3(c) and 3(d) show some examples of partial data where the missing (or occluded) parts are obvious.

The global and parts representation of the partial data are generated in the same manner as those of the complete data as the the full boundary of each of the partial data are available. This is important for valid comparisons of the classification performance of the classifiers on partial and complete data. For example, the Fouriers Descriptors for an object that has a very small part missing would be very similar to that of the complete object. This is also the case for parts representation as most parts of the object would retain their attribute values and only those sections related to the occluded parts would differ.

3   Learning Techniques

In this section, we consider two different learning algorithms, namely C4.5 and CRG where the goal is to obtain a partitioning of the feature space that enables correct classification of seen as well as unseen or novel patterns An issue that is important for robust recognition systems is the amount of information that is needed to represent the patterns in the observed data. The representation of the patterns can be attribute-indexed or part-indexed as previously described. The difference between the two representations lies in the fact that structural pattern information is preserved in the part-indexed representation. In other words, relations between parts are made explicit, for example, 'the angle between part 4 and part 5 is 120°'.

With supervised learning, during the learning phase, each object is presented separately as a number of training views (samples). The training views are presented serially and the order in which they are presented is not important.

The difference between CRG and C4.5 has been investigated by Caelli *et al* [3]. Rules generated by the CRG method are cf (minimal) variable length whereas in C4.5, the dimensionality of the feature space and rule length are fixed. In addition, relational structure has to be encoded implicitly in the attributes extracted from different parts and part relations as C4.5 does not use part-indexing. CRG also differs from relational learners (e.g. ML-SMART [1]) as it uses the ordering of the rules to constrain the search. The method uses knowledge of physical constraints to select the parts and part relations to be added during the rule construction process.

3.1   C4.5 and Global Features

C4.5 [7] constructs rules inductively by generalising from specific examples. There are a number of key requirements for this technique. Firstly, the information about classes is expressed in terms of a fixed set of attributes which may be continuous or discrete. This implies that the attributes used to describe an object may not vary from one object to another. Secondly, categories or classes for the classification of objects must be pre-defined. These classes must be clearly delineated and thus an object can belong to only one class. Thirdly, sufficient data must be available for the construction of the model as the model is constructed inductively by analysing patterns in the data provided. Lastly, the technique constructs classifiers that are expressed as decision trees or as sets of production rules. The description of a class is restricted to a logical expression in which the primitives are statements about the attribute values. The constructed decision tree consists of either a leaf node (indicating a class) or a decision node that specifies some test to be carried out on a single attribute value.

Classification on unseen data is performed by comparing the required attributes at nodes in the tree and following the correct path through the tree until a leaf node is reached or the search terminates with class uncertainities.

3.2   CRG and Relational Features

CRG is a system that is based on the *recognition by parts and relations* (RBP) paradigm. Again it is a rule-based decision tree classifier which has the capability to classify a large number of objects that are either isolated or in a complex scene. This system generates Horn clause descriptions of patterns in terms of labelled parts, their relations and associated attribute bounds. These descriptions can be used to uniquely discriminate pattern identity.

Firstly, a unary feature space, U, is formed by grouping unary attributes from all parts, from all views of all objects. Each single part is then a point in the unary feature space. The unary feature space is partitioned into a number of clusters U_i using the same entropy-based attribute splitting technique as used in C4.5. Some of these clusters may uniquely describe a class to which a part may belong, and some will contain multiple classes. The non-unique (i.e. multi-class) clusters are further processed using the binary features between all parts in that region and other parts in the view where the binary relations are typically restricted to local neighbourhoods of a part. These binary features form a binary feature space, U_iB_i. The conditional binary feature space is then clustered into a number of clusters that may be unique or non-unique with respect to class membership. The non-unique clusters are processed further by including unary features of the second part to form another unary feature space $U_iB_{ij}U_j$. This unary feature space is again clustered and the process continues until all clusters are unique. However, there may be cases where clusters are unresolvable at any level. Such clusters may be refined at some level either by splitting clusters or by reclustering a particular feature space into more clusters (for details see [2]). A point to note is that there must be a balance between tree expansion and cluster refinement. Tree expansion may not always achieve unique clusters and may also reduce the probability of clustering partial data. On the other hand, cluster refinement often results in a more complex feature space partitioning as well as generalising from the presented examples.

During the recognition phase, the same types of features as those used in the training phase are extracted from the scene. Unary and binary attributes are extracted from all the parts (e.g. lines) and the same relational topologies are used as in training. Each part is then classified as a function of the CRG rules that they activate either partially or fully. That is, the evidence vectors of all paths terminating in a particular part will determine the classification of that part. However, the evidence vectors for a given part may be non-unique (i.e. a rule has only been partially instantiated) or incompatible. Techniques are then required to combine the evidence and Bischof and Caelli [3] have investigated a number of methods including the *winner takes all* and relaxation labelling formulations.

4   Experimental Results

The first objective of the experiments was to compare the performances of the attribute-index learning algorithm (C4.5) which uses global feature vectors and the part-indexed learning algorithm (CRG) which uses parts and relationships between parts.

The second objective was to evaluate the effects of introducing perturbed data into the training set. The classifiers were trained using data that consisted of ideal and perturbed data. The performances of these classifiers were then compared with the performances of classifiers trained using only ideal data.

In these experiments, images were obtained using 20 different objects from the three classes. The objects in each class were positioned at various orientations. The data used to generate Classifier 1 and 3 were ideal data of complete objects while Classifier 2 and 4 were trained using ideal and perturbed data. Training and test data were disjoint where the test data consisted of two objects in the category of pliers, four in the category of screwdrivers and eleven in the category of spanners. The test data belongs to one of the following three sets: ideal data of complete objects, perturbed data of complete objects and ideal data of partial objects.

4.1   Classification using Fourier Descriptors and C4.5

Tables 1 and 2 show the results of the classification of the three sets of test data using Classifier 1 and 2 respectively. Both classifiers performed equally well in the case of the ideal data of complete objects giving perfect results (refer to Table 1). Classifier 2, trained using a combination of perturbed and ideal data, performed better in classifying both complete and partial objects as might be expected given that perturbed and partial data are both "imperfect". For example, as shown in Table 1, in the case of $\sigma = 1$, Classifier 1 misclassified three of the spanners as pliers whereas Classifier 2 has classified all the spanners correctly but has misclassified one of the screwdrivers as a spanner. There was also a gradation in the performance of both classifiers as the test data became increasing perturbed as evidenced by the results for $\sigma = 2.5$. This could be due to the fact that as the amount of perturbations increased the boundary between classes can become difficult to resolve.

4.2   Classification using CRG and Unary and Binary Attributes of Parts

A strategy is needed here to classify a given handtool as each handtool is made up of parts (i.e. lines) and the classifier labelled the parts as belonging to a certain class (i.e. pliers, screwdrivers or spanners). We have adopted the *winner takes all* strategy. If most of the parts of an object were classified as belonging to a certain class, then it is more likely that the object is an instance of that class. For example, if nine out of the ten parts making up a spanner were classified as belonging to the class of spanners then the object was considered as belonging to that class.

Tables 3 and 4 indicate the cases where the object has been classified in the right class. A total of 16 cases were used in Table 3 and 36 cases in Table 4. For example, in Table 3, for the case of $\sigma = 1$, only one of the four objects in the class of screwdrivers was correctly classified. The rest of the objects in the class were incorrectly classified. Tables 3 and 4, respectively, show the classification using the two classifiers on complete and partial data. Again, Classifier 4 (the classifier trained using a combination of perturbed and ideal data) performed better on all the test data involving complete and partial objects. This is as expected because the classifier (trained using the combination of perturbed and ideal data) expected the perturbations. Note that the confusion matrices do not represent all the samples because CRG, unlike C4.5, has a *not recognised class.*

4.3   Comparison of classification performance between C4.5 and CRG

An issue of interest here is the classification performance of attribute-indexed (C4.5) and part-indexed classifiers (CRG) on complete and partial data. We have argued that part-indexed representations have a greater representative power than attribute-indexed representations and the experimental results obtained have confirmed this to be the case. Table 5 shows a comparison of the performance of the two classifiers on complete data. The entries in the tables indicate the number of cases of the test data that have been *incorrectly* classified. As discussed earlier, Classifiers 1 and 2 are obtained using C4.5 and Classifiers 3 and 4 are obtained using CRG. It can be seen from this table that the classifiers obtained using the same type of training data perform equally well in classifying ideal data (i.e. none incorrectly classified). In the case of complete data, there is little difference in the performance between Classifiers 2 and 4 or between Classifiers 1 and 3. All four classifiers show a degradation in classifying perturbed data as σ increases.

Table 6 shows a comparison of the classification performance of the four classifiers on partial data based on correctly classified data. In the case of Classifiers 1 and 3, there was little difference in the performance in the classification of pliers and screwdrivers. However, Classifier 3 performed better than Classifier 1 in classifying the spanners. In comparing Classifiers 2 and 4, Classifier 4 again performed better than Classifier 2 in the classification of the spanners.

Although the test (unseen) data set is relatively small, the results indicated that the part-indexed classifiers performed slightly better than the attribute-indexed classifiers when dealing with partial data. This is not so clear in the case of complete data.

5   Conclusion

This paper has described the performance of two different decision tree classifiers that use different types of feature (global *vs* local) obtained from images. An important aspect is concerned with how classifiers can be compared given that they use different features and operate in different ways. A simple decision tree (C4.5) has been used to recognise handtools from global features i.e. Fourier descriptors of the boundary. A more complex decision tree classifier (CRG) requires both unary and binary attributes which are local in extent. Both of these classifiers are trained on a training set that is supposed to be a representative set of patterns i.e. similar results should be expected on unseen data. This is not always the case and perturbations are required on the training set to get more representative data. Analysis has been presented that shows that for the chosen features, perturbations have to be carefully chosen. Training on ideal and *legal* perturbations shows that the classifiers generally improve in performance over those just trained using ideal data. Another issue addressed is which classifier produces the best results i.e. are global features better than combinations of local features or visa versa. From our experiments, the global features give a better performance in cases where the *full* object is available. Global features are poor if occlusion occurs meaning only partial views are available. In the case where only parts of the objects are available, the classifiers obtained from CRG performed slightly better.

References

1. F. Bergadano and A. Giordana. Guiding Induction with Domain Theories. In Y. Kodratoff and R.S. Michalski, editors, *Machine Learning: An Artificial Intelligence Approach*, volume 3, chapter 17, pages 474–492. Morgan Kaufmann, Los Altos, CA, 1990.
2. W. F. Bischof and T. M. Caelli. Learning Structural Descriptions of Patterns: A New Technique for Conditional Clustering and Rule Generation. *Pattern Recognition*, 27(5):689–697, 1994.
3. T. M. Caelli and W. F. Bischof. The Role of Machine Learning in Building Image Interpretation Systems. In T.M.Caelli, H. Bunke, and C. P. Lam, editors, *Spatial Computing: Issues in Vision, Multi-media and Visualisation Technologies*. World Scientific, 1997.
4. A. K. Jain and R. Hoffman. Evidence-Based Recognition of 3-D Objects. *IEEE Transactions on Pattern Analysis and Machine Intelligence*, 10(6):783–802, 1988.
5. David G. Lowe. *Perceptual Organisation and Visual Recognition*. Kluwer Academic, 1985.
6. A. R. Pearce, T. Caelli, and W. F. Bischof. Rulegraphs for Graph Matching in Pattern Recognition. *Pattern Recognition*, 27(9):1231–1247, 1994.
7. J. R. Quinlan. *C4.5: Programs for Machine Learning*. Morgan Kaufmann Publishers San Mareo, California, 1993.
8. M. C. Robey and G. A. W. West. On the Complexity of Parts for Decision Tree Based Recognition. *Fourth International Conference on Control, Automation, Robotics and Vision*, pages 73–77, 1996.

Classifier No.		Ideal data			$\sigma = 1$			$\sigma = 2$			$\sigma = 2.5$		
		P	D	S	P	D	S	P	D	S	P	D	S
1 (C4.5)	P	2			2					2	1		1
	D		4			4			2	2		3	1
	S			11	3		8	1	1	9	2	2	7
2 (C4.5)	P	2			2			2			1		1
	D		4			3	1		2	2		3	1
	S			11			11			11		1	10

Table 1. The table shows the confusion matrices for the classification using C4.5 on ideal and and perturbed data of complete objects. The columns show the chosen classes and the rows show the true classes. P, D and S represents the classes of pliers, screw-drivers and spanners respectively.

Classifier No.	Ideal data of partial objects		
	P	D	S
1 (C4.5) P	3	1	4
D	2	5	4
S	2	5	10
2 (C4.5) P	8		
D		5	6
S		4	13

Table 2. The classification of partial data using the classifier obtained using C4.5.

Classifier No.	Ideal data			$\sigma = 1$			$\sigma = 2$			$\sigma = 2.5$		
	P	D	S	P	D	S	P	D	S	P	D	S
3 (CRG) P	2			1			1			1		
D		4			1			1			0	
S			11			11			11			10
4 (CRG) P	2			2			2			2		
D		4			3			2			2	
S			11			11			11			11

Table 3. The confusion matrices for classification using CRG on ideal and and perturbed data. The number in each column indicates the number of object in a class that has been classified correctly.

Classifier No.	Ideal data of partial objects		
	P	D	S
3 (CRG) P	3		
D		5	
S			17
4 (CRG) P	8		
D		8	
S			17

Table 4. The table shows the classification of partial data using the classifier obtained using CRG.

Classifier	Ideal data	$\sigma = 1$	$\sigma = 2$	$\sigma = 2.5$
1 (ideal training data)	0	3	6	6
3 (ideal training data)	0	4	4	6
2 (ideal + perturbed training data)	0	1	2	3
4 (ideal + perturbed training data)	0	2	2	2

Table 5. Comparing the performances of the classifiers by considering the number of cases in the test data that were *incorrectly* classified.

Partial Ideal Data			
Classifier	P (Total of 8)	D (Total of 11)	S (Total of 17)
1 (C4.5: ideal training data)	3	5	10
3 (CRG: ideal training data)	3	5	17
2 (C4.5: ideal + perturbed training data)	8	5	13
4 (CRG: ideal + perturbed training data)	8	8	17

Table 6. Comparison of the performances of the classifiers by considering the number of cases in the test data consisting of partial objects that were *correctly* classified.

Fig. 1. Image containing three classes of handtools.

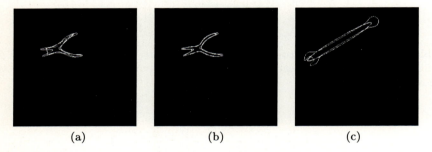

(a)   (b)   (c)

Fig. 2. Ten Fourier descriptors were used in the image in (a) and fifteen Fourier descriptors were used in the image in (b) to reconstruct the boundary of the handtool. The figure in (c) shows the reconstructed ellipse of the spanner using the fundamental frequency. This ellipse is then superimposed on to an image the spanner.

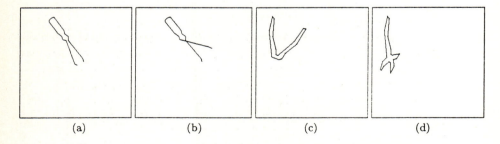

(a)   (b)   (c)   (d)

Fig. 3. Images in (a) and (b) shows the illegal instances of the screwdriver that were reconstructed using the bounds of the attributes associated with the CRG classifier. Images in (c) and (d) are examples of partial objects used for testing the performance of the classifiers.

Development of a Knowledge–Driven Constructive Induction Mechanism

Suzanne Lo[1] and A. Famili[2]

[1] Dept. of Computer Science, University of British Columbia
Vancouver, BC, V6T 1Z1, Canada
[2] Institute for Information Technology, National Research Council Canada
Ottawa, ON, K1A 0R6, Canada, famili@ai.iit.nrc.ca

Abstract. This paper discusses the advantages of knowledge-driven constructive induction (KDCI). Development, testing, and evaluation of a KDCI (or in short CI for constructive induction) system are explained in detail. The objectives of developing this system were to demonstrate the usefulness of the approach and to provide knowledge-driven constructive induction support in our data analysis research. Technical details, particularly the process of building new attributes and changing the representation space, are discussed. Other issues concerning the design and implementation of the CI mechanism are presented. The evaluation process and comparison measures used for evaluation of the system are briefly explained. Experimental results, using 4 data sets from a real-world application, are given.

1 Introduction

Today's world is overwhelmingly dominated by generation and acquisition of large amounts of data in all businesses and industries. Over the last ten years, there have been many advances in data integration, data warehousing, data analysis, and generation of useful knowledge. Examples are several data analysis tools that have become commercially available [9]. Almost all these tools assume that the data representation space is accurate and complete. The data representation space is normally spanned over descriptors such as attributes, variables, terms, relations, or transformations that are used to describe an object. For example, most inductive tools use all initial attributes (numeric and non-numeric) as provided in the data sets, assuming that useful concepts can be generated from these attributes. However, researchers in real-world data analysis applications have experienced that more meaningful concepts can be generated from the data and more accurate knowledge can be discovered if the data is properly transformed to provide a new representation space. This is through constructive induction.

Constructive induction is the process of manually or automatically transforming the data through creating new features from the existing ones and possibly removing the less relevant ones before or during the data analysis. The goal of

X. Liu, P. Cohen, M. Berthold (Eds.): "Advances in Intelligent Data Analysis" (IDA–97)
LNCS 1280, pp. 173–185, 1997.   © Springer–Verlag Berlin Heidelberg 1997

constructive induction is therefore to create a new representation of the domain so that the overall prediction or explanation accuracy of induction based data analysis tools are improved. The prediction accuracy can be estimated through analysing and comparing two randomly selected subsets of a large data set when the representation space is modified on one and not on the other. The explanation accuracy is normally evaluated at the end of the data analysis process by a domain expert.

Generating new features may involve different constructive induction strategies. Three strategies introduced in the literature [16] are: (i) hypothesis-driven in which changes to the data representation space are based on the analysis of hypotheses generated in each data analysis iteration and the discovery of patterns, (ii) data-driven in which data characteristics, such as interrelationships between parameters, are used to generate new data representation space, and (iii) knowledge-driven in which expert provided domain knowledge is applied to modify, construct and/or verify new data representation space. When two or more of the above strategies are combined, it is called multistrategy constructive induction approach.

Two of the most common applications where a knowledge-driven constructive induction approach would be useful are:

- Data analysis applications in which initial parameters have to be corrected through the use of domain knowledge and new features have to be created for an accurate and meaningful data analysis. Examples are industrial operations where environmental conditions affect the operation and performance of a system.
- Data analysis applications in which exploratory research is needed to incorporate qualitative models of a process into the data analysis process and create new features from the existing ones. Examples are applications in which dimensionless terms (p terms) can be introduced to replace all or some of the initial attributes. Each dimensionless term represents two or more of the original attributes. The idea is to transform any number of dimensionally invariant variables (m) represented by n dimensions into m-n dimensionless terms.

The research reported in this paper is focused on knowledge-driven constructive induction. Our goal is to introduce a knowledge-driven construction induction mechanism that we have developed as part of our research in data analysis and demonstrate, using real-world data, how creation of new features could improve the overall performance of this system. In Section 2, we review some of the related work. Section 3 includes description of the problem that motivated this work. In Section 4 we provide an overview of our approach where we introduce technical details of the constructive induction mechanism that we have built and show how new features are generated and used. Section 5 includes testing and evaluation methodology and in Section 6 we provide the results. We conclude the paper in Section 7 and discuss our future research.

2   Related Work

The general form of knowledge-driven constructive induction has been advocated by many researchers [10] and [15]. Most of the related works were motivated by the fact that inductive learning algorithms were sensitive to the original representation space. This section briefly reviews some aspects of knowledge-driven constructive induction research performed by other researchers.

Pagallo [12] developed FRINGE that iteratively constructs new attributes using structural information of the decision tree. Matheus and Rendell [8] and Rendell [14] also proposed a number of frameworks for feature construction based on four main aspects of detection, selection, generalization, and evaluation. In these studies, the process of building new features is at the time of building the decision tree where simple domain knowledge is used for constructive biases. AM (Automated Mathematician), developed by Lenat [7], changes the data representation space by employing pre-defined heuristics for: (i) defining new concepts represented as frames, (ii) creating new slots and their values, and (iii) adapting concept frames developed in one domain to another domain.

Callan and Utgoff [1] used the domain knowledge, in the form of first order predicate calculus, to describe the problem solving operators. This information was then used to generate a better vocabulary by increasing the resolution and by decomposing the description of the goal states into a set of functions that were used to map search states to feature values. On two domains that this method was tested, it was shown that the features it generated, were more effective for inductive learning than the original features. A similar concept has been incorporated into LAIR [3] which is a constructive induction system that acquires conjunctive concepts by applying a domain theory to introduce new features into the evolving concept description. The system works through an iterative process in which it weakens the inductive bias with each iteration of the learning loop.

Most systems that incorporate knowledge into constructive induction use almost complete domain knowledge. Use of fragmentary knowledge in constructive induction, due to varying degrees of completeness, may still result in generation of useful features. Donoho and Randell [2] showed that production of new features, using fragmentary knowledge increases inductive accuracy and it also results in determining knowledge reliability. Use of domain knowledge during the process of building the decision tree was also investigated by Nunez [11] who developed an algorithm to generate more logical and understandable decision trees than are normally generated by learning algorithms like ID3. His algorithm executes various types of generalization and at the same time reduces the classification costs by means of background knowledge.

Constructive induction could also be treated as a preprocess to data analysis. In GALA, developed by Hu [6], a small number of new attributes are generated from the existing nominal or real-valued attributes. GALA was designed for use in preprocessing data sets for inductive algorithms and was tested with C4.5 [13]. Reiger [15] also views constructive induction as a data preprocessing step and proposes a framework in the robotics domain where the numeric data is transformed to logic-based data using domain knowledge.

3   The Problem

Some of the best applications of inductive techniques have been in areas where domain knowledge has been used to generate a new representation space. In some cases, complex domain knowledge, such as causal qualitative models, are used to generate new attributes. The main questions are: (i) where the required knowledge comes from, (ii) how can we appropriately use this knowledge to modify the representation space, (iii) are the new attributes meaningful and (iv) how can we evaluate the results when new attributes are used. This section includes a brief discussion of the above points.

3.1   Types of Knowledge

The knowledge for creating a new representation space could vary from simple concepts of domain knowledge to complex causal models. The best representation space can be created when complete domain theories are used for constructive induction [2]. Two of the most common types of domain knowledge are:

- Process or System Knowledge that represents various aspects of domain theory. This type of knowledge comes from domain experts, empirical results, system design information, and system documentation.
- Data Characteristics Knowledge that is acquired during data preprocessing. Various approaches are taken. Examples are: data visualization, principal component analysis, dimensional analysis and data fusion. The main goal is by understanding the nature of the data (combined with domain theory), one can generate new attributes.

3.2   Validating New Features

The most difficult task in knowledge-driven constructive induction is validating features when a new representation space is introduced since the new representation space has a profound effect on the quality of generated results. One approach is to investigate the relationship between attributes in the new representation space when new features are included. This can be done using design of experiment techniques to evaluate the performance of a system or process represented by the new representation space. The other approach is to evaluate the performance of the new representation space in an induction process and on iterations in which sets of new attributes are added in groups of related features. This approach requires involvement of domain experts [5] who would evaluate the results of induction each time the representation space is modified.

4   Overview of the Approach

Most learning systems have been concerned with learning concepts from examples from a fixed set of attributes. In other words, the attributes relevant to describing the examples are normally provided before the learning process starts, and the resulting concepts are expressed in terms of these attributes. However, inconsistency and incompleteness of data, as well as the need for incorporating the domain knowledge, have shown the importance of learning systems to have the ability to construct new attributes. Our goal in constructive induction is to convert domain knowledge into a set of numeric attributes for which an evaluation function can be generated via existing methods. We have developed a constructive induction facility that is added to our data analysis tool, IMAFO [Intelligent MAnufacturing FOreman]. IMAFO is a data mining tool that is used for failure analysis and process optimization. It helps engineers to discover the reasons for unsuccessful productions or operations [4]. IMAFO has two engines for data analysis. Engine 1 is a variation of Quinlan's ID3 algorithm; Engine 2 uses Quinlan's C4.5 algorithm [13] and [4].

4.1   Design of the CI Mechanism

The CI module was designed as one of the modules of IMAFO. The criteria for designing the CI module was: once new attributes are created, they are added to the measurement space for any application set-ups and data analysis. Creation of new attributes would be based on using domain knowledge and all new attributes will have the capabilities of the existing ones:

- Subspace Definition: New attributes can be used to define new classes (problem definitions).
- Dimension Control: New attributes can be enabled or disabled for certain classes during data analysis.
- Correlation Check: Principal components can be identified using new attributes.
- Graphics: New attributes can be used for data visualization.

Figure 1 shows the interface to the CI facility. The New Variable Names list (top scroll list) contains the names of all the new attributes. The first set of buttons (Insert, Rename and Delete) are used to define new attribute names. The Available Numeric Variables list (bottom scroll list) contains all numeric (integer or real) attributes that can be used in the definition of a new attribute. The Relations selection buttons display the selected relation which can be used to define a new attribute. The New Variable Definition field shows the definition of the new attribute that is highlighted in the New Variable Names list. The Build button is used to build a new attribute definition.

4.2   Technical Details

The relationships between the objects in the interface are illustrated in Figure 2. A new attribute consists of a name and its definition. A definition is a mathematical expression that is built from domain knowledge. When a new attribute name and its definition are entered, they are added to the New Variable Names

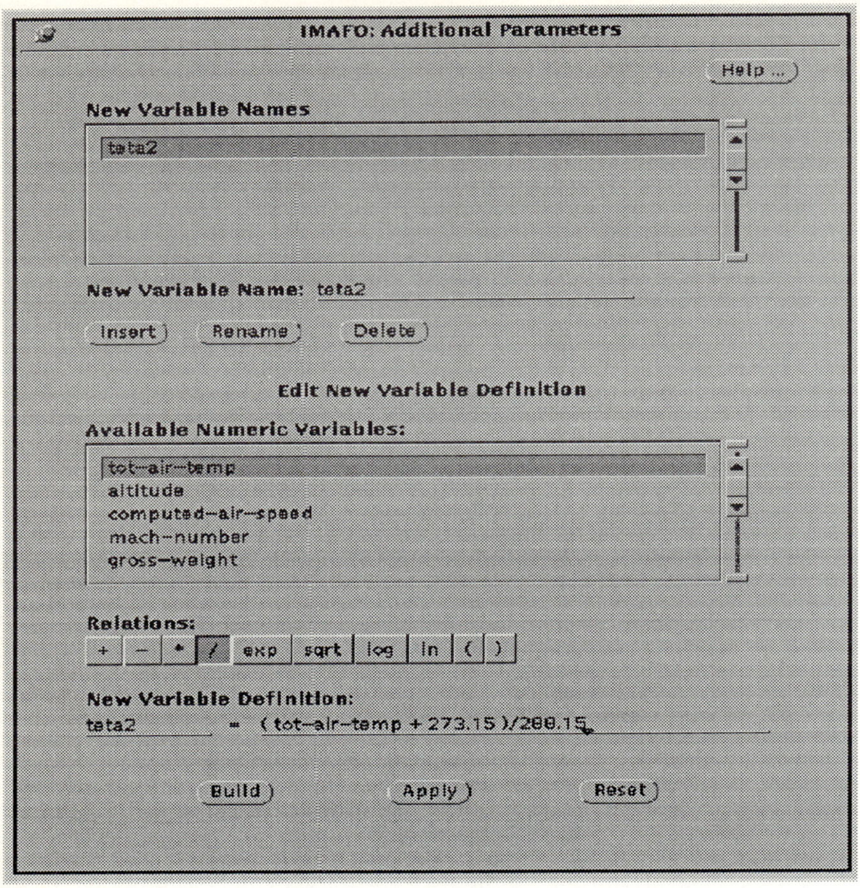

Fig. 1. Interface for the Construcive Induction Facility

list. The most important step in defining a new attribute is the Build process. This process ensures that the expression is syntactically and semantically valid.

The operations in the Relations section for constructing new attributes are based on the generality of the individual relation and completeness of the overall relation set. These relations are relatively common operators that can be used frequently in changing the representation space. It should be noted that the current set of relations in our system can be extended to include operators for strings and logical expressions. They can result in new attributes of type string or boolean.

4.3 Changing the Representation Space

The process of changing the representation space involves three operations: expansion, correction and contraction. Expansion aggrevates the learning process

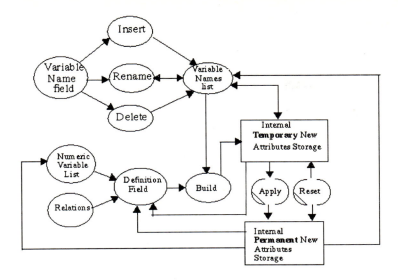

Fig. 2. Interaction between Objects in CI Inetrface

as the representation space is expanded. For example, if two of the original attributes were angular velocity of a rotor blade (ω) and its radius (r), and its linear velocity (v) was also useful for learning a concept, then the linear velocity can be added to the representation space:

$$v = r \times \omega$$

The original representation space can also be corrected for inconsistencies in the values of the attributes to provide a better representation space for the learning process. This can be accomplished by constructing new attributes with more consistent values from existing attributes and eliminating the original attributes. For example, if one of the original attributes were the angular velocity of the rotor bade of an engine (ω), since it is influenced by several operation conditions, it would be corrected by θ:

$$\omega = \frac{\omega}{\theta}$$

where θ is calculated from domain knowledge. wc represents the corrected value of ω.

Contraction decreases the complexity of the learning process as the representation space is restricted. This is performed through abstraction and implies that the learning process can be sped up due to narrowing the search space and decrease in generality. Use of dimensionless terms is a good example of contraction. For example, Reynolds Number (RN) of a moving object is space, a dimensionless term, is derived from air speed (U), air mass density (ρ), length of the moving object (l), and coefficient of air viscosity (μ), as:

$$RM = \frac{\rho \times U \times l}{\mu}$$

The most important effect of changing the representation space is that the overall complexity of the resulting concept description may decrease when domain knowledge is applied. An example is when irrelevant attributes are eliminated from the representation space. Table 1 shows how the existing relations (shown in figure 1) can be used for expansion, contraction, and correction of the original attributes to generate a new representation space.

Table 1. Summary of representation space changes

Operator	Arguments	Notation	Interpretations	E	C	K
+	Attributes x, y	x + y	Sum of x and y	*	*	*
-	Attributes x, y	x - y	Difference between x and y	*	*	*
*	Attributes x, y	x * y	Product of x and y	*	*	*
/	Attributes x, y	x / y	Quotient of x and y	*	*	*
exp	Attributes x, y	x^y	x raised to the power of y	*	*	*
sqrt	Attribute x	sqrt x	Square root of x	*		*
log	Attribute x	logx	Logarithm of x in base 10	*		*
ln	Attribute x	ln x	Natural logarithm of x	*		*
(	-	-	Beginning of sub-expression			
)	-	-	End of sub-expression			

E: Expansion, C: Contraction, and K: Correction

5 Testing and Evaluation Methodology

The overall goal of testing was to focus on the logics of the algorithm that generates the new representation space. Evaluation was performed to investigate improvements in predictive accuracy of the concepts (rules) generated as a result of our data analysis. Predictive accuracy is a criterion that evaluates the performance, consistency, and other behaviours of learning systems, as a measure of success. Having high predictive accuracy means that the learning system learned the correct concept. The correct concept can be represented in different forms or different organizations of a single representation, or different combination of attributes.

The experimentation process consisted of running IMAFO, with and without the new representation space, to evaluate the performance based on improvements on the following comparison measures:

– Coverage: The percentage of records with problems that the rule explains:

$$\text{coverage} = \frac{\text{\# of records with the rule and the problem}}{\text{\# of records with the problem}}$$

– Error rate: The percentage of records without problems that the rule explains:

$$\text{error rate} = \frac{\text{\# of records with only the rule}}{\text{\# of records with the rule and the problem}}$$

– False-negatives rate: The percentage of records with problems that the rule does not explain:

$$\text{false–negatives rate} = \frac{\text{\# of records with the problem but without the rule}}{\text{\# of all records}}$$

– Rule complexity: The number of variables acting together in a rule to characterize a problem.
– Accuracy: The percentage of records with problems that the rule explains and without problems that the rule does not explain:

$$\text{accuracy} = \frac{\text{\# of records with problem and rule} + \text{\# of all records without problem and rule}}{\text{\# of all records}}$$

For experimentation, 46 data sets, representing data collected from an aerospace domain, grouped into four groups, were analyzed. The steps for starting the data analysis were as follows:

1. Filtered data files for corrupt data.
2. Performed data visualization for trend monitoring and class threshold selections.
3. From graphs prepared in step 2, selected proper thresholds for problem (class) definitions.
4. Removed irrelevant attributes from the measurement space.
5. Analyzed all corrected data sets using both engines 1 and 2. For consistency, data analysis was repeated three times.
6. Using the domain knowledge, changed the representation space.
7. Modified all problem (class) definitions to make use of the new attributes, if necessary.
8. Analyzed all data sets using both engines 1 and 2. For consistency, both engines were run three times on each data set.

The following steps were taken for summarizing the results:

1. For each run, the top rule was selected for comparison. The top rule is the rule in which the comparison measures collectively behave better than all the other rules generated (see explanation below).

2. Determined the coverage, error rate, and complexity from the output of all three runs. From the contingency tables of the runs, determined the false-negatives rate and accuracy.

3. Calculated the average of each comparison measures listed in step 2 for the three runs.

4. Repeated steps 1, 2 and 3 for the results, using the new representation space.

5. For each data set, the values of comparison measures without new attributes were subtracted from the values obtained when including new attributes. Thus a positive difference indicates an increase in the comparison measure by using the CI facility, and a negative difference indicates a decrease. Note that an increase does not imply improvement in quality and a decrease does not imply deterioration.

6. The values of differences obtained in step 5 were further averaged for each data group.

For engine 1, the top rule was selected for evaluation because from our experience in the past we have noticed that the first rule had by far the best performance. The first rule usually gives the highest coverage and the lowest error rate. As an example, when we compared two rules generated for the same problem, rule 1 performed better than rule 2 in coverage (64.5% vs. 9.7%), false-negatives rate (7.4% vs. 18.9%), complexity (2 levels vs. 3 levels) and accuracy (92.6% vs. 81.1%). The performance of both engines were the same in error rate. For engine 2, it follows from engine 1 that the top rule is also selected for evaluation for consistency in the overall evaluation process.

6   Results

The data came from an aerospace domain that contained several parameters from the operation of the aircraft main engines and the auxiliary power unit. Results of analyzing four data groups are presented below. Data group 1 consisted of 34 data sets of auxiliary power unit described by 7 attributes; they are divided into individual data sets by aircraft identification number. Data group 2, also represented auxiliary power unit data, consisted of four data sets for four aircraft from data group 1, but described by 55 attributes. Data group 3 consisted of four sets of main engines cruise data for four aircraft. This group consisted of 52 attributes. Data group 4 consisted of four sets of main engines cruise performance data for the four aircraft in data group 3. This group consisted of 53 attributes.

Table 2 summarizes the results of data analysis obtained with the two data analysis engines in IMAFO. This table reports the averages of the differences in the comparison measures obtained for the four data groups. A positive value indicates an overall increase in the comparison measure between the original and the new representation space, for the data set(s) in the data group; a negative value indicates an overall decrease.

For engine 1, the results in Table 2 shows improvements in performance measures with the new representation space. Each comparison measure improved in two to three data groups. In other words, for two data groups, coverage increased (data groups 2 and 4) and error rate decreased (data groups 2 and 4). In addition, false-negatives rate decreased for three data groups (2, 3 and 4) and complexity decreased for three data groups (1, 2 and 3). Finally, accuracy increased in data groups 2, 3 and 4.

For engine 2, the results shows similar improvements. While coverage increased for three data groups (2, 3 and 4), error rate decreased for the same data groups. On the other hand, false-negatives rate decreased for two data groups (2 and 4). As for complexity, it decreased for three data groups (1, 2 and 3). Accuracy increased for three data groups (2, 3 and 4).

Table 2. Averages of Results of Experimentation

Engine	Comparison Measures	Cov(%)	ER(%)	FN(%)	RC	A(%)
Engine 1	Data Group 1	-3.76	1.44	1.8	-0.21	-1.61
	Data Group 2	5.86	-0.94	-1.79	-0.19	2.01
	Data Group 3	-1.04	5.68	-1.14	-0.26	1.04
	Data Group 4	2.39	-2.84	-0.09	0.08	0.09
Engine 2	Data Group 1	-2.92	3.17	0.29	-0.17	-0.49
	Data Group 2	0.49	-0.50	-0.60	-0.29	0.53
	Data Group 3	0.48	-0.48	0.10	-0.44	0.07
	Data Group 4	2.39	-2.93	-0.13	0.13	0.11

Cov=Coverage, ER=Error Rate, FN=False Negatives,
RC=Rule Complexity, and A=Accuracy.

When using engine 1 of IMAFO, changing the representation space showed slightly greater magnitude of improvement than using engine 2. The improvement in coverage using engine 1 ranged from 2.39% to 5.86% while using engine 2, the improvement ranged from 0.48% to 2.93%. The decrease in error rate for engine 1 ranged from 0.94% to 2.84% while for engine 2, the decrease ranged from 0.48% to 2.93%. The change in false-negatives rate using engine 1 ranged from 0.09% to 1.79% while using engine 2, the change ranged from 0.13% to 0.60%. The average ranges of improvement in complexity were 0.19 to 0.35 rules and 0.17 to 0.44 rules using engines 1 and 2, respectively. The increase in accuracy for engine 1 ranged from 0.09% to 2.01% while for engine 2, the increase ranged from 0.07% to 0.53%.

In general, the new representation space resulted in improvements in the data analysis. On average, while it does not always improve analysis results in the comparison meas- ures, no comparison measure has shown constant deterioration with the data groups used in the experimentation.

7   Conclusions

The main goal of a constructive induction approach is to understand what knowledge from the application domain can improve the representation space. One is also interested to investigate the effects of incorporating domain knowledge on the constructed feature space and any improvements in accuracy and predictability of the learned knowledge. Our experiments have shown that knowledge-driven constructive induction helps to incorporate domain knowledge into the measurement space and create a new representation space. However, the procedure for incorporating domain knowledge and the process of evaluating the results may require involvement of domain experts. Although we did not investigate how using a new representation space may speed up the knowledge discovery we found that constructive induction provides means to generate more reliable knowledge.

We have designed and implemented a constructive induction facility into our existing data analysis system. This data analysis tool has been successfully applied in several domains, including semiconductor manufacturing. With the addition of the constructive induction facility, we hope to obtain improvements in predictive accuracy in data analysis. In our experimentation plan, we tested and evaluated the performance of our constructive induction using four groups of data, containing 46 data sets. A significant contribution of this system is that the rules generated can be easily interpreted by the domain expert or can be incorporated into an expert system. The comparison between the existing induction system and the one with constructive induction facility was based on change of the representation space that resulted in improvements in coverage, error rate, false-negatives rate, rule complexity and accuracy.

According to the results based on the five comparison measures, the performance of our data analysis tool for learning new concepts improved with the addition of the constructive induction facility. The user is now able to select the type of relation (operator) to be used to create new attributes and generate a new representation space. This gives engineers and data analysts a way for generating data descriptions that are most suitable based on the available domain knowledge.

We conclude that with knowledge-driven constructive induction and creation of a new representation space we can obtain more accurate and more reliable results in data analysis. There are a number of issues yet to be further investigated: (i) how can generation of new attributes be optimized before the data is analyzed, (ii) is it possible to automate the process of creating new attributes, by taking into account the data characteristics and domain knowledge, and (iii) expansion of our work to include string and logical expressions.

Acknowledgements

The authors would like to thank Riad Hartani and Sylvain Letourneau for providing useful comments on an earlier version of this paper. The preliminary design of our constructive induction mechanism was done by Serge Oliveira.

References

1. Callan, J.P., Utgoff, P.E.: Constructive Induction on Domain Information. Proceedings of AAAI-91, AAAI Press (1991), Vol. 2, 614-619
2. Donoho, S., Rendell, L.: Constructive Induction Using Fragmentary Knowledge, Proceedings of AAAI-91, AAAI Press, (1991), Vol. 2, 614-619
3. Elio R. and Watanabe L.: An Incremental Deductive Strategy for Controlling Constructive Induction in Learning from Examples, Machine Learning, Kluwer Academic Publishers, 7 (1), (1991), 7-44
4. FamiliA., and Turney P.: Intelligently Helping Human Planner in Industrial Process Planning, AIEDAM, 5(2), (1991), 109-124
5. Fawcett T.E.: Knowledge-Based Feature Discovery for Evaluation Functions, Computational Intelligence, 12(1), (1996), 42-64
6. Hu Y-J.: Constructive Induction: A Preprocessor, Proceedings of Canadian AI Conference, Toronto, Canada, Springer-Verlag, (1996), 249-256
7. Lenat D.B.: The Role of Heuristics in Learning by Discovery: Three Case Studies, in R.S. Michalski, J.G. Carbonell, and T.M. Mitchell, eds., Machine Learning: An Artificial Intelligence Approach I, Morgan Kaufmann, Palo Alto, CA, (1983), 243-306
8. Matheus C.J., and Rendell L.A.: Constructive Induction on Decision Trees, Proceedings of IJCAI-89, Detroit, MI, AAAI Press, (1989), Vol.2, 645-650
9. Mena, J.: Automatic Data Mining, PC AI, 10(6), (1996), 16-20
10. Michalski, R.S., et al: The Multi-Purpose Incremental Learning System AQ15 and its Testing Application to Three Medical Domains, Proceedings of AAAI-86, Philadel- phia, PA, AAAI Press, (1986), 1041-1045
11. Nunez, M.: The Use of Background Knowledge in Decision Tree Induction, Machine Learning, Kluwer Academic Publishers, 6 (3), (1991), 231-250
12. Pagallo, G.: Learning DNF by Decision Trees, Proceedings of IJCAI-89, Detroit, MI, AAAI Press, (1989), Vol.2, 639-644
13. Quinlan, J. R.: C4.5: Programs for Machine Learning, Morgan Kaufmann Publishers, San Mateo, CA, (1993)
14. Rendell, L. R.: Learning Hard Concepts Through Constructive Induction: Framework and Rational, Report No. UIUCDCS-R-88-1426, Dept. of Computer Science, Univer- sity of Illinois, Urbana, Il. (1988)
15. Rieger, A.: Data Preparation for Inductive Learning in Robotics, IJCAI Workshop on Data Engineering for Inductive Learning, (1995), 70-78
16. Wnek, J., and Michalski, R.S.: Hypothesis-Driven Constructive Induction in AQ17-HCI: A Method and Experiments, Machine Learning, 14 (2), (1994), 139-168

Oblique Linear Tree

João Gama

LIACC, FEP - University of Porto
Rua Campo Alegre, 823
4150 Porto, Portugal
Phone: (+351) 2 6001672 Fax: (+351) 2 6003654
Email: jgama@ncc.up.pt
WWW: http://www.up.pt/liacc/ML

Abstract. In this paper we present system *Ltree* for proposicional supervised learning. *Ltree* is able to define decision surfaces both orthogonal and oblique to the axes defined by the attributes of the input space. This is done combining a decision tree with a linear discriminant by means of constructive induction. At each decision node *Ltree* defines a new instance space by insertion of new attributes that are projections of the examples that fall at this node over the hyper-planes given by a linear discriminant function. This new instance space is propagated down through the tree. Tests based on those new attributes are oblique with respect to the original input space. *Ltree* is a probabilistic tree in the sense that it outputs a class probability distribution for each query example. The class probability distribution is computed at learning time, taking into account the different class distributions on the path from the root to the actual node. We have carried out experiments on sixteen benchmark datasets and compared our system with other well known decision tree systems (orthogonal and oblique) like *C4.5*, *OC1* and *LMDT*. On these datasets we have observed that our system has advantages in what concerns accuracy and tree size at statistically significant confidence levels.

Keywords: Oblique Decision Trees, Constructive Induction, Machine Learning.

1 Introduction

In Machine Learning most research is related to building simple, small, and accurate models for a set of data. For propositional problems of supervised learning a large number of systems are now available. Most of them use a divide and conquer strategy that attacks a complex problem by dividing it into simpler problems and recursively applying the same strategy to the sub-problems. Solutions of sub-problems can be combined to yield a solution of the complex problem. This is the basic idea behind well known decision tree based algorithms: Quinlan's ID3, Kononenko's et al. Assistant, Breiman's et al. CART[2] and Quinlan's C4.5[16]. The power of this approach comes from the ability to split the hyperspace into subspaces and each subspace is fitted with different functions. The

X. Liu, P. Cohen, M. Berthold (Eds.): "Advances in Intelligent Data Analysis" (IDA–97)
LNCS 1280, pp. 187–198, 1997. © Springer–Verlag Berlin Heidelberg 1997

main drawback of this approach is its instability with respect to small variations of the training set[9].

The hypotheses space of ID3 and its descendants are within the DNF formalism. Classifiers generated by those systems can be represented as a disjunction of rules, each rule being a conjunction of conditions based on attribute values. Conditions are tests held on one of the attributes and one of the values of its domain. These kinds of tests correspond, on the input space, to a hyper-plane that is orthogonal to the axes of the tested attribute and parallel to all other axis. The regions produced by these classifiers are all hyper-rectangles.

A linear discriminant function[1] is a linear composition of the attributes where the sum of squared differences between class means is maximal relative to the internal class variance. It is assumed that the attribute vectors for examples of class A_i are independent and follow a certain probability distribution with probability density function f_i. A new point with attribute vector $\mathbf{x}$ is then assigned to that class for which the probability density function $f_i(x)$ is maximal. This means that the points for each class are distributed in a cluster centered at μ_i. The boundary separating two classes is a hyper-plane and it passes through the mid point of the two centers[5]. If there are only two classes, then one hyper-plane is needed to separate the classes. The general case of $\mathbf{q}$ classes, $\mathbf{q\text{-}1}$ hyper-planes are needed to separate the classes. Linear discriminant is a one step procedure, since it does not make a recursive partition of the input space. The advantage of these kind of systems is their ability to generate decision surfaces with arbitrary slopes. It is a parametric approach: it is assumed that classes can be separated by hyper-planes, and the problem is to determine the coefficients of the hyper-plane. Although parametric approaches are often viewed as being an arbitrary imposition of the model assumptions to the data, discriminant functions due to the small number of free parameters are stable with respect to small variations on the training set[1].

Learning can be a hard task if the attribute space is inappropriate to describe the target concept, given the bias of an algorithm. To overcome this problem some researchers propose the use of *constructive learning*. Constructive induction[11] discovers new features from the training set and transforms the original instance space into a new high dimensional space by applying attribute constructor operators. The difficulty is how to choose the appropriate operators for the problem in question. In this paper we argue that in domains described at least partially by numerical features *discriminant analysis* is a useful tool for constructive induction[18].

The system that we present in this paper, *Ltree*, is in the confluence of these 3 areas. It explores the power of divide and conquer methodology from decision trees and the ability of generating hyper-planes from the linear discriminant. It

---

[1] Throughout this paper we adopt the following notation. The training set consists of $\mathbf{N}$ examples drawn from $\mathbf{q}$ known classes. For each class there are n_i examples, each example is defined by $\mathbf{p}$ attributes and is represented by the vector $x = (x1, ..., xp)$. The sample mean vector for each class is represented by μ_i. The sample covariance matrix for the $\mathbf{i}$ class is S_i, the pooled covariance matrix $\mathbf{S}$ is given by $\sum_{i=1}^{q} \frac{n_i * S_i}{n-q}$.

integrates both using constructive induction.

In the next section we describe the motivation behind *Ltree* and we give an illustrative example using Iris dataset. Section 3 presents in detail the process of tree building and pruning. Section 4 presents related work namely from the area of oblique decision trees. In section 5 we perform a comparative study between our system and other oblique trees on sixteen benchmark datasets from the StatLog and UCI repositories. Last section presents conclusions of the paper.

2   An illustrative example

2.1   Motivation

Consider an artificial two class problem defined by two numerical attributes, which is shown in Fig 1a.

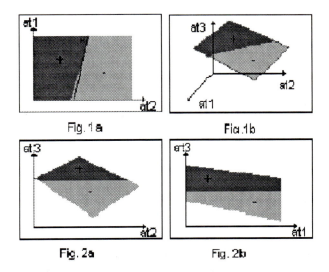

Fig. 1a   Fig. 1b

Fig. 2a   Fig. 2b

Running C4.5, we obtain a decision tree with 65 nodes. Obviously this tree is much more complex than expected. By analyzing paths in the tree we systematically find tests on the same attribute. One of such paths, from the root to a leaf, is given:

IF $at2 \leq 0.398$ **AND** $at1 \leq 0.281$ **AND** $at2 > 0.108$ **AND** $at1 > 0.184$ **AND** $at2 \leq 0.267$ **AND** $at2 > 0.189$ **AND** $at1 \leq 0.237$ **AND** $at2 \leq 0.218$ **AND** $at1 > 0.2$ **THEN CLASS** +

This means that the tree is making an approximation to an oblique region by means of a staircase-like structure. Running a linear discriminant procedure, we get one discriminant:

$$H = 5 - 18 * at1 + 8 * at2$$

Line H is the hyper-plane generated by the linear discriminant. A new attribute (at3) is generated by projecting the examples over this hyper-plane. The new instance space is given in Fig. 1b. In the new instance space a hyper-rectangle orthogonal to the new axis (Fig. 2a e 2b) is able to split classes using only one test. Running C4.5 on this new dataset, we obtain the following tree:

IF $at3 \leq 0$ THEN CLASS -
IF $at3 > 0$ THEN CLASS +

This is the smallest possible tree that discriminates the classes. The test is orthogonal to the axis defined by attribute at3 and parallel to the axes defined by at2 and at1. Rewriting the rule "**IF** $at3 \leq 0$ **THEN** *CLASS* -" in terms of the original space (that is in terms of at1 and at2) we get: "**IF** $8 * at2 \leq 18 * at1 - 5$ **THEN** *CLASS* -". This rule defines a relation between attributes, thus is oblique to the at1 and at2 axes. This simple example illustrates one fundamental point: constructive induction based on combinations of attributes extends the representational power and overcomes limitations of the language bias of decision tree learning algorithms.

We have implemented a two step algorithm, that explores this idea as a *preprocessing* step. Given a dataset, a linear discriminant builds the hyper-planes. New attributes are created by projecting all the examples over the hyper-planes. The transformed dataset is then passed to C4.5. We refer to this algorithm as *C45Oblique*. It was also used in our experimental study with very promising results.

Ltree dynamically explores the constructive step of *C45Oblique*. At each decision node *Ltree* builds new attributes based on linear combinations of the previous ones. If the best orthogonal axis involves one of the new attributes, the decision surface is oblique with respect to the original axes. There are two new aspects in our contribution to the state of the art on oblique decision trees. The first one is that the new attributes are propagated down through the tree. This allows non-linear fittings, because the new attributes build at lower nodes contain terms based on attributes built at previous nodes[2]. The second aspect is that the number of attributes is variable along the tree, including for two nodes at the same level. Another aspect is that this is a probabilistic decision tree in the sense that when classifying an example the tree outputs a class probability distribution that takes into account, not only the distribution at the leaf where the example falls, but a combination of the distributions of each node from the path that the example follows.

---

[2] We call them non linear, in the sense that they can not be discovered using only the primitive attributes.

2.2 The Iris data

To illustrate the basic ideas behind *Ltree*, we will use the well-known Iris dataset. The tree generated by *Ltree* is:

> **IF** *Linear_Test_5* ≤ 0.5 **THEN** Class = Setosa
> **IF** *Linear_Test_5* < 0.5
>   **IF** *Linear_Test_7* ≤ 0.635 **THEN** Class = Virginica
>   **IF** *Linear_Test_7* < 0.635 **THEN** Class = Versicolor

where
$Linear\_Test\_5 = -31.5 - 3.2 * Sepal\_length - 3.5 * Sepal\_width + 7.5 * Petal\_length + 14.7 * Petal\_width$
$Linear\_Test\_6 = -13.3 + 7.7 * Sepal\_length + 16.6 * Sepal\_width - 21.5 * Petal\_length - 24.3 * Petal\_width$
$Linear\_Test\_7 = -16.6 - 3.5 * Sepal\_length - 5.5 * Sepal\_width + 6.9 * Petal\_length + 12.4 * Petal\_width + 0.03 * Test\_5 - 0.03 * Test\_6$

This is a tree with 5 nodes. It misclassifies 2.7[3]. For the same data, C4.5 generates a tree with 11 nodes with 4.7

In this example, *Ltree* proceeds as follows: at the root of the tree, there are 4 attributes and 3 classes. For each class, there are 50 examples. Two hyper-planes are constructed based on linear combinations of the four original attributes. All examples are projected over the two hyper-planes and the instance space is reconstructed by the insertion of two new attributes (*Ltree* calls them *Linear_Test_5* and *Linear_Test_6*). Using the *Gain Ratio* as criteria, the attribute that best splits the data is chosen: in this case the new attribute *Linear_Test_5*. Since the tested attribute is continuous, the original dataset is divided into two distinct datasets. In the first one, all examples belong to the same class: *Ltree* returns a leaf and the splitting process terminates. The second dataset, has 6 attributes and two classes. One hyper-plane is generated, and the examples projected over the hyper-plane. A new attribute is built and the new instance space has 7 attributes and two classes. Now, the attribute that best splits the data is the *Linear_Test_7*, which contains terms based both on the original attributes and on the new attributes built at the root of the tree. This attribute splits the data into two datasets with only one class. The process of generating the tree terminates.

3 Growing the Tree

Ltree is a top down inductive decision tree system. The main algorithm is similar to many other programs from the TDIDT family, except for the construction of new attributes at each decision node.

---

[3] In a 10 Fold Cross Validation.

At each decision point, *Ltree* dynamically computes new attributes. New attributes are linear combinations of all ordered attributes of the examples that fall at this decision node. To build new attributes the system considers at each decision point, different number of classes. Taking into account class distribution at this node, *Ltree* only considers those classes whose number of examples is greater than the number of attributes[4]. Suppose that in a decision node *Ltree* considers q_{node} classes ($q_{node} \leq q$). Applying the linear discriminant procedure described below, we get $q_{node} - 1$ hyper-planes. The equation of each hyper-plane is given by[7]:

$$H_i = \alpha_i + \sum_j \beta_{ij} * x_j$$
$$\text{where}$$
$$\alpha_i = -\tfrac{1}{2}\mu_i^T S^{-1} \mu_i \text{ and } \beta_i = S^{-1}\mu_i$$

Ltree build $q_{node} - 1$ new attributes. All the examples of this node are extended with the new attributes. Each new attribute is given by the projection of the example over the $hyper-plane_i$. The projection is computed as the dot product of the example vector **x** by the coefficients β_i of the $hyper-plane_i$. Because class distribution varies along the tree, the number of new attributes is variable. Two different nodes (also at the same level) may have different number of attributes. As the tree grows and the classes are discriminated, the number of new attributes decreases. New attributes are propagated down the tree. This is the most original aspect of *Ltree*.

It is known that building the optimal tree (in terms of accuracy and size) for a given dataset is a NP complete problem. In this situation we must use heuristics to guide the search. A splitting rule typically works as a one-step lookahead heuristic. For each possible test, the system hypothetically considers the subsets of data obtained. Choose the test that maximizes (or minimize) some heuristic function over the subsets. By default *Ltree* uses *Gain Ratio*[16] as the splitting criteria. A test on a *nominal* attribute will divide the data into as many subsets as the number of values of the attribute. A test on a *continuous* attribute will divide the data into two subsets: *attribute_value > cutpoint* and *attribute_value ≤ cutpoint*. To determine the *cut point*, we follow a process similar to C4.5[16].

The usual stopping criteria for decision trees is: stop building the tree when all the examples that fall at a decision node are from the same class. In noisy domains, more relaxed rules are needed. *Ltree* uses the following rule: stop growing the tree if the percentage of the examples from the majority class is greater than a user defined parameter (by default 95%). If there are no examples at a decision node, *Ltree* returns a leaf with the class distribution of the predecessor of this node.

In spite of the positive aspects of divide and conquer algorithms, one should be concerned about the statistical consequences of dividing the input space. Dividing the data can improve the bias of an estimator, because it allows fine fitting

---

[4] Because data underfits the concept, see Breiman et al. [2]

to the data, but in general increases the variance. The use of soft thresholds is an example of a methodology that tries to minimize the effects of splitting the data[5]. *Ltree* uses a *smoothing* process[4] that usually improves performance of tree based classifiers. When classifying a new example, the example traverses the tree from the root to a leaf. The class attached to the example takes into account not only the class distribution at the leaf, but all class distributions of the nodes in the path. That is, all nodes in the path contribute to the final classification. Instead of computing class distribution for all paths in the tree at classification time, as it is done in [15], *Ltree* computes a class distribution for all nodes when growing the tree. This is done recursively, taking into account class distributions at the current node and at the predecessor of the current node. At each node j, *Ltree* combines both class distributions using the formula:

$$ClDist[i]_{Root} = Freq[i]_{Root}$$
$$ClDist[i]_j = (ClDist[i]_{j-1} + W * Freq[i]_j)/(1 + W)$$

where $Freq[i]_j$ is the relative frequency of examples from class **i** at node **j** and **W** is a weight that the user can set (by default $W = 1$). Classification done using smoothed class distributions is more accurate than when computed from the examples that fall at the leaves[4]. Both processes, pruning the tree and missing values, explore the smoothed class distributions. Another side effect is when considering problems with cost matrices: there are very simple algorithms for minimizing costs that use class probabilities[12].

Pruning is considered to be the most important part of the tree building process at least in noisy domains. Statistics computed at deeper nodes of a tree have low level of significance due to the small number of examples that fall at these nodes. Deeper nodes reflect too much the training set (overfitting) and increase the error due to the variance of the classifier. Several methods for pruning decision trees are presented in literature[2, 6, 16]. The process that we use exploits the way in which *Ltree* computes class probability distributions at each node. Usually, for example in C4.5, pruning is a process that increases the error rate on the training data. In our case, this is not necessarily true. The class that is assigned to an example takes into account the path from the root to the leaf, and is often different from the majority class of the examples that fall at one leaf. At each node, *Ltree* considers the *static error* and the *backed-up error*. *Static error* is the number of misclassifications considering that all the examples that fall at this node are classified from the majority class taken from the class distribution at this node. *Backed-up error* is the sum of misclassifications of all subtrees of the current node. If *Backed-up error* is greater or equal than *Static error* then the node is replaced by a leaf with the class distribution of the node.

When using a tree as classifier, the example to be classified passes down through the tree. At each node a test based on the values of the attributes is performed. If the value of the tested attribute is not known (often in real data some attribute values are unknown or undetermined) the procedure cannot

---

[5] The propagation down of the new attributes is also a more sophisticated form of soft thresholds.

choose the path to follow. Since a decision tree constitutes a hierarchy of tests, the *unknown problem* has special relevance on this type of classifiers. *Ltree* passes the example through all branches of the node where the unknown attribute value was detected[4]. Each branch outputs a class distribution. The output is a combination of the different class distributions that sum to 1.

4   Related Work

In [2], Breiman et al. suggest the use of linear combination of attributes instead of using a single attribute. The algorithm incorporated in CART, cycles through the attributes at each step, searching for an improved linear combination split. Reconstruction of the input space by means of new attributes based on combinations of original ones, appears as a *pre-processing* step in Yip and Webb [18]. Yip's system CAF incorporates new attributes based on canonical discriminant analysis[6]. They show that such techniques can improve the performance of machine learning algorithms. *Ltree* uses a similar strategy at each decision node. The advantage is that for each sub-space local interdependencies between attributes are captured. The FACT system [10] uses a recursive partition of the input space using linear discriminant functions. The advantage of *Ltree* is the downwards propagation of the new attributes through the tree.

Similar to CART, OC1[13], uses a hill-climbing search algorithm. Beginning with the best axes-parallel split, it randomly perturbs each of the coefficients until there is no improvement in the impurity of the hyper-plane. As in CART, OC1 adjusts the coefficients of the hyper-plane individually finding a local optimal value for each coefficient at a time. The innovation of OC1 is a randomization step used to get out of local minima. OC1 generates binary trees. The implementation that we have used in the comparative study accepts numerical descriptors only. Linear Machine Decision Tree (LMDT) of Brodley and Utgoff [3], uses a different approach. Each internal node in a LMDT tree is a set of linear discriminant functions that are used to classify an example. The training algorithm repeatedly presents examples at each node until the linear machine converges. Because convergence cannot be guaranteed, LMDT uses heuristics to determine when the node has stabilized. Trees generated by LMDT are not binary. The number of descendants of each decision node is equal to the number of classes that fall at this node. LMDT uses a randomizing initialization process. Two runs on the same data produce, in general, different results.

Ltree does not need to search for oblique hyper-planes. Conditions based on the new attributes corresponds to oblique surfaces in the original input space. This is the main advantage of *Ltree* over OC1 and LMDT.

---

[6] This is similar to *C45Oblique*

5   Experiments

In order to evaluate our algorithm we performed a 10 fold Cross Validation (CV) on sixteen datasets from the *StatLog repository*[7] and from the *UCI repository*[8]. Datasets were permuted once before the CV procedure. All algorithms were used with default settings. In each iteration of CV all algorithms were trained on the same train partition of the data[9]. Classifiers were also tested on the same test partition of the data.

In the first set of experiences we compared *C4.5* (release 8) with *C45Oblique*. In 11 datasets out of 16 an increase of performance was verified. At a confidence level of 90% from t paired tests, *C45Oblique* performs significantly better on 5 datasets and was significantly worse on 3 datasets (*Iono, Letter* and *Votes*).

In the second set of experiences, we have compared *Ltree* with two oblique decision trees: LMDT and OC1 that are available through Internet, and C4.5 which represents the state of the art in the area of decision trees. The table 1 presents, on the first line, the mean and standard deviation of error rates (in percentages). The second line refers to the tree size (in terms of number of nodes[10]).

In order to have statistical confidence on the differences we have compared the results using t paired tests. The null hypothesis is that *Ltree* is worse than each of the other algorithms. Column ttest refers to the significance between *Ltree* and LMDT, OC1, C4.5 and *C45Oblique* in that order. A sign + or - means that there was a significant difference at 90% of confidence level. A minus sign means that the null hypothesis was verified, that is, *Ltree* was worse than the given algorithm at this confidence level, a plus means that *Ltree* performs significantly better. Our system has an overall good performance. There are two main reasons. The main one is the use of linear combinations of attributes that capture inter-relations between variables and the propagation of these new attributes down through the tree. The second reason is the use of smoothing. However *Ltree* is significantly worse than C45Obl in one dataset (Letter). We think that this is due to the fact that C4.5 is much more sophisticated than *Ltree* in details (like default parameters and internal thresholds). These were verified on a set of experiments, where the constructive step was restricted to the root of the tree. In almost all of the experiences, the performance was verified to be worse than the standard *Ltree*. *Ltree* is a very recent system. Some details need to be refined, in order to take full advantage of the proposed methodology.

For each dataset we consider the algorithm with the lowest error rate. We designate *error margin* as the standard deviation from a Bernoulli distribution for the error rate of this algorithm. The distance of the error rate of an algorithm to the best algorithm on each dataset, in terms of the *error margin* give us

---

[7] http://www.ncc.up.pt/liacc/ML/statlog/index.html

[8] http://www.ics.uci.edu/AI/ML/Machine-Learning.html

[9] Slightly syntatic modifications of the data were necessary to run OC1. On *Letter* data LMDT returns *"Bus error"* on 9 of the ten iterations of CV.

[10] Including leaves.

Table 1. Error rates and Tree sizes

	ttest	*Ltree*	LMDT	OC1	C4.5	C45Obl
Australian	+	**13.9±4**	35.9±12	14.8±6	15.3±6	14.9±5
		14.2	72.6	13.2	35.5	21.1
Balance	+	10.4±4	12.0±4	10.4±3	34.6±4	**9.0±4**
		6.9	59.6	20.8	43.5	7.6
Breast(Wis)	+ + +	**3.1±2**	5.1±4	6.4±5	6.1±6	3.3±3
		3.0	19.2	19.4	29.0	**3.0**
Diabetes	+ +	**24.4±5**	31.9±7	28.0±8	24.7±7	24.9±7
		43.2	139.8	33.0	42.6	**20.2**
German	+	26.4±5	39.5±10	25.7±5	29.1±4	**25.1±4**
		18.5	85.0	24.4	151.1	110.8
Glass		34.7±8	37.0±10	32.8±11	**32.3±12**	34.6±10
		34.8	52.0	**16.4**	44.4	48.4
Heart	+ + +	**15.5±4**	25.9±5	29.3±7	21.1±8	19.2±10
		4.2	29.0	15.6	37.4	20.8
Hepatitis	+	**18.9±9**	20.3±13	25.5±13	24.9±13	24.2±15
		10.6	**6.6**	7.8	17.0	7.6
Iono	+	9.4±4	18.8±11	11.9±3	**9.1±5**	13.1±6
		15.0	11.0	**9.8**	27.4	19.6
Iris	+ +	**2.7±3**	6.0±7	7.3±6	4.7±5	3.4±4
		5.0	7.0	6.2	8.4	**5.0**
Letter	- -	18.2±1		16.7±1	**13.7±1**	15.6±1
		1522.0		2349.0	1962.6	1823.6
Segment	+	**2.9±1**	3.6±1	4.5±1	3.6±2	3.4±1
		54.6	50.9	**37.2**	81.2	77.2
Vehicle	+ +	22.5±5	21.8±5	34.1±5	28.8±4	**21.5±4**
		97.4	74.1	120.0	139.0	**57.0**
Votes	+	3.5±4	22.3±12	6.7±6	**2.8±3**	3.9±4
		4.4	11.6	**3.0**	10.8	4.0
Waveform	+ + +	17.8±2	20.2±3	22.1±1	23.9±2	**16.9±2**
		167.2	115.8	**61.8**	309.2	128.0
Wine	+ +	**2.8±3**	3.4±4	9.0±6	6.7±8	2.8±5
		5.0	**4.0**	9.2	9.6	5.0

an indication about the algorithms performance taking into account problem difficulty. The next table, presents the average of distances over all datasets where all algorithm runs.

	Ltree	*Lmdt*	*OC1*	*C4.5*	*C45Obl*
Average	0.39	5.57	3.46	3.68	0.75

There is strong evidence that if we need to use a learning classifier on new data, and do not have further information, we shall first use *Ltree*.

Ltree, LMDT, and OC1 generate trees with multivariate tests on its decision nodes. Tree size between these systems could be compared directly. Comparisons with univariate tests are more problematic because we must take into account the increase in complexity of combinations of attributes. Oblique trees (namely *Ltree*) are more compact than univariate ones. Although some nodes contain multivariate tests that are more difficult to understand, the overall tree is, on the whole, substantially simpler.

The training time for divide and conquer is often in orders of magnitude faster than gradient based algorithms[8]. This is one of the aspects why *Ltree* is significantly faster than other oblique decision trees like OC1 and LMDT that use gradient descendent approaches to determine hyper-planes. We verify, for example on Letter data, that OC1 takes about 10 hours on a Sun Sparc10 to run 1 fold of cross validation. Similar time was needed for LMDT. *Ltree* needs 7,25 hours to run the complete 10 folds of CV.

In [10] the authors of CART, presents a discussion on the use of linear combination of attributes in tree learning. They refer that *"Although linear combination splitting has strong intuitive appeal, it does not seem to achieve this promise in practice...."* This was confirmed in our experiments when comparing systems such as LMDT and/or OC1 with C4.5. The use of linear combination of attributes increases the number of degrees of freedom in order to obtain better fit to the data, but also increases the variance of the classifier. The process of downwards propagation of new attributes through the tree is a process that reduces the variance (see section 3). The increase of performance observed with *Ltree* is strongly due to this process.

6 Conclusions

We described a new method for the construction of oblique decision trees, combining a decision tree with a linear discriminant by means of constructive induction. There are two main features in our system. The first one is the use of constructive induction: when building the tree, new attributes are computed as linear combinations of the previous ones. As the tree grows and the classes are discriminated, the number of new attributes decreases. The new attributes that are created at one node are propagated down the tree. This allows non-linear fittings, because the new attributes built at lower nodes contain terms based on attributes built at previous nodes. The second aspect is that this is a probabilistic decision tree in the sense that, when classifying an example, the tree outputs a class probability distribution that takes into account not only the distribution at the leaf where the example falls, but a combination of the distributions of each node on the path that the example follows. In problems with numerical attributes, attribute combination extends the representational power and relaxes the language bias of univariate decision tree algorithms. In an analysis based on Bias-Variance error decomposition[9] *Ltree* combines a linear discriminant which is known to have high bias, but low variance with a decision tree known to have low bias but high variance. This is the desirable composition of algorithms. We use constructive induction as a way of extending bias. Using Wolpert's terminology [17] the constructive step performed at each decision node is a bi-stacked generalization. We have shown that this methodology can improve both accuracy and tree size comparatively to other oblique decision trees' systems without increasing learning time.

Acknowledgments: I thank P.Brazdil, L.Torgo, A.Jorge, and the two anonymous reviewers for useful suggestions. Gratitude is expressed to financial support under PRAXIS XXI project and Plurianual support attributed to LIACC.

References

1. Breiman, L.: Bias, Variance and Arcing Classifiers, Technical Report 460, Statistics Department, University of California
2. Breiman, L., Friedman, J., Olshen, R., Stone, C.: *Classification and Regression Trees*, Wadsworth International Group, 1984.
3. Brodley, C., Utgoff, P.: Multivariate Decision Trees, in *Machine Learning, 19*, Kluwer Academic Press, 1995.
4. Buntime,W,: A theory of learning Classification rules, PhD thesis, University of Sydney, 1990
5. Dillon, W., Goldstein,M.: *Multivariate analysis, Methods and Applications*, John Willey & Sons, 1984
6. Esposito, F., Malerba, D., Semeraro, G.: Decision Tree Pruning as a Search is the State Space, in *Machine Learning: ECML93*, Ed. Pavel Brazdil, 1993
7. Henery, B.: FORTRAN programs for Discriminant Analysis, Internal report, Dep. Statistics and Modelling Science, University of Strathclyde, 1993
8. Jordan, M. Jacob, R.: Hierarchical mixtures of experts and the EM algorithm, *Neural Computing, n.6*, 1994
9. Kohavi, R., Wolpert, D.: Bias plus variance decomposition for zero-one loss functions, in *Proceedings of 13 International Conference on Machine Learning - IML96*, Ed. Lorenza Saitta, 1996
10. Loh W., Vanichsetakul N.: Tree-Structured Classification Via Generalized Discriminant Analysis, *Journal of the American Statistical Association*, 1988
11. Matheus,C., Rendell, L.: Constructive Induction on Decision Trees, in *Proceedings of IJCAI 89*, 1989
12. Michie, D., Spiegelhalter,J. Taylor,C.: *Machine Learning, Neural and Statistical Classification*, Ellis Horwood, 1994
13. Murthy, S., Kasif, S., Salzberg, S.: A system for Induction of Oblique Decision Trees, *Journal of Artificial Intelligence Research*, 1994
14. Press, W., Teukolsky, S., Vetterling, W. and Flannery, B.: *Numerical Recipes in C: the art of scientific computing*, 2 Ed. University of Cambridge, 1992
15. Quinlan R.: Learning with continuous classes, in *Proceedings of AI92*
16. Quinlan, R.: *C4.5: Programs for Machine Learning*, Morgan Kaufmann, 1993
17. Wolpert, D.: Stacked Generalisation, *Neural Networks Vol.5*, 1992
18. Yip, S., Webb, G.: Incorporating canonical discriminant attributes in classification learning, in *Proceedings of the tenth Canadian Conference on Artificial Intelligence*, Morgan Kaufmann, 1994

Feature Selection for Neural Networks through Functional Links Found by Evolutionary Computation

S. Haring & J.N. Kok & M.C. van Wezel

Department of Computer Science
Leiden University, The Netherlands

Abstract. In this paper we describe different ways to select and transform features using evolutionary computation. The features are intended to serve as inputs to a feedforward network. The first way is the selection of features using a standard genetic algorithm, and the solution found specifies whether a certain feature should be present or not. We show that for the prediction of unemployment rates in various European countries, this is a succesfull approach. In fact, this kind of selection of features is a special case of so-called functional links. Functional links transform the input pattern space to a new pattern space. As functional links one can use polynomials, or more general functions. Both can be found using evolutionary computation. Polynomial functional links are found by evolving a coding of the powers of the polynomial. For symbolic functions we can use genetic programming. Genetic programming finds the symbolic functions that are to be applied to the inputs. We compare the workings of the latter two methods on two artificial datasets, and on a real-world medical image dataset.

1 Introduction

Feedforward networks are widely applied algorithms for function approximation and pattern classification. The most commonly applied neural network is the multi-layer feedforward network that is trained with the back-propagation learning rule. In this paper, we will perform *feature selection* (i.e. select and transform input features) for neural networks using evolutionary computation, in order to improve the performance of the neural networks.

Evolutionary computation (EC) is the generic term for a number algorithms that are based on an evolutionary concept, examples are genetic algorithms [Hol75] and genetic programming [Koz92]. EC is based on a set of potential solutions; in our application a solution is a configuration of functional links. In the evolutionary process repeatedly candidate solutions are recombined to new candidates to seek for good solutions. We will use both genetic algorithms and genetic programming to find good functional links.

Feature selection seems to be well suited to be tackled with evolutionary computation. In the first place, there is a straightforward fitness value for the

X. Liu, P. Cohen, M. Berthold (Eds.): "Advances in Intelligent Data Analysis" (IDA–97)
LNCS 1280, pp. 199–210, 1997. © Springer–Verlag Berlin Heidelberg 1997

strings. Each string corresponds to a configuration of functional links that is to be used for input transformation. The error rate of the neural network that is achieved on the transformed problem is a measure for the quality of the functional links, and can therefore be used as fitness value. In the second place, we are seeking functions, that is symbolic expressions with different functional components and continuous parameters; we might thus view our problem as a mixture of a discrete and a continuous optimization problem. EC is especially suited for that kind of optimization problems.

Many combinations of neural networks and evolutionary computation can be found in the literature; an extensive overview is given in [Bra95]. Evolutionary computation has two major applications for neural networks: (i) it has been used as an alternative to learning rules like back-propagation to find appropriate connection weights [MD89, TSVdM93, YHLK94]; (ii) evolutionary computation has been used to determine good network connection schemes [MTH89, Kit90, Man93]. There are also studies that have a closer relation to our approach. Examples are the studies of [KS91] and [Rog91]. The paper [Rog91] combines Multivariable Adaptive Regression splines with genetic algorithms, in which incremental search is replaced by genetic search. No neural networks are used. The paper [KS91] enhances the GMDH method with the search for good nodes. Use is made of the minimal description length principle for the fitness of nodes, and each layer is separately optimized using multi-modal genetic optimization. We only select and transform features, and we do not try to find the network structure.

The remainder of this paper is structured as follows: in Section 2 a more precise description of the concept 'functional link' is given, and the three types of functional links that we use are described. In Section 3 results of our methods are given on artificial and real data, and Section 4 gives a short discussion.

2   Functional Links

To increase the capabilities of networks so-called *functional links* can be added [Pao89]. With functional links the input pattern space is transformed to a new pattern space. If the input layer of a perceptron consists of n functional links f_i, $i = 1, \ldots, n$, then input pattern x will be transformed to $(f_1(x), \ldots, f_n(x))$. Transfer functions can thus be seen as a special case of functional links.

The transformed problem may be simpler, such that for example it can successfully be solved with a single layer network. However, the choice of the functional links is crucial for this approach to succeed.

We propose to find good functional links automatically making use of *evolutionary computation*. To integrate functional links in the evolutionary computation concept a functional link is to be encoded. We study three encoding mechanisms: *binary encoding* for functional links that are either $f(x) = 0$ or $f(x) = x$, *polynomial encoding* for functional links that are polynomials, and *symbolic encoding* for functional links that are symbolic expressions. For each encoding mechanism, the class of the functional links is fixed (e.g. whether the functional link is a polynomial or a symbolic expression), but the exact form of the functional link is to be found by the evolutionary algorithm. We discuss the three kinds of encodings in turn.

It is straightforward to use binary encoding for the functional links that are $f(x) = 0$ or $f(x) = x$. In many regression and classification problems, the possible number of indicators (input variables) is too big for the amount of data that is available. Using all possible input features would not leave enough degrees of freedom in the model, and hence, could lead to over-fitting and poor generalization. So, in many applications, it is desirable to reduce the number of input indicators. This is not an easy task because the number of possible feature subsets grows exponentially with the available number of features. Using the functional link with binary encoding, we hope to arrive at a near-optimal feature subset. This method was originally proposed in [CL91].

Next we discuss the so-called polynomial encoding. A functional link has as input a pattern $x = (x_1, \ldots, x_D)$, and as output a scalar. Consider a functional link f that is a product of powers of the separate components x_i of the input pattern:

$$f(x) = \prod_{i=1}^{D} x_i^{n_i},$$

where n_i is an arbitrary integer. An example of such a functional link applies the function $x_1^2 x_2$ to input pattern (x_1, x_2). If we consider a simple perceptron, that has m of such functional links, then the separation plane that it constructs has the following form:

$$w_1 \prod_{i=1}^{D} x_i^{n_{i1}} + \ldots + w_m \prod_{i=1}^{D} x_i^{n_{im}} = 0,$$

where n_{ij} is the power that is applied to x_i in functional link j. The components of the polynomial are the functions of the functional links, and the coefficients of the polynomial are the weights of the network. We call these functional links *polynomial functional links*. Using polynomial functional links a simple perceptron constructs separation planes that are polynomials in the pattern space. Therefore polynomial functional links increase the capabilities of simple perceptrons.

It is relatively easy to encode a polynomial functional link. We only need to store the powers of the polynomial. We use an array of length D of integers, in which the i-th integer corresponds to n_i, the power of x_i, and $i = 1, \ldots, D$. In our algorithm we only consider polynomials with a maximum order of six. An string in the population encodes a transformation of a pattern space; that is a configuration of functional links. As each functional links is encoded by an array, an string is the concatenation of such arrays. In Figure 1 we illustrate a network with polynomial functional links, and its encoding as an arrays with integers. The strings that are normally used in genetic algorithms are binary strings; however, the concept of genetic algorithms can also be applied to strings of integers.

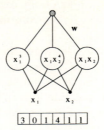

Fig. 1. Simple perceptron with functional links

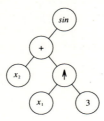

Fig. 2. Tree of $\sin(\text{plus}(\text{power}(x_1, 3), x_2))$

We put the restriction that all strings in a population have the same length. Hence, as a string represents a transformation of a pattern space, the dimension of the transformed pattern spaces is equal for all strings.

Next we come to the symbolic encoding. The concept of *genetic programming* was introduced by [Koz92]. In genetic programming each string in the population is a symbolic expression, which can be seen as a tree built of a number of basic functions. In this manner it is possible to represent complex programs by a single string.

We use symbolic expressions to encode functional links; we call links with such a coding scheme *symbolic functional links*. The basic functions that we use are: plus, minus, times, divide, power, exp, ln, sin, arcsin, cos, arccos, tan, and cotan. With these components we can build complex functions. Recall that a functional link acts on a multidimensional input pattern and outputs a scalar. An example of a symbolic expression that encodes a functional link is: $\sin(\text{plus}(\text{power}(x_1, 3), x_2))$; this expression is illustrated as a tree in Figure 2. The nodes in the tree are the functions, and the leaves are components of the input pattern. The basic functions that we use are unary and binary. The operands of the functions are again symbolic expressions or input pattern components. The power function is implemented as an binary operation of which the second operand is the power to which the operand is raised; the power is not an expression but is a number from a limited the set of valid powers.

Symbolic encoding yields functional links with more expressive power than polynomial functional links, but symbolic functional links give a larger search space. To limit the size of the search space, we have bound the depth of symbolic expression to a maximum of four.

An string encodes a pattern transformation; that is, it encodes configuration of functional links. Therefore in our algorithm an string is a series of symbolic expressions. Our method has the restriction that all strings consist of an equal number of symbolic expressions. Furthermore, each string has an additional functional link that always outputs value one; such a functional link thus acts as a bias term.

It might be so that symbolic expressions arise with no mathematical meaning, e.g. $\arcsin(x_1)$ is undefined for $x > 1$. The prevent the algorithm to get stuck in such inconsistencies, we have protected the basic functions, such that if an inconsistency arises, the functions return a zero. We have not included schemes that simplify the symbolic expressions.

3 Results

We have tested our method to find functional links on three kinds of datasets. First we applied the feature selection by binary encoding to a real-world problem: prediction of unemployment rates in various European countries with a multilayer perceptron neural network. A multilayer perceptron neural network is better suited for this task than linear methods because there is theoretical reason to believe that there exist non-linear relationships between various input features we used and the unemployment rate [PS56, vWH95]. The task of the GA was to select the most indicative subset of 11 possible input features.

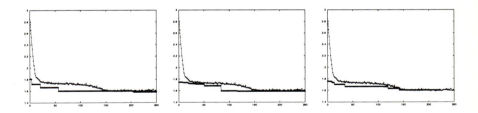

Fig. 3. Error curves during the evolution process. The lower line indicates the error of the best string in the population, the upper line the average error. From left to right: run 1, run 2 and run 3.

The GA used two crossover operators and one mutation operator. The first crossover operator used by the GA chooses a segment which is swapped between the two parent strings. The second crossover operator chooses a random number of cross-sites and swaps the values of the parent strings at these cross sites. The mutation operator mutates only one position in a parent string.

The error development of three arbitrary runs is shown in Figure 3. In each of the runs, the population size was 75 strings and the evolution went on for 250 generations. Both crossovers were applied once per generation. The mutation operator was applied twice per generation. From all graphs, it can be seen that the quality of the model improves during the evolution process. The best string resulting from each run is shown in Table (1). From this table, it can be seen that variables MONEYS (money stock), RLC (real labour cost), GCPCT (government consumption as a percentage of the gross national product), PCQ (private consumption quota), LF (labour force), and ET (total employment) are indicative for the unemployment rate because they are always selected by the GA, whereas PSB (profit share business sector), ICE-IP (increase in real total compensation per employee minus increase in productivity-index), and AIUdivWR (average income of the unemployed divided by the wage rate) are less indicative, because they are never selected.

	GPDV	MONEYS	RLC	GCPCT	PSB	ICE-IP	WSRE	PCQ	LF	ET	AIUdivWR
run1	1	1	1	1	0	0	1	1	1	1	0
run2	1	1	1	1	0	0	0	1	1	1	0
run3	0	1	1	1	0	0	1	1	1	1	0

Table 1. Best strings found by the GA in the various runs.

We obtain a significant decrease in error levels. Also other applications have shown the potential of this method (see for example [BSvW95, BOSvW94]).

Then we considered two artificial pattern classification problems. For both problems we know the optimal functional links; it is our goal to investigate whether an evolutionary algorithm can find these functional links. We also investigate whether one of both coding schemes is to be preferred over the other.

A third series experiments concern real-world medical images. We investigate if the proposed algorithm is of use for classifying image pixels.

For the second and third series of experiments we use populations containing one thousand strings, which we evolve a hundred generations. We apply tournament selection with four candidates per tournament; the crossover rate was 0.6, and the mutation rate was 0.1. Such a parameter setting is common in evolutionary computation; tests showed that this setting gave good results for our application.

The first experiment concerns a two-dimensional pattern classification problem with two classes (see Figure 4A). The two classes are distributed in clusters that are separable with a sinusoidal. The horizontal direction in the figure is

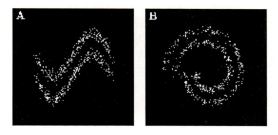

Fig. 4. Artificial classification problems.

the x_1-direction, and the vertical direction is the x_2-direction. The set of functional links $\{\sin x_1, x_2\}$ linearizes the classification problem. In the image only the training patterns have been shown; the test patterns are different but have been generated using the same protocol as the training patterns.

We test whether the optimal functional links can be found if symbolic encoding is used. We use three functional links per string (that includes the bias functional links that always output value one), i.e. the two-dimensional input patterns are transformed to three-dimensional patterns. The result of typical run is illustrated in Table 2. The table shows that initially the best string in the population corresponds to a rather unsuccessful configuration of functional links. In later generations the fitness value increases (or the error rate decreases), and eventually the optimal pattern transformation is found. Note that the weights of the simple perceptron indicate that the bias term is of no use for the considered pattern classification problem.

generation	error	functional links	weights links
1	0.234	$\cos x_1^2 + \arcsin(\log(\arcsin x_2)) - \arcsin x_2 x_1 - x_2$	0.343
		$\tan x_1$	6.64e-4
		1	-0.0805
10	0.142	$\exp(2x_1)$	0.510
		x_2	-0.861
		1	-0.502
100	0.001	x_2	-1.57
		$\sin x_1$	2.57
		1	0.000

Table 2. The result of a typical run on the "sinus data-set" if symbolic encoding is used. For each considered generation the best configuration of functional links is shown, together with the corresponding perceptron weights and the classification error on test patterns.

In a second experiment, which concerns the same data-set, we apply polynomial encoding. Again we set the dimension of the transformed pattern space to three. After a hundred generations we obtain a classification error of 0.007 on the test patterns (only slightly more than in the previous experiment). The functional links that give these results are the following: $\{x_2^3, x_2, x_1\}$. With these functions it is possible to construct exactly the first order Taylor expansion of the sinus function.

Next, we consider a two-spiral problem, of which the training patterns are illustrated in Figure 4B. A polar transformation transforms both spirals to straight lines. Such a transformation can be done with the following functional links: $\{\arctan(x_2/x_1), \sqrt{x_1^2 + x_2^2}\}$. We test whether with genetic programming these links can be found.

In the left half of Figure 5 we have plotted the average classification error rate and the minimum error rate in the population. Note that the fitness of an string equals one minus the classification error rate. Clearly, the average fitness gradually increases, and the best fitness value approximates one after a hundred generations. The functional links that yield the best transformation indeed carry out a polar transformation.

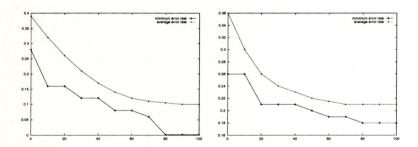

Fig. 5. The average and minimum classification error rate of a population during evolution. (The fitness of an string is one minus the classification error rate.) The left graph depicts the error development using symbolic encoding, the right graph depicts the error development using polynomial encoding.

We repeated the experiment with the two-spiral data-set, using polynomial encoding. We present results for the case that the two-dimensional input patterns are transformed to five-dimensional patterns. The right half of Figure 5 shows the course of the average and minimum error rate in the population during evolution. Although the error rates decrease in the evolution process, the eventual results are considerably worse than in the case of symbolic encoding. Apparently, polynomial functional links do not form a successful pattern transformation, in case of the considered pattern classification problem.

The reason behind the discrepancy in results is probably that it is practically impossible to do a polar transformation with polynomial functional links. This is evident form the fact that the function $\arctan(x_2/x_1)$ has a singularity around the origin: therefore it is hard to approximate with a polynomial around that point.

Now that we have tested our algorithm on artificial data-sets, we attempt real-world data. We use patterns from a medical image segmentation problem, where individual pixels have to be assigned to classes; each class corresponds to an object in the image. Training patterns and test patterns are derived from the images in Figure 6. On the left side of this figure, the original image is shown, on the right side the desired segmentation. For initial feature extraction we use Gaussian filters at multiple scales. From that point of view we apply a *multi-scale* or *multi-resolution* approach; a pixel is represented by features that are extracted with filters of various sizes. It has been shown that multi-scale approaches are successful for image segmentation [HVK94]. We convolve the input image with five Gaussian filters with scales of one, two, four, eight, and sixteen pixels. The resulting feature values, with inclusion of the original intensity, make up the feature patterns that we use in our algorithm. Hence, pixels are represented by six-dimensional feature pattern $(x_1, \ldots, x_6)$, which are transformed by the functional links.

It is not our aim to study if the algorithm gives better results than other pattern classification algorithms; we only use the medical segmentation problem as a test case for our algorithm. The algorithm tries to find functional links that successfully simplify the resulting pattern classification problem. We investigate several issues: (i) the influence of the encoding scheme, (ii) the influence of the dimension of the transformed pattern space (i.e. the number of functional links per string); and (iii) the nature of the evolved functional links (i.e. do they give insight in the problem?). For the parameters of the evolutionary algorithm we use the same values as in the experiments on artificial data.

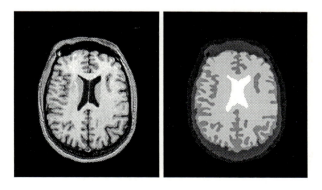

Fig. 6. Left: input image, right: desired segmentation.

Classification results on the test image for both encoding schemes are presented in Figure 8. With symbolic encoding we achieve surprisingly good results, which are competitive to results that we obtained in [HVK94]. Increase of the number of functional links per string has no effect from a certain point; an eight-dimensional transformed pattern space gives good results for the considered problem. (Note that the eighth dimension is the bias term.) The following set of seven functional links have been found by the algorithm:

$$\left\{ \begin{array}{c} \tan x_2 \\ \arcsin(\arcsin(\arccos x_4)) \\ \arccos(\cos x_3)/\sin(x_1 \times x_6) \\ \arcsin(\cos x_1) \\ \log(\tan x_3) - x_1 \\ \sin(\arctan(\arcsin x_1)) \\ \arctan(\cos^4 x_6) \end{array} \right\}$$

Figure 7 shows the original image shown in Figure 6 transformed through three functional links found by the evolutionary algorithm. These transformed images each highlight very specific features of the original image, making the segmentation process easier.

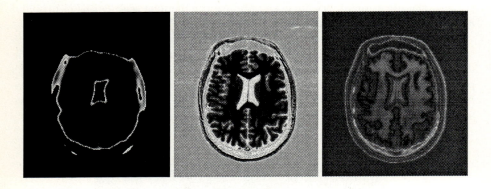

Fig. 7. Left: $\arcsin(\arcsin(\arccos x_4))$, middle: $\arcsin(\cos x_1)$, right: $\log(\tan x_3) - x_1$.

Figure 8 also shows that the results of polynomial encoding are not as good as the results of symbolic encoding.

4   Conclusions & Discussion

With evolutionary computation it is possible to find good functional links. We have applied the algorithm to two artificial problems and to two real-world datasets: one on unemployment rates and one on medical images. Concerning the

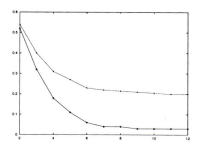

Fig. 8. The number of functional links per string versus the achieved error rate for both encoding schemes; the top graph refers to polynomial encoding, and the bottom graph refers to symbolic encoding. Both graphs concern the medical image segmentation problem.

two artificial data sets we conclude that, using symbolic encoding, the appropriate pattern transformations could be found in both cases. These transformations allowed for transformed pattern distributions that are separable with a simple perceptron. Polynomial encoding was less successful; apparently the approach depends on the used encoding scheme. Results on a medical image segmentation application suggest that symbolic encoding is more generally applicable.

We expect that the algorithm is especially applicable to areas where there is an (unknown) physical model behind the data. As the evolutionary algorithm searches for the appropriate model, the parameters of the model are being tuned with the neural network.

References

[BOSvW94] B. Back, G. Oosterom, K. Sere, and M. van Wezel. A comparative study of neural networks in bankruptcy prediction. In *Proceedings of the 10th Conference on Artificial Intelligence Research in Finland, Turku, Finland, 1994*, pages 140–148. Finnish Artificial Intelligence Society, 1994.

[Bra95]   J. Branke. Evolutionary algorithms for neural network design and training. In *Proceedings of the First Nordic Workshop on Genetic Algorithms and its Applications*, 1995.

[BSvW95] B. Back, K. Sere, and M. van Wezel. Bankruptcy prediction (choosing the best set of bankruptcy predictors. In *Proceedings of the Nordic Workshop on Genetic Algorithms, Waasa University, Finland, 1995*, pages 295–299, 1995.

[CL91]   Eric I. Chang and Richard P. Lipmann. Using genetic algorithms to improve pattern classification performance. In *Advances in Neural Information Processing Systems*, volume 3, pages 797–803. Morgan Kaufmann, 1991.

[Hol75]   J.H. Holland. *Adaptation in Natural and Artificial Systems*. The University of Michigan Press, Ann Arbor, 1975.

[HVK94]   S. Haring, M.A. Viergever, and J.N. Kok. Kohonen networks for multiscale image segmentation. *Image and Vision Computing*, 6(12):339–344, 1994.

[Kit90]   H. Kitano. Designing neural networks using genetic algortithms with graph generation. *Complex Systems*, 4:461–476, 1990.

[Koz92]   J. Koza. *Genetic Programming*. MIT Press, 1992.

[KS91]   H. Kargupta and R.E. Smith. System identification with evolving polynomial networks. In *Proceedings of the Fourth International Conference on Genetic Algorithms*, pages 370–376. Morgan Kaufmann, 1991.

[Man93]   M. Mandischer. Representation and evolution of neural networks. In *Proceedings of the Conference on Artificial Neural Networks and Genetic Algorithms*, pages 643–649, Heidelberg, 1993. Springer.

[MD89]   D.J. Montana and L. Davis. Training feedfoward networks using genetic algorithms. In *Proceedings of the International Joint Conference on Artificial Intelligence*, pages 762–767, 1989.

[MTH89]   G.F. Miller, P.M. Todd, and S.U. Hegde. Designing neural networks using genetic algorithms. In *Proceedings of the Third International Conference on Genetic Algorithms*, pages 379–384. Morgan Kaufmann, 1989.

[Pao89]   Y. Pao. *Adaptive Pattern Recognition and Neural Networks*. Addison Wesley, 1989.

[PS56]   T. Parsons and N.J. Smelser. *Study in the Integration of Economic and Social Theory*. Routledge and Keganpaul ltd., 1956.

[Rog91]   D. Rogers. G/SPLINES: A hybrid of friedman's multivariate adaptive regression splines (MARS) algorithm with Holland's genetic algorithm. In *Proceedings of the Fourth International Conference on Genetic Algorithms*, pages 384–391. Morgan Kaufmann, 1991.

[TSVdM93]   D. Thierens, J. Suykens, J. Vandewalle, and B. de Moor. Genetic weight optimization of a feedforward network controller. In *Proceedings of the Conference on Artificial Neural Networks and Genetic Algorithms*, pages 658–663, Heidelberg, 1993. Springer.

[vWH95]   J.A.M. van Wezel and M. Havekes. *Economie en Samenleving: een internationaal vergelijk van het arbeidsbestel (in Dutch)*. Lemma publishers BV, 1995.

[YHLK94]   B. Yoon, D.J. Holmes, G. Langholz, and A. Kandel. Efficient genetic algorithms for training layered feedforward networks. *Information Sciences*, 76:67–85, 1994.

Building Simple Models:
A Case Study with Decision Trees

David Jensen, Tim Oates, and Paul R. Cohen

Department of Computer Science
University of Massachusetts
Amherst, MA 01003
{jensen|oates|cohen}@cs.umass.edu

Abstract. Building correctly-sized models is a central challenge for induction algorithms. Many approaches to decision tree induction fail this challenge. Under a broad range of circumstances, these approaches exhibit a nearly linear relationship between training set size and tree size, even after accuracy has ceased to increase. These algorithms fail to adjust for the statistical effects of comparing multiple subtrees. Adjusting for these effects produces trees with little or no excess structure.

1   Introduction

Many induction algorithms construct models with unnecessary structure. These models contain components that do not improve accuracy, and that only reflect random variation in a single data sample. Such models are less efficient to store and use than their correctly-sized counterparts. Using these models requires the collection of unnecessary data. Portions of these models are wrong and mislead users. Finally, excess structure can reduce the accuracy of induced models on new data [8].

For induction algorithms that build decision trees [1, 7, 10], *pruning* is a common approach to remove excess structure. Pruning methods take an induced tree, examine individual subtrees, and remove those subtrees deemed unnecessary. Pruning methods differ primarily in the criterion used to judge subtrees. Many criteria have been proposed, including statistical significance tests [10], corrected error estimates [7], and minimum description length calculations [9].

In this paper, we bring together three threads of our research on excess structure and decision tree pruning. First, we show that several common methods for pruning decision trees still retain excess structure. Second, we explain this phenomenon in terms of statistical decision making with incorrect reference distributions. Third, we present a method that adjusts for incorrect reference distributions, and we present an experiment that evaluates the method. Our analysis indicates that many existing techniques for building decision trees fail to consider the statistical implications of examining many possible subtrees. We show how a simple adjustment can allow such systems to make valid statistical inferences in this specific situation.

X. Liu, P. Cohen, M. Berthold (Eds.): "Advances in Intelligent Data Analysis" (IDA–97)
LNCS 1280, pp. 211–222, 1997.   © Springer–Verlag Berlin Heidelberg 1997

2   Observing Excess Structure

Consider Figure 1, which shows a typical plot of tree size and accuracy as a function of training set size for the UCI `australian` dataset.[1] Moving from left-to-right in the graph corresponds to increasing the number of training instances available to the tree building process. On the left-hand side, no training instances are available and the best one can do with test instances is to assign them a class label at random. On the right-hand side, the entire dataset (excluding test instances) is available to the tree building process. C4.5 [7] and error-based pruning (the C4.5 default) are used to build and prune trees, respectively.

Note that accuracy on this dataset stops increasing at a rather small training set size, thereafter remaining essentially constant.[2] Surprisingly, tree size continues to grow nearly linearly despite the use of error-based pruning. The graph clearly shows that unnecessary structure is retained, and more is retained as the size of the training set increases. Accuracy stops increasing after only 25% of the available training instances are seen. The tree at that point contains 22 nodes. When 100% of the available training instances are used in tree construction, the resulting tree contains 64 nodes. Despite a 3-fold increase in size over the tree built with 25% of the data, the accuracies of the two trees are statistically indistinguishable.

Under a broad range of circumstances, there is a nearly linear relationship between training set size and tree size, even after accuracy has ceased to increase. The relationship between training set size and tree size was explored with 4 pruning methods and 19 datasets taken from the UCI repository.[3] The pruning methods are error-based (EBP – the C4.5 default) [7], reduced error (REP) [8], minimum description length (MDL) [9], and cost-complexity with the 1SE rule (CCP) [1]. The majority of extant pruning methods take one of four general approaches: deflating accuracy estimates based on the training set (e.g. EBP); pruning based on accuracy estimates from a pruning set (e.g. REP); managing the tradeoff between accuracy and complexity (e.g. MDL); and creating a set of pruned trees based on different values of a pruning parameter and then selecting the appropriate parameter value using a pruning set or cross-validation (e.g. CCP). The pruning methods used in this paper were selected to be representative of these four approaches.

Plots of tree size and accuracy as a function of training set size were generated for each combination of dataset and pruning algorithm as follows. Typically,

---

[1] All datasets in this paper can be obtained from the University of California–Irvine (UCI) Machine Learning Repository.
http://www.ics.uci.edu/ mlearn/MLRepository.html.

[2] All reported accuracy figures in this paper are based on separate test sets, distinct from any data used for training.

[3] The datasets are the same ones used in [4] with two exceptions. The `crx` dataset was omitted because it is roughly the same as the `australian` dataset, and the `horse-colic` dataset was omitted because it was unclear which attribute was used as the class label. Note that the `vote1` dataset was created by removing the `physician-fee-freeze` attribute from the `vote` dataset.

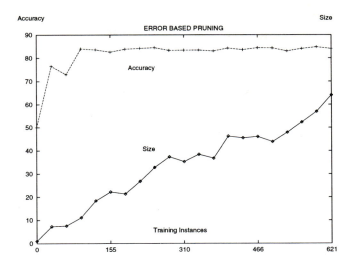

Fig. 1. Tree size and accuracy as a function of training set size for the australian dataset. C4.5 and error-based pruning were used to build and prune trees.

k-fold cross-validation is used to obtain estimates of the true performance of decision tree algorithms. A dataset, D, with n instances is divided into k disjoint sets, D_i, each containing n/k instances. Then for $1 \le i \le k$, a tree is built on the instances in $D - D_i$ and tested on the instances in D_i, and the results are averaged over all k folds [2]. That procedure was augmented for this paper by building trees on subsets of $D - D_i$ of various sizes, and testing them on D_i. Specifically, 20 subsets were created by retaining from 5% to 100% of the instances in $D - D_i$ in increments of 5%; standard k-fold cross-validation corresponds to the case in which 100% of the instances in $D - D_i$ are retained. The order of the instances in D was permuted prior to creating the $k = 10$ folds, and the instances to be retained were gathered sequentially starting with the first instance in $D - D_i$ for each level of data reduction. In this way, 10-fold cross-validated estimates of tree size and accuracy as a function of training set size were obtained.

This procedure was performed twice for each combination of dataset and pruning method, generating complete size and accuracy curves for two different permutations of the data, and the results were averaged. The goal was to reduce the inherent variability of cross-validated estimates of size and accuracy. Note that the same divisions of a given dataset were used for all of the pruning methods. With 19 datasets, 4 pruning methods, 20 levels of training set size, and 2 runs of 10-fold cross-validation at each level of training set size, the results reported in this section involved running C4.5 30,400 times.

For each plot generated by this procedure, the training set size at which accuracy ceased to grow was found by scanning the accuracy curve from left to right, stopping when the mean of three adjacent accuracy estimates was no more than 1% less than the accuracy of the tree based on all available training data (the right-most point on the accuracy curve). Averaging three adjacent accuracies makes the stopping criterion robust against random variations in the accuracy curve.[4] Bounding the absolute change in accuracy from below by 1% ensures that any reduction in tree size costs very little in terms of accuracy. Then, a linear regression of tree size on training set size was performed on the points in the tree size curve to the right of the training set size at which accuracy ceased to grow.

In general, additional tree structure is welcome as long as it improves classification accuracy, and it is unwelcome otherwise. Ideally, there will be no correlation between tree size and training set size once classification accuracy stops increasing. The results of a linear regression of tree size on training set size reports the probability, p, of incorrectly rejecting the null hypothesis that there is no such correlation (that the slope of the regression line is zero), the estimated slope of the regression line, and the amount of variance in tree size accounted for by training set size, R^2. When p is significant and R^2 is high, changes in training set size have predictable effects on tree size. When the slope of the regression line is large, the effects are strong.

For each combination of dataset and pruning method, we recorded the percentage of available training instances at which accuracy ceased to grow, results of the linear regression of tree size on training set size (p, slope, and R^2), the percentage decrease in tree size (Δ size) and the absolute difference in accuracy (Δ accuracy) between the tree built from all available training instances and the tree built from the number of instances at which accuracy ceased to grow. A summary of that information is given in Table 1, which shows for each pruning method the number of datasets for which accuracy peaked prior to seeing 100% of the available training instances (% Kept < 100), the number of datasets for which the relationship between tree size and training set size is statistically significant ($p < 0.1$), the number of datasets for which the relationship is both statistically significant and strong ($slope > 0.1$), and the means of R^2, Δ size and Δ accuracy for those datasets with significant p values.

For EBP, accuracy peaked prior to seeing 100% of the available training instances for 16 of the 19 datasets. Every one of those 16 datasets exhibited a significant relationship between tree size and training set size beyond the point at which accuracy stopped growing, and 12 of them were highly significant (at the 0.001 level). In 13 of the 16 datasets that exhibit a significant relationship, the slope of the regression line exceeds 0.1, indicating that at least one node is added for every 10% increase in the size of the dataset.

In spite of the fact that accuracy remains basically constant, tree size continues to grow as training set size does (the slope of the regression line is positive in all cases). The most remarkable feature of the EBP row in Table 1 is the R^2 column. Recall that $100 * R^2$ is the percentage of variance in tree size accounted

---

[4] We did not use the mean of the final three points on the accuracy curve minus 1% as the accuracy threshold because those points represent different training set sizes, and their mean is therefore not an estimate (robust or otherwise) of the accuracy of trees built on all available training instances.

Pruning Method	% Kept < 100	p < 0.1	Slope > 0.1	Mean R^2	Mean Δ size	Mean Δ accuracy
EBP	16	16	13	0.90	38.29	-0.14
REP	17	17	11	0.75	39.32	-0.32
MDL	18	17	13	0.88	44.03	-0.37
CCP	19	10	4	0.62	30.11	-0.06

Table 1. Summary of the effects of random data reduction for all of the pruning methods.

for by training set size. Across 16 datasets, the average R^2 is 0.90. This result is interesting for two reasons. First, it says that training set size has an extremely predictable effect on tree size. Increasing training set size invariably leads to larger trees; decreasing training set size invariably leads to smaller trees. Second, this effect is robust over a large group of datasets with widely varying characteristics. Regardless of the default accuracy, the number and types of attributes, the presence or absence of class and attribute noise, and differences in a number of other features along which the datasets vary, EBP does not appropriately limit tree size as training set size increases.

The results for REP and MDL are qualitatively the same as those for EBP. For REP, 17 datasets show a significant relationship between tree size and training set size (12 at the 0.001 level), 11 of the 17 datasets had a slope greater than 0.1, and the mean R^2 is 0.75. The average reduction in tree size obtainable via random data reduction is 39.32% with an average loss in accuracy of less than four tenths of one percent. Accuracy was higher with reduced training sets in 12 of the 17 cases. For MDL, 17 datasets had significant p values (14 at the 0.001 level), 13 of the 17 datasets had a slope greater than 0.1, and the average R^2 was 0.88. Trees based on reduced training sets were on average 44.03% smaller and less than four tenths of one percent less accurate. Note that for one dataset, **hypothyroid**, there is no significant relationship between tree size and training set size past the point at which accuracy stopped growing. In this one case, MDL appropriately limits tree size by not adding structure to the tree unless a concomitant increase in classification accuracy occurs.

The results for CCP indicate that it appropriately limits tree growth much more frequently than the previous three pruning methods. Accuracy peaked for all 19 datasets prior to seeing 100% of the available training instances. However, only about half of the time (10 out of 19 datasets) was there a significant relationship between tree size and training set size after accuracy stopped growing. CCP appropriately limits tree growth for 9 datasets, whereas EBP and REP never did so, and MDL did so once. For the 10 datasets that exhibited significant relationships between tree size and training set size, random data reduction still leads to substantially smaller trees (30.11% on average) with little loss in accuracy (less than one tenth of one percent on average).

3   Explaining Excess Structure

Why do three common pruning methods retain excess structure? Error-based pruning (EBP), reduced-error pruning (REP), and minimum description length pruning (MDL) almost invariably produce trees whose size increases as a function of training set size. Cost complexity pruning (CCP) produces this pathology as well, but less frequently and to a substantially lesser degree than the three other methods.

3.1   Selecting Among Multiple Trees

The three pruning methods which retain the most excess structure (EBP, REP, and MDL) share a general approach. For each node N_T that forms the root of a subtree, the methods calculate the value of an evaluation function f based on a sample of data S. We call the value of f the *score*, $x_T = f(N_T, S)$. The methods compare the score for the subtree rooted at N_T to a threshold value T. If $x_T > T$, the subtree is retained, otherwise it is pruned.[5] The threshold T could be determined in several ways; for EBP, REP, and MDL, the threshold is the score that N_T would receive if it were converted into a leaf node N_L, that is $T = x_L = f(N_L, S)$.

EBP, REP, and MDL can each be described in terms of this general approach. For example, EBP uses the training sample to calculate an adjusted error rate for a subtree and compares it to the adjusted error rate for a leaf. REP uses a pruning sample to calculate accuracy for the subtree and compares it to the accuracy of the subtree when it is converted into a leaf. MDL uses the training sample to calculate a description length for the subtree and compares it to the description length for the leaf.

Given this approach, when will pruning retain excess structure? Pruning will fail if, for a large proportion of subtrees, $x_T > T$ when the subtree does not improve accuracy. This can occur if the threshold is set too low. With an incorrectly low threshold, subtrees will be judged to be useful when they actually are not, and they will be retained.

Algorithms that use EBP, REP, and MDL all produce the score x_T in a way that almost guarantees T will be too low. They: 1) generate n possible subtrees; 2) produce a score for each subtree based on a data sample S; and 3) select the subtree with the maximum score. The pruning methods compare the maximum score $x_{max} = x_T$ to the threshold T.

The procedure above is used to both grow and prune trees. During the *growing* phase, algorithms select each decision node in a subtree by generating and evaluating many possible decision nodes. Each decision node uses a different attribute and a different partitioning of the values of that attribute. For a subtree with D total nodes, a data sample with A attributes, and attributes with an

---

[5] In many cases, the score x_T measures error, the inverse of accuracy, and a subtree is retained only if its score is *less than* a threshold. This transformation can be made to the discussion with no loss of generality.

average of P possible partitionings, algorithms select from approximately DAP possible subtrees. During the *pruning* phase, algorithms select each subtree by generating and evaluating many possible pruned subtrees. EBP, REP, and MDL each examine subtrees only after attempting to prune their constituent nodes. For subtrees with D' non-leaf nodes, algorithms select from $2D'$ possible subtrees during pruning, up to and including the entire subtree.

3.2   Why Selection Affects Scores

Why should selecting the maximum score affect the threshold T? Recall that any score x results from applying an evaluation function f to a tree and a data sample S. Suppose an algorithm examines n subtrees with scores $x_1, x_2, \ldots, x_n$. Each score x_i is the value of a random variable. That is, f is a function whose value is a real number determined by a specific sample S. An algorithm examines n subtrees and selects the one with the score $max(x_1, x_2, \ldots, x_n)$.

For simplicity and concreteness, consider a tree-building algorithm that examines two subtrees, with scores x_1 and x_2, and assume that their scores are random variables whose values are drawn from independent uniform distributions of integers $(0, \ldots, 6)$. The distribution of $max(x_1, x_2)$ is shown in Table 2. Each entry in the table represents a joint event with the resulting maximum score; for example, $(x_1 = 3 \land x_2 = 4)$ has the result, $max(x_1, x_2) = 4$. Because x_1 and x_2 are independent and uniform, every joint event has the same probability, $1/49$, but the probability of a given maximum score is generally higher; for example, $Pr(max(x_1, x_2) = 6) = 13/49$.

		x_1					
	0	1	2	3	4	5	6
0	0	1	2	3	4	5	6
1	1	1	2	3	4	5	6
2	2	2	2	3	4	5	6
x_2 3	3	3	3	3	4	5	6
4	4	4	4	4	4	5	6
5	5	5	5	5	5	5	6
6	6	6	6	6	6	6	6

Table 2. The joint distribution of the maximum of two random variables, each of which takes integer values (0...6).

For independent and identically distributed (i.i.d.) random variables $x_1, \ldots, x_n$, it is easy to specify the relationship between cumulative probabilities of individual scores and cumulative probabilities of maximum scores:

$$\text{If } Pr(x_i < T) = q, \text{ then } Pr(max(x_1, x_2, \ldots, x_n) < T) = q^n. \qquad (1)$$

For example, in Table 2, $Pr(x_1 < 4) = 4/7$ (and $Pr(x_2 < 4)$ is identical, because x_1 and x_2 are i.i.d.), but $Pr(max(x_1, x_2) < 4) = (4/7)^2 = 16/49$. It is also useful to look at the upper tail of the distribution of the maximum:

$$\text{If } Pr(x_i \geq \mathcal{T}) = p, \text{ then } Pr(max(x_1, x_2, \ldots, x_n) \geq \mathcal{T}) = 1 - (1 - p)^n. \quad (2)$$

These expressions and the distribution in Table 2 make clear that the distribution of any individual random variable x_i from i.i.d. variables $x_1, x_2, \ldots, x_n$ underestimates the distribution of the maximum of all the variables $x_{max} = max(x_1, x_2, \ldots, x_n)$. $Pr(x_i \geq \mathcal{T})$ underestimates $Pr(max(x_1, x_2, \ldots, x_n) \geq \mathcal{T})$ for all values $\mathcal{T}$ if the distributions are continuous. Said differently, the distribution of x_i has a lighter upper tail than the distribution of x_{max}.

This disparity increases with the number of random variables, $x_1, x_2, \ldots, x_n$. Consider three variables distributed in the same way as the two in Table 2. Then,

$$Pr(x_i \geq 4) = 3/7 = 0.43$$
$$Pr(max(x_1, x_2, x_3) \geq 4) = 1 - (1 - 3/7)^3 = 0.81.$$

The distribution of $Pr(x_i \geq 4)$ underestimates x_{max} by almost half its value. In comparison, $Pr(max(x_1, x_2) \geq 4) = 0.67$.

We have examined this relationship, and how it leads to excess structure, in greater detail elsewhere [3].

3.3   What About Cost-Complexity Pruning?

This analysis applies to EBP, REP, and MDL. However, cost-complexity pruning (CCP) is unaffected by the difference in the distributions of x_{max} and x_i. Like the other pruning methods, CCP selects among multiple trees: for each of ten cross-validation test sets, it creates a large number of pruned trees (where each tree is characterized by a pruning parameter a) and selects the one with the maximum score. However, CCP does not compare the maximum score to any threshold. Instead, it combines the values of a from each of the cross-validation folds and uses that parameter value to prune the tree grown on the entire dataset. Trees are not selected based on their values of a, nor is the selected tree's a compared to some threshold. Therefore, the difference in the distributions of a single score and a maximum score do not affect CCP.

4   Controlling Excess Structure

Section 3 suggests a method for controlling excess structure. Equation 2 can be used to set a threshold $\mathcal{T}$ such that $max(x_1, x_2, \ldots, x_n) \geq \mathcal{T}$ with a specified probability α_{max}, given that the null hypothesis is true. Changing notation slightly:

$$Pr(max(x_1, x_2, \ldots, x_n) \geq \mathcal{T}) = \alpha_{max} = 1 - (1 - \alpha_i)^n. \tag{3}$$

Given a specified value of α_{max} and the number of independent scores n, we can determine $\alpha_i = Pr(x_i \geq \mathcal{T})$. Given α_i and a reference distribution for a single score, we can determine $\mathcal{T}$. Alternatively, we can use equation 2 to adjust probability values derived from the reference distribution for a single score and then compare the adjusted probability value directly to α_{max}.

Equation 3 is known as a *Bonferroni equation*. Applying it is referred to as Bonferroni adjustment. The experiments in this section test the utility of Bonferroni adjustment in the same way as EBP, REP, MDL, and CCP were evaluated in section 2.

4.1 Building and Pruning Trees with Bonferroni Adjustment

We developed an algorithm — Tree-building with Bonferroni Adjustment (TBA) — that grows and prunes decision trees by using Bonferroni-adjusted significance tests. TBA resembles nearly all other algorithms for top-down induction of decision trees [10]. It differs in its evaluation function, its use of Bonferroni-adjusted significance tests to select attributes and partitions during tree construction and to select subtrees during tree pruning, and how it handles missing values. All of TBA's significance tests depend on the adjusted significance level α_{max}. For all experiments reported here, $\alpha_{max} = 0.10$.

Evaluation function: TBA uses the G statistic to evaluate contingency tables during both tree construction and pruning. The G statistic is used because it has a known reference distribution, a requirement for using Bonferroni adjustment. G is computed for contingency tables as follows:

$$G = 2 \sum_{cells} f_{ij} \ln \left(\frac{f_{ij}}{\hat{f}_{ij}} \right), \tag{4}$$

where f_{ij} is the number of occurrences, or frequency, in the cell i, j of the table and $\hat{f}_{ij}$ is the expected value of that cell. In this case, the expected value is $f_{i.} f_{.j} / f_{..}$, where $f_{i.}$ is the total frequency in row i, $f_{.j}$ is the total frequency in column j, and $f_{..}$ is the total of all cells in the table.

Selecting partitions: During tree construction, attribute partitions are selected using an approach suggested by Kass [5] and Kerber [6]. For each attribute, a contingency table is constructed with a row for each class value and a column for each of k attribute values — every possible value for discrete attributes or every unique interval for discretized continuous attributes. Then, the pair of columns in the table with the least significant difference is merged. The merging process is repeated until all columns are significantly different. For continuous attributes, only adjacent columns, corresponding to adjacent numeric intervals, can be merged. The result is a node that partitions the sample into j subsamples, where $1 \leq j \leq k$.

The test that determines whether pairs of columns are significantly different uses a Bonferroni adjustment. The exponent n is the number of column pairs. Without this adjustment, an inappropriately large number of partitions would be produced.

Selecting attributes: TBA selects attributes based on probability values. These values are calculated by comparing the G value for the merged contingency table to an appropriate reference distribution, and applying a Bonferroni adjustment. For a table with r rows and c columns, the appropriate reference distribution is chi-square with $(r-1)(c-1)$ degrees of freedom. Following Kass, the Bonferroni exponent n is the number of possible ways of combining k initial categories into j final categories. This calculation depends on the type of attribute. All possible pairs of categories can be merged in a discrete attribute, while only adjacent pairs of categories can be merged in continuous attributes:

$$n_{discrete} = \sum_{i=0}^{k-1}(-1)^i \frac{(k-i)^j}{i!(k-i)!}; \quad n_{continuous} = \binom{j-1}{k-1} \qquad (5)$$

While these estimates of n are approximate, at best, they provide a rough balance between the total number of possible tables (certainly an overestimate because the tables are highly correlated) and an exponent of 1 (certainly an underestimate). In Kass' experiments with randomly-generated data, they adjusted appropriately for the bias introduced by the merging process.

TBA forms a decision node using the attribute with the lowest probability value, regardless of whether that value falls below some threshold α. The algorithm uses the decision node to partition the sample into j subsamples based on the attribute's values, and repeats the attribute selection process for each subsample. Tree growth stops when no partition can improve accuracy on the training set.

Pruning: After constructing a tree, TBA prunes the tree by examining the probability values calculated during tree construction. Recall that those probability values are adjusted to account for multiple comparisons *within* an attribute. However, they are not yet adjusted to account for multiple comparisons *among* the many attributes that could be used at an individual node, nor are they adjusted to account for the many possible nodes available in other parts of the tree. The latter two adjustments are made at this stage, where the Bonferroni exponent n is the number of attributes considered at that node and the total number of decision nodes at the same tree depth as that node, respectively.

TBA examines each frontier node of the tree — decision nodes that have only leaf nodes as children. Frontier nodes where $p \leq \alpha_{max}$ are retained; frontier nodes where $p > \alpha_{max}$ are converted to leaf nodes and labeled with the majority class of the appropriate training subsample. The process continues until all frontier nodes are significant. Note that this process cannot eliminate non-frontier nodes for which $p > \alpha_{max}$. This could potentially "trap" insignificant nodes in the interior structure of the tree, but it guards against eliminating potentially useful subtrees.

Handling Missing Values: TBA handles missing values by assigning a default class to each decision node (the majority class of the training instances at that node). When the decision tree is used to classify instances, this class is assigned to any instance that reaches a decision node for which it lacks a value for the appropriate attribute. While this approach is easy to implement, it lacks the sophistication of the approaches in other algorithms. For example, C4.5 sends instances that lack attribute values down *all* relevant branches of a node, weighted according to overall frequency in the training set, and then makes a prediction based on the weighted class labels of all branches. In principle, nothing prevents such an approach from being implemented in TBA.

4.2   Results Using TBA

We repeated the experiments from section 2 on TBA. The results are summarized in table 3 along with results reproduced from table 1 for comparison.

Pruning Method	% Kept < 100	p < 0.1	Slope > 0.1	Mean R^2	Mean Δ size	Mean Δ accuracy
EBP	16	16	13	0.90	38.29	-0.14
REP	17	17	11	0.75	39.32	-0.32
MDL	18	17	13	0.88	44.03	-0.37
CCP	19	10	4	0.62	30.11	-0.06
TBA	**16**	**11**	**2**	**0.68**	**36.72**	**-0.18**

Table 3. Summary of the effects of random data reduction for TBA and the other pruning methods.

Based on the results in table 1, TBA performs better than EBP, REP, and MDL and performs similarly to CCP. Its accuracy peaked prior to seeing 100% of the available training instances for 16 of the 19 datasets. Eleven datasets exhibited a significant relationship between tree size and training set size beyond the point at which accuracy stopped growing. However, the slope of the regression line exceeds 0.1 in only two of those datasets. While TBA still exhibits a significant relationship between training set size and tree size in many datasets, the relationship is a relatively weak one in all but two of those cases.

5   Acknowledgments

The authors would like to thank Donato Malerba, Floriana Esposito, and Giovanni Semeraro of the Dipartimento di Informatica, Università degli Studi, Bari Italy for supplying their implementations of reduced error pruning and cost-complexity pruning. M. Zwitter and M. Soklic of the University Medical Centre,

Institute of Oncology, Ljubljana, Yugoslavia provided the breast cancer and lymphography datasets, and Dr. William H. Wolberg of the the University of Wisconsin Hospitals provided the breast-cancer-wisc dataset.

This research was supported by Sterling Software, Inc. subcontract #7335-UOM-001 (DARPA F30602-95-C-0257), and by a National Defense Science and Engineering Graduate Fellowship. The U.S. Government is authorized to reproduce and distribute reprints for governmental purposes not withstanding any copyright notation hereon. The views and conclusions contained herein are those of the authors and should not be interpreted as necessarily representing the official policies or endorsements either expressed or implied, of the Advanced Research Projects Agency, Rome Laboratory or the U.S. Government.

References

1. L. Breiman, J. Friedman, R. Olshen, and C. Stone. *Classification and Regression Trees*. Wadsworth International, 1984.

2. Paul R. Cohen. *Empirical Methods for Artificial Intelligence*. The MIT Press, Cambridge, 1995.

3. Paul R. Cohen and David Jensen. Overfitting explained. In *Preliminary Papers of the Sixth International Workshop on Artificial Intelligence and Statistics*, pages 115–122, 1997.

4. George H. John. Robust decision trees: Removing outliers from databases. In *Proceedings of the First International Conference on Knowledge Discovery and Data Mining*, 1995.

5. G.V. Kass. An exploratory technique for investigating large quantities of categorical data. *Applied Statistics*, 29(2):199–127, 1980.

6. Randy Kerber. Chimerge: Discretization of numeric attributes. In *Proceedings of the Tenth National Conference on Artificial Intelligence*. MIT Press, 1992.

7. J. R. Quinlan. *C4.5 : programs for machine learning*. Morgan Kaufmann Publishers, Inc., 1993.

8. J. Ross Quinlan. Simplifying decision trees. *International Journal of Man-Machine Studies*, 27:221–234, 1987.

9. J. Ross Quinlan and R. Rivest. Inferring decision trees using the minimum description length principle. *Information and Computation*, 80:227–248, 1989.

10. J.R. Quinlan. Induction of decision trees. *Machine Learning*, 1(1):81–106, 1986.

Exploiting Symbolic Learning in Visual Inspection

M. Piccardi, R. Cucchiara, M. Bariani, P. Mello

Dipartimento di Ingegneria, University of Ferrara
Via Saragat 1, 44100 Ferrara, Italy
E-mail {MPiccardi, RCucchiara, MBariani, PMello}@ing.unife.it

Abstract. The paper describes the use of data analysis techniques in the computer-vision inspection of industrial workpieces. Computer-vision inspection aims at accomplishing quality verification of fabricated parts by means of automated visual procedures. Gathering the visual information into models proves a critical task, especially when subjective judgement is involved in quality verification. In this work, intelligent data analysis techniques based on symbolic learning by examples have been explored in order to automatically devise and parametrize effective quantitative models. The paper reports and discusses the experimental results achieved in an industrial application.

1 Introduction

Inspection of fabricated components is a fundamental part of a production process, aiming at the respect of predefined quality levels in final products. Since human inspection carries high costs and scarce reliability, automated techniques are required to substitute for or integrate the human work.

An important role in this field is played by computer vision, that provides fast, non-destructive procedures for visual inspection [1]. Approaches adopted are often classified as quantitative vs. qualitative: a quantitative inspection is allowed when precise models of defective and non defective objects are available; for instance, this is the case of geometrical measures such as object widths, lengths, areas, automatically extracted from camera-acquired images. Instead, qualitative inspection mainly relies on expert judgement: in these cases, the acceptance/rejection criterion is not well-defined and often depends on human experience in accordance with very low-constrained issues.

When approaching the latter class of problems, inducing models from pre-classified examples by machine learning can prove an attractive solution [2]. Many works in the area of computer vision exploited machine learning as a powerful approach to support various phases of the vision process. In particular, symbolic learning by examples has been recently used to infer sets of rules or decision trees in order to assess effective and reliable object classification [3]. Symbolic learning, unlike other techniques, as neural networks, provides rules

X. Liu, P. Cohen, M. Berthold (Eds.): "Advances in Intelligent Data Analysis" (IDA–97)
LNCS 1280, pp. 223–234, 1997.

that can be easily interpreted by humans, thus allowing expert validation of the decision criteria.

In this paper we present an approach to a visual inspection task based on adoption of symbolic learning by examples. The application goal is quality evaluation of industrial workpieces, and in particular discrimination of defective exemplars from non defective ones. The components to be inspected are metallic pieces presenting surface defects due to thermal and mechanical treatments. Defects can be qualitatively described as objects with rather high luminosity, mainly elongated shape and nearly rectilinear edges. These properties can not be rigorously quantified with precise values; moreover, they can be partially shared by other spurious objects which are to be classified as "appearing" defects, thus resulting in a non trivial classification task.

The main goal of this paper is to evaluate the effectiveness of supervised, symbolic learning-by-examples tools for the defect detection application. In particular, we focus on: the impact of the training set composition on the classifier's structure; the error rates achieved using different sets of visual primitives; the trade-off between classifier's complexity and error rate attainable by constraining the classifier's structure.

The paper is divided as follows: in the next section the data analysis techniques are described together with the machine learning tools; in section 3 the visual items of the application are explained in detail, and the adopted visual primitives discussed. Section 4 presents the experimental results achieved with two different approaches: in the former, machine learning is explored on the basis of a set of general-purpose visual primitives; in the latter, a specific visual primitive is introduced; performance of the two approaches are compared in terms of error rate. Eventually, conclusions summarize work guidelines and discusses the achieved results.

2   Data analysis techniques

The quality inspection application aims at detecting the presence of defects in workpieces: thus, two classes of interest have been defined, namely Defect and NonDefect. Objects to be classified are extracted from images and are submitted to classification in form of tuples of attribute values, with each attribute corresponding to one visual primitive [4]. The classifiers used to assess the presence of defects have been constructed by using a symbolic learning-by-examples approach [2].

In a preliminary phase of the work, we have experimented different machine learning tools, including C4.5, a system developed at the Irvine University of California by Quinlan, and FOIL, a tool for inductive logic programming form Quinlan and Cameron-Jones[5, 6]. Among the explored tools, C4.5 proved the most adequate for the application [7].

C4.5 generates a classification system represented in form of a decision tree: each leaf indicates a class while each decision node specifies a single or a set of tests to be carried out on a single attribute value with one branch and subtree for each possible outcome of the test. C4.5 belongs to the class of supervised algorithms: it requires a training set of correctly pre-classified examples from the entire population in order to generate the classifier.

The result of running C4.5 is an initially un-pruned tree. Often, the resulting decision tree can be over-specialized, in the sense that it overfits the data by inferring more structure than it is justified by the training cases [5]. The limit case carries one leaf for each classified example of the training set; that is a sort of simple storage of the training set without any generalization inferred. In order to avoid this possibility, several parameters of C4.5 can be tuned, for instance by operating a further pruning phase or by indicating the minimum number of examples which must be used for generating a new leaf. Obviously these operations provide both generalization and simplification of the classification process and thus may increase the error rate. C4.5 provides two primal ways to evaluate the effectiveness of the generated classifier: the former is the number of examples from the training set that are misclassified (i.e., incorrectly classified) by the classifier, separately for each class; the latter is an estimate of the error rate on the entire population, based on heuristics. In addition, several tests can be applied to achieve statistical validation of classifier performance. When the set of available data is small (as in our case), cross-validation techniques can be exploited [8]: they consist of randomly dividing the set of available data into k disjoint subsets of equal size, and performing k trials (k-fold cross-validation). In each trial one of the subset is used as testing set at a time and the union of the other ones as training set; errors are averaged over all the performed trials. The main advantage of cross-validation is that testing sets are independent, allowing to explore variation of classifier's performance with the testing data; instead, training data substantially overlap: for instance, in the case of 10-fold cross validation, each pair of training sets overlap by the 80 %, and this may partially prevent form generalizing the error rate to the whole population. Another test, suggested by Quinlan in [5], is the N-fold cross validation, with N the number of available data (leave-one-out cross-validation). This technique doesn't claim to explore dependence from the training set, but rather to evaluate performance of the classifier which is the most similar to the one generated from the whole available data set. In our work, we have used both 10-fold cross-validation and leave-one-out to compare the error rates of the different classifiers.

C4.5 maps instances into hyperrectangles of the attribute space (i.e.: domain regions with boundaries parallel to the coordinate axes) [5]. Although simple, this approach can lead to very complex trees even in presence of regular distributions of instances into domain regions of shape different from hyperrectangles. A possible way for simplifying the classifier is to increase the number of attributes by adding redundant ones with the only goal of guiding the construction of the decision tree. Therefore, we have introduced a set of redundant attributes, functions of the initial ones, but holding still a semantic value with respect to the qualitative model of defects.

3   The application of visual inspection

The application of visual inspection aims at verifying fabricated workpieces: these are metallic objects that may present surface or subsurface defects. Typical inspection techniques make use of magnetoscopy: first, the piece is magnetized and dipped in a water suspension of fluorescent ferromagnetic particles; then it is exposed to UV light and inspected [9]. Since defects causes discontinuities in the magnetic field, the ferromagnetic particles are attracted on the piece surface in correspondence of the defect locations. As particles are previously pigmented with fluorescent material, when exposed to UV light their fluorescence is enhanced and defects can be detected by human inspectors. Defects shape appears to the human eye as elongated, roughly straight and thin, very bright with respect to the neighbouring background.

The initial step of automated inspection consists of acquiring images of pieces under UV light and using sets of standard image analysis operators to extract salient features. These computer-vision primitives account for shape features like edge rectilinearity, gradient magnitude, and shape thickness. In particular, the Hough transform (HT) for lines gives evidence to rectilinear shapes by transforming the image space into a parametric space, where line segments are mapped into single points [10]. Each point in the Hough space has two coordinates, (ρ, θ): ρ is the distance of the line from the origin, and θ is the angle that the normal to the line forms with the x axis. The point's value is proportional to the number of image points in a same line (called voting points); when this number is high, it leads to formation of a peak in the Hough space. Moreover, gradient-based versions of the Hough transform account also for gradient information: contributions from image points to a peak value are weighted by the magnitude of the luminosity gradient, thus giving more evidence to highly contrasted objects; the gradient orientation distributes votes in $\theta \in [0, 2\pi]$, thus distinguishing rising from falling edge slopes [10].

The detection criterion can be expressed by looking for two associated peaks in the Hough space, corresponding to the two (straight and parallel) main edges of a defect. In the case of an ideally-straight shape, the two peaks are found at (ρ, θ) and (ρ', θ') coordinates respectively, with $|\rho' - \rho|$ measuring object thickness and $\theta' = \theta + \pi$, because of the opposing slope of edges. Since defects are thin, an upper bound to $|\rho' - \rho|$ can be considered. Nevertheless, deciding how to use the computed values in order to classify shapes as defects or non defects is a critical task, since spurious objects are present in images, sharing some features with real defects. For this reason, we decided to perform data analysis with automatic learning-by-examples tools.

A set of visual primitives was defined in order to compute tuples of values to be evaluated by C4.5. The set of visual primitives include:

1. peak1: the value of a Hough peak, $H1(\rho, \theta)$, corresponding to a rising edge;
2. peak2: the value of a Hough peak, $H2(\rho', \theta')$, corresponding to a falling edge;
3. votes: the number of image points that were transformed into $H1(\rho, \theta)$ or $H2(\rho', \theta')$;

4. distance: the Euclidean distance between Hough points (ρ, θ) and $(\rho', \theta' - \pi)$;
5. delta_rho: the $|\rho' - \rho|$ value;
6. delta_theta: the $|\theta' - \theta - \pi|$ value.
7. gradient: the average value of the gradient magnitude in the image.

Since high peak values represent well-formed straight line segments (i.e. clear defects), features 1. and 2. are expected to be used by the inferred classifier as figure of merits; instead, a large number of votes at a parity of Hough values could reveal a non clean-cut shape that may arise from spurious objects; all of the 4., 5., and 6. values should be low in the case of a real defect, and in particular 6. should be zero in the case of an ideally-straight defect. Anyway, in the case of approximately vertical edges, two peaks detecting a real defect can have ρ and ρ' of opposing sign and delta_theta close to π. Moreover, since Hough values are gradient-based, the average gradient of the image 7. is added in the tuples, in order to let the machine learning tool explore a possible gradient influence.

As explained before, C4.5 maps examples into hyperrectangles of the attribute space: this leads to complex decision trees when examples show a different distribution. Therefore, we added other attributes, obtained by combining the previous ones, exploring if they could be more effectively mapped into intervals:

8. product: the product $H1 \times H2$;
9. ratio: the ratio $H1/H2$;
10. sum: the sum $H1 + H2$;
11. difference: the difference $H1 - H2$;
12. delta_product: the product delta_rho $\times$ delta_theta.

8. and 10. are expected to behave as figures of merit, claiming for the presence of a shape with two relevant straight edges; instead 9. and 11. reveal the presence of a straight object which edges differ significantly either in length or gradient magnitude (thus improbably ascribable to a defect); low values of 12. accounts for thin shape and parallel edges together.

4   Experimental results

Experiments were carried out by first pre-classifying a large number of examples in order to compose an adequate training set for the learning phase. Several images were inspected, with both real and appearing defects. The inclusion criterion was based on Hough values: the HT is computed and values of the Hough space ranked; the highest values in the two Hough semispaces (with $\theta \in [0, \pi]$ and $\theta \in]\pi, 2\pi]$ respectively) form the couples $(H1(\rho, \theta), H2(\rho', \theta'))$ that are used for classification. The resulting training set includes N = 2700 pre-classified examples, 311 of which belonging to the Defect class (actual defects) and the remaining part classified as NonDefect (appearing defects).

When generating the decision tree, C4.5 gives the possibility of exploring a trade-off between tree complexity and error rate, by constraining the minimum number of cases per leaf, n. In an unconstrained running, C4.5 produced

```
Simplified Decision Tree:
distance <= 8.60233 :                                              (1)
|   peak2 <= 1.12634 :
|   |   gradient > 0.0456994 : NonDefect (69.0/1.4)
|   |   gradient <= 0.0456994 :
|   |   |   peak1 <= 0.662371 : NonDefect (16.0/1.3)
|   |   |   peak1 > 0.662371 :
|   |   |   |   peak1 <= 0.74439 : Defect (10.0/3.5)                (7)
|   |   |   |   peak1 > 0.74439 : NonDefect (21.0/8.0)
|   peak2 > 1.12634 :
|   |   delta_theta <= 6 :
|   |   |       peak2 > 1.62996 : Defect (174.0/12.9)              (11)
[...]
distance > 8.60233 :                                              (22)
|   delta_theta <= 164 :
|   |   delta_theta > 16 : NonDefect (1600.0/2.6)
|   |   delta_theta <= 16 :
|   |   |   votes > 14 : NonDefect (401.0/5.0)
|   |   |   votes <= 14 :
|   |   |   |   delta_rho > 21 : NonDefect (162.0/10.7)
[...]
|   |   |   difference <= 0.387529 : NonDefect (26.0/1.3)         (41)
|   |   |   difference > 0.387529 : Defect (18.0/10.0)


Evaluation on training data (2700 items):
  Before Pruning              After Pruning
 ----------------    ----------------------------

Size     Errors     Size      Errors    Estimate
 51     71 (2.6%)    43      70 (2.6%)    (4.0%)    <<
 (a)    (b) <-classified as
 ---- ----
 276    35 (a): class Defect
  35  2354 (b): class NonDefect
```

Fig. 1. Decision tree with $n = 10$ minimum examples per leaf.

a pruned decision tree with 105 nodes and an error rate on the training set of 0.8 % (21 examples misclassified); the error rate for the classifier is estimated in 3.3 %. However, many of the tests in the tree lack generalization, especially those leading to leaves covering very few examples. For this reason, we examined also trees generated with a minimum number of cases per leaf.

Fig. 1 shows the pruned decision tree generated with $n = 10$; each line associated with a leaf reports two numbers in parentheses: the number of cases of the training set classified by that leaf, and a pessimistic estimate of the number of them which are wrongly classified. In the decision tree of Fig. 1, the initial test is based on the Euclidean distance between the two peaks of the couple: the

```
Simplified Decision Tree:
distance <= 8.60233 :
|   peak2 <= 1.12634 : NonDefect (116.0/17.1)
|   peak2 > 1.12634 :
|   |   delta_theta <= 6 : Defect (195.0/21.5)
|   |   delta_theta > 6 :
|   |   |   peak2 > 4.35773 : Defect (24.0/3.7)
|   |   |   peak2 <= 4.35773 :
|   |   |   |   distance <= 5.38516 : Defect (36.0/12.5)
|   |   |   |   distance > 5.38516 : NonDefect (43.0/13.6)
distance > 8.60233 :
|   delta_theta <= 164 : NonDefect (2222.0/35.6)
|   delta_theta > 164 :
|   |   delta_rho <= 313 : Defect (20.0/1.3)
|   |   delta_rho > 313 : NonDefect (44.0/12.6)


Evaluation on training data (2700 items):
  Before Pruning             After Pruning
 -----------------   --------------------------

Size      Errors    Size      Errors    Estimate
  17    98 (3.6%)      15    96 (3.6%)    (4.4%)    <<
  (a)   (b) <-classified as
 ----  ----
  245    66 (a): class Defect
   30  2359 (b): class NonDefect
```

Fig. 2. Decision tree with $n = 20$ minimum examples per leaf.

attribute seems to be used in the right way, since the largest part of the Defect class is classified below the computed threshold (see lines 1 and 11) while most NonDefect instances are classified above (see lines 22 and following). Instead, lines 7-8 report a test that is difficult to justify: if peak1 - which should be used as a figure of merit - is below a threshold, a defect is stated; an appearing defect, otherwise. Moreover, this test leads to a large value of the estimate of misclassified examples (8 of 21). Another test of difficult interpretation is reported in lines 41-42: the difference attribute should be low in the case of a defect with main edges of similar length and gradient, while in the decision tree it is used in the opposite way; this test, too, carries a high misclassification estimate (10 of 18).

Since the tree still has a rather complex structure, we tried to constrain further the tree generation with $n = 20$. In effect, all of the tests shown in the corresponding Fig. 2 seem to reflect physical properties of the qualitative model of defects. In this case, although the estimated error rate is only slightly worse than the previous case (4.4. % vs. 4.0 %), the number of misclassified examples substantially increases (3.6 % vs. 2.6 %). The poorest performance concerns the error rate on the Defect class, which is 21 % (66 misclassification

```
Simplified Decision Tree:
distance <= 11.4018 :
|    gradient > 0.0502131 : Defect (200.0/11.8)
|    gradient <= 0.0502131 :
|    |    gradient <= 0.0470579 : Defect (59.0/5.0)
|    |    gradient > 0.0470579 :
|    |    |    votes <= 15 : Defect (29.0/10.3)
|    |    |    votes > 15 : NonDefect (19.0/1.3)
distance > 11.4018 :
|    gradient <= 0.0456994 : Defect (31.0/1.4)
|    gradient > 0.0456994 : NonDefect (284.0/15.1)


Evaluation on training data (622 items):
 Before Pruning              After Pruning
----------------     --------------------------

Size      Errors    Size      Errors    Estimate
  15    31 (5.0%)     11     32 (5.1%)    (7.2%)   <<
  (a)    (b)  <-classified as
 ----  ----
  299     12  (a): class Defect
   20    291  (b): class NonDefect
```

Fig. 3. Decision tree with $n = 10$ minimum examples per leaf, balanced training set.

of 311 examples). This behaviour is even more unsuitable in the context of the inspection application, since it leads to validate defective workpieces. This asymmetry between the NonDefect and Defect classes could be due to over-description of the former class with respect to the latter in the training set (88.5 % vs. 11.5 %).

For this reason, we derived a balanced training set from the initial one, including all the 311 Defect instances and as many of the NonDefect class. Figg. 3 and 4 show the generated trees in the case of $n = 10$ and $n = 20$, respectively. The estimated error rates for the generated trees are greater than in the previous cases (7.2 % for $n = 10$ and 8.4 % for $n = 20$), but the number of misclassified Defect examples is substantially less (12 vs. 35 for $n = 10$ and 20 vs. 66 for $n = 20$).

In general, by using the balanced training set, we have observed that the number of misclassified examples from the training set is nearly equal for the two classes. Instead, in the case of the unbalanced training set, the number of misclassified defects tends to grow fast when we constrain the minimum number of cases per leaf, n. In a primal analysis, we have seen that, when n grows, small leaves of the Defect class tend to be merged in the larger leaves of the NonDefect one; this leads to considering a rather high number of defects as non defects. This behaviour may be ascribed to two causes: first, the Defect class may be more spread in the attribute space, thus resulting in a large number of

```
Decision Tree:
distance <= 11.4018 :
|    delta_theta <= 12 : Defect (282.0/22.0)
|    delta_theta > 12 : NonDefect (25.0/8.0)
distance > 11.4018 :
|    gradient <= 0.0456994 : Defect (31.0)
|    gradient > 0.0456994 : NonDefect (284.0/12.0)

Evaluation on training data (622 items):
 Before Pruning              After Pruning
 ---------------       -------------------------

Size       Errors    Size       Errors    Estimate
  7     42 (6.8%)      7      42 (6.8%)      (8.4%)   <<
 (a)     (b) <-classified as
 ----   ----
 291      20 (a): class Defect
  22     289 (b): class NonDefect
```

Fig. 4. Decision tree with $n = 20$ minimum examples per leaf, balanced training set.

small leaves; second, the number of the Defect examples is too low with respect to the NonDefect ones. In both cases, constraining with the same n value the two classes will result in decreasing the number of Defect leaves more than NonDefect ones, with greater error rates on the Defect class.

Having a small number of misclassified defects is highly preferable for the specific application, and suggests that the training-set composition should match the application requirements: since the automatic tool doesn't associate class-dependent parameters to each class (for instance, weighted error rates), an important goal of the trainer is to compose the training set in accordance with application requirements.

The approach described until now is based on the use of a set of well-known computer vision primitives; then, we introduced one more specialized visual primitive in the set for purpose of comparison. The visual primitive, called Correlated Hough transform, is based on the gradient Hough transform, and was specifically proposed for detecting bar-shaped objects [11]. It exploits explicitly the mutual presence of two associated peaks $H1(\rho, \theta)$ and $H2(\rho', \theta')$, by computing correlation between $H1(\rho, \theta)$ and the Hough values in a neighbourhood of $(-\rho, \theta + \pi)$. In addition to the CHT peak, the ratio CHT/H1 was also included in the visual primitive set, measuring the amount of correlation. The same images used to devise the initial training set were analyzed again in order to detect examples on the basis of CHT values; Fig. 5 shows the tree generated without constraining the minimum number of cases per leaf.

The generated decision tree uses only three attributes, namely Correlated-Hough (the CHT peak's value), gradient, and CHT/H1; CorrelatedHough is the first attribute tested, and is used correctly as the most salient figure of merit:

```
Decision Tree:
CorrelatedHough > 78.6283 : Defect (29.0)
CorrelatedHough <= 78.6283 :                                        (2)
|  CorrelatedHough <= 8.86457 : NonDefect (133.0)
|  CorrelatedHough > 8.86457 :                                      (4)
|  |  gradient > 0.0470579 : NonDefect (98.0)                       (5)
|  |  gradient <= 0.0470579 :
|  |  |  CorrelatedHough > 17.7267 : Defect (18.0)
|  |  |  CorrelatedHough <= 17.7267 :
|  |  |  |  gradient > 0.0456994 : NonDefect (13.0)
|  |  |  |  gradient <= 0.0456994 :
|  |  |  |  |  gradient <= 0.0446796 : NonDefect (3.0)               (11)
|  |  |  |  |  gradient > 0.0446796 :
|  |  |  |  |  |  CHT/H1 > 12.1436 : Defect (15.0)
|  |  |  |  |  |  CHT/H1 <= 12.1436 :
|  |  |  |  |  |  |  CHT/H1 <= 11.4278 : Defect (5.0)
|  |  |  |  |  |  |  CHT/H1 > 11.4278 : NonDefect (3.0)              (16)
```

```
Evaluation on training data (317 items):
  Before Pruning              After Pruning
----------------        --------------------------

Size     Errors     Size     Errors    Estimate
 17      0 (0.0%)     17      0 (0.0%)    (3.6%)    <<
(a)     (b)  <-classified as
----    ----
 67            (a): class Defect
        250 (b): class NonDefect
```

Fig. 5. Decision tree (unconstrained) with use of the Correlated Hough transform.

values above 78.6283 indicate Defect, below 8.86457 indicate NonDefect, and require further testing inside this range. The gradient attribute seems to be used as a normalization term: if the CorrelatedHough attribute belongs to the intermediate interval (lines 2,4), highly-contrasted images indicate NonDefect (line 5); this can be explained since all Hough related values are gradient-based: if the average gradient is high, higher peak values are expected in order to state a defect. Lines 11 and 16 report two tests of difficult validation, because they define two small intervals in the gradient and CHT/H1 attributes in which 6 NonDefect instances are classified (3 and 3 respectively).

The generated tree shows the best performance of all: no error is made on the training set, even with a tree of compact size (17 nodes), and the estimated error rate is 3.5 % only. In case of pruning the two difficult-to-validate leaves (11 and 16), the error rate on the training set would be 1.9 %; however no error on the Defect class would be introduced.

Since the error estimates provided by C4.5 are mainly based on heuristics, we have performed also cross-validation analysis of the generated classifiers. Table 1

Training Set	10-fold CV			Leave-one-out CV		
	free	$n = 10$	$n = 20$	free	$n = 10$	$n = 20$
Unbalanced	4.1	5.0	4.7	4.6	4.4	4.1
Balanced	5.3	9.2	9.0	5.1	6.1	6.8
With CHT	4.4	3.7	8.8	2.8	3.2	11.7

Table 1. Error rate (%) in cross validation tests.

reports the error rates in the case of 10-fold and leave-one-out cross-validation, in the cases of a tree with $n = 1$, 10, 20 minimum number of examples per leaf; the table compares the error rates achieved with the unbalanced and balanced training sets, and the classifier using the CHT attribute. The balanced training set shows higher error rates than the unbalanced one, especially when n is constrained: in fact, since it has a smaller training set, a same n constraint results in eliminating a large number of leaves, thus over-simplifying the tree. Anyway, we could expect better classification of the Defect class from the balanced training set, as it was shown by misclassifications on the training examples; a future analysis will try to assess separate error rates for the two classes. Best performance in classification is achieved by using the CHT attribute, since error rates are less both for the 10-fold and leave-one-out tests; also in this case, the n parameter must be carefully tuned to avoid excessive tree simplification.

The overall experiment with symbolic learning techniques embodies some final considerations:

1. Selection of the training set should reflect the expected target of the classifier more than the actual frequency in the classes. This can be observed by comparing results of the two training sets (see Figg. 1,2 and 3,4) the former reflecting the frequency of actual and appearing defects in images, with a substantial lower percentage of defective images, and the latter balancing actual and appearing defects. In both cases the learning tools generate decision trees of moderate complexity, with satisfactory overall performance, since the error rates are generally low. Nevertheless, best performance for the Defect class were achieved with the balanced training set, and may be further improved with over-description of this class; this is important for the semantic of the application, where the incorrect classification of a Defect instance can lead to validation of a defective workpiece.

2. Symbolic learning systems are powerful tools for classification since they help in selecting the minimal and reliable set of symbolic features actually required. In this application, using specialized visual primitives in addition to standard ones resulted in an effective approach, since it led to simpler mapping of the training data into the decision tree, as proved by the experimental results. Moreover, the preliminary analysis shows that the achievable error rates may improve with use of specialized visual primitives.

5   Conclusion

Automatic visual inspection is a critical task when the data models guiding the inspection are only qualitative. In this case, machine learning by examples can substantially support the phase of model devising, by automatically analyzing the set of available data.

In this work, we have evaluated an approach to visual inspection making use of symbolic learning-by-examples techniques. Symbolic learning provides decision tree that can be easily interpreted by human experts, thus enabling for a-posteriori validation. In the paper we have described a visual inspection application, aimed at detecting small-sized defects on the surface of metallic workpieces. The substantial use of symbolic learning allowed for developing a reliable, fast and efficient visual system that is currently used for a real industrial inspection problem. Depending on the application requirements, the effort of devising more specialized primitives may be justified, or standard visual primitive sets may provide adequate accuracy. In both cases, combining the computed values into quantitative models and tuning thresholds in the data set are critical tasks; in our application, automatic data analysis by symbolic learning proved an adequate solution in terms of flexibility, expressive power and precision. Finally, since the explored quality-inspection application embodies a substantial asymmetry between classes, in future work we intend to evaluate other machine learning tools that explicitly adopt class-dependent parameters, as for instance weighted error rates and class-dependent constraints on the classifier structure.

References

1. T.S. Newman, A.K. Jain: *A survey of automated visual inspection*, Comp. Vision and Image Understanding, **6**, n. 2, pp. 231-262.
2. R.S. Michalski, J.G. Carbonell, T.M. Mitchell, editors: Machine Learning - An Artificial Intelligence Approach. Springer-Verlag, Berlin, 1984.
3. K. Cho, S.M. Dunn: *Learning shape classes*, IEEE Trans on PAMI, **16**, n. 9, pp. 882-893, 1994.
4. R.M. Haralick, L.G. Shapiro: Computer and Robot Vision. Addison-Wesley, 1992.
5. J.R. Quinlan: C4.5: Programs for Machine Learning. Morgan Kaufmann Publishers, San Mateo, California, 1993.
6. J.R. Quinlan, R.M. Cameron-Jones: *Induction of Logic Programs: FOIL and related systems*, New Generation Computing, **13**, pp. 287-312, 1995.
7. M. Bariani, R. Cucchiara, P. Mello, M. Piccardi: *Data mining for automated visual inspection*, Proc. of PADD97, London, UK, 1997, pp. 51-64.
8. T. G. Dietterich: *Statistical tests for comparing supervised classification learning algorithms*, Pre-print, submitted, Dept. of CS, Oregon State University, OR, 1996.
9. R. Mason (ed.): *Magnetic Particle Inspection*, Nondestructive Testing, **33**, pp. 6-12.
10. J. Illingworth, J. Kittler: *A survey of the Hough transform*, Comp. Vision Graphics, Image Process, **43**, 221-238.
11. R. Cucchiara, F. Filicori: *A highly selective HT based algorithm for detecting extended, almost rectilinear shapes*, Computer Analysis of Images and Patterns: LNCS n. 970 Springer Verlag, pp. 692-698, 1995.

Forming Categories in Exploratory Data Analysis and Data Mining

P. D. Scott, R. J. Williams, K. M. Ho

Dept of Computer Science, University of Essex, Colchester CO4 3SQ, UK

Abstract. This paper describes the techniques used for categorizing variables in SNOUT an intelligent assistant for exploratory data analysis of survey and similar data sets that is currently under development. We begin by reviewing existing work on category formation in data mining which has been mainly concerned with enabling decision tree programs to handle numeric variables. It is argued that there are other important but neglected aspects of category formation, notably the formation of new categorizations of nominal variables. We report the limited success achieved in categorizing variables from survey data using either endogenous methods or exogenous methods that maximise the association with only one dependent variable. We then describe the categorization technique used in SNOUT: a procedure that selects a partition that both maximises the number of variables associated with the partitioned variable and maximises the strength of those associations. We report on the success achieved using this procedure in exploring real survey data.

1 Introduction

We are currently engaged in developing an intelligent assistant, called SNOUT [20], for exploratory data analysis. The goal of this work is to construct a system that will be provide users with a wide variety of tools for exploratory analysis and data mining including both established EDA techniques ([21], [5]) and methods using machine learning procedures in a single integrated system. During the course of the project it has become increasingly apparent that the success of the system is going to depend on our ability to provide SNOUT with effective procedures for automatic category formation.

Category formation is the most fundamental technique that people use to manage the otherwise overwhelming variety of their experiences. If every event, object, or phenomenon observed were regarded as unique, bearing no relationship to any previous observation, it would be impossible to make any sense at all of the world. Segmenting experiences into groups that are in some sense of the same kind or type is the simplest possible form of organization that an observer can impose. Furthermore such segmentation is an essential first step towards any more sophisticated structuring of experience. It is therefore hardly surprising that a system that attempts to discover order in complex domains requires some means of finding useful ways to sort the entities it encounters into classes.

X. Liu, P. Cohen, M. Berthold (Eds.): "Advances in Intelligent Data Analysis" (IDA–97)
LNCS 1280, pp. 235–246, 1997.   © Springer–Verlag Berlin Heidelberg 1997

In section 2 we review existing work on category formation in data mining which has been mainly concerned with enabling decision tree programs to handle numeric variables. We argue that there are other important aspects of category formation, notably the formation of new categorizations of nominal variables. In section 3 we discuss category formation using only the values of the partitioned variable before turning to systems that create partitions that maximise the association with one or more other variables. Section 4 examines partitioning using one other variable while in section 5 we consider systems in which partitioning is determined by many variables and describe this paper's main novel contribution to the field: a procedure for dichotomizing all levels of variable in a manner that maximises their association with other variables. Finally we summarise the main conclusions and give some indication of profitable directions for further work.

2 Category Formation in Data Exploration

Category formation could be defined succinctly as follows:

> Given a set of entities X, find an equivalence relation E_X that partitions X into equivalence classes.

However, something very important is missing from such a definition: not all equivalence relations would generate a partition that we would recognise as a sensible or useful set of categories. Worthwhile classification systems typically have the property that members of a given category resemble each other rather more than they resemble the members of other categories. Thus the notion of category formation implies the existence of some form of similarity metric that would enable judgements to be made about resemblances within and between groups. Different similarity metrics could be expected to give rise to different sets of categories: classifying animals by where they live would give rise to a very different taxonomy than that derived from consideration of their anatomical structure. Thus we might refine our original definition thus:

> Given a set of entities X and a similarity metric S, find an equivalence relation $E_{X,S}$ that partitions X into equivalence classes such that similarity within equivalence classes is maximised and similarity between equivalence classes is minimised.

It is clear therefore that the choice of similarity metric is at the heart of category formation. It includes the aspects of the entity that will be considered in assessing resemblance, the way in which this resemblance will be measured and the way in which such measurements can be extended to measure the resemblance of groups rather than individuals.

2.1 Discretization of Numeric Features for Classification Learning

There are two reasons why category formation is a useful step in exploring a dataset: as an end in itself and as part of some other method for discovering regularities.

Existing work on categorisation in data mining has concentrated on forming categories as a means to an end rather than as an end in itself. Most of it has been motivated by the desire to extend classification tree induction methods ([1],[16]) to handle numeric variables. Recent research on discretization of numeric features for classification learning procedures has been reported by

Catlett [2], Kerber [15] Fayyad & Irani [8], [9], Holte [14], Richeldi & Rossotto [19], and Quinlan [18] : Dougherty, Kohavi & Sahami [4] includes a useful systematic overview of this work.

ID3 [16] incorporated a simple but computationally expensive method for dichotomising numeric variables: as each numeric variable becomes a candidate for the construction of a subtree, the information gain for each possible partition point is considered. Since each variable may be a candidate many times in the course of tree construction this method can be very time consuming, particularly for large datasets that give rise to many possible partition points. Catlett [2] developed a modified version of ID3 called D-2 that partitions each numeric variable once before tree construction begins. The resulting categories may be suboptimal but a considerable amount of effort is saved.

D-2 and ID3 thus exemplify two contrasting strategies for creating categories to be used by another procedure: D-2 uses an early binding time strategy to form a *global* categorization in which all the partitions are constructed in advance of running the procedure; ID3 uses a late binding time strategy in which *local* partitions are constructed or re-constructed at the point where they are needed. In theory this should be a classic computing trade-off: the global method should require less computational effort but the local method should give better results.

Dougherty *et al.* [4] report some comparative studies of alternative methods using sixteen real world datasets. Fayyad & Irani's entropy based global method [8] was the most successful but the advantage over C4.5 [17], which uses a local method, was slight. Further research in which the same discretization technique is used to build both global and local categorizations is needed to establish whether the latter offers any advantages to justify its greater cost.

2.2   Forming Categories Using Nominal Features

Almost all existing work on category formation for data mining has addressed the problem of partitioning continuous features. As noted above the motivation has been to allow techniques developed for categorical models to be used when the data is numerical. Consequently the problem of partitioning nominal variables has been relatively neglected.

This is regretable because deriving new categorizations by combining existing categories is one of the simplest yet most powerful ways of building useful models of the world: for example, it is obviously useful to divide the numerous species of plants into those that we are good to eat and those that are not. Consequently developing new techniques for partitioning nominal variables has been an important part of our work on SNOUT.

2.3   Issues in Category Formation

Developing systems that automatically form categories involves a number of important issues, many of which can only be briefly mentioned in this paper. There is a fundamental distinction between procedures that depend only on the values of the variable to be partitioned and those that also use information about the corresponding values of one or more other variables. We term the former *endogenous* and the latter *exogenous* methods. Dougherty *et al.* [4] use the terms *unsupervised* and *supervised* to draw a a similar but not identical distinction.

Another important aspect of category formation is the problem of measures of success. This is a major topic well beyond our present scope. We shall assume one or both of two criteria: that the partitioning improves our ability to make predictions about the value of one or more other variables, or that it corresponds to the commonsense division that a person might make because it reflects some *fundamental* property of the feature partitioned. Both of these are closely associated with another important notion: the *fertility* of a variable. Fertile variables are those that have strong associations with many other variables, typically because of some direct or indirect causal relationship. Category formation techniques that favour the formation or identification of fertile variables are particularly useful.

3   Endogenous Category Formation

Endogenous category formation procedures use only information concerning the distribution of values of the variable to be partitioned. There are two basic approaches: percentile methods and clustering methods.

3.1   Percentile Methods

Percentile methods, often termed "binning", operate by dividing the range of values into contiguous subranges and assigning all the values in each subrange to a distinct category. In order to take account of the distribution of values actually occuring in the dataset it is common to arrange that each subrange should contain a certain proportion of the total sample. The simplest example of such a technique is to dichotomise an interval variable at the medial value into high and low categories. Alternatively, a trichotomy can be formed comprising the lowest quartile, the middle two quartiles, and the highest quartile. The computational cost of such a partitioning method is typically $O(n \log n)$ since it is dominated by the sorting operation necessary to determine the subrange boundaries

Percentile methods are conceptually simple but unfortunately they are not very sensitive to the actual distribution of values. In particular they often make a poor job of separating the two underlying populations in a variable that has a bimodal distribution. Furthermore decisions about the number and size of each partition must be made in advance and are often arbitrary.

3.2   Clustering Methods

In contrast clustering methods([6], [7], [11]) are much more sensitive to the distribution of values and provide a strong indication of the number of categories that is most appropriate. Such techniques normally operate by constructing a tree whose leaf nodes hold the individual items in the dataset: subtrees whose roots lie immediately below the root of the tree correspond to major subgroups of the variable. The tree construction procedure minimises the similarity the children of each node. A large number of clustering procedures of this type have been developed ([6], [7], [11]) to partition numeric variables. They differ chiefly in two respects: whether the tree is built bottom up (usually termed agglomerative hierarchical methods) or top down, and what metric is used to determine the similarity of a pair of individuals or groups.

A number of endogenous partitioning procedures were evaluated with a view to providing SNOUT with a simple method for forming categories. These included a percentile method, four agglomerative hierarchical methods using different similarity metrics, and one top down method. Tests were carried out using two types of data: artificial data comprising one, two, or three overlapping normal distributions; and real data taken from widely accessible datasets [1].

Limitations of space preclude a detailed discussion of our results so we confine ourselves to reporting the principle findings. None of the methods always found the 'right answers' for the artificial data. Ward's agglomerative method [22], which minimises the internal variances of the groups constructed, was the most successful: unfortunately it is also computationally expensive. More disturbingly the various methods showed little consensus on the real world data where there was no obvious correct answer. We concluded that the categories formed on the type of data that SNOUT is intended to investigate would be very sensitive to the choice among these partitioning methods and that none was particularly trustworthy. Thus it appears it would be unwise to rely on endogenous techniques for partitioning numeric variables.

All the techniques discussed in this section are designed primarily for numeric variables. Most of them can readily be extended to accommodate ordinal variables but none of them can be adapted to handle nominal variables. Indeed the authors are unaware of any useful endogenous method for partitioning nominal variables.

4   Exogenous Category Formation: One Variable

Having concluded that endogenous methods will not provide an effective way of forming categories using any type of variable, we turn our attention to exogenous techniques. These operate by partitioning a variable so as to maximize its

---

[1] Ward's 1985 Automotive Yearbook, Wisconsin Breast Cancer, and Credit Card Application Approval all from the Machine Learning Database Repository held at University of California, Irvine, CA.; and the British Household Panel Survey available through The Data Archive, University of Essex, UK.

association with some other categorical variable. For example, given a dataset whose variables included income as an interval variable and house ownership as a dichotomous nominal variable, one could partition income into two categories at the point which maximises some measure of association between income and house ownership. In such a case the existing nominal variable may be said to *induce a partitioning* of the partioned variable.

Almost all of the numeric feature discretization methods for classification learning that were discussed above are examples of this type of category formation. They are intended for use as part of a supervised learning process in which the training examples are classified. Hence an obvious way of categorizing a numeric variable is to partition it in a way that maximizes its association with the classification to be predicted.

Unfortunately many of these methods are ineffective in partitioning a variable when the association between it and a classification is weak, and some of them of computationally expensive. We have developed a method for dichotomising numeric variables [13] by maximizing either *lambda* [12], a measure of the proportionate reduction in prediction error achieved by forming the categorization, or *zeta* a measure of the prediction accuracy achieved. This procedure, like the percentile methods, has $O(n \log n)$ complexity. It has been tested on the same datasets as the endogenous methods discussed in the preceding section. For most of the datasets the technique found partitioning points that greatly increased prediction accuracy. Work is currently in progress to incorporate this method into an decision tree algorithm, thus permitting direct comparison with the results reported by Dougherty *et al.* [4].

However, the results obtained with the British Household Panel Survey (BHPS) dataset were disappointing: the procedure was unable to find division points which significantly increased prediction accuracy. The BHPS dataset is part of a very large annual survey that collects a lot of data concerning individuals' socio-economic situation, family circumstances and opinions on social and political issues. It is intended to allow social scientists to investigate social change over a period of years. The dataset used for our experiments was collected in 1991 and contains about 3500 sets of responses to about 100 questions.

Datasets such as the Breast Cancer and Credit Approval databases have obvious dependent variables. In contrast the BHPS data contains numerous relationships but there is no obvious candidate for the variable to be predicted.

We expected that if we partitioned certain obviously fertile variables, such as age or income, using a range of nominal variables we would find that divisions were repeatedly made at similar points. For example, age might be repeatedly split at around retirement age because this is associated with numerous changes in lifestyle. In fact we completely failed to discover such recurrent breakpoints. The problem is that although retirement is a socially significant event, its effect on any one variable is very modest and hence it is impossible to precisely locate the partitioning point.

The obvious conclusion is that if SNOUT is to be able to partition data of this type, it must use category formation techniques that simultaneously take account of relationships with many variables.

5   Exogenous Category Formation: Many Variables

The techniques discussed in the preceding section maximize the association between a single existing categorical variable and the partitioned variable. An obvious extension of this approach is to consider whether a group of variables can collectively induce a partitioning. Such a technique would be useful when the investigator is not concerned with identifying good predictors of particular variables within the dataset but is more concerned with identifying *fertile variables*: that is, those variables that have significant associations with many other variables.

Once a variable has been partitioned it too can induce partitionings in other variables. This suggest the possibility of category formation methods in which every variable could contribute to the partitioning of all the others.

5.1   Exogenous Category Formation in SNOUT

SNOUT currently includes a module providing this type of partitioning using a two phase method to dichotomize all the variables in a dataset. It has its origins in exploratory data analysis techniques advocated by Davis [3] based on dichotomizing all variables so a common measure of association can be used.

During the first phase SNOUT attempts to dichotomise all the variables using simple endogenous techniques adapted from Davis (*op. cit.*):

Phase 1: Initial Dichotomisation

> *Interval Variables:*
>> Divide at the median
>
> *Ordinal Variables:*
>> Divide at point minimizing difference in size of the two resulting groups.
>
> *Nominal Variables:*
>> If largest existing category contains at least 35% of
>> total, assign this to one category and all remaining examples to another.
>> Otherwise set this variable aside for refinement in phase 2.

During the second phase SNOUT constructs revised dichotomies for all the variables using a method which allows all the other dichotomised variables to "vote" for a prefered partitioning. For ordinal and interval variables, the other dichotomised variables vote for a single partitioning point. For nominal variables, the other variables vote for their preferred categorization of each of its values.

Phase 2: Refining the Dichotomisation

> *Ordinal variables:*
>> 1. Find optimal partition point induced by each of the other dichotomised variables D_i thus:

(a) For every possible partitioning point measure the association with D_i by calculating Cramer's V [12]

(b) Use Chi-Square test to check that association using partitioning point p yielding largest value of Cramer's V is statistically significant.

(c) If so record one vote for partitioning at p; otherwise record no vote.

2. Dichotomise variable at point receiving most votes.

Interval Variables:

Divide interval into octiles; then apply method used for ordinals.

Nominal Variables:

1. Create two new categories, A and B; Place the largest existing category of the nominal variable in A.

2. Find optimal partitioning of the nominal variable V induced by each of the other dichotomised variables D_i thus:

(a) Form the $2 \times n$ cross tab table for V and D_i:

(b) Use Chi-Square test to check that there is a statistically significant association between V and D_i.
If the test fails abandon attempt to use D_i and proceed to use next dichotomised variable D_{i+1}.

(c) Determine which cells in the $2 \times n$ cross tab table exceed the counts expected if V and D_i were independent.

(d) Find the row r in the cross tab table in which the actual count for the largest category of V exceeds the expected value.

(e) For each count in row r that exceeds the expected value, record a vote for assigning the corresponding value of V to class A.
For all other counts in row r record a vote for assigning the corresponding value of V to class B

3. Assign each possible value of the nominal variable V to A or B on the basis of the number of votes cast by all the dichotomised variables.

In summary, this procedure selects a partition that both maximises the number of variables associated with the partitioned variable and maximises the strength of those associations.

Note that it would be possible to repeat Phase 2 either a preset number of times or until some stability criterion was satisfied. We have considered but not yet implemented this possibility because Phase 2 is computationally expensive: it takes about 10 minutes to dichotomise all the variables in the British Household Panel Survey dataset using a 66MHz 486 PC. Since SNOUT is intended as an interactive tool this type of delay would often be unacceptable. Fortunately faster machines are likely to bring the response time down to something more acceptable but further delays for additional iterations of Phase 2 could only be justified if they were likely to provide the investigator with new insights. More experimental work will be needed to establish if this is the case. A single iteration of Phase 2 has proved very capable of exposing interesting relationships; and this is the fundamental purpose of SNOUT.

5.2   Results

This method was used to dichotomize all the variables in the British Household Panel Survey 1991 dataset described above. The following examples illustrate the results obtained:

SNOUT partitioned the responses to the question "Which daily newspaper do you purchase or read first?" as follows:

Group A: Mirror, Star, Sun
Group B: Express, Financial Times, Guardian, Independent, Mail, Telegraph, Times, Today

Anyone familiar with the British press will immediately recognise that SNOUT has identified the most fundamental distinction: Group A contains the tabloid papers while Group B contains the broadsheets plus the "middlebrow" Express and Mail. Note that this dichotomy is based entirely on the attributes of the readers: SNOUT has no other information concerning the properties of these newspapers.

Another nominal variable indicates the part of the UK where the respondent lives. SNOUT dichotomised this thus:

Group A: Inner London, Outer London, Rest of South East, South West, East Anglia, North Yorkshire and Humberside, Rest of North West.
Group B: East Midlands, West Midlands, West Midlands Conurbation, Greater Manchester, Merseyside, South Yorkshire, West Yorkshire, Tyne and Wear, Rest of North, Wales, Scotland.

This is basically a North/South division that coincides with many traditional stereotypes. It is interesting to note that Group A comprises those areas where support for the Conservative or Liberal Democrats is strongest while Group B is mainly Labour territory.

Both of these examples concern the dichotomy of nominal variables. The dichotomising procedure also behaves sensibly with interval variables. For example, the original medial partition of age at 42 is moved to 61; close to a typical value for retirement age.

5.3   Additional Information for the Exploratory Data Analyst

While such dichotomies are often interesting in isolation the serious investigator will normally want to to know more about how they were derived. To facilitate this SNOUT also provides the user with a list of all those other variables that contributed to a dichotomy. In the examples discussed above, 16 variables gave rise to the newspaper groupings while 45 were used in partitioning the regions. The number of contributing variables provides some indication of the fertility of the partitioned variable.

In addition, SNOUT tells the user how each of the contributing variables was associated with the categories formed. For example, the information supplied with the dichotomy of region includes the findings that people living in Group A regions tend to have more years of education, and are less likely to smoke, read a tabloid newspaper, live in a council house or regard themselves as working class. On the basis of this array of information, the investigator is likely to identify relationships that are of particular interest and investigate them more thoroughly using conventional statistical methods.

5.4   Relationship to Alternative Clustering Techniques

It is interesting to compare SNOUT's dichotomisation procedure with other techniques for forming categories in data sets containing nominal variables. Many of these are basically techniques for forming clusters of similar data items. For example, COBWEB [10] partitions the set of data items into classes in such a way as to optimise category utility: that is, to maximise the number of variable values that can be correctly predicted given knowledge of the class to which a data item belongs.

SNOUT's dichotomisation procedure forms groups of the values a variable may take rather than groups of data items. Hence direct comparison of the two approaches is not possible. Indirect comparison would be possible since any partitioning of data items induces a partitioning of a given variable's values and vice versa. However, SNOUT dichotomises all the variables and thus many distinct partitionings of the data items may be induced. Consequently it is not clear how its performance should be compared with a system like COBWEB.

The difference between SNOUT and COBWEB can thus be summarised: SNOUT forms clusters of variable values while COBWEB forms clusters of data items. This suggests a third possibility: a procedure that forms clusters of variables. In fact SNOUT incorporates just such a module that makes use of the results of dichotomisation. The degree of association between every pair of dichotomised variables is determined using either Yule's Q or Cramer's V. An agglomerative hierarchical clustering technique [6] is then applied to the variables, rather than the data items, using the degree of association as the measure of similarity. In this way SNOUT builds a variable similarity tree in a bottom up fashion thus identifying groups of similar variables. (For further discussion of this component of SNOUT see [20]).

We thus have three distinct approaches to category formation in data sets comprised of nominal variables. Each can be viewed as a clustering process, but each forms clusters of different types of entity: data items, variables, or variable values. Hence each does a different job and all can be regarded as useful tools to be put at the disposal of the exploratory data analyst.

6   Conclusions

There is a considerable body of work concentrated on one aspect of category formation in data mining: partitioning numeric variables for use in decision tree induction systems. Nevertheless there is still much scope for further work, particularly to determine whether local partitioning can bring benefits to offset its greater cost, and to find the most computationally efficient global and local methods.

There is very little work on other aspects of category formation. We have described a new procedure that dichotomises all levels of variable and produces sensible results. However there is much scope for further work. For example, the program in its present form only generates dichotomies. While this may be appropriate for many variables, three or possibly even more may be better in many cases. An improved program would be able not only to produce such alternative partitionings but decide which is best without using inordinate amounts of computing resources.

Acknowledgments

Work on the development of SNOUT has been funded by the Economic and Social Research Council's programme on the Analysis of Large and Complex Datasets under grant number H519255030.

References

1. L. Breiman, J. H. Friedman, R. A. Olshen, and C. J. Stone. *Classification and Regression Trees*. Wadsworth, Pacific Grove, CA., 1984.
2. J. Catlett. On changing continuous attributes into ordered discrete attributes. In Y. Kodratoff, editor, *EWSL-91. Lecture Notes in Artificial Intelligence 482*, pages 164–178. Springer-Verlag, Berlin – Heidelberg – New York, 1991.
3. J. A. Davis. *Elementary Survey Analysis*. Prentice-Hall, Englewood Cliffs, New Jersey, 1971.
4. J. Dougherty, R. Kohavi, and M. Sahami. Supervised and unsupervised discretisation of continuous features. In *Proc. Twelfth International Conference on Machine Learning*, Los Altos, CA, 1995. Morgan Kaufman Publ. Inc.
5. B.H. Erickson and T.A. Nosanchuk. *Understanding Data*. The Open University Press, 1979.
6. B. S. Everitt. *Cluster Analysis*. Heinemann, London, 2nd edition, 1980.
7. B. S. Everitt and G. Dunn. *Applied Multivariate Statistical Analysis*. Edward Arnold, London, 1991.
8. U. M. Fayyad and K. B. Irani. On the handling of continuous-valued attributes in decision tree generation. *Machine Learning*, 8:87–102, 1992.
9. U. M. Fayyad and K. B. Irani. Multi-interval discretization of continuous-valued attributes for classification learning. In *Proc. Thirteenth International Joint Conference on Artificial Intelligence*, pages 1022–1027, Los Altos, CA, 1993. Morgan Kaufman Publ. Inc.

10. D. H. Fisher. Knowledge Acquisition Via Incremental Clustering. *Machine Learning*, 2:139–172, 1987.
11. D. H. Fisher and P. Langley. Conceptual clustering and its relation to numerical taxonomy. In W. A. Gale, editor, *Artificial Intelligence and Statistics*, pages 77–116. Addison-Wesley, Reading, Mass., 1986.
12. J. Healey. *Statistics: A Tool For Social Research*. Wadsworth, Belmont, CA., 1990.
13. K. M. Ho and P. D. Scott. Discretization of continuous variables in bivariate relationships. In *Proceedings of KDD-97, The Third International Conference on Knowledge Discovery and Data Mining, Newport Beach, CA.*, Menlo Park, CA., August 1997. AAAI Press.
14. R. C. Holte. Very simple classification rules perform well on most commonly used datasets. *Machine Learning*, 11:63–91, 1993.
15. R. Kerber. Chimerge: Discretisation of numeric attributes. In *AAAI-92 Proceedings of the Tenth National Conference on Artificial Intelligence*, pages 123–128, Cambridge, Mass., 1992. The MIT Press.
16. J. R. Quinlan. Induction of decision trees. *Machine Learning*, 1:81–106, 1986.
17. J. R. Quinlan. *Programs for Machine Learning*. Morgan Kaufman Publ. Inc., Los Altos, CA, 1993.
18. J. R. Quinlan. Improved use of continuous attributes in c4.5. *Journal of Artificial Intelligence Research*, 4:77–90, 1996.
19. M. Richeldi and M. Rossotto. Class-driven statistical discretisation of continous attributes (extended abstract). In *ECML-95: Proceedings of the European Conference on Machine Learning, Lecture Notes in Artificial Intelligence*, volume 914, Berlin – Heidelberg – New York, 1995. Springer-Verlag.
20. P. D. Scott, A. P. M. Coxon, M. H. Hobbs, and R. J. Williams. Snout: An intelligent assistant ofr exploratory data analysis. In *Lecture Notes in Artificial Intelligence: Proceedings of PKDD-97, The First European Symposium on Principles of Data Mining and Knowledge Discovery, Trondheim.*, Berlin – Heidelberg – New York, June 1997. Springer-Verlag.
21. J. W. Tukey. *Exploratory Data Analysis*. Addison-Wesley, Reading, Mass., 1977.
22. J. H. Ward. Hierarchical grouping to optimize an objective function. *Journal of the American Statistical Association*, 58:236–244, 1963.

A Systematic Description of Greedy Optimisation Algorithms for Cost Sensitive Generalisation

Maarten van Someren[1], Cristina Torres[1,2] and Floor Verdenius[2]

[1] **Department of Social Science Informatics (SWI)**
Faculty of Psychology, University of Amsterdam
Roetersstraat 15, 1018 WB Amsterdam, The Netherlands
email: maarten@swi.psy.uva.nl
[2] **AgroTechnological Research Institute (ATO-DLO)**
Postbox 17, 6700 AA Wageningen, The Netherlands
email: F.Verdenius@ato.dlo.nl

Abstract. This paper defines a class of problems involving combinations of induction and (cost) optimisation. A framework is presented that systematically describes problems that involve construction of decision trees or rules, optimising accuracy as well as measurement- and misclassification costs. It does not present any new algorithms but shows how this framework can be used to configure greedy algorithms for constructing such trees or rules. The framework covers a number of existing algorithms. Moreover, the framework can also be used to define algorithm configurations with new functionalities, as expressed in their evaluation functions.

1 Introduction

There is a wide range of practical problems in which a general model must be constructed from a sample of observations and background knowledge about the domain. The purpose of such models can be predictions (e.g. of class membership) or decisions (e.g. on the best action to take). In this paper we discuss algorithms that apply earlier observations and knowledge about costs and benefits to construct a model of a domain. Then this model is used to derive predictions or decisions in later contexts.

Suppose that a set of records of patient descriptions is given with their diagnoses. One problem then is to discover how in general a patient can be classified as having a disease. The goal in this case is to maximise the accuracy of future classifications. Although appropriate in many circumstances, application of such a model in practice is limited when its application generates high costs. This is for example the case when the costs of obtaining attribute-values vary substantially over the attributes. This is for instance the case in medical diagnosis, where doctors balance the effectiveness of various possible tests with the costs of doing those tests. Expensive tests should only be performed when inexpensive

X. Liu, P. Cohen, M. Berthold (Eds.): "Advances in Intelligent Data Analysis" (IDA–97)
LNCS 1280, pp. 247–257, 1997. © Springer–Verlag Berlin Heidelberg 1997

tests prove insufficient to make a diagnosis. The goal in this case is to maximise future accuracy but also to minimise *measurement costs*.

A diagnosis for a patient will lead to a specific treatment of that patient. Incorrect diagnosis, i.e. misclassification, may have costs associated with it. These costs will not be uniformly distributed over all possible misclassifications. For example, classifying a pneumonia patient as healthy will lead to different costs (no treatment, serious health problems) than in the opposite case (treatment, patient is cured, possibly minor side effects of treatment). Similarly, classifying a pneumonia patient as suffering from bronchitis leads to different costs than classifying the same patient as healthy. This adds another aspect to the problem: the problem may now be to optimise future accuracy and misclassification costs.

A type of problem that does not involve future classification is to **characterise** a class. For example, from a database of patients who are classified by their final diagnoses, a scientist may want to find the properties that characterise a particular disease. The goal is then to find a description of a class that covers as many instances of a class as possible. This description does not need to distinguish between the class it described and other classes but only to give a description that covers it well. Riddle et. al. [Riddle et al., 1994] point out that there are situations in which cases belonging to a target class must be identified, but in which characterisation of the whole input space is not required. They give the example of a manufacturing process where people are interested in predicting whether a given case belongs to a special class representing an anomaly to be monitored. In this *characterisation* problem *measurement costs* and *misclassification costs* are irrelevant.

If it is possible to characterise a class in terms of attribute-value pairs, the next step is to achieve that class by manipulating the values of attributes. This constitutes the third problem type: finding an optimal set of attribute-value pairs for **achieving** a class. Here the problem is to find a model that is optimal in achieving a particular class, given a set of cases. An example in a medical context is treatment planning. From a set of patients who are treated with different combinations of medicine, the goal is to find which combination is optimal in leading to a particular class (recovery) and minimises costs of the medication. There are two variants of the **achievement** problem: a value must be found for *all* attributes (a complete vector) or only for a *subset* of attributes, because not all attributes may be relevant. Optimality refers to a property of the cases, in particular to the costs associated with attribute-value pairs and costs of *unwanted achievements* (cf. misclassifications), rather than attributes or classes as above. The costs of realising specific attribute-value pairs accumulate over all decisions in a rule. The goal is to find a set of attribute-value pairs that minimises attribute-value and misclassification costs and maximises the probability of achieving a required class.

We now end up with four dimensions along which problems can be perceived:

- classification accuracy
- measurement costs

— misclassification costs
— achievement costs

In the problems we are interested in the first item always plays a role. Wether other costs become important depends on the problem at hand. Achievement costs behave as *dynamical misclassification costs.* In evaluation criteria achievement costs are treated similarly to misclassification costs [Turney, 1995] .

We present a language that systematically describes these inductive optimisation problems. To see why a systematic description of inductive techniques is useful, consider the following problem, taken from our experience. Ceramic objects are produced in a baking process that has numerous parameters, such as baking temperature and duration of the baking process. Data representing several different parameter settings of the baking process and also different products were available. Moreover, the result of quality control (accept or reject) were available. Associated with different values of the parameters are costs (e.g. higher temperatures are more expensive). The company involved wants to know the optimal parameter settings and also the relation between parameter settings and passing quality control. The question is: which (if any) of the methods described here is appropriate for this problem?

Our goal is to acquire a model that gives for a given task the optimal classification result. Optimal is expressed in terms of the dimensions as presented above. Many different types of model representations can be constructed for these tasks. Here we discuss two types of representation: decision trees and rules. Both can be combined with measurement and misclassification costs. In section 2 we discuss the difference between these representations. Section 3 details the notion of model accuracy. Then, in section 4 the learning goals associated with the different problem types are formulated in one criterion, and in section 5 we show how these descriptions can be used to configure algorithms that can be used to solve inductive optimisation problems in the style of [Langley, 1996]. Section 6 shows how a number of existing techniques are covered by the framework, and the paper ends with some concluding remarks.

2   Inductive optimisation tasks

Inductive optimisation problems concern the construction of a model from a sample of data. The data consist of a set of *cases*. A case is an attribute-value vector. The values may be continuous or nominal. Each case is associated with a *class*.

In addition to cases some problems involve *background knowledge*. Background knowledge is not specific for particular cases but it is general information related to attributes, values or classes.

Attribute measurement costs (attCosts(A)): costs associated with an attribute, e.g. for measuring it

Attribute value costs (attValCosts(A, V)): associated with the value of an attribute, e.g. for realising that attribute-value

Misclassification costs (misClassCosts(Class1, Class2): costs of classifying an object as class Class2 when it in fact belongs to Class1. The misclassifications costs are given in a matrix of size NrCl x NrCl, where NrCl is the total number of classes.

The goal of an inductive optimisation problem is to derive a model that is both a good generalisation and that is also optimal with regard to costs. The model can take the form of a *decision tree*, a *rule set*, a *rule* or a *vector*. These languages are equally expressive but there is a difference in how they are acquired (see section 5) which leads to different models.

A *decision tree* [Quinlan, 1986] consists of nodes and branches. Each node corresponds to a test on the value of an attribute. The starting node is called the root of the tree. The branches represent the possible values of the attribute. The *leaves* of the tree are associated with classes. Given a case to classify, the test specified by the root is applied to the case and according to the result, a branch is chosen. This process is repeated at each non-terminal node until a leaf is reached, indicating the class of the case.

A *rule* and a *vector* consist of a conjunction of attribute-value pairs and a class. A rule or vector covers a case if the case satisfies all the rule conditions. If so, the case is classified as belonging to the class of the rule or vector. The difference between rule and vector is that a vector includes *all* attributes. A *rule set* is a set of rules. When these are used for classification, all rules are applied to a case and the results are combined to decide on the class.

The goal of an inductive optimisation problem is to find a model in the form of a decision tree, a rule, a ruleset or a vector that also minimises costs, in particular, *expected costs*. As we discussed in the introduction, there are several different types of (expected) costs. Here we define accuracy and the different forms of (expected) costs for the decision trees, rules/vectors and rule sets.

path(Leaf): set of attribute-value pairs on the path from the root of a tree to a Leaf

probability(Leaf): probability that a case ends up in a leaf. Estimated by the relative frequency

$$probability\,(Leaf) = \frac{\#\,(Leaf)}{\#\,(Total)} \tag{1}$$

where $\#\,(Leaf)$ is the total number of cases covered by the Leaf and $\#\,(Total)$ is the total number of cases.

probabilityCorrect(Leaf): probability that a case that ends up in a leaf is classified correctly, assuming that it is classified as the most frequent class in the cases at the leaf. This is therefore equal to the proportion of the most frequent class in the leaf.

$$probabilityCorrect\,(Leaf) = \frac{\#\,(MaxClass\,(Leaf))}{\#\,(Leaf)} \tag{2}$$

where $\#MaxClass\,(Leaf)$ is the class that has the highest frequency for Leaf, $\#\,(Leaf)$ is the total number of cases covered by the Leaf.

classCoverage(Leaf, Class): the probability of ending up in Leaf if an instance belongs to Class

$$p\left(Leaf/Class\right) = \frac{\#\left(Class/Leaf\right)}{\#\left(Class\right)} \tag{3}$$

where $\#\left(Class/Leaf\right)$ is the number of cases in Leaf that belong to the required Class, and $\#\left(Class\right)$ is the total number of cases.

pathMeasuringCost(Leaf): the sum of the measuring cost of the attributes that appear in the path from the root to Leaf. If an attribute appears more than once in that path, its cost is considered only once.

$$pathMeasuringCost\left(Leaf\right) = \sum_{node=root}^{Leaf-1} Cost\left(Attrib_{node}\right) * ff\left(Attrib_{node}\right)$$
$$\tag{4}$$

where $ff\left(Attrib_{node}\right) = 1$ if $Attrib_{node}$ is encountered for the first time in the path, 0 otherwise, and $Cost\left(Attrib_{node}\right)$ is given in a measurement cost table. $Attrib_{node}$ is the attribute tested upon in the decision node.

expected measuring costs(Tree): the sum of (probability(Leaf) * pathMeasuringCost(Leaf) for all leaves.

$$\frac{\sum probability\left(Leaf\right) * pathMeasuringCost\left(Leaf\right)}{N(Leaves)} \tag{5}$$

where N(Leaves) is the number of leaves.

misclassificationCost(Leaf): all cases at this leaf will be classified as the most frequent class and therefore all cases with different classes will be misclassified (as defined by [Roberts et al., 1995]). The **misclassficationCost(Leaf)** is the sum of misclassification costs for all misclassified cases.

$$misclassificationCost\left(Leaf\right) = \sum_{Classj} p\left(j/Leaf\right) * C\left(i,j\right) \tag{6}$$

where i and j are classes, C is the Cost matrix for classifying a case class j as being of class i.

expected misclassification costs(Tree): average product of probability(Leaf) and misclassificationCost(Leaf) for all leaves:

$$\frac{\sum\left(probability\left(Leaf\right) \times misclassificationCost\left(Leaf\right)\right)}{N} \tag{7}$$

where N is the number of leaves.

Rules correspond closely to *paths* through a tree. If we call the *end* of a rule a *leaf*, we can directly apply the definitions that we gave above for trees to properties of rules and rule sets.

3 Accuracy

One of the main goals is to obtain models that will give accurate classifications for future cases and accurate characteristic descriptions. There is of course no obvious way to assess the accuracy of a given model. The three approaches to this problem use an independent test set, confidence intervals and *comparisons*. The latter is possible because criteria can be given that allow comparison of two models on accuracy without giving an *absolute* value. Here we want to use accuracy in learning goals and in heuristic learning methods. Therefore we do not discuss details and value of these approaches but instead show how they can be used to define learning goals and greedy algorithms.

The *testset* approach evaluates possible (extensions of) models on a test set that is independent of a learning set. This method can be easily included in a learning goal and an algorithm because applying a (partial) model to a testset gives a percentage correct that can be interpreted as an estimate of accuracy. This method does not give an unbiased estimate of accuracy because the testset is used in the construction of the hypothesis. The results will therefore be higher than the actual accuracy.

The *confidence interval* approach is to construct a confidence interval for the proportion correct predictions at a node, based on the subset of the original data that is associated with this node (e.g. see [Quinlan, 1992]). It has two problems. The first is that it requires a parameter for the probability of an error and the second is that the amount of data is usually too small to give reliable intervals (Quinlan exploits this in the C4.5 system because larger samples lead to smaller intervals; this favours larger samples and thereby gives a criterion for pruning). The statistical basis of this approach is weak, mainly because comparing intervals based on different sized samples is not very meaningful.

The comparison approach is based on the idea that accuracies of two models can be compared even if they cannot be assessed absolutely. Comparison uses a dimension of models and data that is related to accuracy. Frequently used measures are:

Proportion correct predictions: Although at first it may seem that a model that is completely correct with respect to the set of cases is optimal, this is not so because of the problem of *overfitting*. This means that there is not a monotonic relation between the actual accuracy and the proportion of correct predictions. At some point actual accuracy is likely to decrease with increasing proportion correct predictions. If this point can be detected then proportion correct predictions can be used.

Size: Many learning systems are based on Occams razor: if two models explain the same data then the simpler one is better.

Amount of information: This is the minimal number of bits that is needed to transmit a model. If it is possible to find a minimal coding then this gives a better basis for comparison than *size* because it abstracts from the language of the model. Unfortunately, finding such an encoding is difficult. In practice an ad hoc encoding is used.

Size of information residual A different but related measure is used in ID3, one of the roots of decision tree learners. The amount of information in the dataset can be estimated from marginal frequencies of the classes. This is then used to compute how much information is *extracted* from the data into a decision tree by adding the *amounts of information* in the subsets that arise after applying the new decision tree. This measure can be interpreted as a prediction of how many nodes the tree will need to fully cover the data. In this sense it is closely related to proportion correct. In fact it is a direct function of proportion correct (propCorrect(Node)) in the case of 2 classes:

$$Information(Set) = -\left(propCorrect(Set) \times \log propCorrect(Set)\right)$$

$$-\left((1 - propCorrect(Set)) \times \log\left(1 - propCorrect(Set)\right)\right) \qquad (8)$$

Total description length: This is a more sophisticated approach than *amount of information*. It takes the *actual* model and the cases that are **not** explained by the model together as the model and minimises the description length of this *compound model*. This is then used as measure of the accuracy of the *actual* model.

The key question is of course how to trade-off simplicity of the model against the extent to which a model explains the data. If two models are equally simple or explain an equal amount of data, the choice is easy but in the general case more complex models explain more data. The total description length is the only criterion that defines a combined function: the sum of the two encodings. It is not clear yet if this is the right way to make the trade-off.

A difficulty with the use of the accuracy measures in learning algorithms is that they are used both to evaluate accuracy and to *select* the sample (in the next cycle). Therefore they are not *unbiased* estimators of accuracy and in general they will overestimate accuracy.

4   Learning Goals

The learning goals that we mentioned in the introduction can now be stated more precisely in terms of the properties of decision trees. In general the goal of an inductive optimisation problem is to construct a model from data that is maximally accurate and that is optimal with respect to one ore more cost criteria. There are two standard ways to combine multiple criteria. The first is to use weights and the second is to order them. We shall include both ways. This gives the following language for inductive optimisation problems:

Find a model represented as {*tree* | *rule* | *vector* | *ruleset*} that maximises:

$$w_1 \times C_1 + w_2 \times C_2 + w_3 \times C_3 + w_4 \times C_4 \qquad (9)$$

where criteria C_i are `accuracy, -expected measurement costs, -expected misclassification costs, class coverage.`

Now we can accurately formulate the problem of baking ceramic products, as described in the introduction. The company is interested in finding one or more rules that minimise the costs of achieving the specific parameter values. In other words:

> Find a model represented as a ruleset that maximises the value of the criterion as formulated in 9 with $C_{1...4}$ defined as above, $w_1 = 0$, and $w_{2..4} \neq 0$. Proper values for $w_{2..4}$ may be determined experimentally, or they may be defined a priori on the basis of client requirements.

5   Greedy top down algorithms

From an algorithmic viewpoint we note that all these tasks are forms of the general task of finding a model that optimises a function on the data, in particular, once an operational criterion is chosen for *best generalisation*. We also note that these criteria can be applied to *incomplete* decision trees, rules and vectors. The problem definitions can therefore be used as evaluation function in a search technique that searches for an optimal model. Here we show how our systematic description can be used to automatically configure greedy algorithms for solving inductive optimisation problems.

Greedy algorithms construct a model step by step; at each step possible model extensions are assessed and the best is selected. The generic algorithm is:

```
DecisionTreeInduction(SetOfCases, Node, Value)

IF termination criterion(SetOfCases, Node, Value)
THEN return leaf labelled with most frequent class in SetOfCases
ELSE generate all possible extensions(Node);
     evaluate extensions with SetOfCases;
     findBestTreeExtension;
     IF value of best-extension  > value of Node
     THEN extend Node with best extension;
         split  SetOfCases into subsets SubsetOfCases(i)
                according to best-extension
         FOR each new node in extension:
                DecisionTreeInduction(SubsetOfCases(i),
                         new node, value of best extension)
     ELSE return leaf labelled with
                most frequent class in SetOfCases
```

Algorithm 1. Generic Top-down Induction of a Decision Tree.

Here the function `findBestTreeExtension` uses the learning goal as evaluation function. In this case, this function *optimises the gain over all nodes of the subtree*. The induction process stops when no extension of the tree gives an improvement. It is possible to specify a `termination criterion(SetOfCases, Node, Value)` that will cause an earlier stop.

The underlying assumption when generating decision trees is that the different values of attributes give information pertaining to class membership. However, it can occur that one specific combination of attribute/value is characteristic for a class while other values of that attribute give no information about class membership. In such situations, a Top Down Induction of Decision Tree (TDIDT) algorithm can fail to extract those characteristic attribute/value combinations from the data.

An analogous generic algorithm can be given for constructing rules (see Algorithm 2). Instead of a subtree this constructs a rule. To construct a set of rules, a new rule is started when a rule has stopped growing and when there are still uncovered cases. The generic algorithm for rule induction is:

```
RuleInduction(SetOfCases, Node, Value)

IF termination criterion(SetOfCases, Node, Value)
THEN return leaf labelled with most frequent class in SetOfCases
ELSE generate all possible extensions(Node);
        evaluate extensions with SetOfCases;
        findBestRuleExtension;
        IF value of best-extension  > value of Node
        THEN extend Node with best extension;
             Derive optimal subset SubsetOfCases(i) of SubsetOfCases
                    according to best-extension
             RuleInduction(SubsetOfCases(i), new node,
                    value of best extension)
        ELSE return leaf labelled with
                    most frequent class in SetOfCases
```

Algorithm 2.   Generic Top-down Rule Induction.

The structure of this algorithm is very similar to that of Algorithm 1. Again, learning goal is used in the evaluation function, now in `findBestRuleExtension`. In this case however, this function *optimises the gain over the best node in the extension, and only this node is further expanded.*

Both algorithms 1 and 2 show how greedy inductive optimisation algorithms can be configured by using the learning goal as evaluation function. Background knowledge can be used in this function. Because the algorithms construct their models *from small to large*, there is an implicit preference for smaller models.

Non-linear optimisation problems are usually expensive (both NP-complete and large) which means that they need heuristic techniques that only approximate the best solution. One type of algorithm is the class of greedy construction techniques: conjunctions, trees or DNF formulae are constructed greedy with a small lookahead. Partial models are evaluated with the optimisation function. Alternative solutions to similar problems are offered by local search techniques, equipped with similar heuristic evaluation functions. Genetic algorithms for example are less sensitive to properties of the distribution of the evaluation function (such as the presence of nonmonotonocity, local maxima, plateaus). On the other hand they are more expensive. Greedy algorithms have a good track record as offering a good trade-off between computation costs and quality of the result.

6   Related systems

Our framework gives a systematic description that covers a number of existing systems. EG2 [Nunez, 1991] is a TDIDT algorithm that uses attribute measuring costs in generating a decision tree. EG2 generates a decision tree with the following selection function:

$$measurementCost\,(Attribute) + (w \times informationGain) \qquad (10)$$

Roberts et al. [Roberts et al., 1995] propose an approach for integrating misclassification costs in a tree induction algorithm. We refer to the method as AFMC, that stands for Accounting for Misclassification Costs. This system uses misclassification costs in evaluating possible decision trees. Similar notions are also worked out in [Schiffers, 1997].

Gold-digger [Riddle et al., 1994] is an algorithm that generates predictive rules, called nuggets, by looking for test/outcome combinations that are characteristic for the target class. Several nuggets are obtained by applying the algorithm iteratively. The selection function is maximal classCoverage. It was successfully applied to find the cause of faults in a production process.

The MICO system [Verdenius, 1991] is based on the idea of finding a rule that maximises probability of a class while minimising attribute-value costs. This system was used to find optimal values for parameters of a production process from experiments with different parameter settings.

All these systems can be described in our general framework. They can be seen as special configurations of the learning goal as presented in section 4. Moreover, if the individual algorithms are seen as instantiations in the three dimensional space spanned by *classification accuracy*, *measurement costs* and *misclassification costs*, then new functionalities can be defined by adapting the weights in the goal function as introduced in section 4 ([Torres and Verdenius, 1996]).

7   Conclusion

This paper gives a systematic description of inductive optimisation problems in terms of properties along which models are optimised. Inductive optimisation problems frequently occur in practice. We show how algorithms can be constructed from the systematic descriptions to solve particular inductive optimisation problems. The choice (or configuration) of an algorithm depends on: the purpose of the target knowledge (*classification, characterisation* or *achievement*), variation in predictive value between values of attributes (large differences suggest rules instead of trees) and the role of costs associated with misclassification, measurement or value settings.

Several of the algorithms that can be configured in a principled way correspond to existing systems that were published before. Moreover, new functionalities can be derived in the same framework, by expressing the learning goal as weight vector in the goal function.

References

[Langley, 1996] Langley, P. (1996). *Elements of machine learning*. Morgan Kaufmann, San Francisco.

[Nunez, 1991] Nunez, M. (1991). The use of background knowledge in decision tree induction. *Machine Learning*, 6:231–250.

[Quinlan, 1986] Quinlan, J. (1986). Induction of decision trees. *Machine Learning*, 1:81–106.

[Quinlan, 1992] Quinlan, J. (1992). *C4.5: Programs for Machine Learning*. Morgan Kaufmann, San Francisco.

[Riddle et al., 1994] Riddle, P., Segal, R., and Etzioni, O. (1994). Representation design and brute-force induction in a boeing manufacturing domain. *Applied Artificial Intelligence*, 8:125–147.

[Roberts et al., 1995] Roberts, H., Denby, M., and Totton, K. (1995). Accounting for misclassification costs in decision tree classifiers. In Lasker, G. and Liu, X., editors, *Proceedings of the International Symposium on Intelligent Data Analysis, IDA-95*, pages 149–156.

[Schiffers, 1997]
Schiffers, J. (1997). A classification approach incorporating misclassification cost. *Intelligent Data Analysis*, Vol. 1, No. 1, http://www.elsevier.com/locate/ida

[Torres and Verdenius, 1996] Torres, C. and Verdenius, F. (1996). Selecting decision-tree based learning algorithms. In van der Herik, H., van den Bosch, A., and Weijters, T., editors, *Proceedings Benelearn-1996*, Maastricht. RU Maastricht.

[Turney, 1995] Turney, P. D. (1995). Cost sensitive classification: Empirical evaluation of a hybrid genetic decision tree induction algorithm. *Journal of AI Research*, 2:369–409.

[Verdenius, 1991] Verdenius, F. (1991). A method for inductive cost optimization. In Kodratoff, Y., editor, *Proceedings of the Fifth European Working Session on Learning EWSL-91*, pages 179–191, Berlin. Springer Verlag.

Dissimilarity Measure for Collections of Objects and Values

Petko Valtchev and Jérôme Euzenat

INRIA Rhône-Alpes
ZIRST, 655 av. de l'Europe, 38330 Montbonnot Saint-Martin, France
phone + 33 (0)4 76 61 53 75, fax + 33 (0)4 76 61 52 07
{Petko.Valtchev, Jerome.Euzenat}@inrialpes.fr

Abstract. Automatic classification may be used in object knowledge bases in order to suggest hypothesis about the structure of the available object sets. Yet its direct application meets some difficulties due to the way data is represented: attributes relating objects, multi-valued attributes, non-standard and external data types used in object descriptions. We present here an approach to the automatic classification of objects based on a specific dissimilarity model. The *topological* measure, presented in a previous paper, accounts for both object relations and the variety of available data types. In this paper, the extension of the topological measure on multi-valued object attributes, e.g. *lists* or *sets*, is presented. The resulting dissimilarity is completely integrated in the knowledge model TROPES which enables the definition of a classification strategy for an arbitrary knowledge base built on top of TROPES.

1 Introduction

The global aim of our study is the development of a strategy for automatic taxonomy building within object knowledge bases. Methods for inferring taxonomic structures, or classifications, first appeared within the numerical taxonomy paradigm, in statistics. Statistic classification is aimed at detecting regularities in sets of feature-described individuals. Feature values, mainly numerical, are used to establish a proximity function on individuals. The classification methods tend to group highly similar individuals into clusters and, in some cases, organize clusters hierarchically. The automatic classification may be used to discover the conceptual structure of the specific domain where data comes from [19].

Such structure-detecting methods may be useful for domains where large amounts of data are processed, like databases and knowledge-based systems. In fact, the extraction of structural knowledge from databases, in particular by means of clustering, is one of the goals of the recently emerged *data mining* field [14].

Our own concern is the introduction of such techniques within object formalisms. The classification task has to be carried out within the knowledge base, i.e. in the context in which the data is stored and manipulated. Therefore,

X. Liu, P. Cohen, M. Berthold (Eds.): "Advances in Intelligent Data Analysis" (IDA–97)
LNCS 1280, pp. 259–272, 1997. © Springer–Verlag Berlin Heidelberg 1997

the complexity of the object description languages has to be successfully dealt with.

The necessity of classifying more complex data motivated the constitution of the *conceptual clustering* paradigm as an extension of numerical taxonomy in machine learning [13]. In contrast to the statistical methods conceptual clustering ones work on symbolic and structured data and put the emphasis on the constitution of intentional descriptions, a concept, for each cluster. Various conceptual clustering methods on different kinds of data description formalisms have been reported since: attribute-value-like [8], first order logic [2] and graph formalisms [11]. Other approaches like *concept formation* [10] or Bayesian classification [6] rely on probabilistic considerations about feature values when grouping individuals.

For the purposes of object taxonomy inference, a proximity-based approach seems to be well suited [3]. Therefore, a key problem to address is the definition of a proximity model which fits object descriptions. This means, in particular, that the model should be able to process the variety of object features and inter-object relations admitted by the concrete formalism.

We propose a generic dissimilarity model, the *topological* measure, which is universally applicable both on features and relations. Its basic principle is the use of the hierarchical structure of a domain to assess the proximity between domain elements. The model has been presented in a previous paper [18].

In the present paper, the extension of the topological measure on multi-valued object attributes is discussed. The paper starts by a motivating example of a domain which illustrates some specific features of the object formalisms like complex objects, rich data type sets and multi-valued attributes (Section 2). Then, for self-containment purposes, we recall the definition of the topological dissimilarity (Section 3). We also discuss the concrete functions in some typical cases of domain structure like *nominal, ordinal,* etc. Next, the extension of the topological dissimilarity on multi-valued attributes, *sets* and *lists*, is introduced (Section 4). Finally, the integration of the measure within a concrete object model, TROPES, and its taxonomy building tool T-TREE is presented (Section 5).

2   Motivating example

Electromyography is a set of electrophysiological technics which allow the neuromuscular diseases to be diagnosed. The electromyographic diagnosis is carried out from a systematic acquisition of numeric and symbolic data. It is decomposed into a set of well defined steps: formulation of hypotheses and specialized examination procedure suited to the patient treated, evaluation of procedure results, validation or questioning of the current hypothesis, elaboration of a conclusion. The domain of electromyography (EMG) is broad, covering more than a hundred existing diagnoses, and about four thousand tests of nervous or muscular structures.

The complexity of EMG examination procedure and the specific conditions in which it is carried out (tests are painful and unpleasant for the patient) make the implementation of a decision support system for the physician quite useful. MYOSYS [20] is a knowledge-based decision supporting system on EMG built on top of an object formalism. The system has been designed to assist the physician in different tasks ranging from symptom evocation to test choice. A base of already resolved EMG cases is integrated into MYOSYS. The way cases are modeled as complex objects is described below.

2.1   EMG case model

When a real-world domain is modeled with object knowledge formalisms the domain entities are represented as objects. Entity features are modeled by object attributes. Attribute values belong to a specific domain, data type or object set.

In the EMG field, entities are divided into several *concepts*: EMG case, clinical data, hypothesis, test, EMG conclusion, etc. The objects that represent entities of the same concept are described through a fixed set of attributes. Thus, they form homogeneous groups, we shall further call those groups *object sorts*. EMG object sorts define attributes of various domain structure. For example, test results are mainly numerical values: floats or integers, but may also be expressed as ordinals. Nominal features are used to describe the general state of the examined anatomic structures, while anatomic structures, i.e. muscles and nerves, themselves constitute hierarchical domains. Moreover, normal values for tests are introduced in form of intervals.

The existing relations between domain entities are modeled through *object-valued* attributes, as opposed to *primitive* attributes, describing features. For example, each EMG case is characterized by its clinical data. In the model, the `cl-data` attribute of the EMG case sort takes its values in Clinical data. Objects-valued attributes give rise to *complex* objects. A sub-set of the EMG domain concepts together with their relations are shown on Fig. 1. As it is shown on the figure, relations may associate an entity to a group of other entities. Thus, an EMG examination includes several tests and may lead to a set of final conclusions. One-many relations are modeled through multi-valued object attributes which may be defined on primitive features as well (see Section 4).

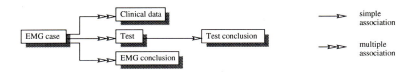

Fig. 1. Some EMG domain concepts and relations among them.

Objects are organized in class taxonomies. A class represents a group of objects, its instances. It is characterized in terms of the same attribute set as its instances. Class attributes describe sub-domains to which the attribute values of instances should belong. Usually, each object sort is assigned a class taxonomy. In the case of EMG, some of the domain concepts like hypothesis, clinical data, conclusion are subject to standardization so the underlying hierarchical structure is well known (see Fig. 2). In contrast, the sets of EMG tests and EMG cases admit no hierarchical structure a priori, since they are specific to a given physician or a hospital department. We tend to see here, the utility of clustering methods: they may be applied to extract a meaningful taxonomy from a set of objects, say EMG cases. The obtained structure can be useful both for extracting the expert knowledge from data and for optimizing the object storage.

In the following section, issues related to automatic classification of objects are discussed.

2.2   Classification-related problems

The construction of meaningful object taxonomies requires an efficient means for assessing objects. However, most of the numerical and conceptual clustering methods are defined upon data representations which are much simpler than the above object model.

First, only a small set of data types, though variable, are admitted in the description of individuals. Compared to this, object formalisms use the whole variety of data types they inherited from programming languages: integers, reals, strings, Boolean; as well as other, less common types like date for example. Even more, some concrete object models [5] do not limit the set of the admitted data types: new types may be imported from outside. External types are introduced through an abstract data type (ADT), which represents a minimal interface of type management primitives (identity predicate, order predicate, etc.).

In addition, most of the existing clustering methods admit only primitive attributes in the description of individuals. However, the relations between objects represent an important part of the domain model. Thus, an object is characterized by the total set of its attributes, both primitive and object-valued. Clustering with relational data within a logical formalism has been studied in machine learning [2]. First attempts to adapt the method on objects have been reported in [3], but no universal approach exists so far.

Finally, objects are often described by means of multi-valued attributes. Such an attribute is defined upon a basic domain, object sort or primitive type, through a collection constructor, *set* or *list*. As collections may have variable cardinality, they cannot be compared directly. Some work on multi-valued features has been done in different fields within the frame of machine learning: in [12] the multiple associations between individuals has been studied for concept formation purposes, whereas the utility of set-valued primitive attributes in decision tree induction is discussed in [7].

For the purposes of object taxonomy building, a proximity-based strategy seems to be a reasonable choice. It requires, however, the definition of proximity functions which satisfies the following criteria. First, proximity between objects should depend on each of their attribute values. In other words, the proximity function should take into account all the kinds of object attributes, in particular the object-valued and multi-valued ones. Next, each data type used in object descriptions has to be processed with the highest possible precision. Finally, the overall object proximity should remain of a low computational cost.

In the following, a dissimilarity model which meets the above requirements is presented. First, a generic function for single-valued object attributes is introduces. Then, the function is extended on collections of both primitive values and objects.

3   Topological dissimilarity

Providing a set of individuals I with a dissimilarity measure means defining a function $d : I \times I \longrightarrow \mathbb{R}_0^+$, which satisfies, for arbitrary $a, b \in I$: (i) $d(a,b) \geq 0$ (*positiveness*), (ii) $d(a,a) = 0$ (*minimalness*) and (iii) $d(a,b) = d(b,a)$ (*symmetry*). Most often, the dissimilarity measures are calculated on the features of the individuals. For this purpose, each feature is provided with a function to assess resemblances between values. Usually, these are *ad hoc* functions tied to the feature types (e.g. *nominal* or *ordinal*).

In the context of an object formalism with an extensible type system, i.e. where user-defined types are possible to import, such an *ad hoc* approach fails. A possible remedy could be to include a primitive for value resemblance computation in the mandatory interface for external types. For a representation formalism, this solution seems to be rather restrictive. A reasonable alternative consists to define a generic function which applies to all data types, both built-in and external, admitted by the formalism. The function could be then overridden by a user-provided primitive, which fits better a particular data type. The topological dissimilarity model represents such a universal means for comparing members of a given domain. It is based on the fact that all domains share a common structure. In fact, for a given object attribute of a primitive domain D, the restrictions imposed by object classes on that attribute, let us call them *type expressions* like in [5], define sub-domains of D. The set of type expressions on D are naturally provided with an inclusion relationship called *sub-typing*. Sub-typing induces a partial order structure on D which is quite similar to the class taxonomy on an object set (see [18]). We use the classification scheme (CS) model to provide a formal description of the structural analogy between domains.

3.1   Classification Schemes

For the sake of compactness, we shall only insist on model's basic components (see [5] for details).

A classification scheme (CS) is defined over a domain, say D, provided with two languages: L_I, of individuals, and L_C, of categories. Individuals are interpreted as domain entities, whereas the categories have two different interpretations. The first one, called *abstract* interpretation (I_A), is the set of all entities

the category may potentially represent. The second one, namely the *real* interpretation (I_R), includes only elements which are currently represented by the
category. A *taxonomy*, with respect to one of the interpretations, is a partially
ordered set of categories whereby the order respects the inclusion of interpretations. Now, a *classification scheme* $S = \langle L_C, C, \ll, \leq \rangle$ is composed of two
taxonomies:

- $\langle L_C, \ll \rangle$ respects I_A; $\ll$ is called *sub-categorization criterion.*
- $\langle C, \leq \rangle$ respects I_R whereby $C \subseteq L_C$ and $\leq$ is called *sub-categorization relation.*

Both object sorts and primitive types may be seen as classification schemes.
For example, in the EMG domain the integer type, used to encode the age of a
patient, can be seen as a classification scheme. The admitted integer numbers
constitute the language of individuals whereas categories correspond to integer
intervals. On Fig. 2, the taxonomy of the classification scheme on the Clinical
data sort is given. Categories representing object classes are given their standard
names. Individuals, i.e. objects of the Clinical data sort, are drawn as rectangles
and are attached to their most specific categories.

A classification scheme of an object sort may be obtained as *product* of the
classification schemes of the object attributes. Thus, the EMG test sort may
be seen as the product of all its attributes like test results, tested anatomic
structure, test conclusion, etc. The same holds for EMG case, EMG conclusion
and Clinical data. The reverse operation of the *CS product* is *projection*; it allows
the separation of a sub-set of the product factors.

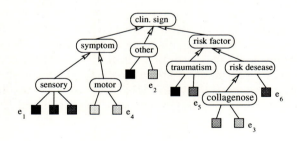

Fig. 2. A possible taxonomy on Clinical data.

The uniform representation of both object sorts and primitive types offered
by the CS model may be used in the definition of a proximity function. In the
next section, a dissimilarity measure ($d : L_I \times L_I \longrightarrow \mathbb{R}_0^+$) is presented which
is entirely based on the taxonomy structure of a CS.

3.2   Basic model

The *topological* dissimilarity is made up of of two functions: a *high level* function, d^t, computed on the objects and *low level* function, $\bar{\delta}$ that accounts for the field values.

The *low level function* δ is measured on the taxonomy of a given classification scheme $S = \langle L_C, C, \ll, \leq \rangle$. It may be roughly described as the shortest path between two individuals in the taxonomy. In fact, what is measured is the minimal sum of path lengths over paths leading to a common category. Thus, given e_1 and e_2 in L_I with $C^* = \{c | e_1, e_2 \in I_R(c)\}$:

$$\delta(e_1, e_2) = \min_{c \in C^*} [dist(e_1, c) + dist(e_2, c)]$$

where $dist(e, c)$ is the number of intermediate categories between individual e and the category c. The above function is further normalized:

$$\bar{\delta}(e_1, e_2) = \frac{\delta(e_1, e_2)}{\max_{x_1, x_2 \in \cup_C I_R(c)} \delta(x_1, x_2)}$$

The way $\bar{\delta}$ works is illustrated on the Clinical data taxonomy on Fig. 2. For example, the topological dissimilarity between the couple of objects $e1$ and $e2$ which represent symptoms of **sensory** and **motor** disorder respectively is $\delta(e_1, e_4) = 4$ since the shortest path between them is of length four. Since the maximal path-length in the above taxonomy is seven ($\delta(e_1, e_3) = 7$), the normalized topological dissimilarity is $\bar{\delta}(e_1, e_4) = 4/7$. Here are some other examples: $\bar{\delta}(e_1, e_2) = 5/7$, $\bar{\delta}(e_1, e_3) = 1$ and $\bar{\delta}(e_2, e_3) = 6/7$.

Both δ and $\bar{\delta}$ are valid dissimilarity indices. Moreover, δ may be extended to a category dissimilarity ($\delta_c : C \times C \longrightarrow \mathbb{R}_0^+$).

The *high level function* d^t is defined on a CS product, i.e. object sort, K of n direct factors (attributes). For individuals, say e and e', given with their corresponding unit projections (attribute values) e_i and e_i' ($i = 1 \ldots n$):

$$d^t{}_K(e, e') = Aggr_{i=1}^n \lambda_i \bar{\delta}(e_i, e_i')$$

where the $Aggr$ is a generic aggregation operator and λ_i is the weight assigned to the i-th attribute of K. Various instantiations are imaginable for this operator, for example City bloc or other Minkowski metrics.

3.3   Concrete functions

Although defined on a graph structure, most often the computation of the topological measure does not require extensive graph search. In fact, for the most common domain structures, the generic model coincides with well known functions which are easy to compute directly on primitives included in the ADT. These functions are be implemented to speed-up the computation.

For instance, on a *nominal* domain D, like strings or symbols, the topological dissimilarity equals the reverse of the identity ($\bar{\delta} = 1 - id_D$). In fact, the classification scheme of such a domain is made up of all possible sets of elements, i.e. $L_C = 2^D$, since there is no reason to distinguish some of them. Thus, for a couple of individuals e_1 and e_2, the nearest common category corresponds to the set $\{e_1, e_2\}$. Consequently, all possible values for δ are 0 and 2 which yields after normalization $1 - id_D$.

When an *ordinal* domain D is considered, the categories can be intervals. Thus, the taxonomy is made up of all intervals on D with interval inclusion as sub-categorization relation. For a couple of values, δ accounts for the number of intervals between each value and the most specific common category. The latter is exactly the interval where those values are bounds. With some elementary computation one may show that $\delta(e_1, e_2) = 2 \times abs(ord(e_1) - ord(e_2))$. Consequently,

$$\overline{\delta}(e_1, e_2) = \frac{abs(ord(e_1) - ord(e_2))}{ord(max(D)) - ord(min(D))}$$

Since based on discrete structures, the topological dissimilarity is impossible to measure directly on a *continuous domain*. Therefore, we assume in this case $\overline{\delta}$ coincides with the normalized real number subtraction.

The low-level topological function applies to object sorts provided a taxonomy is available on them. In this case, the proximity between objects is assessed with respect to their mutual position within the taxonomy. In other words, $\overline{\delta}$ deals with objects as if they were atomic and no more products of attribute values. For example, when d^t is computed on a couple of EMG cases, $\overline{\delta}$ will be applied on each attribute. For a couple of Clinical data objects associated to the EMG cases, the graph distance between them within the taxonomy will be extracted by $\overline{\delta}$ as shown previously.

Disregarding object attributes means, in particular, that $\overline{\delta}$ explores the relational structure of the object sort set at depth one. Thus, the value of d^t on EMG cases depends on the respective EMG tests, but not on the Test conclusions.

We deliberately chose a measure based on the taxonomy. As a matter of fact, the existing taxonomy, created either manually or automatically, is a synthetic expression of the object sort (conceptual) structure. If it is meaningful (if it is not, then the whole modeling and/or clustering is meaningless), then a good dissimilarity approximates the object proximities induced by the taxonomy.

4   Multi-valued types

Multiple associations between entities lead to characterizations in terms of value collections instead of single values. For instance, chemical elements have several possible valences. When elements are modeled, the set of valences is to be associated to each of them. The nucleotide sequence associated to a gene is just another example.

Multiple associations are usually modeled through multi-valued object attributes. A multi-valued attribute is defined on a *basic* type, by means of a constructor: *set* or *list*. Constructors apply to both primitive and object-valued attributes. For example, in the EMG domain, EMG case is assigned a set of tests.

When processing a multi-valued attribute difficulties arise due to the variable length of the collections. In machine learning, for example, set-valued features are often encoded through a set of single-valued ones and only rarely processed in a direct way (see [7] for a discussion).

4.1 Comparing collections

A generic dissimilarity model on multi-valued attributes should account for both the pairwise member dissimilarity and cardinality differences. One may imagine an exhaustive computation, leading to the average of all member pairwise dissimilarities. However, the obtained measure is not *minimal*, i.e. its values on identical collections are strictly non-negative.

In order for *minimalness* to be guaranteed, some preliminary selection of member pairs is necessary. More precisely, the set of selected pairs must satisfy: (*i*) a collection member may take part in at most one pair and (*ii*) the number of pairs is maximal. Those conditions define a *matching* between collections which is maximal in cardinality. In addition, we require the matching to minimize the total dissimilarity of the selected pairs.

4.2 Set-valued attributes

Let $S = \{e_1, e_2, ...e_k\}$ and $S' = \{e'_1, e'_2, ...e'_l\}$ be two sets over a basic domain D. Let also δ_D be a normalized dissimilarity measure on D.

Matching S and S' in the way described above means resolving the problem of an optimal matching in a weighted bipartite graph. Algorithms of $O(n^3)$ complexity for the problem have been reported in [1].

Let $M_{opt}(S, S') = \{(e_i, e'_j)\}$ be an optimal matching. When at least one of the sets is non empty, the set dissimilarity between S and S' may be defined as the average over the total dissimilarity of the matching and the unmatched elements taken with a maximal dissimilarity, i.e. 1.:

$$\delta_s(S, S') = \frac{\sum_{(e_i, e'_j) \in M_{opt}(S, S')} \delta_D(e_i, e'_j) + |l - k|}{\max(l, k)}.$$

If both S and S' are empty, then we set the result of the function to zero: $\delta_s(\emptyset, \emptyset) = 0$. The obtained measure is a valid dissimilarity index, since it is positive, symmetric and minimal.

For example, let $S = \{2, 7, 4, 3, 9\}$ and $S' = \{6, 4, 8, 1\}$ be sets constructed over an integer domain $D = [0, 10]$. An optimal matching is $M_{opt}(S, S') = \{(2, 1), (7, 6), (4, 4), (9, 8)\}$, consequently $\delta_s(S, S') = (3/10 + 1)/5 = 0, 26$.

The δ_s function applies successfully to object sets as well. Indeed, let $S = \{e_1, e_2, e_6\}$ and $S' = \{e_4, e_5\}$ be two sets of EMG tests (see Fig. 2) associated to a couple of EMG cases. Their dissimilarity is $\delta_s(S, S') = (8/7 + 1)/3 = 5/7$ obtained with $M_{opt}(S, S') = \{(e_4, e_1), (e_5, e_6)\}$.

4.3   List-valued attributes

In the case of lists, the collection is provided with a sequential structure. The new structure implies some extra constraints for the matching procedure. Thus, for lists $L = \langle e_1, e_2, ...e_k \rangle$ and $L' = \langle e'_1, e'_2, ...e'_l \rangle$, a matching $M_l(L, L') = \{(e_i, e'_j)\}$ should preserve the order induced by the lists. In other terms, $M_l(L, L')$ should satisfy:

$$\forall (e_i, e'_j), (e_m, e'_n) i \leq m \Rightarrow j' \leq n'$$

The new kind of matching is more complex than the previous one. In fact, the sequential structure implies stronger dependencies between member pairs than in the previous case. For instance, $(e_1, e'_2) \in M_l$ implies $(e_2, e'_2) \notin M_l$ but also $(e_2, e'_1) \notin M_l$.

The task may be evaluated in the following way. First, with no loss of generality we may suppose $k < l$, the case $k = l$ being trivial. Then, for lists of different length the matching we are looking for may be seen as a mapping from the shorter list, L, to the longer one, L'. The number of all maps that preserve the list-induced order is C_l^k. In the worst case, this number is an exponential function of l. We use therefore a branch-and-bound algorithm exploring the space of all possible matchings, i.e. k-tuples on $[1, l]$, in lexicographic order.

The initial solution is provided by a greedy heuristic algorithm which implements a recursive divide-and-conquer strategy. At each step, it chooses the best pair, i.e. the one of lowest dissimilarity, between all possible pairs. For a given element of the shorter list, only matchings are considered which do not prevent other elements of the same list to be further matched. Thus, for e_i in L, only $e'_i, e'_{i+1}, .., e'_{l-k+i}$ will be taken into account. Once the best pair is fixed, it is added to the matching, its elements are extracted from their respective lists and each list is splited in two sub-lists: one to the left and one to the right of the extracted element. The algorithm is recursively applied on both pairs of respective lists; it stops with empty list. In the worst case, the complexity of the above procedure is $O(n^3)$.

For example, let $L_1 = \langle 5, 3, 6 \rangle$ and $L'_1 = \langle 2, 4, 6, 2, 7 \rangle$ be lists on the integer domain $D = [2, 12]$. For this couple of lists, the heuristic algorithm will provide a matching $M_l = \{(5, 2), (3, 4), (6, 6)\}$ with total dissimilarity between matched elements of 4/10. Now, when the *branch-and-bound* algorithm is applied with this initial solution, it will rapidly find the optimal matching $M_{lopt} = \{(5, 6), (3, 2), (6, 7)\}$ with total dissimilarity of 3/10.

Thus, the matching computed by the heuristic algorithm provides quite a high bound for the following search. Finally, the total dissimilarity between lists is completed in order to take into account the unmatched elements. Let $M_{opt}(L, L') = \{(e_i, e'_j)\}$ be an optimal matching obtained by the above procedure. With at least one non empty list, the list dissimilarity between L and L' will be:

$$\delta_l(L, L') = \frac{\sum_{(e_i, e'_j) \in M_l(S, S')} \delta_D(e_i, e'_j) + |l - k|}{\max(l, k)}.$$

Should both L and L' be empty, their dissimilarity is zero: $\delta_l(\emptyset, \emptyset) = 0$.

In the above example of integer lists, $\delta_l(L, L') = (3/10 + 2)/5 = 0,46$ with M_{lopt} and $\delta_l(L, L') = (4/10 + 2)/5 = 0,48$ with M_l.

With object lists $L = \langle e_1, e_6 \rangle$ and $L' = \langle e_3, e_4, e_2 \rangle$ (see Fig. 2), and matching $M_{lopt}(L, L') = \{(e_1, e_4), (e_6, e_2)\}$, the dissimilarity is $\delta_l(L, L') = (9/7 + 1)/3 = 16/21$.

5   Classification strategy

The topological measure represents an efficient tool for building a taxonomy within an object formalism. However, in case of multiple object sorts and several taxonomies to infer, the application of the measure, requires a specific strategy. In the following, we describe the strategy we developed for the case of the TROPES knowledge model [15].

In a TROPES knowledge base, objects are instances of disjoint *concepts*. TROPES concepts correspond to what we called *object sorts*: their instances share the same set of attributes. Values of a given attribute belong to a specific data domain, either a primitive data type or an object concept. Furthermore, the type system integrated to TROPES supports encapsulated external types, introduced via abstract data types [4], as well as multi-valued types.

The model has been provided with a taxonomy building tool, T-TREE [9] which implements some numerical classification algorithms. Enhanced with the topological measure T-TREE is able to process objects with no restriction and thus can infer several taxonomies on disjoint concepts in the knowledge base.

In doing that, the set of concepts is considered as a graph. In fact, the knowledge base may be considered as a graph where concepts and ADT are vertices and attributes are edges (see Fig. 1 for a partial view on that structure). The obtained structure is a directed acyclic graph since for the time being we excluded mutual dependencies between concept characterizations. When classification has to be carried out on several concepts the global principle is to process each concept only after all its subordinated concepts, i.e. those related by object-valued attributes. In the case of the EMG domain, this means that if EMG cases have to be classified, then the EMG test concept should first be provided with a suitable taxonomy. This amounts to exploring the graph structure in a bottom-up manner, at each step inferring a taxonomy on a concept by referencing the taxonomies on the concept attributes.

Such an exploration has the following advantages. First, both object features and relations are taken into account, whereby the greatest attention is paid to the domain structure of each attribute. Next, object relations are dealt with at a reasonable cost. In fact, only direct attribute values are processed, their possible structure and further relations remaining hidden. For example, when classifying EMG cases, each EMG tests will be considered only as a member of its class in the test concept taxonomy. Finally, the taxonomic structure discovered at a particular level is reused on higher levels.

6   Related works

A dissimilarity measure based on graph distance has been discussed in [16]. The underlying measure is defined on a semantic net and accounts for the shortest path between a couple of nodes. Link directions and nature are not considered. Compared to that model, ours focuses on taxonomy, i.e. specialization links, and considers only up-going paths.

In [3], a possible way to compare complex objects has been presented. The proposed similarity function considers all relations between individuals to be reflexive and transitive. Thus, the similarity of a couple of objects is assumed to depend on the similarity of all related objects. When the model is applied to the EMG domain, for example, the proximity of a couple of EMG tests is computed with respect to proximity of the whole EMG cases, the EMG conclusions, etc. This additional information is not unlikely to disturb the EMG test proximity assessment. It seems, therefore, that the topological measure is better adapted to real-world domains where relations are mainly non-reflexive. Yet both models should be compared experimentally in order to find out which one is better.

Issues on inter-individual associations, matching and clustering with multiple individual sets have been addressed for the first time in [17]. The paper presents an extension of the concept formation algorithm COBWEB [10] for structured domains. A possible way to further extend the basic concept formation approach on multiple associations between individuals is described in [12].

7   Conclusion

The automatic inference of object taxonomies is a special kind of analyzing data and extracting implicit knowledge from it. A straightforward way to build object taxonomies is to use an automatic classification method on object sets. The detection of meaningful object clusters requires a proximity measure which completely fits object descriptions.

We described here such a measure, called *topological* dissimilarity. The *topological* dissimilarity is a generic model which applies to any data domain used in object descriptions. It allows to handle a variety of data types: nominal, ordinal, continuous, etc. as well as to successfully explore the inter-object relations during classification.

Furthermore, we proposed an extension of the topological dissimilarity for multi-valued attributes, sets and lists. which utility has been exemplified in the EMG domain. The computation of a dissimilarity between collections requires a preliminary step of matching between collection members. Strategies for matching sets and lists have been discussed. The extended model is able to process primitive and object-valued attributes as well as single and multi-valued ones.

Finally, we presented a strategy for taxonomy inference based on the extended *topological* dissimilarity. The advantages of the described strategy are multi-fold: (*i*) classification is carried out directly within the knowledge base, (*ii*) the domain structure of different data types is respected and (*iii*) the existing taxonomies on object sorts are reused in the construction of new taxonomies on other sorts.

This new measure has been integrated into the taxonomy building module of the TROPES system and is currently under evaluation. Its comparison with other measures is a subject of future works and shall include studies on several fields of application.

References

1. R.K. Ahuja, T.L. Magnanti, and J.B. Orlin, *Network Flows: Theory, Algorithms and Applications*, Prentice Hall, 1993.
2. G. Bisson, 'Conceptual clustering in a first order logic representation', in *Proceedings of the 10th European Conference on Artificial Intelligence, Vienna, Austria*, pp. 458–462, (1992).
3. G. Bisson, 'Why and how to define a similarity measure for object-based representation systems', in *Towards Very Large Knowledge Bases*, ed., N.J.I. Mars, pp. 236–246, Amsterdam, (1995). IOS Press.
4. C. Capponi, *Identification et exploitation des types dans un modèle de connaissances à objets*, Ph.D. dissertation, Joseph Fourier, Grenoble (FR), 1995.
5. C. Capponi, J. Euzenat, and J. Gensel, 'Objects, types and constraints as classification schemes', in *Proceedings of the 1st KRUSE symposium*, pp. 69–73, Santa Cruz (CA US), (1995).
6. P. Cheeseman, J. Kelly, M. Self, J. Stutz, W. Taylor, and D. Freeman, 'Autoclass: A bayesian classification system', in *Proceedings of the 5th Internatinal Conference on Machine Learning, Ann Arbor, MI*, pp. 54–56, (1988).
7. W. Cohen, 'Learning trees and rules with set-valued features', in *Proceedings of the 13th AAAI and 8th IAAI*, (1996).
8. F. Esposito, 'Conceptual clustering in structured domains: a theory guided approach', in *New Approaches in Classification and Data Analysis*, eds., E. Diday, Y. Lechevallier, M. Schader, P. Bertrand, and B. Burtschy, pp. 395–404, Berlin, (1994). Springer Verlag.
9. J. Euzenat, 'Brief overview of t-tree: the TROPES taxonomy building tool', in *Proceedings of the 4th ASIS SIG/CR classification research workshop*, pp. 69–87, Columbus (OH US), (1993).
10. D.H. Fisher, 'Knowledge acquisition via incremental conceptual clustering', *Machine Learning*, **2**, 139–172, (1987).
11. R. Godin, G.W. Mineau, and R. Missaoui, 'Incremental structuring of knowledge bases', in *Proceedings of the 1st KRUSE symposium*, pp. 179–193, Santa Cruz (CA US), (1995).
12. A. Ketterlin, P Gançarski, and J.J. Korczak, 'Hierarchical clustering of composite objects with variable number of components', in *Proceedings of the 5th International Workshop on Artificial Intelligence and Statistics*, eds., D. H. Fisher and P. Lenz, Fort Lauerdale (FL USA), (1995).
13. R. Michalski and R. Stepp, *Machine learning: an Artificial Intelligence approach*, volume I, chapter Learning from observation: conceptual clustering, 331–363, Tioga publishing company, Palo Alto (CA US), 1983.
14. G Piatetsky-Shapiro and W. Frawley, *Knowledge discovery in databases*, AAAI Press, 1991.

15. Sherpa project, *Tropes 1.0 reference manual*, INRIA Rhône-Alpes, Grenoble (FR), 1995.

16. R. Rada, H. Mili, E. Bicknell, and M. Blettner, 'Development and application of a metric on semantic nets', *IEEE Transactions on Systems, Man and Cybernetics*, **19**(1), 17–30, (1989).

17. K. Thompson and P. Langley, *Knowledge and experience in unsupervised learning*, chapter Concept formation in structured domains, 127–161, Morgan Kaufman, San Mateo (CA US), 1991.

18. P. Valtchev and J. Euzenat, 'Classification of concepts through products of concepts and abstract data types', in *Ordinal and symbolic data analysis*, eds., Y. Lechevallier E. Diday and O. Opitz, pp. 3–12, Heildelberg (DE), (1996). Springer Verlag.

19. B. van Cutsem, *Classification and dissimilarity analysis*, Lecture notes in statistics, Springer Verlag, New York, 1994.

20. D. Ziébelin, A. Vila, and V. Rialle, 'Neuromyosys a diagnosis knowledge based system for emg', in *Proceedings of the 12th International Congress of Medical Informatics in Europe*, Lisboa (PT), (1994).

Section III:

Medical Applications

ECG Segmentation Using Time–Warping

H.J.L.M. Vullings, M.H.G. Verhaegen, H.B. Verbruggen

Delft University of Technology, Department of Electrical Engineering, Mekelweg 4,
2628 CD Delft, The Netherlands, E-mail: E.Vullings@et.tudelft.nl

Abstract. We present a method to segment the electrocardiogram (ECG)
using time-warping, a technique commonly used in speech recognition.
First, the ECG is transformed to a piecewise linear approximation. Next,
the slope amplitude is used to cut the ECG into distinct periods (R-R
interval). These periods are then compared to each other using time-
warping, and the pair which is most similar is selected. Finally, this pair
is segmented into the different subpatterns usually encountered in the
ECG, such as the QRS complex, the T wave, and the P wave.

1   Introduction

The electrocardiogram (ECG) is an important physiological signal for the evaluation of the cardiac status of patients. ECG-based diagnosis can be performed
on patients at rest, during exercise, in the intensive care, or during an operation.
The important features are examined by a physician, who visually inspects the
signal. This is a time-consuming task and great relief in the inspection would result from an accurate automatic assessment of the important features. Figure 1
shows an ideal cardiac cycle with the most important subpatterns: the QRS
complex, the P wave, and the T wave. From these, the QRS complex is the most
important and most dominant feature, and several computer algorithms exist
to detect its presence [5, 15, 16, 17]. However, the durations and amplitudes of
the other subpatterns also contain vital information for the cardiologist, so it is
important to develop methods that automate the assessment of the ECG. Or,
put differently, we need intelligent methods to extract information from data.

To automate the assessment of the ECG, an algorithm must detect the presence of the above mentioned subpatterns, and calculate the durations and amplitudes as accurately as possible. Some available methods are based upon syntactic
methods [3, 7, 10, 17], which need to label the slopes of a linear approximation.
In [8] the slopes are used directly to detect the presence of the subpatterns. Here,
we are interested in analyzing the ECG using a dynamic-programming technique
which also uses the slopes explicitly, without labels: time-warping [13]. Time-
warping is often used in speech recognition for the recognition of phonemes,
words, or sentences [11]. However, we will adapt it for the recognition of a cardiac cycle, and for a decomposition in subpatterns. In the discussion, we will
treat the differences in more detail.

X. Liu, P. Cohen, M. Berthold (Eds.): "Advances in Intelligent Data Analysis" (IDA–97)
LNCS 1280, pp. 275–285, 1997. © Springer–Verlag Berlin Heidelberg 1997

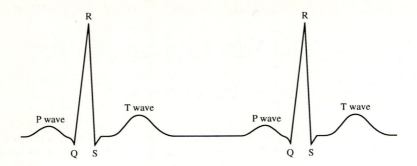

Fig. 1. Ideal cardiac cycle with the most important subpatterns.

2   Analysis and Segmentation

To analyze and to segment the ECG, we propose a three stage approach. First, we approximate the ECG via piecewise linear approximation (PLA), where the original signal is represented using straight line segments (§2.1). Next, we collect several normal, patient-specific ECGs, which we can compare to the current ECG cycle using time-warping (§2.2). Finally, we select the best ECG cycle based on the comparisons done in the second stage, and use it to segment the current ECG cycle in the above mentioned subpatterns (§2.3).

2.1   Approximation of the ECG via PLA

The recorded ECG is often contaminated with noise. Also, it is computational intensive to classify a signal sampled at 500 Hz, as is the case here. For these two reasons, it is recommended to transform the recorded ECG to a domain more suitable for analysis. Although it is feasible to transform the signal to the frequency domain, we want to develop a method that analyzes the ECG in a way which is familiar to cardiologists and anesthetists: in the time domain. However, even in the time domain there exist many methods. The most frequently used methods are based on line segments [1, 6, 9, 14], but it is also possible to use peaks and waves [17], or energy primitives [2]. As these latter transformations are more difficult to obtain, we propose to approximate the ECG using line segments. However, this is not essential to our approach, and it is quite easy to adapt our method to incorporate a different transformation.

To approximate the ECG with a series of line segments (a PLA), we use a slightly adapted version of the algorithm proposed by Koski [7]. The ECG is regarded as a vector $\mathbf{x} = [x(1), \ldots, x(n)]$, where $x(i)$ $(1 \leq i \leq n)$ is the voltage of the ECG at time i. The first proposed segment consists of the first s $(2 \leq s \leq n)$ samples of $\mathbf{x}$. We approximate this segment with a straight line connecting the first and last sample. As long as this line approximates the original segment with an acceptable error e, s more samples are added to the segment. To calculate the error e, consider figure 2, where j samples are approximated by the straight line $y(i) = ai + b$. The error $e(i)$ for sample i $(1 < i < j)$ should never exceed an empirical determined threshold ϵ:

$$e(i) = \frac{|x(i) - y(i)|}{\sqrt{a^2 + 1}} < \epsilon \qquad (1)$$

Now, we sequentially add s more samples to the segment, until (1) does not hold. Then, we start shrinking the segment in order to obtain a segment which does not exceed ϵ. The new end-point of the segment is the point i for which $e(i)$ is maximal. If the error on the new segment remains below ϵ, the line segment to approximate a part of the ECG is found. The new segment will start at the end point of the previous segment. However, if the error is still not below ϵ, we shrink the segment again and again, until (1) holds. The algorithm to segment the ECG is given in box 2.1.

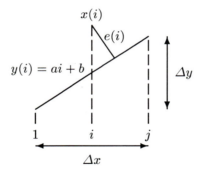

Fig. 2. The error $e(i)$ for a segment.

A good value of the parameters in our case (500 Hz sampling, 10 bits quantization, recorded during operations in the Amsterdam Medical Center using the Hewlett Packard Merlin anesthesia monitor) is $s = 16$ and $\epsilon = 8$. The advantage of a large step size s is a speedup when compared to algorithms which only increment one sample at a time. As we always break our segment where we have found the largest error, we are still capable of finding the important (fiducial) points, which are needed for an acceptable detection of the subpatterns (see figure 3).

A complete cardiac cycle consists of a sequence of lines which can be displayed in string format. One line can be represented as a combination of a slope and a horizontal length, i.e. $(\Delta y/\Delta x, \Delta x)$ (see figure 2), which enables us to describe

Box 2.1: PLA algorithm

$j = 1, \ e(l_{max}) = 0$
while $\{\max(e(l_{max})) < \epsilon\}$
       $j = j + s$
       $a = \frac{x(1) - x(j)}{1 - j}$
       $b = \frac{x(j) - jx(1)}{1 - j}$
       $l_{max} = \max \left\{ \frac{x(i) - ai - b}{\sqrt{a^2 + 1}} \right\}_{1 < l < j}$
while $\{\max(e(l_{max})) > \epsilon\}$
       $j = l_{max}$
       $a = \frac{x(1) - x(j)}{1 - j}$
       $b = \frac{x(j) - jx(1)}{1 - j}$
       $l_{max} = \max \left\{ \frac{x(i) - ai - b}{\sqrt{a^2 + 1}} \right\}_{1 < l < j}$
$\Delta x = j - 1$
$\Delta y = x(j) - x(1)$

the cardiac cycle in figure 3 by the string

$$\mathbf{a} = \{ \ (0.10, 60), (1.00, 25), (0.05, 19), (-1.23, 26), (0.37, 33),$$
$$(-2.33, 9), (11.78, 9), (-21.00, 9), (6.29, 14), (0.53, 30), \qquad (2)$$
$$(0.00, 65), (0.69, 55), (-1.04, 25), (-0.20, 70), (0.08, 50)\}$$

After transforming the ECG to a sequence of lines, we need a way to divide this quasi-periodic signal into its cycles. For this, we use the QR line which is easy to detect. If the slope of a line exceeds a predetermined threshold (slope>4), we consider it a possible QR line. The lines generated between two consecutive QR lines are the ECG cycles we will examine. In case of a possible QR line which was caused by an artefact, the ECG cycle will be rejected, as the difference between a normal ECG cycle and the current cycle will be large.

2.2   Selecting the best string

Before we can analyze an ECG, we first need to know whether the current cycle is a normal ECG, or whether it is deformed beyond recognition, e.g. by noise interference or patient movement. To examine this, we compare the current cycle with normal cycles obtained from the same patient at the beginning of the recording. If the computed distance between the string of the current cycle and a string of a previously recorded cycle is not too large, we can start with the segmentation of the current cycle. The distance between two strings is calculated using time-warping [13]: compress $(k + 1)$ adjacent lines into 1 line, or expand 1 line into $(k + 1)$ adjacent lines. Consider two strings $\mathbf{a}$ and $\mathbf{b}$, where $\mathbf{a} = \{a_1, a_2 \ldots a_m\}$ and $\mathbf{b} = \{b_1, b_2 \ldots b_n\}$. Every element of a string is a pair $(\Delta y / \Delta x, \Delta x)$, e.g. $a_1 = (0.10, 60)$ in (2). Let $\mathbf{a}^i$ be the substring $\{a_1 \ldots a_i\}$ and $\mathbf{b}^j = \{b_1 \ldots b_j\}$. The total distance $\mathbf{d}(\mathbf{a}, \mathbf{b})$ between $\mathbf{a}$ and $\mathbf{b}$ is calculated in a recursive matrix as shown in figure 4.

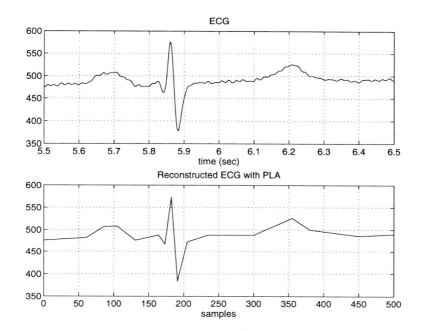

Fig. 3. Example of a filtered ECG and its approximation for one cycle.

To calculate $\mathbf{d}(\mathbf{a}, \mathbf{b}) = \mathbf{d}(\mathbf{a}^m, \mathbf{b}^n) = \mathbf{d}_{m,n}$ the elements of the matrix are calculated recursively, starting from cell $(0,0)$ in the upper-left corner, and moving toward the cell (m,n) in the lower right. Cell (i,j) contains the distance $\mathbf{d}_{i,j}$ which is computed from the previous obtained distances as follows:

$$
\mathbf{d}_{i,j} = \min \begin{cases} \mathbf{d}_{i-1,j} & + \frac{1}{2}w(i,j)t(a_{i-1}) \\ \mathbf{d}_{i-1,j-1} & + \frac{1}{2}w(i,j)(t(a_{i-1}) + t(b_{j-1})) \\ \mathbf{d}_{i,j-1} & + \frac{1}{2}w(i,j)t(b_{j-1}) \end{cases} \tag{3}
$$

where $t(a_{i-1})$ is the time feature a_{i-1} is active, and $w(i,j) = w(a_i, b_j)$ is the distance between two string elements. However, not all transitions are allowed (see figure 5).

Further, to align two strings with minimum distance (the trace), a trace matrix $\mathbf{r}$ is also computed. This matrix contains pointers of the route the algorithm took to arrive at a certain point in $\mathbf{d}_{i,j}$:

$$
\mathbf{r}_{i,j} = \left\{ \begin{matrix} (i-1,j) & or \\ (i-1,j-1) & or \\ (i,j-1) & \end{matrix} \right\} \text{ depending on the choice of (3)} \tag{4}
$$

To display the best trace, we start in $\mathbf{r}_{m,n}$ and trace the route back to $\mathbf{r}_{1,1}$.

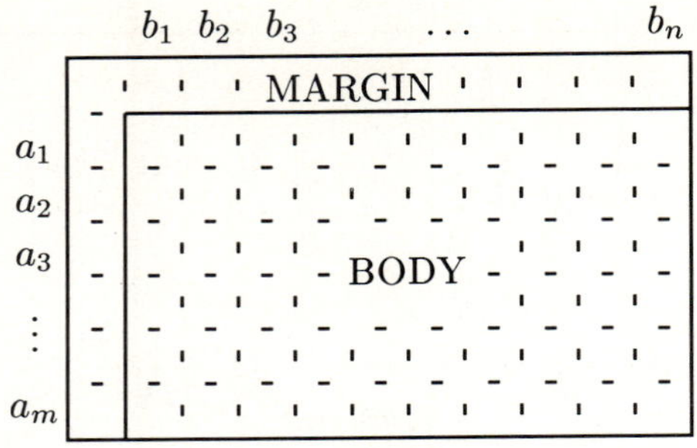

$$b_1 \quad b_2 \quad b_3 \qquad \cdots \qquad b_n$$

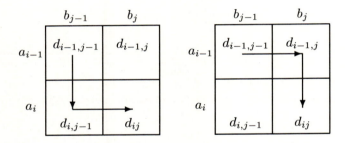

Fig. 4. Computational array for the computation of the distance $\mathbf{d}(\mathbf{a}, \mathbf{b})$ between two strings.

Fig. 5. Some two-step patterns are not allowed in time-warping.

In our case, the strings $\mathbf{a}$ and $\mathbf{b}$ are cardiac cycles, where every element consists of two parts: a slope, and a horizontal length (2). The slope is used to calculate the distance $w(i, j) = |\mathrm{slope}(a_i) - \mathrm{slope}(b_j)|$ between two elements. The horizontal length is equivalent to the time t a feature is active, e.g. $t(a_1) = 60$ in (2). Now it is possible to compute the distance $\mathbf{d}(\mathbf{a}, \mathbf{b})$ between the current string and an old valid string, derived from the beginning of the recording and validated by a physician. As long as the computed distance is less than twice the computed average distance between valid strings, we accept the string as valid and we can start the segmentation.

2.3 Segmentation of the current ECG cycle

After determining the best-matching pair of strings (strings with minimum distance), we are now ready to determine which lines constitute the different sub-patterns. To accomplish this task, we use a three dimensional distance matrix $\mathbf{D}$ which consists of several layers. Every layer is a two dimensional matrix $\mathbf{d}$ as used in the previous subsection. However, if the slope of the lines is not within a predefined interval for a certain layer, a penalty is added to the calculated distance. Further, if the distance in a previously computed layer is smaller than the distance computed in the current layer, the value of the previous layer is taken. The rationale behind this is that every slope will be placed in the most optimal layer, where optimal is in the sense of minimum total distance for the matching of two strings. The distance between $\mathbf{a}^i$ and $\mathbf{b}^j$ in layer k is expressed as:

$$\mathbf{D}_{i,j,k} = \min \begin{cases} \mathbf{D}_{i-1,j,k} & + \frac{1}{2}(w(i,j) + p_1(a_i,k))t(a_{i-1}) \\ \mathbf{D}_{i-1,j-1,k} & + \frac{1}{2}(w(i,j) + p_1(a_i,k) + p_1(b_j,k))(t(a_{i-1}) + t(b_{j-1})) \\ \mathbf{D}_{i,j-1,k} & + \frac{1}{2}(w(i,j) + p_1(b_j,k))t(b_{j-1}) \\ \mathbf{D}_{i,j,k-1} & + p_2(k) \qquad\qquad\qquad\qquad\qquad \text{if } k > 1 \end{cases}$$
(5)

where $p_1(s,k)$ (with $s \in \{a_i, b_j\}$) and $p_2(k)$ are the penalty functions:

$$p_1(s,k) = \begin{cases} 0 & \text{if } B_l(k) < |s| < B_u(k) \\ 20 & \text{otherwise,} \end{cases}$$

$$p_2(k) = \begin{cases} 200 & \text{if we skip a layer} \\ 0 & \text{otherwise,} \end{cases}$$

Here, $B_u(k)$ and $B_l(k)$ are, respectively, the upper and lower boundary of the slope in layer k. Thus, if the slope of the compared lines is not within a predefined boundary, we add an empirically determined penalty to the calculated distance. Further, as we skip a layer by moving upwards more than once, we also add a penalty. As a consequence, as long as the slopes of the lines remain within the layer's boundary, it is not necessary to add a penalty. However, as soon as this does not hold, the penalty is added, and the distance will probably become smaller in a different layer. In figure 6 the different layers are displayed, together with the empirically determined thresholds for every layer. From this figure, it is also clear which lines agree with which subpattern. However, to calculate the beginning and ending of each subpattern, we need a trace matrix like the one in (4), with the only difference that this matrix now has three dimensions:

$$\mathbf{R}_{i,j,k} = \begin{cases} (i-1,j,k) & or \\ (i-1,j-1,k) & or \\ (i,j-1,k) & or \\ (i,j,k-1) \end{cases} \text{depending on the choice of (5)} \qquad (6)$$

Finally, all we have to do to get the subpatterns is trace back $\mathbf{R}$ and log the points where we switch of layer.

3   Results

In this section, some results are shown using the proposed algorithm to compute the different subpatterns in the ECG. For three patients (figures 7, 8, and 9), we show several seconds of the ECG, segmented in cycles and subpatterns, where we used the boundaries shown in figure 6 for all patients.

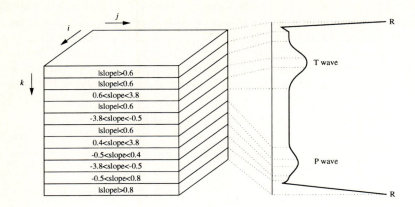

Fig. 6. Three dimensional distance matrix **D** and the slope boundaries for the different layers.

We examined 50 seconds of ECG data for each of the three above mentioned patients, consisting of a total of 172 cycles. Of these, we found 6 cycles (3.5%) with an incorrect classification of the P wave, the T wave, or the QRS complex. Most errors were due to a junction of the P wave and the QRS complex.

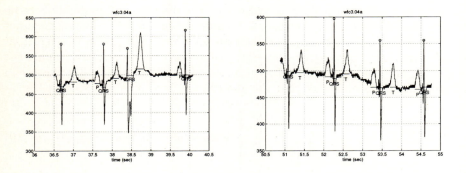

Fig. 7. Patient 1: Extrasystole (left) and baseline wander (right).

One should pay special attention to figure 9, where we incorrectly classified one of the ECG cycles. The problem associated with this method is that when one line is incorrectly classified, the complete classification can fail. In this case, it was more optimal to put the P wave in the T wave segment, and add the P wave to the QRS complex. However, we think this problem can be diminished,

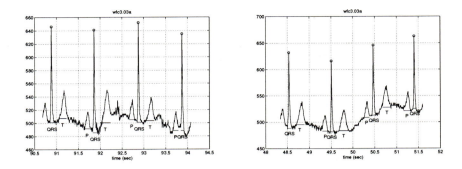

Fig. 8. Patient 2: Noise interference (left) and baseline wander (right).

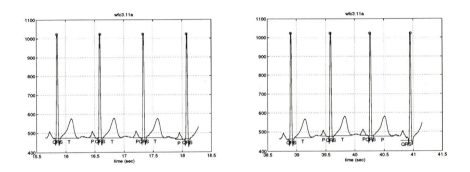

Fig. 9. Patient 3: Correct (left) and incorrect classification (right).

if not annihilated, by using more explicit information of the ECG, such as the moment in time a certain subpattern can occur within a cycle.

4   Discussion

If we compare the use of time-warping in speech recognition to our approach, there are three kind of differences: first, time-warping in speech recognition is often based on linear predictive coding of the signal, whereas we take a linear approximation of the signal as input to the algorithm. Second, instead of just using one distance function in the time-warping algorithm, we combine several distance functions to obtain a segmentation of the ECG into its subpatterns. A final difference is that in speech recognition, time-warping is often based on an asymmetrical distance function (or $\mathbf{d}(\mathbf{a}, \mathbf{b}) \neq \mathbf{d}(\mathbf{b}, \mathbf{a})$), whereas our approach is symmetrical (see also [12] for a comparison of symmetrical and asymmetrical distance functions).

Although the proposed algorithm works quite well, it should be evaluated further using data from more patients. Also, a number of improvements have to be made. First, the erroneous detection of the P wave has to be corrected. Second, the upper and lower boundaries of a layer should be optimized using information obtained after initial segmentations, as one setting probably will not work for all patients. Third, we would like to use this algorithm also for other quasi-periodic signals, such as the blood pressure, and the capnogram (CO_2 signal). In the latter case, we are not so much interested in obtaining a proper segmentation of the signal, but we want to be able to validate the measured signal for use in an anesthesia monitor [4]. Finally, the execution time would become smaller and the segmentation would become better if we calculate $\mathbf{D}$ only partially, i.e. only that part of $\mathbf{D}$ is computed which is relevant. For example, looking for a RS segment when you are already in the P layer is meaningless, so computing time can be saved, and incorrect classification avoided.

References

1. J.P. Abenstein and W.J. Tompkins. A new data-reduction algoritm for real-time ECG analysis. *IEEE trans. on Biomedical Eng.*, 29(1):43–48, 1982.
2. G. Belforte, R. De Mori, and F. Ferraris. A contribution to the automatic processing of electrocardiograms using syntactic methods. *IEEE trans. on Biomedical Eng.*, 26(3):125–136, 1979.
3. A. Cohen. *Biomedical signal processing*, volume 2. CRC Press, Inc., 1986.
4. P.M.A. de Graaf, G.C. van den Eijkel, H.J.L.M. Vullings, and B.A.J.M. de Mol. A decision-driven design of a decision support system in anesthesia. *Artificial Intelligence in Medicine*, Admitted to the special issue on "Decision support in the operating theatre and intensive care", 1997.
5. G.M. Friesen, T.C. Jannett, M.A. Jadallah, S.L. Yates, S.R. Quint, and H.T. Nagle. A comparison of the noise sensitivity of nine QRS detection algorithms. *IEEE trans. on Biomedical Eng.*, 37(1):85–98, 1990.
6. B. Furht and A. Perez. An adaptive real-time ECG compression algorithm with variable treshold. *IEEE trans. on Biomedical Eng.*, 35(6):489–494, 1988.
7. A. Koski, M. Juhola, and M. Meriste. Syntactic recognition of ECG signals by attributed finite automata. *Pattern Recognition*, 28(12):1927–1940, 1995.
8. P. Laguna, R. Jané, and P. Caminal. Automatic detection of wave boundaries in multilead ECG signals: validation with the CSE database. *Computers and biomedical research*, 27:45—60, 1994.
9. T. Pavlidis and S.L. Horowitz. Segmentation of plane curves. *IEEE transacions on Computers*, c-23(8):860–870, 1974.
10. E. Pietka. Feature extraction in computerized approach to the ECG analysis. *Pattern Recognition*, 24(2):139–146, 1991.
11. L.R. Rabiner and B-H. Juang. *Fundamentals of speech recognition*. Signal Processing series. Prentice Hall, 1993.
12. H. Sakoe and S. Chiba. Dynamic programming algorithm optimization for spoken word recognition. *IEEE trans. on Acoustics, Speech, and Signal Processing*, 26(1):43–49, 1978.
13. D. Sankoff and J.B. Kruskal. *Time warps, string edits, and macromolecules: the theory and practice of sequence comparison*, chapter 1 & 4. Addison-Wesley, 1983.

14. E. Skordalakis. Recognition of the shape of the ST-segment in ECG waveforms. *IEEE trans. on Biomedical Eng.*, 33(10):972–974, 1986.

15. L. Sörnmo, O. Pahlm, and M-E. Nygårds. Adaptive QRS detection: A study of performance. *IEEE trans. on Biomedical Eng.*, 32(6):392–401, 1985.

16. Y. Suzuki. Self-organizing QRS-wave recognition in ECG using neural networks. *IEEE trans. on Neural Networks*, 6(6):1469–1477, 1995.

17. P. Trahanias and E. Skordalakis. Syntactic pattern recognition of the ECG. *IEEE trans. on pattern analysis and machine intelligence*, 12(7):648–657, 1990.

Interpreting Longitudinal Data through Temporal Abstractions: An Application to Diabetic Patients Monitoring

Riccardo Bellazzi[1], Cristiana Larizza[1] and Alberto Riva[2]

[1] Dipartimento di Informatica e Sistemistica, Università di Pavia, via Ferrata 1, 27100 Pavia, Italy
[2] I.R.C.C.S. Policlinico S. Matteo, P.le Golgi 2, 27100 Pavia, Italy

Abstract. In this paper we present a new approach for the intelligent analysis of longitudinal data coming from diabetic patients home monitoring. This approach consists in exploiting temporal abstractions to preprocess the raw data and to obtain a new time series of abstract episodes, whose features are then interpreted through statistical and probabilistic techniques. We finally show the application of this methodology on the data of two diabetic patients monitored for six months.

1 Introduction

Temporal Abstraction (TA) techniques are methods that can be used to derive an abstract description of the course of longitudinal data by extracting their most relevant features. In expert systems they have been exploited as the first step of the inference process, both in diagnostic and monitoring problems [1, 2]. In particular, in medical applications, the interpretation of the time series data coming from patient monitoring is performed on the basis of a two-step procedure: *first* the data are filtered using standard quantitative techniques and *then* they are summarized using TAs. The result of this process is a set of abstractions that represent the relevant features of the monitoring period. This set is used to highlight dangerous situations and to generate suggestions through Artificial Intelligent (AI) techniques [3].

In this paper we will present a slightly different use of TAs: given a set of longitudinal data coming from patient monitoring, *first* we exploit TAs to preprocess the raw data and to obtain a new time series of abstract episodes, whose features will *then* be analyzed through quantitative methods for interpretation purposes.

This approach is motivated by the particular medical problem we are facing: the interpretation of data coming from home monitoring of Insulin-Dependent Diabetes Mellitus patients. Insulin-Dependent Diabetes Mellitus is a chronic disease that forces patients to measure their Blood Glucose Levels (BGL) several times a day (e.g. three times a day), in order to control their glucose metabolism.

X. Liu, P. Cohen, M. Berthold (Eds.): "Advances in Intelligent Data Analysis" (IDA–97)
LNCS 1280, pp. 287–298, 1997. © Springer–Verlag Berlin Heidelberg 1997

This home-monitoring activity generates a huge amount of data, that the physician periodically revises to assess the quality of the patient's metabolic control. Unfortunately, to completely reconstruct the BGL daily pattern, at least eight measurements per day are required, while no more than three/four measurements per day are usually available. This means that the BGL signal is under-sampled, and it is therefore very difficult to distinguish abnormal values from measurement errors: it is hence in general not correct to filter the data *before* performing TAs. On the contrary, it is more convenient to move from the original time series to a higher-level description of the patient's behavior, and then to analyze it, resorting to statistical and probabilistic techniques. The TA analysis generates a multi-dimensional sequence of episodes that contains, as a by product, information on the *persistence* of the abstract state of the variables. In this paper we propose a method for exploiting this information to evaluate the patient response to the current therapeutic protocol. Moreover, we apply probabilistic techniques to derive a concise description of the patient behavior that can be also used to forecast the metabolic response to the current therapeutic protocol.

2   The Clinical Application

The application domain in which we are applying TA techniques is the interpretation of data coming from the home monitoring of Insulin-Dependent Diabetes Mellitus (IDDM) patients. IDDM is caused by the destruction of the pancreatic cells producing insulin, the main hormone that regulates glucose metabolism.

IDDM patients control their glucose metabolism through the delivery of exogenous insulin several times a day. The patients perform self-monitoring of BGL and glycosuria (glucose in the urine) at home, usually before meals, and report the monitoring data and the therapy followed in a diary. The accuracy of the patients' self-care is very important, since the onset and development of diabetic complications is strictly related to the degree of metabolic control that can be thus achieved. Recent studies [4] show that a good metabolic control can significantly delay or prevent the development of long-term complications. Unfortunately, tight metabolic control involves 3 to 4 insulin injections per day or continuous sub-cutaneous injections, accurate home BGL monitoring, and leads to an increase in the probability of hypoglycemic events.

Current information technologies may be exploited to improve the quality of patient monitoring by providing patients and physicians with support tools for making proper decisions. In particular, several authors have developed methods and systems for helping the physician in data interpretation. The methodologies used to this purpose range from simple statistical analysis and graphical representation of the raw data, to more complex techniques, like time-series analysis [5], causal probabilistic networks [6] and temporal abstractions [2]. In the following sections, we will present our approach to data analysis, based on a combination of TA methods and statistical and probabilistic techniques.

3 The TA Methods in the Diabetes Domain

3.1 Definition

TAs are methods that can be used to obtain an abstract description of the course of multi-dimensional time-series by extracting their most relevant features. Hence, in patient monitoring, TA methods provide a useful instrument to transform the fragmentary representation of the patient's history into a more compact one.

The basic principle of TA methods is to move from a time-point to an interval-based representation of the data. Given a sequence of time stamped data (*events*), the adjacent observations which follow meaningful patterns are aggregated into intervals (*episodes*).

In particular, to conceptualize the TA problem solving method, we defined an ontology which distinguishes two main classes of abstractions: BASIC abstractions for detecting predefined courses in a time series and COMPLEX abstractions for investigating specific temporal relationships between intervals.

BASIC abstractions extract TRENDS (increase, decrease or stationarity patterns), or STATES (e.g. low, normal, high values) from a uni-dimensional time series. TREND abstractions allow a flexible definition of "fast" or "slow" trends, by specifying the minimum slope and the minimum temporal extension of the pattern to be detected. The STATE pattern detection of numerical variables requires a preliminary qualitative abstraction [8]. In the diabetes domain, the qualitative abstractions can be context-dependent: for example, a BGL of 160 mg/dl may be classified as "normal" or "high" in dependence of the measurement time and of the patient's age. COMPLEX abstractions search for specific temporal relations between episodes which can be generated from a basic abstraction or from other complex abstractions. The relation between intervals can be any of the temporal relations defined by Allen [7]. The COMPLEX abstraction method can be exploited to detect multi-dimensional patterns or to extract compound data patterns from a uni-dimensional time series.

3.2 Data Pre-processing

In order to allow a proper interpretation of the data, we subdivided the 24-hour daily period into a set of consecutive non-overlapping time slices. This process is a first type of abstraction, that generates a qualitative time scale on the basis of the information about the patient's life-style, in particular the meal times. The relationships between actions (insulin intakes) and effects (BGL measurements) are derived using the concept of *competent time slice*: an action in a certain time slice will be competent for the BGL measurements in the time slices that it directly affects. For example, an intake of regular insulin will be competent for the time slices that cover the subsequent six hours.

The five time slices used in this work are shown in the following:

Breakfast	Mid-Morning	Lunch	Afternoon	Dinner	Bed-Time	Night-Time

Usually diabetic patients collect a measure of BGL, the glycosuria level and the insulin dosages for each time-slice. Further information about the patient's life-style, physical exercise and diet are very frequently missing. In our work, we have hence performed the analysis only on the time series of BGL, insulin and glycosuria. When detecting STATE patterns in time series of numerical variables a preliminary qualitative abstraction is carried out [8]. The following table shows the qualitative abstraction exploited in our work:

Table 1. Qualitative Abstractions

variable	qualitative level
BGL	Low
	Normal
	High
INSULIN	Low
	Medium
	High
GLYCOSURIA	Absent/Traces
	Present

The mapping between the qualitative abstractions and the quantitative levels of each numerical variable depends on the time-slice and on the specific patient's characteristics. For example, the BGL normal range is wider in the morning than around lunch and it is wider in pediatric patients than in adult ones.

Once the time slices have been generated, following the ontology defined for the TAs problem solving method, we have defined a set of BASIC and COMPLEX abstractions for each time slice. For example, the BASIC TAs defined for the *Breakfast* time slice are shown in Table 2.

Table 2. BASIC Temporal Abstractions.

TA type	finding	Temporal Abstractions
STATE	BGL	Hypoglycemia
		Normal BGL
		Hyperglycemia
	Insulin	Low Insulin
		Medium Insulin
		High Insulin
	Glycosuria	Glycosuria Absent/Traces
		Glycosuria Present
TREND	BGL	BGL Increase
		BGL Stationarity
		BGL Decrease
	Insulin	Insulin Increase
		Insulin Stationarity
		Insulin Decrease

Exploiting the BASIC abstractions and the above introduced definitions, we can characterize the patient behavior through the concept of ABSTRACT STATE (AS), that corresponds to the combination of the TAS that are *true* in that period. We can construct the time series of the ASs along the daily time axis for each time-slice.

The general form of the abstract state in the i−th day for the j−th time-slice AS_{ij} is:

$AS_{ij} = \{$ BGL$_{ij}$, BGL-TREND$_{ij}$, COMPETENT INSULIN (C-I)$_{ij}$, C-I TREND$_{ij}$, GLYCOSURIA$_{ij}\}$

To explain the process of deriving the AS time series from the raw data, let us suppose to have the following data for the breakfast time slice, extracted from the daily diary of an 8 years old pediatric patient:

monitoring day	BGL (MG/DL)	C-I (NPH) (U/kg))	GLYCOSURIA
6	72	0.2	Absent
7	208	0.21	Traces
8	206	0.23	Absent
9	112	0.2	Present
10	106	0.17	Absent
11	238	0.17	Absent
12	193	0.2	Traces
13	238	0.2	Absent

NPH is a kind of insulin having a delayed effect on the glucose metabolism, so that the insulin competent for the breakfast time slice is the NPH insulin taken at dinner.

The following tables show the values of the parameters used to define the TAS on the BGL and the NPH variables of the same patient in the breakfast time slice. The *granularity* parameter specifies the maximum gap that allows two measurements to be aggregated in the same episode. The *minimal extent* parameter specifies the minimal time-span of the considered TA.

STATE TA definitions

variable	granularity (days)	qualitative level	range	minimal extent (days)
BGL (mg/dl)	1	Low	0 - 80	1
		Normal	80 - 180	1
		High	180 - 400	1
NPH (Units/Kg)	1	Low	0 - 0.25	1
		Medium	0.25 - 0.35	1
		High	0.35 - 0.8	1

TREND TA definitions

variable	granularity (days)	trend	slope (per day)	minimal extent (days)
BGL	1	Dec	≤ -50	3
(mg/dl)		Stat	< +/-50	3
		Inc	≥ +50	3
NPH	1	Dec	≤ -0.015	2
(Units/Kg)		Stat	< +/-0.015	2
		Inc	≥ +0.015	2

On the basis of these values, the following episodes are detected:

TA	episodes
Glycosuria Absent/Traces [6,8]	[10,13]
Glycosuria Present	[9,9]
Normal BGL	[6,6] [9,10]
Hyperglycemia	[7,8] [11,13]
Low NPH	[6,13]
NPH Decrease	[8,10]
NPH Stationarity	[6,7] [10,11] [12,13]
NPH Increase	[7,8] [11,12]
BGL Decrease	[8,9]
BGL Stationarity	[7,8] [9,10] [11,13]
BGL Increase	[6,7] [10,11]

From these episodes we can derive the following AS time series for the breakfast time slice:

day	BGL	BGL-TREND	NPH	NPH trend	GLYCOSURIA
6	Normal	Dec	Low	Stat	Absent/Traces
7	High	Inc	Low	Inc	Absent/Traces
8	High	Stat	Low	Inc	Absent/Traces
9	Normal	Dec	Low	Dec	Present
10	Normal	Stat	Low	Dec	Absent/Traces
11	High	Inc	Low	Stat	Absent/Traces
12	High	Stat	Low	Inc	Absent/Traces
13	High	Stat	Low	Stat	Absent/Traces

The above presented methodology proposes a two-step procedure for data-preprocessing: i) the raw time-series is analyzed using TAs, moving from the original time scale to a new scale derived from the sequence of relevant patterns detected in the data; ii) a new time series of abstract episodes is obtained, going back from the new scale to the original one. The new time series contains more meaningful information than the raw one, incorporating temporal trends and, in general, it may include also multi-variable patterns.

3.3　Data Interpretation

Once that the AS time series is defined, a number of analysis can be performed. The more interesting ones are described in the following:

Modal Day Extraction

A major goal of monitoring diabetic patients is to verify if their glucose metabolism follows a daily *cyclo-stationary* pattern, modulated by meal intakes and physical activity. To test this hypothesis, several authors suggest the extraction of the BGL *modal day* (BG-MD).

BG-MD is a characteristic daily BGL pattern that summarizes the typical patient's response to the therapy in a specific monitoring period. It can be used to evaluate the protocol performance over the selected time interval, even when the information is poor (e.g. data on meals missing). Several approaches for deriving the BG-MD have been presented in the literature, from simple statistics to time series analysis [1, 5].

In our approach it is easy to derive the BG-MD by calculating the marginal probability distribution of the BGL from the AS time series. In particular, in this context it is possible to apply a Bayesian technique described in [9] that is able to deal with incomplete probabilistic models.

At the beginning of the monitoring period, an equal prior probability is assigned to the occurrence of each BGL qualitative value (e.g. Hypoglycemia, Normoglycemia, Hyperglycemia). As the monitoring process proceeds, the posterior probability distribution of BGL is updated according to the available evidence. If some data are missing, the point posterior probability estimate turns into an interval, which is specified through a lower and a higher *posterior probability bound*. So, at the end of the learning process we obtain, for each time slice, an interval probability distribution over the BGL states [10]. The modal day is then extracted taking the BGL states with the highest probability in each time slice. The "imprecision" of the distributions (proportional to the width of the intervals) will increase with the amount of missing data, and therefore the derived modal day will be more or less reliable, in dependence of the available data set. By using the same procedure it is possible to extract the typical insulin regimen that is followed by the patient, called *control actions modal day* (CA-MD).

The Blood Glucose Modal Daily Pattern

Although the BG-MD is a widely accepted instrument for summarizing the patient's metabolic control over a given monitoring period, it represents a "vertical" temporal abstraction [2] which misses the information on the actual daily pattern, i.e. on the sequence of consecutive measurements in one day. As a matter of fact, the BG-MD only reflects the frequencies of measurements in each time-slice, and not the frequencies of patterns of consecutive measurements. Nevertheless the information derived on the BG-MD may be used as the basis to search for the most frequent daily pattern in the AS time series.

To take into account the "horizontal" dependencies of the TAs extracted in each time slice, we define the *blood glucose modal daily pattern* (BG-MDP) as the most frequent sequence of abstract states of the BGL variables in the different time slices.

We have implemented a technique that is able to derive the most typical daily patterns starting from the information contained in the modal day and the

TAs performed on the daily basis. More details on this technique are contained in [12].

Persistence Analysis

Other interesting results can be obtained through the time span distribution of the episodes. For example, if the normal level of the BGL variable has an exponential time span distribution, it is clear that the patient is not able to control its glucose metabolism for a long period. This kind of information is important for understanding the patient response to a certain therapeutic protocol, as well as for classifying the quality of his/her self-management.

In our approach, it is very easy to derive such information from the AS time series, including univariate and multivariate time span distributions.

Causal Models for Explanation

Starting from the AS time series, it is in general of interest to derive the cause-effect relationships underlying the individual patient's response: the detection of situations that happened to be dangerous in the past may allow to define *context-based* therapeutic rules tailored to each patient. Several techniques may be applied to this purpose, from statistical analysis to machine learning [13]. Here we resort to a probabilistic technique, which derives the probability of the AS at a certain monitoring time given the past history of abstract states.

The general problem of deriving the needed conditional probability distributions from the data is hampered by the well-known "curse of dimensionality" problem, since the number of conditionals grows combinatorially with the number of variables involved in the model and the number of their levels. It is hence necessary to derive a simplified model, that conjugates the need to explain the patient's behavior with the possibility to derive the needed probability distributions from the available data.

A possible solution is to try to derive a simple causal model that provides the probability distribution of the next BGL state given the AS at a certain time point. Unfortunately, the probabilistic model underlying the AS evolution is intrinsically non-Markovian, since the state variables are represented with intervals whose different extents implicitly contain information on their past course.

A Markov model may be obtained by augmenting the AS, including the information on the extent of each state variable calculated at that time. In order to obtain a manageable model, also the extent of the variables must be abstracted into few qualitative levels, automatically derived from the time-span distribution analysis.

In this way, we obtain a new time series of AS*; the AS* of the i-th day for the j-th time slice can be formalized as follows:

$$AS^*{}_{ij} = \{ \text{BGL}_{ij}, \text{BGL TIME-SPAN}_{ij}, \text{BGL-TREND}_{ij}, \text{BGL-TREND TIME-SPAN}_{ij}, \text{C-I}_{ij}, \text{C-I TIME-SPAN}_{ij}, \text{C-I TREND}_{ij}, \text{C-I TREND TIME-SPAN}_{ij}, \text{GLYCOSURIA}_{ij}, \text{GLYCOSURIA TIME-SPAN}_{ij}\}$$

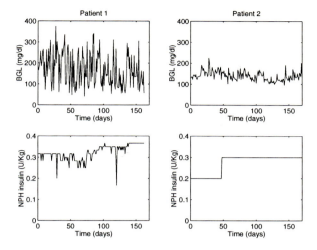

Fig. 1. BGL and C-I time series for two patients.

where the variable time spans represent the current time extent of each episode. If we want to calculate the transition probability distributions, it is possible to perform a mapping of these values into qualitative ones. Also this solution requires a very high number of conditionals. So, it is necessary to resort to marginal models, that forecast the next BGL given a subset of the extended AS*.

4   A Data Interpretation Example

In order to better explain the methods proposed in the previous section, we will show some results obtained on 2 patients from a population of 6 pediatric patients monitored for periods lasting from 6 months to one year. These patients underwent Intensive Insulin Treatment (IIT) with three injections a day for the overall period. The time-slices have been identified for all patients in dependence of their life-style, and the TAs have been derived as described in the previous section. In Fig. 1 the raw BGL and C-I of the BREAKFAST time-slice of two patients are shown. Patient 1 has a worse BGL control than patient 2; moreover (and consequently) the C-I for patient 1 is continuously modified, while patient 2 follows a nearly constant insulin protocol.

Modal Day Extraction

The first step of the analysis involved the calculation of BG-MD. Fig. 2 shows the results obtained for the breakfast time-slice. This figure summarizes in a very intuitive way the metabolic response of the different patients.

Persistence Analysis

Once the BG-MD has been derived, it is important to model the persistence of

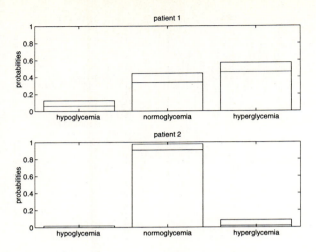

Fig. 2. Probability bounds for the bgl state variable for patient 1 and 2 (see Fig. 1), in the breakfast time-slice.

the different episodes, in order to derive an indication of the relative stability of the metabolic control. A way to perform this analysis is to test whether the time-span distribution of the different BGL episodes (hypo, normo and hyperglycemia) follows an exponential decay. In particular, if the normoglycemia episodes time-span show an exponential shape, an immediate conclusion is that the patient is not able to obtain a desirable control, since episodes of persistent normoglycemia are not likely to happen. Fig. 3 shows the probability distribution of the nor-moglycemic episodes time-span for patient 1 and 2 in the breakfast time-slice. While patient 1 presents an exponential probability distribution, confirming the instability of the metabolic control, the distribution of patient 2 shows a signifi-cant discrepancy from the exponential one, with a high probability of persistent normoglycemic episodes.

Causal Models for Explanation

Finally, for the two patients we have derived a very simple causal model, with which we try to explain the BGL state using the information on its past level and on the episode persistence. The model, consisting of a set of conditional probability distributions, can be specified, for each time-slice as:

$$p(\mathrm{BGL}_{i+1} \mid \mathrm{BGL}_i, \mathrm{BGL\ TIME} - \mathrm{SPAN}_i).$$

In order to test the explanation performance of this simple model, we have chosen the cross-validation method: the probability distributions were learned from a half of the available data and the prediction on the next qualitative level was performed on the remaining half. Predictions were derived using the Maxi-mum a Posteriori Method. The results obtained show 50% of correct predictions for patient 1, and 96% of correct predictions for patient 2. Such results have a straightforward interpretation: patient 2 has a good quality of metabolic control,

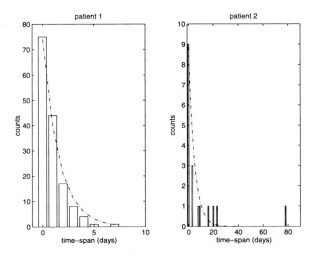

Fig. 3. Probability distribution of the normoglycemic episodes time-span for patient 1 and 2, in the breakfast time-slice and the relative exponential fitting (dashed line).

and hence it is possible to explain the next BGL qualitative levels just by analyzing the past measurement history. Patient 1 has a worse degree of metabolic control, showing a highly unstable time series; a simple model is hence inadequate to interpret the data in a reliable way.

A more systematic approach may involve the utilization of statistical techniques for variable selection, like regression trees [14], applied on the overall augmented abstract state AS*. In the example, while for patient 2 the model including BGL_i, BGL TIME $-$ $SPAN_i$ as explanatory variables is sufficient to obtain a good prediction performance, for patient 1 we must consider also the variable $BGL - TREND_i$ to slightly decrease the forecasting error rate (60% of correct predictions).

5   Conclusions

In this paper we have described an approach to pre-process and interpret biomedical time series using Temporal Abstraction techniques. The basic idea underlying the paper is to filter the original time series using TAs and hence to interpret the derived new time series resorting to both statistical and AI techniques. Moreover, in this paper we faced the problem of modeling episodes persistence and of exploiting causal relationships between episodes to derive suitable explanation models. The proposed approach is particularly useful in application contexts in which a large amount of data is present, with a low information content. The analysis of data coming from home-monitoring of diabetic patients falls within the above mentioned class of problems: intelligent data analysis may greatly enhance the quality of monitoring performed by patients and physicians.

Acknowledgments. This work is part of the EC project HC-1024, T-IDDM, Telematic Management of Insulin-Dependent Diabetes Mellitus.

References

1. Kahn, M.G., Abrams, C.A., Orland, M.J., Beard, J.C., Cousin, S.B., Miller, J.P., Santiago, J.V.: Intelligent computer-based interpretation and graphical presentation of self-monitored blood glucose and insulin data. Diab. Nutr. Metab. **4** (1991) 99-107.
2. Shahar, Y., Musen, M.A.: Knowledge-Based Temporal Abstraction in Clinical Domains. Artificial Intelligence in Medicine **8** (1996) 267–298
3. Miksch, S., Horn, W., Popow, C., Paky, F.: Therapy Planning Using Qualitative Trend Descriptions. Lecture Notes in Artificial Intelligence, P. Barahona, M. Stefanelli and J. Wyatt eds., Springer Verlag, Berlin (1995) 197–208
4. The Diabetes Control and Complication Trial Research Group: The effect of intensive treatment of diabetes on the development and progression of long-term complications in insulin-dependent diabetes mellitus. The New England Journal of Medicine **14-329** (1993) 977–986
5. Deutsch, T., Lehmann, E.D., Carson, E.R., Roudsari, A.V., Hopkins, K.D., Sönksen, P.: Time series analysis and control of blood glucose levels in diabetic patients. Comp. Meth. and Programs in Biomed. **41** (1994) 167–182
6. Andreassen, S., Benn, J., Hovorka, R., Olesen, K.G., Carson, E.R.: A probabilistic approach to glucose prediction and insulin dose adjustment: description of metabolic model and pilot evaluation study. Computer Methods and Programs in Biomedicine 41 (1994) 153-165
7. Allen, J. F.: Towards a general theory of action and time. Artificial Intelligence 23 (1984) 123–154
8. Larizza, C., Bernuzzi, G., Stefanelli, M.: A General Framework for Building Patient Monitoring Systems. Lecture Notes in Artificial Intelligence, P. Barahona, M. Stefanelli and J. Wyatt eds., Springer Verlag, Berlin (1995) 91–102
9. Ramoni, M., and Sebastiani, P.: Robust Learning with missing data. Knowledge Media Institute Report, The Open University, KMI-TR-28 (1996)
10. Riva, A., Bellazzi, R.: Intelligent Analysis Techniques for Diabetes Data Time Series. In: Advances in Intelligent Data Analysis, G. Lasker and X. Liu eds., IIAS Press, Germany (1995) 144-148
11. Riva, A., Bellazzi, R.: Learning Temporal Probabilistic Causal Models from Longitudinal Data. Artificial Intelligence in Medicine **8** (1996) 217–234
12. Bellazzi, R., Larizza, C., Riva, A.: Temporal abstractions for pre-processing and interpreting diabetes monitoring time series. in: Intelligent Data Analysis in Medicine and Pharmacology, IJCAI workshop, Nagoya, Japan (1997)
13. Advances in Knowledge Discovery and Data Mining. U. Fayad, G. Piatetsky-Shapiro, P. Smyth and R. Uthurusamy eds., AAAI Press, Menlo Park, California (1996)
14. Breiman, L., Friedman, J.H., Olshen, R.A., Stone, C.J.: Classification and regression Trees. Wadsworth, California, 1984

Intelligent Support for Multidimensional Data Analysis in Environmental Epidemiology

Vera Kamp[1] and Frank Wietek[2]

[1] University of Oldenburg, FB Informatik
Escherweg 2, D-26121 Oldenburg, Germany
kamp@informatik.uni-oldenburg.de
[2] Institut OFFIS, Escherweg 2,
D-26121 Oldenburg, Germany
wietek@OFFIS.uni-oldenburg.de

Abstract. Within the scope of the project CARLOS (Cancer Registry Lower–Saxony), a software system — CARESS (CARLOS Epidemiological and Statistical Data Exploration System) — was developed to support modeling and conducting of descriptive epidemiologic studies. The fundamental idea was to implement a powerful core of a system for statistical analysis, which is easily extensible with regard to both data types and algorithms for processing the data. We followed a knowledge-based approach, i. e. a strict separation of data and knowledge on the one hand and the control cycle processing this knowledge on the other. The main concepts concerning data structures, methods, and data processing are presented. Special emphasis is put on the underlying data analysis model and the user interface, namely a visual workbench providing easy access to the whole trail of a study and all relevant data and knowledge. CARESS aims at novel techniques for analysing cancer clustering using advanced database technology to support multidimensional analysis.

1 Application Scenario

Apart from establishing a population–based cancer registry in Lower–Saxony, a federal state of Germany, the project CARLOS also aims at providing software support for all steps of cancer registration [1]. Especially, novel techniques for analysing cancer clustering using an advanced analysis system and database technology are being developed and implemented. A database documents cancer cases in a predefined area. Incoming data is stored and epidemiologists may use it for describing the distribution of disease or testing hypotheses on cancer clusters and their determinants. Additional, e. g. spatial data is related to data about cancer cases to support epidemiologists in generating hypotheses about cancer clustering. Events in time and space are used to trigger rules evaluating time–space–configurations constituting candidates for such clusters.

To overcome cancer is not only still an immense medical task but also an increasing interdisciplinary task. The main aim of investigating causality is to

X. Liu, P. Cohen, M. Berthold (Eds.): "Advances in Intelligent Data Analysis" (IDA–97)
LNCS 1280, pp. 299–310, 1997. © Springer–Verlag Berlin Heidelberg 1997

improve prevention. To achieve this, computer supported population–based cancer registries form an important foundation. They improve methodical founded cancer research providing the possibility to manage huge and differentiated data about cancer diseases, death and complementary health-related and environmental data [8]. Besides, they offer new chances and methods to support epidemiological research for monitoring the development of diseases, especially cancer, by modern analysis and database technology [2] and appropriate interactive visualisation mechanisms.

Indispensable for modern health reports are descriptive epidemiological studies used to evaluate the data. Descriptive epidemiology is population–based and not related to individual cases. It deals with the occurrence of diseases, their respective accumulations, especially their spatial and/or temporal distributions in comparison to standardised populations. A continuous description of the development of cancer diseases in the population depends on an operating epidemiological cancer registry and is based on the periodical computation of quantified indicators. Indicators like mortality describe the health state of the population and are strongly related to the living environment and therefore can be influenced by collective and social activities. Analysing the spatio–temporal distribution of regional health data is a common way to detect possible health risks within a population [15]. Cluster analysis belongs to this set of statistical methods used to group variables or observations into strongly interrelated subgroups. To detect regional patterns in health data it is necessary to define measures of spatial clustering and to use significance tests evaluating the similarity of adjacent data values. The patterns may represent various environmental effects like relations to socioeconomic status and urbanisation. In general, the interpretation of health data affords additional knowledge. Therefore, environmental data has to be integrated, like sociodemographic data, geographic data and data describing health risks resulting from the technical and social environment. For example, the event of an increased cancer rate around an incineration plant for garbage can only be estimated correctly if data concerning the pollutants, the kind of disease and the age composition of the population are available. Interrelation between the data serves to generate hypotheses concerning striking temporal or spatial situations and their causalities. Often the hypotheses are already known and are verified by the data. To explore relations between cases and environmental influences the database has to manage geographic data as well. Geo–operators, for example *around*, have to be defined on top of geographic information in order to specify spatial events like in the previous example.

The cancer registry of Lower–Saxony determines the main application area of the system presented in this paper. In particular, we want to support[20] all steps of descriptive and exploratory statistical analysis for incidence monitoring and health reporting [8], [15], [16]. Besides, the system is supposed to be also capable of supporting further, similar fields of application, e. g. processing simple analytical models or dealing with different kinds of population-based aggregate data in general. Thus, our approach is application–driven, but the developed concepts should not and will not be restricted to the specific application. In contrast, they will fit for a huge class of scientific data processing and analysing applications having similar requirements in common.

2   Requirements

The description of the application above gives an impression of the need for improving analysis and database system support for multidimensional data analysis problems and applications. The characteristics of the application can be summed up as follows:

1. Large multidimensional data sets categorized by different parameters describing the study population are processed.
2. Normally, these studies are carried out in an iterative cycle, in the course of which parameters are modified and different data sets and results of various methods are compared. Intermediate results serve as a foundation of new hypotheses and new calculations. This can be summarized in the term exploratory data analysis.
3. Often non–expert users with regard to the statistical background of all the methods used participate in the enforcement of a study — especially as rather simple descriptive statistics are concerned. (Of course, they have to be familiar with at least the basic concepts.)

Conventional systems for statistical analysis often do not support all steps necessary to carry out descriptive epidemiologic studies. As a consequence different systems for data storage, data management, statistical analysis and visualization of data have to be used. Our objective is to take into consideration all aspects listed above and to design a user-friendly, flexible system for data analysis, which allows to integrate all methods required in a single system. In particular, the following requirements have to be met:

1. The core of the system should be independent of specific data types and methods. Easy incorporation of data types and methods — at least from the programmers point of view — is of very great importance.
2. To support data exploration, mechanisms for book–keeping, i. e. storing the trail of a study, are necessary. The user has to be able to modify and re-execute his calculations carried out so far.
3. The user interface has to provide help facilities, making available information about both usage of the system as well as data and methods employed in the current study. The user needs an overall view of the whole study network, comprising all methods applied to the data base and intermediate results respectively. Furthermore, all elements of the study should always be accessible for information and modification.

An appropriate management of scientific data is important to meet the requirements of an intelligent multidimensional data analysis. It is based on using spatial, temporal and statistical data which occurs in a lot of scientific applications (geographic and environmental information systems, decision support systems, data mining systems). Therefore, new approaches have to provide an adequate modeling, special storage and access structure for efficient data management, and an extended (descriptive) query language for an integrated view and multidimensional data processing. Commercial systems, database systems or geographic information systems support only special fragments and there exist only a few research approaches dealing with an integrated solution. But for an operationalisation it is not sufficient to integrate the various aspects within the data model. Comfortable querying and special implementation techniques have to be considered. Our approach is based on the interoperability of analysis and database systems. The evaluation of the results will guide the developement of more adequate and efficient variants. For this purpose, we are implementing our system CARESS, which will be evaluated in the scientific application of analysing spatio–temporal distribution of regional health data — especially of cancer data in Lower-Saxony.

3   Basic Concepts

As shown above, analysing multidimensional health data is a complex problem and shares some characteristics of OLAP (Online Analytical Processing) applications, geographic or environmental information systems and scientific data analysis in general. In this section, first the underlying multidimensional data analysis model is described. This model guided the design of our system architecture which is presented in Sect. 3.2 and our approach for data integration. We will concentrate in Sect. 3.3 and Sect. 3.4 on the core of our system, a graphical network editor which provides a visual workbench for intelligent data analysis.

3.1   Multidimensional Data Analysis Model

The model for data analysis underlying the design of CARESS is based on several layers: data integration, querying, analysing, visual presentation and interactive exploration as it is shown in Fig. 1. Data integration means that an integrated view onto the different kinds of data has to be provided. Health data, e. g. cancer cases with a spatio-temporal relation, has by itself a complex structure with additional statistical aspects. For example, every data item, a cancer case, is described by the location of occurence, the geographic coordinates, and the time the event occured, the date of diagnosis. In the process of analysis this item has to be combined with geographic data, e. g. data describing the political borders of administrative units. The main view consists of spatio–temporal aggregation of health data. Besides the view must provide access to additional information sources, environmental data and pure statistical data like demographic population facts. The user must not be bothered by integrating this information by himself. It has to be achieved by the interface to the underlying database system.

Following the argumentation above, the interface to the database system has to provide suitable querying techniques. The selection of data requires facilities to perform visual queries and the existence of especially spatial and temporal

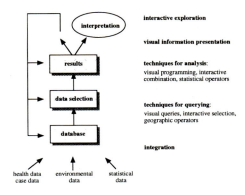

Fig. 1. Data analysis model

operators which can be used in long interactive and iterative data selections. The user determines the data selection which serves as a basis for the following analysis. In general the selection will be related to spatio-temporal aspects. The consequences for data modeling and the required extension of existing query languages become evident. The process of data selection is an iterative process and as the whole iterative analysis process it is motivated by interactive user requirements.

This becomes much clearer regarding the performed techniques of analysing the query results of data selection. The techniques applied are determined by visual combination of predefined operators, especially statistical operators. Complex operation can be constructed and combined interactively. In this context, a lot of the analysis is based on aggregation and summation of data. Because of the interactive query model, a certain part of the selected and analysed information will be reused in the same analysis process but also in other data explorations, at another time or by different people. So new techniques have to be developed to fulfill those requirements and to improve performance aspects of interactive analysis. Interactive data exploration is mainly based on visual presentation of analysis results and an online interpretation which triggers new analysis steps. This iterative process requires selected data and results to be reused and combined. The response time of the analysis system has to be kept low to be acceptable for the user.

3.2   System Architecture

Derived from the presented model for data analysis and the application–based requirements, Fig. 2 depicts the system architecture of CARESS. Based on a multidimensional data (MDD) mapping layer for data integration three different analysis layers, user interfaces, are provided for different user groups:

1. a graphical network editor (visual workbench) to easily combine different calculation and visualization procedures, to compare algorithms and datasets

in an exploratory data analysis, and even facilitate integration of external tools,

2. a tool to define and execute sequences of analysis steps for report generation and data export, and

3. a menu–based user interface providing an easy–to–use access to routine calculations and graphics.

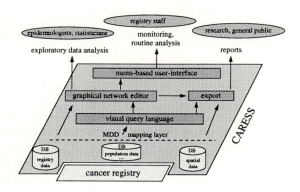

Fig. 2. Prototypical system architecture

As the graphical network editor forms the core of our system, we will concentrate on the concepts of this editor. But first we will sum up some important aspects concerning the data integration.

Data Integration. The MDD mapping layer builds a database interface which integrates datasets from different data sources into a unified view onto the database. Currently CARESS, our data analysis system, uses a special layer which maps the object oriented (extended) analysis access onto the SQL interface of the relational database system ORACLE (cf. Fig. 2) In this context extended access means that the layer provides extended query facilities like spatial and statistical operators. Additionally, it provides an object puffer for caching determined results for reuse. Epidemiologic analysis data is represented by so–called statistical objects ([12] ,[13] ,[17]). All data sets processed by the system are aggregate (summary) data, i. e. multidimensional data about subpopulations of a study population classified by different attributes, e. g. counts of incidence cases or standardized rates by age, sex, time, and district. Formally, a statistical object is defined as a triple (S, C, m) of one summary attribute S, a set of category attributes C, and a multidimensional data array m. The summary attribute provides the metadata associated with the data set. It describes, for instance, type (like count, rate, test statistic etc.), domain, and statistical features of the data set and gives access to its creation history. The category attributes define the data sets' dimension and size by describing the parameters dividing the population under study into subpopulations (e. g. region, time (interval or point), age, sex, or kind of disease). Finally, the data array contains the data — possibly with additional reference data (e. g. population data along with counts of cases) — and provides a variety of functions for a comfortable access to single values and for iteration over its elements. Thus, data sets from different data sources are modelled and implemented as data cubes classified and aggregated along different dimensions. Efficient data access and manipulation (*roll up, drill down, slice and dice*), as demanded in currently discussed OLAP applications and Data Warehouse environments ([4], [3], [19], [6], [5]) combined with application specific data analysis procedures are provided. Different types of metadata associated with the data sets will guide search and data processing.

For selecting and combining data, (cf. Fig. 2) on top of the MDD mapping layer there will be a visual query language interface which makes formulating complex textual SQL–queries unnecessary. This language should comprise statistical, spatial, and temporal operators as well as data management facilities for multidimensional data, especially aggregation. Such an interface improves the data access support of the visual data analysis workbench which will be described in Sect. 3.4. The next section introduces the underlying knowledge–based analysis approach representing the central idea of the visual workbench.

3.3   A Knowledge–based Analysis Approach

Our considerations described in the previous sections lead to a knowledge–based ([14], [11], [7]) architecture, i. e. a separation of control cycle on the one hand and processed data and knowledge about data and methods on the other [18]. The system clearly defines interfaces to incorporate new data types and methods for integration of various data sources, for data management, statistical analysis, and visualization. We follow an object–oriented approach. Thus, both data and methods are modeled as objects, which carry information about themselves and knowledge about their usage. This knowledge is used explicitly within a help system (describing the methods used and features of the data sets processed), and implicitly by tests for suitability (appliance of methods to data sets).

Methods and their Suitability. The arrangement of data types and methods in a class hierarchy facilitates the extension of the system. Figure 3 shows a part of the extensible class hierarchy comprising all kinds of methods. Up to now, methods providing interfaces to data sources, data management routines, standardization procedures, algorithms for cluster analysis, and methods for visualization of data are incorporated into the system. Standardization procedures, for instance, are split into methods for direct and indirect standardization: Direct method death rates or cumulative mortality figures are examples for the first and the standardized mortality ratio is one for the latter. Each method or class of methods respectively

1. defines the type of data and its category attributes processed by the method.
2. implements inheritable procedures and functions for processing the data.

3. provides the possibility to attach user-defined tests for suitability to the method. By inspecting the respective statistical objects in the course of a session, these tests may control:
   - correctness of data type, dimensions of data sets, or type of parameters. A method for drawing thematic maps e. g. checks whether the data being processed is spatial.
   - statistical features of the data. A test of clustering e. g. may claim at least a certain number of cases in each subregion or a special distribution of the data.
   - whether a method matches the problem the user wants to solve by using the respective method.

In particular the first kind of tests is of great importance as we will see later, whereas the other two types of tests are optional and their usage is strongly influenced by the users intention.

Usage and Interchange of Knowledge. Figure 3 summarizes and illustrates the knowledge-based system architecture, especially the interchange of knowledge between data, methods, tests for suitability and the user of the system. The user is able to manipulate the metadata attached to statistical objects and may specify his intentions as an input of tests for suitability. In addition, he has access to data descriptions (the metadata), background knowledge about methods and their classes, and may let the system explain results of tests. Methods process and create statistical objects and thus also define the metadata associated with the data sets. They also control the application of tests and provide them with the required metadata of the statistical objects. The control cycle of

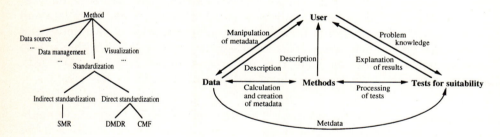

Fig. 3. The class hierarchy of methods and the interchange of knowledge

the knowledge interchange is reflected in our implementation by the graphical network editor, a visual workbench, which is described in the following section.

3.4   Visual Workbench

The system's main component is a visual workbench for describing the study of interest. A study is modeled as a network of methods, which constitute the nodes of the network. Each node represents the application of one method of a certain method class to one or more data sets, which may be processed in parallel (independently) or in combination as a joint input of one calculation. At present, five classes of methods are available, namely

1. data sources (interfaces to different data bases),
2. data management routines (restriction and aggregation of data sets),
3. (especially age- and sex-) standardization of rates,
4. algorithms for spatial cluster analysis (tests for e. g. heterogenity) , and
5. visualization of a single or a set of statistical objects.

The connections between nodes represent the data flow from out–ports of one method to in–ports of another. Each port of a node provides access to a statistical object. By selecting a certain mode of data propagation the user defines which

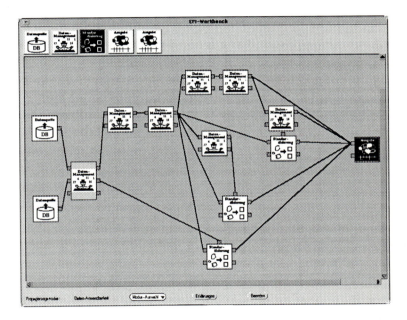

Fig. 4. Visual Workbench

kind of tests for suitability should be applied to the data propagated along the connections, before a method calculates a new statistical object as a new output from the data serving as input. As the user is free to connect each node with each other, especially basic tests for suitability are required to avoid senseless networks. Figure 4 shows an example of a network used to produce a table from mortality data containing five columns: number of cases, population, crude rate and two rates standardized by age and sex. The left part of the network selects age groups, sex and time interval, whereas the right part calculates the different measures. Each port of a node in a network offers the possibility to inspect the data stored in the port just by clicking with the mouse into the port. The data description may comprise e. g. a name of the data set, a general comment and information about data source, data type, category attributes, or the reference data stored along with the data values. Similarly, the user is provided with information about the method applied in a node by a mouse–click in this node. The menu–based user interface mentioned in Fig. 2 constitutes just a special–

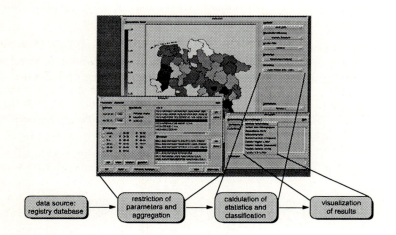

Fig. 5. Translation of menu–based parametrizations to analysis networks

purpose view onto the visual workbench for routine calculations. Each menu selection can be translated into a network of single analysis steps (outlined in Fig. 5), which may be modified and extended in the network editor for further, more detailled calculations.

4   Summary and Future Work

So far, a prototype of a data analysis system to support descriptive epidemiologic studies has been designed and implemented following a knowledge–based approach. This prototype is already employed in analysing the database of the cancer registry in Lower–Saxony and is being evaluated also in a joint project with the Hamburg cancer registry. Still a lot of work has to be done to transform this prototype into a powerful tool, which is able to fulfill what we claimed to achieve, namely to provide a user-friendly, flexible tool that supports the whole process of descriptive epidemiologic studies. In particular, the following topics have to be elaborated:

1. So far, menu-based user interface and graphical network editor are still isolated tools. The next step will be to integrate these two systems into a homogeneous environment as described above.

2. The user interface has to be improved, especially to provide the user with comfortable access to all relevant knowledge. To enable the system to process studies of typical size, abstraction mechanisms for large study networks have to be developed, which group the application of a number of methods. Furthermore, the system might be extended to support interactive statistical graphics.

3. The interfaces defined by the graphical network editor may also serve as a basis for interchanging data and knowledge with other systems. A special class of network nodes might provide access to external data sources and statistical data analysis packages.

4. Up to now, the user is restricted to implementing new C++–classes if he wants to expand the pool of methods, data types and tests for suitability. We do not want to design a new language for statistical analysis, which would be a very complex task, but an important objective is to provide a simple declarative programming language used to describe new data types and to build an interface between the data structures defined by the system on the one hand and the users algorithms on the other. It could serve as a tool to easily describe the knowledge base, which is relevant for the studies to be carried out. An interpreter might enable the user to incorporate new data types and methods dynamically during runtime.

5. Finally, according to the idea of realizing a generic core of a data analysis system, the module has to be evaluated by incorporating various methods for data management, data analysis and visualization.

Furthermore, different database system support capabilities have to be evaluated. Based on the experience we gained implementing the mapping layer, we examine some other approaches based on advanced database system technology like extensible and object–relational database systems. The evaluation of these will help on the developement of an adequate architecture. One approach considers the use and extension of multidimensional and statistical databases [10]. Another one is based on incorporating spatio–temporal aspects [9] like a spatio–temporal query language extension and the use of special indexing techniques. Problems in selecting and combining information from different sources for the analysis process are analogous to already known problems in heterogenous database systems. Currently we are reimplementing CARESS on top of the object-relational database system ILLUSTRA. We are working on specifying a *minimal data model* which covers the basic requirements of our application, environmental epidemiology. A formalisation in terms of this model will help on developing a global support approach for the combination of statistical and spatio-temporal aspects.

References

1. H.-J. Appelrath. CARLOS Tätigkeitsbericht 1996. Technical report, OFFIS, Oldenburg, 1996.

2. H.-J. Appelrath, H. Behrends, H. Jasper, and V. Kamp. Active Database Technology Supports Cancer Clustering. In *Proceedings of the first International Conference on Applications of Databases*, number 819 in LNCS, 1994.

3. S. Chaudhuri and U. Dayal. An Overview of Data Warehousing and OLAP Technology. *SIGMOD Record*, 26(1):65–74, 1997.

4. G. Colliat. OLAP, relational and multidimensional database systems. *SIGMOD Record*, 25(3):64–69, September 1996.

5. J. Gray, A. Bosworth, A. Layman, and H. Pirahesh. Data Cube: A relational operator generalizing group-by, cross-tab, and sub-totals. In *Proceedings of the 12th International Conference on Data Engineering*, pages 152–159, 1996.

6. A. Gupta, V. Harinarayan, and D. Quass. Aggregate-Query Processing in Data Warehousing Environments. In *Proceedings of the 21nd International Conference on Very Large Data Bases, Zürich, Switzerland*, pages 506–521, 1995.

7. R. Haux and K.-H. Jöckel. Database Management and Statistical Data – The Need for Integration and for Becoming More Intelligent. In *Proceedings in Computational Statistics, CompStat, Heidelberg*, 1986.

8. O. Jensen, D. Parkin, R. MacLennan, C. Muir, and R. Skeet. *Cancer Registration: Principles and Methods, IARC Scientific Publications No. 95*. International Agency for Research on Cancer, Lyon, 1991.

9. V. Kamp, M. Grawunder, R. Grupe, M. Hinrichs, F. Oldenettel, S. Weidlich, and L. Zachewitz. Spatio-Temporale Erweiterung eines DBS zur Unterstützung wissenschaftlicher Anwendungen. In *Proceedings des 10. Symposiums Informatik für den Umweltschutz, Hannover, Deutschland*, pages 277–285, 1996.

10. F. Laskowski. Relationale Verwaltung statistischer Daten im Rahmen von CARLOS. Master's thesis, Carl-von-Ossietzky Universität Oldenburg, Fachbereich Informatik, 1995.

11. J.L. McCarthy. Metadata Management for Large Statistical Databases. In *Proc. of the 8th International Conference on Very Large Data Bases (London, UK)*, 1982.

12. Z. Michalewicz, editor. *Statistical and Scientific Databases*. Ellis Horwood, London, 1991.

13. M. Rafanelli and F. Ferri. VIDDEL - An Object Oriented Visual Definition Language for Statistical Data. In *Proceedings of the 6th International Working Conference on Scientific and Statistical Database Management*, Department of Computer Science ETH Zuerich, 1992.

14. E. Rödel and R. Wilke. A Knowledge Based System for Testing Bivariate Dependence. *Statistical Newsletter*, 16(1), 1990.

15. K. Rothman. *Modern Epidemiology*. Little, Brown and Company, Boston, 1986.

16. T. Schäfer and H.-W. Wachtel. *Umweltbezogene Gesundheitsberichterstattung*. Asgard Verlag, Sankt Augustin, 1989.

17. A. Shoshani. Statistical Databases - Characteristics, Problems and Some Solutions. In *Proceedings of the 8th International Conference on Very Large Data Bases*, Morgan Kaufman, 1982.

18. K. Sundermeyer. *Knowledge Based Systems - Terminology and References*. BI Wissenschaftsverlag, Mannheim, 1991.

19. J. Widom. Research problems in Data Warehousing. In *Proceedings of the 4th International Conference on Information and Knowledge Management, Baltimore, USA*, 1995.

20. F. Wietek. Eine wissensbasierte generische Statistikkomponente für die Epidemiologie. Master's thesis, Carl-von-Ossietzky Universität Oldenburg, Fachbereich Informatik, 1994.

Section IV:

Soft Computing

Network Performance Assessment for Neurofuzzy Data Modelling

Steve R. Gunn, Martin Brown and Kev M. Bossley

ISIS group, Dept.of Electronics and Computer Science
University of Southampton, U.K.
E–Mail: {srg, mqb, kmb93r}@ecs.soton.ac.uk

Abstract. This paper evaluates the performance of ten significance measures applied to the problem of determining an appropriate network structure, for data modelling with neurofuzzy systems. The advantages of Neurofuzzy systems are demonstrated with application to both real and synthetic data interpretation problems.

1 Introduction

Neurofuzzy systems have recently received an intensified research effort [1, 2], as they combine the learning ability of neural networks with a fuzzy representation, to provide an enhanced linguistic representation. Conventional neural networks are successful at approximating continuous multivariate functions from a supervised training set, but, with the exception of trivial cases, it is difficult for a designer to interpret the knowledge that is stored within such a network. This research focuses on the ability of a learning system to automatically configure its own structure, so that it models the training data and explains the stored knowledge in a transparent fashion. A network structure which encapsulates the principle of transparency is the class of additive B-spline fuzzy networks [1], which are members of the group of associative memory networks. They have the benefit that only the output layer weights are adjusted which allows established linear training algorithms to be employed, such as conjugate gradient (CG) or singular value decomposition (SVD).

Conventional multi-layer perceptron (MLP) networks [3] use sigmoidal ridge functions to decompose the input space. These networks employ a projection pursuit type learning for hidden layer identification, but they are not transparent. However, they can produce smooth models but are often difficult to train. Radial basis function (RBF) networks [4] can be constructed using orthogonal least squares. They can have their nodes placed anywhere in the input space and hence they have a semi-transparency. However, the lack of structured placement makes them difficult to interpret and sometimes their learning is badly conditioned. The B-spline networks considered here are based on an additive tree structure which uses a forward selection and backward elimination algorithm to decompose the input space. Their disadvantage is in application to strongly coupled functions

X. Liu, P. Cohen, M. Berthold (Eds.): "Advances in Intelligent Data Analysis" (IDA–97)
LNCS 1280, pp. 313–323, 1997. © Springer–Verlag Berlin Heidelberg 1997

in a high dimensional input space, and here MLPs have an advantage. Their strength is a comparatively simple local representation which can be used to explain the knowledge extracted from the training data, with the advantages of a fuzzy rule base interpretation. The fundamental question with all these networks is how to select the size and structure of the network for a particular problem. This paper addresses this issue by evaluating some performance functions for assessing the significance of a network with respect to a set of training data. Additionally the technique of regularisation is discussed as a method for post-processing the network to provide further evidence about the performance of the model and its sub-networks. This is preceded by an introduction to the construction algorithms employed within the neurofuzzy networks.

1.1 B-Spline Neurofuzzy Networks

When neurofuzzy systems are applied to applications they tend to suffer from the curse of dimensionality [5]. To address this issue an additive decomposition is employed, which can exploit redundancy in the training data [1]. Here a function is expressed as a sum of simpler sub-functions,

$$f(\mathbf{x}) = f_0 + \sum_{i=0}^{n-1} f_i(x_i) \sum_{i=0}^{n-1} \sum_{j=i+1}^{n-1} f_{i,j}(x_i, x_j) + \cdots + f_{0,1,\ldots,n-1}(\mathbf{x}),$$

where f_0 represents the bias and the other terms represent the univariate, bivariate etc. additive components. The benefit of this representation is that often many of the sub-functions are redundant providing a more compact description.

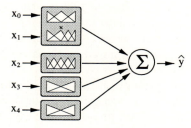

Fig. 1. A B-spline Neurofuzzy network composed of four sub-networks.

Fig. 1. illustrates an additive network, with its fuzzy B-spline basis functions shown for each input. The B-spline basis functions can be of arbitrary order dependent on the application domain, and are illustrated here for the piecewise linear case. The introduction of additional knots within the basis functions enables increasingly complex functions to be approximated, whilst an increase in their order enables smoother functions to be obtained.

1.2   Construction Algorithm

The purpose of data modelling is to understand the variables and relationships within the data. To achieve this a method of model selection is required which can search through the potentially high dimensional model space for an appropriate data representation. The question of what is an appropriate network will be returned to in section 2. The process of model selection is achieved here by an evolutionary search algorithm. An initial model is given, which can be empty if no knowledge is available about the network structure, or the designer can introduce information within this network by means of an initial structure, using fuzzy rules. This model is then updated by evaluating the best refinement from a list of potential refinements. Current refinements include: univariate addition, tensor product, tensor split, knot insertion, knot deletion, sub-network deletion, reduce order and regularise. The B-spline basis functions are chosen from a set of candidate basis functions. The refinements are collected into passes to provide a restricted and coherent method of searching the model space. A typical pass structure is,

Pass 1 Univariate addition, Tensor Product, Tensor Split.
Pass 2 Sub-network deletion.
Pass 3 Knot insertion.
Pass 4 Knot deletion.
Pass 5 Reduce order.
Pass 6 Regularise.

This forward selection and backward elimination approach allows the model to increase and decrease in complexity to adapt the network structure to model the data.

1.3   Termination Criterion

To terminate the model selection at each pass a termination criterion, Fig. 2, is used which embodies two rules. The first rule allows the refinement to escape from local minima in the refinement process by allowing the model to look ahead in the tree structure. The second rule places an emphasis on the parsimony of the network, by requiring a new refinement which increases the network size to reduce the significance measure by a certain percentage, f_{tol}. Similarly, a refinement which reduces the network size is allowed to increase the significance measure by a certain percentage, b_{tol}. This hysteresis enables superior refinement termination. (A high significance measure corresponds to a poor network). Typically $f_{tol} = 3\%$ and $b_{tol} = -1.5\%$.

2   Significance Measures

What is a significant model of a set of data? Four characteristics that are required of a network are:

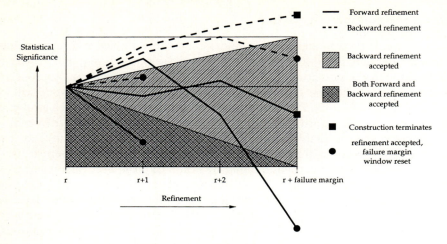

Fig. 2. Construction termination criterion.

Parsimonious The network is as simple as possible to model the data.

Transparent The network is interpretable by the designer allowing the knowledge stored in the network to be understood, and when necessary, modified by the designer.

Generalisation The network should produce good models outside the limits of the training data set.

Accuracy The network should accurately model the data.

Transparency is implicit within the B-spline neurofuzzy networks at four levels; the fuzzy rule base, visualisation of sub-network outputs, linear relationships and the ability to reject redundant inputs. The accuracy, generalisation, and parsimony of the network are closely related. The following two sections considers two ways of measuring these properties; the first primarily measures the accuracy of the network, coupled with an a estimate of the networks complexity, the second additionally addresses the question of generalisation by providing an estimate of the noise on the training data. Finally, the method of regularisation is introduced as a post-processing technique to provide additional insight into the validity of the data model.

2.1   Mean Square Error based Measures

The *MSE* is an estimator of the accuracy of the model, and as such is often employed within significance measures. However, as a biased estimator [6] it is impractical because a network can always reduce the *MSE* by introducing additional degrees of freedom. The *MSE* is often utilised as an unbiased estimator by considering the degrees of freedom of the model. There are many such functions in the literature of the form,

$$ss = MSE.f(n_w, n_p)$$

where n_w is the number of weights and n_p is the number of training patterns. To illustrate the behaviour of some of these significant measures, Fig. 2 shows the equi-potential functions, $ss = constant$, plotted against the number of weights, for a fixed training pattern size, $n_p = 100$; the measures are: bayesian information criterion (BIC), generalised cross validation (GCV), unbiased estimate of variance (UEV), final prediction error (FPE), Akaike information criterion (AIC), unbiased Akaike information criterion (UAC), full generalised cross validation (FGV), structural risk management (SRM), and minimum descriptor length (MDL) [4, 7, 8, 9]. From inspection it is evident that the measures can be grouped into threes classes: Class I: $\{AIC, FGV, MDL\}$, Class II: $\{BIC, GCV, UEV, FPE, UAC\}$ and Class III: $\{SRM\}$. All the functions are monotonically increasing, with respect to nw. Class I functions place no upper limit on the number of weights and potentially allow over fitting of the data to occur. Class II functions have an asymptote at $n_w = n_p$ and hence limit the number of weights in the network to be less than the number of training patterns. They can be ordered such that $UEV < FPE < (GCV, BIC, UAC)$. The Class III function places an upper limit on n_w which is dependent upon its parameter K_1; for $K_1 = 1.0$ the maximum number of weights is $\approx 0.37 n_p$ $(n_p > 40)$. Additionally, it is a rapidly increasing function and it will have a tendency to under-fit the data with respect to the functions of class I and II. The MDL, BIC and SRM measures are dependent upon the number of training pairs whereas the other measures are solely dependent upon the ratio n_w/n_p. To limit the testing of these functions, a representative of each class is used for comparison on the data sets. The measures that were chosen limit the tendency to overfitting from their respective class. They are,

$$MDL(MSE, n_w, n_p) = MSE.\exp\left(n_w \frac{\ln(n_p)}{n_p}\right),$$

$$BIC(MSE, n_w, n_p) = MSE.\left[\frac{n_p + (\ln(n_p) - 1)n_w}{n_p - n_w}\right]_\infty,$$

$$SRM(MSE, n_w, n_p) = MSE.\left[\frac{1}{1 - K_1\sqrt{\frac{(1+n_w)\ln(2n_p) - \ln((1+n_w)!) + K_2}{n_p}}}\right]_\infty$$

where

$$[x]_\infty = \begin{cases} x & x \geq 0 \\ \infty & x < 0 \end{cases}.$$

The disadvantage of these measures is that they are inherently dependent upon the MSE as a measure of model suitability. The training data is often sparse and in order to avoid relying on one particular instance it can be advantageous to divide the training set in different ways to provide several different training and test sets. There are two common frameworks for achieving this, bootstrapping [10] and cross validation [4]. Here we consider the method of cross validation.

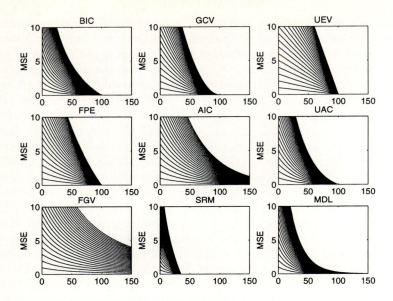

Fig. 3. Equi-potential curves for *MSE* based measures ($n_p = 100$).

2.2 Cross Validation

To address the problems of model significance our research has focused on the method of cross validation as an alternative to the above *MSE* based measures. Cross validation allows all the training patterns to be used for both training and testing of the network, by partitioning the data into packets of size m, and training the network on all but one of these, which is then used to measure the error. This is repeated for all the packets in turn to provide a mean estimate. When $m = 1$ the technique is referred to as leave one out cross validation (*LOOCV*), which is the method considered here due to its applicability to networks which are linear in their parameters. The increased computation of *LOOCV* is minimised in such cases and it may be calculated from the projection matrix, **P** [4],

$$LOOCV = \frac{\hat{\mathbf{y}}^T \mathbf{P} \left(\text{diag}(\mathbf{P}) \right)^{-2} \mathbf{P}\hat{\mathbf{y}}}{n_p}$$

The projection matrix is calculated via a SVD of the auto-correlation matrix, which can be used to directly solve for the network weights. This typically is about three times slower than the *MSE* based measures which can exploit the increased speed of the CG method for weight training.

2.3 Regularisation

Regularisation of a network smoothes the output surface and enhances the models interpolation and its extrapolation capabilities, particularly where training data is sparse. It achieves this by reducing the models sensitivity to individual data sets. It has been found [8] that second order regularisation, which introduces a soft prior smoothness constraint, is relatively simple to implement and gives good results, effectively giving more reliable fuzzy rules. The complexity of the resulting model is controlled by the regularisation coefficients which can be determined using a re-estimation formula derived from bayesian inferencing.

2.4   Summary

To evaluate the significance measures considered in this section, a representative set containing, *MDL*, *BIC*, *SRM*, and *LOOCV* is applied to real and simulated data sets in the next section. The method of second order regularisation is demonstrated as an aid to model interpretation.

3   Performance Measure Evaluation

To evaluate the performance of the four measures two problems where chosen. The first data set is taken from [11] and concerns the modelling of an additive function. This example was used to investigate the behaviour of the performance measures as the training data size varies. The second example was taken from [12] and concerns the modelling of automobile MPG data. In both experiments the models were initialised with an empty network structure, and the refinements were chosen by minimising the current performance measure; piecewise linear B-splines were used. The model refinements were stopped according to the termination criterion of section 1.3, with a failure margin of three and $f_{tol} = 3\%$, $b_{tol} = -1.5\%$. The coefficients in the *SRM* measure were taken from [7], $K_1 = 1.0$, $K_2 = 4.8$.

3.1   Example 1: Additive Data Modelling

The model considered is a ten input function, five of which are redundant, given by,

$$f(x_0, x_1, \ldots, x_9) = 10\sin(\pi x_0 x_1) + 20\left(x_2 - \frac{1}{2}\right)^2 + 10x_3 + 5x_4 + \mathcal{N}(0, 1)$$

where $\mathcal{N}(0, 1)$ is zero mean additive Gaussian noise, corresponding to approximately 20% noise, and the inputs were generated independently and randomly from a uniform distribution in the interval [0,1]. The trials were performed for five different training data sizes, {50, 100, 200, 500, 1000}, with ten independent data sets for each size. The results are presented in Fig. 4, showing the sample mean and sample standard deviation of the network size for each of the ten data sets.

The function can be modelled well with four sub-networks corresponding to a network in Fig. 1, with 16 weights. It is evident that the *BIC*, *MDL* and *LOOCV*

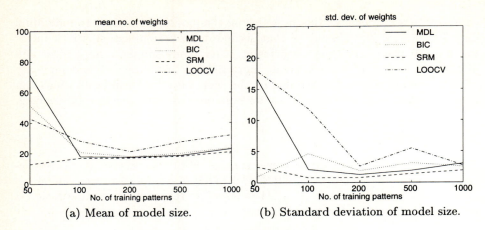

(a) Mean of model size.   (b) Standard deviation of model size.

Fig. 4. Model size vs. training pattern size.

measures generate over parameterised models for the 50 sample case, whereas the *SRM* measures generates a slightly undersized model. The *MDL* measure produces a mean model with $n_w > n_p$ and a high variation in model size, the *BIC* measure produces a mean model with $n_w \approx n_p$ and a low variation in model size, and the *LOOCV* measure produces a mean model with $n_w < n_p$ and a high variation in model size. The *SRM* measure is relatively successful, with a mean model size of 13 weights and a small variation in model size, because the cut-off for the *SRM* measure corresponds to 18 weights, just allowing the model to be approximated. When the sample size is increased above 100 all measures provide a similar mean model size that increases slightly with an increase in n_p. The standard deviations of the model size for the *BIC* and *SRM* measures are approximately constant across the different training sizes, whereas *MDL* and *LOOCV* initially have a large variation that reduces as n_p increases. Table 1. illustrates the number of correct network structures. It can be seen that the *SRM* measure provides the best models for small n_p.

n_p	50	100	200	500	1000
MDL	0	6	10	10	10
BIC	0	4	10	10	10
SRM	6	8	10	10	10
LOOCV	0	4	9	10	10

Table 1. Number of correct network structures for the 10 trials.

Fig. 5 illustrates the method of regularisation applied to a model obtained using *LOOCV* for $n_p = 50$. Fig. 5(a) shows the sub-network outputs before regularisation and Fig. 5(b) show the sub-network outputs after regularisation. These serve to demonstrate that the redundant inputs can be identified by regularisation even when the model is structure is over complex.

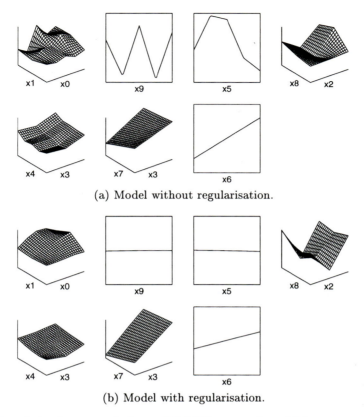

(a) Model without regularisation.

(b) Model with regularisation.

Fig. 5. Regularisation example for an $LOOCV$ measured model (x_{5-9} are redundant).

3.2   Example 2: Automobile MPG Data Modelling

The automobile MPG training set contains the following data: no. of cylinders, displacement, horsepower, weight, acceleration, year and the mpg for 392 cars. The performance measures were used to produce four respective models. Fig. 6 illustrates the resulting models, with the model structure on the left and the output of the sub- networks on the right. All four networks choose a structure consisting of the three inputs, weight, horsepower and year. However, the $LOOCV$ measure produced a tensor product between the year and horsepower properties, and for the purposes of comparison the remaining measures are displayed in the same manner. The difference of the $LOOCV$ model is the plateau around the region, 75 hp./1972, that is consistent with the data in this region, which is dominated by VW cars! In contrast, the other measures predict an increase in MPG with an increase in horsepower, which is due to the less flexible

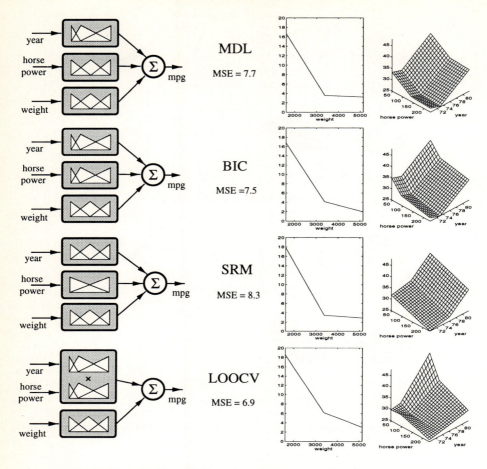

Fig. 6. Automobile MPG network models and their sub-networks.

models that were chosen. However, the important feature here is that overall the models have identified similar structures from the data.

4   Conclusions

The transparency of additive B-spline fuzzy networks has been demonstrated by their ability to describe the data in a form that the designer can inspect; the outputs of the sub-networks can be visualised to provide good understanding of the data structure and reasoning can be done using fuzzy rules. In order to form these models ten performance measures were considered. These were reduced to four representative measures which were evaluated on a synthetic modelling problem and an example to model automobile MPG data. The *SRM* measure showed the best performance on the synthetic modelling problem, providing good results over all data sizes considered, and out performing the other measures significantly for small data sizes. The automobile data demonstrated that for larger data sizes $LOOCV$ can provide a better interpretation of the data. However, it may advantageous for a designer to form models using different performance measures to determine the sensitivity of the models to the performance measure used.

References

1. M. Brown and C. J. Harris: Neurofuzzy Adaptive Modelling and Control. Prentice Hall, Hemel Hempstead, (1994)
2. J-S. R. Jang, C-T Sun and E. Mizutani: Neuro-fuzzy and Soft Computing: A computational approach to learning and machine intelligence. Prentice Hall, Upper Saddle River, NJ 07458, (1997)
3. D. J. C. Mackay: Bayesian Methods for Neural Networks: Theory and Applications. Neural Networks Summer School, Cambridge, (1995) http://wol.ra.phy.cam.ac.uk/mackay/abstracts/cpi4.html
4. M. J. L. Orr: Introduction to Radial Basis Function Networks, Technical Report, (1996) http://www.cns.ed.ac.uk/people/mark/intro/intro.html
5. R. E. Bellman: Adaptive Control Processes. Princeton University Press, Princeton, NJ, (1961)
6. S. Geman, E. Bienenstock and R. Doursat: Neural Networks and the Bias/Variance Dilemma. Neural Computation, 4:1-58, (1992)
7. T. Kavli and E. Weyer: On ASMOD - an algorithm for building multivariable spline models. Neural Networks Engineering in Control, Springer-Verlag, 83-104, (1995)
8. K. M. Bossley: Neurofuzzy modelling approaches in systems identification. Ph.D. Thesis, Faculty of Engineering, University of Southampton, (1997)
9. B. Droge: Some comments on Cross-validation. Technical Report No. 7, Humboldt-Universtt, Berlin, (1994)
10. B. Efron and R. J. Tibshirani: An Introduction to the Bootstrap. Chapman and Hall, (1993)
11. J. H. Friedman: Multivariate Adaptive Regression Splines. The Annals of Statistics, 19(1): 1-141, (1991)
12. Automobile Database, University of California Machine Learning Repository. http://www.ics.uci.edu/ mlearn/MLSummary.html

A Genetic Approach to Fuzzy Clustering with a Validity Measure Fitness Function

Susana Nascimento and Fernando Moura–Pires

Departamento de Informática
Faculdade de Ciências e Tecnologia
Universidade Nova de Lisboa, Portugal
e–mail: snt@di.fct.unl.pt, fmp@di.fct.unl.pt

Abstract. This paper presents an extension to the genetic fuzzy clustering algorithm proposed by the authors. The original algorithm, which combines the powerful search technique of genetic algorithms with the *fuzzy c-means* (FCM) algorithm, is extended such that the FCM algorithm was totally embedded in the genetic operators design. Two objective functions are applied as fitness functions: the *performance index* of a P fuzzy c-partition $J_m(P)$, used on the FCM algorithm, and the *partition coefficient* $F_c(P)$, a function commonly used as a measure of cluster validity.

The fuzzy c-means and the new proposal for the genetic fuzzy clustering algorithm were compared on generating multiple prototypes. The experimental results show that the use of genetic search improves the quality of the clustering solutions and that the *partition coefficient* $F_c(P)$ is a better measure for clustering than the *performance index* $J_m(P)$.

1   Introduction

Cluster analysis is a powerful tool for data analysis and pattern recognition. Its goal is to find the best partition of n entities into c classes. One of the most popular algorithms in fuzzy cluster analysis is the fuzzy c-means (FCM) algorithm [2]. However, the family of fuzzy c-means algorithms correspond to local search techniques that search for the optimum by using an hill-climbing approach. Thus, they often fail in the search for the global optimum. Some efforts have been made for the FCM approach to avoid getting trapped in local minimum [13].

Genetic Algorithms (GAs) [10],[9],[8] are adaptive techniques which may be used to solve search and optimization problems. These algorithms have been shown to be capable of adaptive and robust search over a wide range of search space topologies. Even though GAs do not guarantee finding the global optimum of a problem, they are usually good at finding good solutions quickly. Further, when specialized techniques exist for solving particular problems, some improvements can be made by hybridizing them with GAs. This is one major motivation of our work in applying GAs to fuzzy clustering problems.

X. Liu, P. Cohen, M. Berthold (Eds.): "Advances in Intelligent Data Analysis" (IDA–97)
LNCS 1280, pp. 325–335, 1997.   © Springer–Verlag Berlin Heidelberg 1997

There exists already some applications of genetic algorithms to fuzzy clustering, in the literature. Bhuyan et al.[5] had applied GAs to find a good ordered representation (permutation) of the elements of a data set to cluster. However, the genetic operators proposed seems quite complex. Buckles et al. [7] use genetic search to discover dynamically the best number of clusters for a clustering problem. For that, it was proposed a variable length genotype model. The use of genetic algorithms in optimal solution search for fuzzy clustering was also applied in [3] and [13]. However, in these works the operators applied are the ones generally used in genetic algorithms, while our approach uses genetic operators taking into account the existing knowledge about the context. Yuan et al. [16] have been explored GAs from an interesting point of view. The authors present some distance measures and then apply genetic algorithms to obtain the best partition in a set of examples, in conjunction with the distance function best suited to the problem. The optimization function is the partition coefficient, a validity criteria for fuzzy clustering described in [6].

Nascimento and Moura-Pires [15] proposed an *hybrid genetic fuzzy c-means algorithm* (GFC, for short), which combines the powerful search technique of GAs with the FCM algorithm. The representation of the search space and the proposed genetic operators, were designed based on specific features of the FCM algorithm. In this paper the new proposal for the GFC algorithm is an extension in two ways. First, the fuzzy c-means was totally embedded on crossover and mutation operators, that is, an hill-climbing approach was taken. Second, two clustering measures were applied as fitness functions on the GFC algorithm: the performance index of a fuzzy c-partition P, $J_m(P)$ used on the FCM algorithm, and the partition coefficient $F_c(P)$, a function commonly used as a measure of cluster validity.

Another aspect explored on this work is the multiple prototypes approach. There has been some research on schemes of editing training data sets. The basic idea of these schemes is to select or derive a smaller set of labeled data from a training data set X_{tr}, that can be used as a substitute for X_{tr}, without appreciable degradation of some performance index of classification. A description of the most common editing schemes can be found in [4]. The multiple prototypes approach have been tested with the classical FCM and the proposed extension of the GFC algorithm, and the obtained results were compared with the test results presented by Bezdek and Reicherzer [4].

The paper is organized as follows: section 2 introduces the FCM algorithm and the cluster validity function, $F_c(P)$. Section 3 describes the proposed GFC algorithm, in particular the genetic operators it uses. Section 4 gives a brief description of the multiple prototypes approach. Section 5 presents a comparative experimental study between fuzzy c-means and the GFC algorithm with each one of the fitness functions applied to the multiple prototypes approach. Finally, section 6 summarizes the main conclusions and discusses further work.

2 The FCM Algorithm and a Cluster Validity Measure

Clustering is an important technique used in searching for structures in data. Given a finite set of data, X, the problem of clustering in X is to find a collection of cluster centers which can properly characterize relevant classes of X. In classical cluster analysis, these classes are required to form a partition of X, such that the degree of association is considered strong for data lying in the same blocks of the partition and weak for data in different blocks. If this requirement is relaxed, a crisp partition on X is replaced with a weaker requirement of a fuzzy c-partition on X, and therefore it falls in the area of *fuzzy clustering*.

One technique of fuzzy clustering is the fuzzy c-means algorithm (FCM, for short) which is based on fuzzy c-partitions, where c designates the number of fuzzy classes in the partition. The purpose of FCM algorithm is to find an optimal fuzzy partition for which a given objective function is minimized.

Let $\mathbf{X} = \{\mathbf{x}_1, \ldots, \mathbf{x}_n\}$ be a set of given data, where $\mathbf{x}_k$ $(1 \le k \le n)$ is a vector in $\mathbb{R}^p$. A fuzzy c-partition of X is a family of fuzzy subsets of X, denoted by $P = \{A_1, \ldots, A_c\}$, which satisfies the two conditions:

$$\sum_{i=1}^{c} A_i(\mathbf{x}_k) = 1 \qquad \text{for all } k \in \mathbb{N}_n, \tag{1}$$

$$0 < \sum_{i=1}^{c} A_i(\mathbf{x}_k) < n \qquad \text{for all } i = 1, \ldots, c. \tag{2}$$

The main problem of fuzzy clustering is to find a fuzzy c-partition and the associated cluster centers by which the structure of the data is represented as best as possible. Given a c-partition $P = \{A_1, \ldots, A_c\}$, the c cluster centers, $\mathbf{v}_1, \ldots, \mathbf{v}_c$ associated with the partition are calculated by

$$\mathbf{v}_i = \frac{\sum\limits_{k=1}^{n} [A_i(\mathbf{x}_k)]^m \, \mathbf{x}_k}{\sum\limits_{k=1}^{n} [A_i(\mathbf{x}_k)]^m} \qquad \text{for all } i \in \mathbb{N}_c, \tag{3}$$

where $m \in [1, \infty)$ is a real number that influences the membership grades. By turn, the memberships functions of fuzzy c-partitions are calculated by:

$$A_i(\mathbf{x}_k) = \left[\sum_{j=1}^{c} \left(\frac{\|\mathbf{x}_k - \mathbf{v}_i\|^2}{\|\mathbf{x}_k - \mathbf{v}_j\|^2} \right)^{\frac{1}{m-1}} \right]^{-1}. \tag{4}$$

In solving the problem of fuzzy clustering it is necessary to define a criterion which expresses strong associations inside clusters and weak associations between clusters. For a fuzzy c-partition P, this criterion is formulated by the performance index $J_m(P)$, as follows:

$$J_m(P) = \sum_{k=1}^{n} \sum_{i=1}^{c} A_i(\mathbf{x}_k)^m \|\mathbf{x}_k - \mathbf{v}_i\|^2 \tag{5}$$

where $\|\mathbf{x}\|$ denotes the Euclidean norm of a vector $\mathbf{x}$.

The aim of FCM algorithm is to find a fuzzy c-partition P that minimizes $J_m(P)$ and it is described by an iterative procedure based on equations (3) and (4). One chooses a value for c and m and randomly generates a fuzzy c-partition P_0. Using these memberships and (3), one computes cluster centers; then using these centers and (4) recomputes memberships and so forth, until the memberships or cluster centers for successive iteration differ by no more than a given small positive number ε. The algorithm is guarantee to converge to a local minimum for the objective function $J_m(P)$ [2].

A difficulty with $J_m(P)$ is that this kind of objective functions usually have multiple local stationary points at fixed c and global extreme are not necessarily the best c-partitions of the data [2]. Some heuristic methods exist to measure the quality of a cluster, which can be used as an objective function for clustering problems [16]. These methods are based on the idea of measuring the amount of fuzziness in a c-partition of n data points and presume the least fuzzy partitions to be most valid. The partition coefficient was the first functional designed as a cluster validity measure and is defined as follows.

Let P be a fuzzy c-partition of n data points. The partition coefficient of P is the scalar:

$$F_c(P) = \sum_{k=1}^{n} \sum_{i=1}^{c} \frac{A_i(\mathbf{x}_k)^2}{n}. \tag{6}$$

A crisp partition is obtained for $F_c(P) = 1$, which is considered the ideal result. When a fuzzy partition is obtained $F_c(P) < 1$. In general, the larger $F_c(P)$ the better is the partition. An interesting property of $F_c(P)$, is that this function is uniformly continuous and strictly convex on the fuzzy c-partition space. Since this space is convex, it is guaranteed a unique global minimum for $F_c(P)$ [2].

To compare the performance of the two clustering measures $J_m(P)$ and $F_c(P)$, each of them had been used as a fitness function on the proposed GFC algorithm described in next section.

3   An Hybrid Genetic Fuzzy C-Means Algorithm

The success of a GA for a particular problem is strongly dependent on the coding mechanism, the fitness evaluation function and the genetic operators used for manipulating individuals. When a GA has to compete with other search and optimization techniques, the incorporation of domain knowledge often makes sense. In this section it is proposed an extension for the original GFC algorithm [15]. The coding mechanism and the genetic operators are described pointing out how the domain knowledge has been explored.

Since the main problem of fuzzy clustering is to find a fuzzy c-partition and its associated cluster centers, where the structure of the data is represented as best as possible, the cluster centers are the key issue to represent a solution to the problem. Thus, in the GFC algorithm each chromosome, corresponding to a c-partition P, represents the c clusters $\mathbf{v}_1, \ldots, \mathbf{v}_c$ associated with P. This means that each chromosome has c genes, each one corresponding to a $\mathbf{v}_i$, a vector of p reals.

Two alternative objective functions have been applied as fitness functions in the GFC algorithm: the performance index $J_m(P)$ (5) of a fuzzy c-partition, which corresponds to a minimization approach for a clustering problem; and the partition coefficient $F_c(P)$ (6), corresponding to a maximization approach for a clustering problem. Thus, two versions of the GFC (2-GFC, for short) algorithm were implemented, each for each one of the two objective functions. The GFC algorithm is described in pseudo-code[1] bellow.

GFC_Algorithm
Input: Crossover Probability, P_c;
          Mutation Probability, P_m;
          Population Size, $NPOP$;
          Number of Generations, $NGEN$;
          Number of Cluster Centers V;
Output: Final Pool;
{/* *Initialize the population of chromosomes, Chroms*/* }
for(p=1; p≤NPOP; p++)
  {As= RandomizeBelongs();
   cr_p=CalcClusterCenters(As);} /*p^{th} *string in population Chroms*
   FitnessValuesOf(Chroms);
   for(g= 1; g≤NGEN; g++)
   { p=0;   /**Generate NPOP candidates, Cands* */
     while (p < NPOP)
     {cr_i= RouletteSelect();
     cr_j= RouletteSelect();
     if (randFraction()< P_c) **then**
      Crossover(cr_i,cr_j,cand$_p$,cand$_{p+1}$);
     else
     {cand$_p$ = CopyChrom(cr_i);
      cand$_{p+1}$= CopyChrom(cr_j);
     }**end if**
     cand$_p$ = Mutation(cand$_p$);
     cand$_{p+1}$= Mutation(cand$_{p+1}$);
     p = p+2; }[2] /* *based on SUGAL* */
   FitnessValuesOf(Cands);
   ReplaceBy(Chroms, Cands);
   }}

---

[1] The implementation was conducted using the SUGAL genetic algorithm package [11].

[2] Note that in the last iteration when there is an odd number of chromosomes, one chromosome is chosen from the population and the other one is randomly generated.

Concerning the genetic operators, they can be summarized as follows:

• The proposed *Crossover* operator is based on the classical single point crossover (4). Since in the proposed representation of chromosomes the *locus* has no meaning (i.e., the representation does not impose any position for a cluster center), it could happen that distinct chromosomes correspond to the same solution (e.g. $cr_1 = \mathbf{v}_1\mathbf{v}_2\ldots\mathbf{v}_c$ and $cr_2 = \mathbf{v}_c\mathbf{v}_1\ldots\mathbf{v}_2$). To prevent this situation, each pair of chromosomes selected to crossover is subjected to a reordering distance based operation. Therefore, the reordering *Cdistance* operator is defined as follows: suppose that the two chromosomes selected to crossover are $cr_1 = \mathbf{v}_{11}\mathbf{v}_{12}\ldots\mathbf{v}_{1c}$ and $cr_2 = \mathbf{v}_{21}\mathbf{v}_{22}\ldots\mathbf{v}_{2c}$; one of them is fixed, say cr_1, and the other, cr_2, is reordered becoming $cr_2' = \mathbf{v}_{21}'\mathbf{v}_{22}'\ldots\mathbf{v}_{2c}'$ such that:

$$\text{Letting } U = \{\mathbf{v}_{21},\ldots,\mathbf{v}_{2c}\}$$
$$S^0 = \{\}$$
$$S^i = S^{i-1} \cup \{\mathbf{v}_{2i}'\}$$
$$\mathbf{v}_{2i}' = \mathbf{v}_{2j} : \|\mathbf{v}_{1i} - \mathbf{v}_{2i}\|^2 \leq \|\mathbf{v}_{1i} - \mathbf{v}_{2k}\|^2 \wedge \mathbf{v}_{2j}, \mathbf{v}_{2k} \in U - S^i.$$

The *Crossover* operator is thus defined as follows:

```
Crossover(cr_i, cr_j, cand_1, cand_2)
    {cr'_j = Cdistance(cr_i, cr_j);
    OnePointCrossover(cr_i, cr'_j, cand_1, cand_2)
    As_1 = CalcBelongs(cand_1);
    cand_1 = Fuzzy_CMeans(As_1);
    As_2 = CalcBelongs(cand_2);
    cand_2 = Fuzzy_CMeans(As_2);
    }
```

• The *Mutation* operator is based on the mutation operator proposed by Krishna and Murty [12], here designated as *DistanceBased Mutation*. The basic idea of this latter operator is to change the center of a cluster to which a data point belongs, depending on the distances of all cluster centers to that data point. Specifically, the probability of changing the membership of a data point belonging to a certain cluster is higher if the corresponding cluster center is closer to the data point. The Mutation operator uses the *DistanceBasedMutation* operator to determine new memberships for a chromosome, and is defined as follows:

```
Mutation(cand_p)
    {A_s = CalcBelongs(cand_p);
    A_s'= DistanceBasedMutation(A_s);
    cand_p = Fuzzy_CMeans(A_s');
    return cand_p;
    }
```

• The operation *Fuzzy_CMeans* corresponds to the FCM algorithm. In particular, the operation *CalcBelongs* calculates equation (4), while *Randomize Belongs* randomly generates the initial fuzzy c-partition $P(0)$. These operations are introduced based on the assumption that genetic operators must be meaningful with respect to the problem. Moreover, these operations ensure the convergence of the problem.

• *Fitness ValuesOf* calculates the fitness of all elements of a population and normalizes those fitness.. For the objective function $J_m(P)$ (5) the fitness function is defined as the inverse of $J_m(P)$. For the partition coefficient $F_c(P)$, the fitness function corresponds to $F_c(P)$.

• The *RouletteSelect* operation corresponds to the roulette wheel proportional selection [9] and is based in the normalized fitness..

• The *ReplaceBy()* operation replaces the current population by candidates as follows: if the best individual of the current population is better than the best of the candidates, then this individual is selected to the new population plus $NPOP - 1$ of the candidates. Otherwise, the new population will be defined by the whole set of candidates.

4   A Multiple Prototypes Approach

When a data set is large in dimension (i.e. number of features) and/or number of examples, clustering algorithms can require too much storage and CPU time for efficient deployment. To circumvent these problems, there has been some research on schemes of editing training data sets. The basic idea of these schemes is to select or derive a smaller set of labeled data from a training data set X_{tr}, that can be used as a substitute for X_{tr}, without appreciable degradation of some performance index of classification. A description of the most common editing schemes can be found in [4]. As described on section 2, the FCM algorithm is an optimization algorithm for generating a set of centers (or prototypes) for clusters in a data set X. An input parameter of FCM is the number of prototypes $|v|$ that one wants to generate and which usually corresponds to the number of classes $|c|$ of the original labeled data set X. If $|v|$ is set greater than $|c|$ then one or more of the generated prototypes are implicitly assigned to each one of the $|c|$ classes. At this point the question is: which prototypes correspond to which class? The idea is to assign a label to each prototype using the original class label of the training data set. In [4] it is discussed a labeling algorithm that uses the original class labels to attach the most likely (as measured by a simple percentage of the labeled neighbors) original class label to each v_i.

Let c' be the number of classes in X_{tr} labeled by crisp vectors $(\mathbf{e}_1, \mathbf{e}_2, ..., \mathbf{e}_{c'})$. Define p_{ij}, $i = 1, \ldots, c'$; $j = 1, ..., c$ to be the percentage of training data from class i closest to v_j via a distance rule, and define the matrix $P = [p_{ij}]$. Label e_i is assigned to v_j such as:

$$p_{ij} = \max_k (p_{kj}) \ k = 1, \ldots, c'. \tag{7}$$

Finally, the solution of labeled vectors is evaluated by computing an error rate for the classification of the original data set used as test set (X_{te}). The error rate is defined using a matrix $C = [c_{ij}] = (\#labeled\ class\ j|\ but\ were\ originally\ on\ class\ i)$. The error rate (in percent) is defined as [4]:

$$Er(X_{te}|X_{tr}) = 100 \times \left(1 - \frac{\#right}{|X_{te}|}\right) = 100 \times \left(1 - \frac{tr(C)}{|X_{te}|}\right). \qquad (8)$$

Based on these ideas, the FCM and 2-GFC algorithms have been used to test the multiple prototypes approach. The procedure follows three steps:

step 1 Generation of V prototypes from a training data set (for which the original class labels were removed), by running one of the FCM or 2-GFC algorithms.

step 2 Labeling of each of the V prototypes. The A's (5) had been used as a distance measure to determine the percentage of training data from a class i closest to a prototype v_j.

step 3 Evaluation of the solution of labeled prototypes by calculating the error rate.

5   Experimental Study

The main goal of the experimental study is threefold. Firstly, to compare the FCM and the 2-GFC algorithms. Secondly, to compare the performance of the evaluation functions $J_m(P)$ and F_c. Thirdly, to analyse the multiple prototypes approach comparing the results of the FCM and the 2-GFC algorithms with the results presented in [4].

The experimental evaluation of the proposed approach was conducted using the Iris data set [1], [14]. Iris contains 150 plants described by four-dimensional vectors each of which gives some measures of $|c| = 3$ subspecies of Iris. There are 50 plants from each subspecies and the train set is labeled. One interesting property of the Iris data set is that class 1 is very well separated from classes 2 and 3 (this is true for third and fourth features).

The study was performed considering $Iris = X_{tr} = X_{te}$ and using the following parameters: number of prototypes $|v| = 3, 4, \ldots, 9$; $m = 1.25$, $m = 1.5$ and $m = 2.0$. The FCM algorithm was run with termination criteria $\varepsilon = 0.001$ and maximum iterations $t \leq 200$ (all of the runs were completed in less than 130 iterations). The 2-GFC algorithms (GFC$_{J_m}$ and GFC$_{F_c}$) were run for 100 generations with a population of 100 individuals. Crossover and mutation rate were fixed at 0.9 and 0.001, respectively.

Typical (re-substitution) error rates for supervised designs on Iris are between 0 and 5 mistakes; for unsupervised designs, around 16 mistakes for classifiers that are subsequently designed with clustering outputs (usually prototypes) [4]. Table 1 reports the best case results (as number of re-substitution errors) after running each of the algorithms for the various values of $|v|$ and m.

Table 1. Number of re-substitution errors. $^*\backslash^{**}$- Classification with two\three proto-types mapped to class 1, respectively.

m	1.25			1.5			2.0		
$\lvert v\rvert$	FCM_m	GFC_{Jm}	GFC_{Fc}	FCM_m	GFC_{Jm}	GFC_{Fc}	FCM_m	GFC_{Jm}	GFC_{Fc}
3	16	16	16	17	17	17	16	16	16
4	23	24	24	24	24	24	22	22	22
5	24*	14	14	24*	14	14	22*	13	13
6	14*	14*	14	14	14*	14	14*	13	5
7	8*	5*	5	5	5*	5	5*	5*	4
8	5*	5*	4	5*	5*	5*	4**	4*	5
9	8*	5**	4	4*	5**	4*	4**	4**	5

As can be seen from Table 1 each one of the 2-GFC algorithms has better performance than the FCM algorithm, in particular for $\lvert v\rvert = 5, 7, 8$ and 9. By comparing the results of the two versions of GFC it is clear that the results are more or less equivalent considering the number of errors (except for $\lvert v\rvert=6$ and m= 2.0). However, notice that GFC_{Fc} almost never places two (or more) prototypes in class 1. This is a good result concerning a priori knowledge we have about the original classification of Iris. Thus, for the Iris data set GFC_{Fc} performs better than GFC_{Jm}. As to the best result of these experiments, it is for $\lvert v\rvert= 6$ and m= 2.0 with 5 errors.

The experimental results of the FCM and GFC_{Jm} algorithms also show that the function $J_m(P)$ presents better values for worst partitions. Therefore as stated in [2] it is confirmed that global extremes of $J_m(P)$ does not correspond necessarily to the best c-partitions of the data.

It is also interesting to compare these results with the ones obtained by Bezdek and Reicherzer [4] where three prototype generation schemes had been applied: Kohonen's learning vector quantization (LVQ); a family of fuzzy LVQ models due to Karayiannis and co-workers, called GLVQ-F and the *dog-rabbit* (DR) model of Lim and co-workers (see [4]). The DR model provides the best results for every value of $\lvert v\rvert$ ($\lvert v\rvert = 3, 4, 5, ...9, 15$ and 30). The best solution of DR is $\lvert v\rvert = 5$ prototypes with 3 errors. Even though none of the FCM or 2-GFC results is better than the ones of the DR model, several of GFC_{Jm} and GFC_{Fc} results are better than the corresponding ones for LVQ or GLVQ-F. In particular, for $\lvert v\rvert = 6$ LVQ, GLVQ-F and DR presents 14, 14, and 3 errors respectively, while GFC_{Fc} presents 5 errors. This is an encouraging result for future tests of the proposed algorithms with other test problems.

6   Conclusions and Further Work

The 2-GFC algorithms proposed for fuzzy clustering analysis are an hybridiza-tion of GAs and the fuzzy c-means algorithm. On designing the genetic operators, the incorporation of domain knowledge prevents unfit chromosomes (in particu-lar those which violate problem constraints) from being generated. This process also avoids introducing poor individuals into the population. The introduction of the *CalcBelongs* and *Fuzzy-CMeans* local improvement operations provide a more efficient exploration of the search space around good points.

The results obtained by the GFC algorithm using $F_c(P)$ as fitness function are better than the corresponding ones for GFC using $J_m(P)$. Notice that this observation concerns solely the obtained results for the *Iris* data set. It would be a mistake to generalize this conclusion without more experimental evidence with other data sets. Since the partition coefficient $F_c(P)$ presents some disadvantages, such as its monotonic tendency and lack of direct connection to some property of the data themselves [2], it is our intention to study other measures for clustering, and apply them as fitness functions on the GFC algorithm.

With respect to the multiple prototypes approach, one objective for further work is to take a comparative study with DR model in particular, and to study other models for multiple prototypes generators.

Acknowledgments This research has been carried out in the Department of Computer Science of FCT- Universidade Nova de Lisboa, and is supported by JNICT, the Portuguese Council for Science and Technology, under PRAXIS XXI program. The work was conducted using the SUGAL Genetic Algorithm package, written by Dr. Andrew Hunter at the University of Sunderland, England.

References

1. E. Anderson. The irises of the gaspe peninsula. *Bull. Amer. IRIS Soc.*, 59:2–5, 1935.
2. J. C. Bezdek. *Pattern Recognition with Fuzzy Objective Function Algorithms.* Plenum Press, New York, 1981.
3. J. C. Bezdek and R. J. Hathaway. Optimization of fuzzy clustering criteria using genetic algorithms. In *First IEEE Conference on Evolutionary Computation (EC-IEEE'94)*, pages 46–50, Orlando, June 1994.
4. J. C. Bezdek and T. R. Reichherzer. Multiple prototype classifier design. In *NATO- Advanced Study Institute on Soft Computing and its Applications*, pages 1–24, Manavgat, Antalya, Turkey, Aug. 1996.
5. J. N. Bhuyan, V. Raghavan, and V. Elayavalli. Genetic algorithm for clustering with an ordered representation. In K. B. Richard and B. Lashon, editors, *Fourth International Conference on Genetic Algorithms)*, pages 408–415, University of California, San Diego, 1991. Morgan Kaufmann Publishers, San Mateo, CA.
6. L. Bobrowski and J. C. Bezdek. c-means clustering with l1 and l norms. *IEEE Trans. on Syst. Man and Cyb. Journal*, 21(3):545–553, 1991.
7. B.P.Buckles, F. Petry, D. Prabhu, R. George, and R. Srikanth. Fuzzy clustering with genetic search. In *First IEEE Conference on Evolutionary Computation (EC-IEEE'94)*, pages 46–50, Orlando, June 1994.
8. L. Davis. *Handbook of Genetic Algorithms.* Van Nostrand Reinhold, New York, 1991.
9. D. E. Goldberg. *Genetic Algorithms in Search, Optimization and Machine Learning.* Addison-Wesley, 1989.
10. J. H. Holland. *Adaptation in Natural and Artificial Systems.* MIT Press, 1993.

11. A. Hunter. *SUGAL Programming Manual*. University of Sunderland, England, 1995.

12. K. Krishna and M. N. Murty. Genetic k-mean algorithm. *Communicated to IEEE Transactions on SMC*, 1995.

13. J. Liu and W. Xie. A genetics-based approach to fuzzy clustering. In W. V. Oz and M. Yannakakis, editors, *Fourth IEEE International Conference on Fuzzy Systems (FUZZY-IEEE '95)*, pages 2233–2240, Yokohama, Mar. 1995.

14. C. J. Merz and O. Murphy. Uci repository of machine learning databases, 1996.

15. S. Nascimento and F. Moura-Pires. A genetic fuzzy c-means algoritm. In *Information Processing and Managment of Uncertainty in Knowledge-Based Systems (IPMU'96)*, number 2, pages 745–750, Granada Spain, July 1996.

16. B. Yuan, G. J. Klir, and J. F. Swan-Stone. Evolutionary fuzzy c-means clustering algorithm. In *Fourth IEEE International Conference on Fuzzy Systems (FUZZY-IEEE '95)*, pages 2221–2226, Yokohama, Mar. 1995.

The Analysis of Artificial Neural Network Data Models

C.M. Roadknight[1], D. Palmer–Brown[1] and G.E. Mills[2]

[1] Department of Computing, The Nottingham trent University
Burton Street, Nottingham NG1 4BU
[2] Department of Life Sciences, The Nottingham Trent University
Clifton Lane, Nottingham NG11 8NS

Abstract. Artificial neural networks are good non-linear function approximators but their multi-layer, non-linear form gives little immediate indication of the features they have learnt. Several methods are put forward in this paper that reduce the complexity of the network or give simplified equations that are easier to interpret. Relative weight analysis and equation synthesis are summarised while correlated activity pruning is introduced and explained in detail. The former techniques use the weights of a trained network to assign importance to inputs or groups of inputs. The latter algorithm reduces complexity of a network by merging hidden units that have correlated activations. This procedure also allows the relationship between detected features to be evaluated. Data from pollutant impact studies are used but the techniques developed are applicable to many scientific data modelling environments.

1 Introduction

1.1 The data to be analysed

Artificial neural networks (ANN's) are used at Nottingham Trent University for modelling the complex climatic and pollutant interactions that damage crops in this and other European countries [1, 2]. ANN's are trained using daily climatic and pollutant conditions to predict development of crop injury or yield reduction. These effects have traditionally been modelled using standard statistical techniques such as dose response curves [3]. While the techniques covered in this paper were developed and tested using data from this domain, they are applicable to any 3 layer, feed forward networks.

1.2 ANN's for data analysis

ANN's have been shown to cope well with data containing non-linearity and to model non-linear relationships [4]. While ANNs have outperformed many statistical techniques for theoretical problem solving, their application to real world problems is the true test of their usefulness. They have been successfully

X. Liu, P. Cohen, M. Berthold (Eds.): "Advances in Intelligent Data Analysis" (IDA–97)
LNCS 1280, pp. 337–346, 1997.   © Springer–Verlag Berlin Heidelberg 1997

applied in a number of areas, including the fields of medicine [5, 6] and finance [7, 8]. In this application of ANN's, the extent to which the training data is of sufficient quantity and quality is detected by R.M.S error measures and R^2 coefficients on randomly selected test data.

1.3   Current rule extraction and optimisation techniques

The connection weights of a trained ANN contain all the information required to carry out the desired task (prediction, classification etc.), but because the knowledge acquired is represented in a complex, non-linear form, it is difficult to interpret in all but the most trivial of networks. The ANN acts as a 'Black Box' taking in and giving out information with no explanation of the relationship between them.

Efforts have been made to dissect ANNs into a set of rules [9, 10]. These methods usually aim for a set of conjunctive rules based on a 'decomposisional' approach at the individual unit level.

There are several reasons why the demystifying of an ANN is desirable:

— The ability to explain how a solution is arrived at is essential for any safety critical' systems (ie. Controlling temperature regulation in a nuclear power station).

— Symbolic AI systems (ie. Expert Systems) declare their knowledge explicitly. This ability to explain their decisions can mean inferior systems are chosen ahead of more accurate ANN.

— Rules synthesised from an ANN could be used to construct a knowledge base for an expert system, the most difficult and time consuming part of building an Expert System [11].

— When small data sets are used, extracting meaning can allow an expert to decide under what conditions the ANN will not be able to generalise.

— There is evidence that rules extracted from the network sometimes give more accurate solutions than the source ANN [12] though this is not usually the case.

Transforming the network into a form that is more interpretable, however, is of maximum value if the network architecture has been optimised. Pruning can help to achieve optimum network architecture. Such methods benefit from the learning advantages of larger networks in terms of speed and aquisition of features while reducing the amount of overtraining or memorisation within these networks.

Weight pruning is most commonly used. This involves the removal of connections based on the value of the connecting weights and these methods can be divided into two groups:

- Sensitivity Calculation. Once the full sized network is trained the sensitivity of the error function to zeroing of a weight is estimated and weights with a low impact are removed [13].

- Penalty-Term Methods. These involve the introduction of a new cost function to enforce weights of small magnitude to converge to zero during training, so can then be removed with no effect [14].

2   Basic Approaches to Weight Analysis

In this paper, two approaches are taken to explain the rules by which an ANN makes its decisions; relative weight analysis [15] [16] and equation synthesis [17] [18].

2.1   Relative weight analysis

The weights of any ANN can be followed from input to output, the product of each route giving an indication to the relative influence of the chosen input. This can only be a coarse measure of an input's affect due to the non linear nature of ANN neuron activation functions.

This method can be used to prioritise environmental and pollutant factors that affect crop injury [16]. For example, a weight analysis from a network trained to predict the onset of injury in clover (*Trifolium subterraneum* and *T. repens*) showed 4 variables to have the following total effect on the output (development of leaf injury) (table 1).

2.2   Equation Synthesis

Once an ANN has been successfully applied, the solution manifests itself as a set of activation functions and connection weights, but the full network equation will only be comprehensible for trivial networks.

To synthesise useful equations from non-trivial networks some rationalisation is required to keep the size and form of the equation manageable. This is achieved by removing all connections with relatively low weights and testing the remaining, partially connected network. More connections may be included until this minimised network performs to an acceptable standard [17] [18]. It is important that the model is found in this way and not by recursively removing

Table 1. Impact of environmental parameters on leaf injury

Variable	Relative Influence on Output
Ozone	12.8
Vapour Pressure Deficit	4.6
Plant Age	3.8
Solar Radiation	-2.20

small weighted connections from a complete network. The relationship between number of connections and network accuracy is a non-linear one so the simplest accurate model will not in general be found by succesive removal of connections.

The performance of partially connected networks was usually worse than the fully connected network, but the relationship between number of connections, described by terms in an equation, and performance was non-linear. This shows that increasing the number of terms in the equation did not necessarily increase the accuracy of predictions the associated ANN would make. One reason for this may be that beyond an essential number of generalising terms, the introduction of new inputs only brings in nodes that learnt pattern specific noise. It also underlines the importance of seeking a minimal model ie. the simplest accurate model possible, since any unnecessary terms are extraneous and serve to both obfuscate the interpretations of the model and to increase the error.

Once the minimal network has been found, it can be written in the form of an equation. From the weights of a network trained to predict whether plant injury would follow a three day ozone episode (0 errors from 37 training and test patterns), the following equation was derived:

Occurrence of injury =
  F[Day2(7hmean + Max) - Day3(AOT40+MaxOzone)] +
  F[Day2(MaxOzone-AOT40) - Day1(AOT40 + 7hmean)
      + day3(7hmean)] - 1

The terms in this equation are pollutant factor levels eg. Mean of hourly readings of ozone levels between 10am and 5pm during day 2 is **Day2(7hmean)**. *Where* **AOT40** = *accumulated ozone above 40 parts per billion.*

This equation contains the primary inputs for 2 hidden units, contained within the activation function (F), which can be approximated as a straight line between 0 and 1, for a linearised model.

The partially connected network described by this equation performed equally well in terms of number of erroneous outputs. The first, and most influential, node appears to show that a fall from high to low levels of ozone precedes the onset of injury. This is apparent because ozone levels on day three are used in a negative way, ie higher levels of ozone give a lower output. Ideal conditions for a positive prediction of injury are therefore high levels of ozone on day 2 and low levels on day 3. This equation can be further simplified to:

Occurrence of injury =
  F[rise in ozone levels] + F[fall in ozone levels]

3   Correlated Activity Pruning

The generalisation ability of an Artificial Neural Network (ANN) is dependent on its architecture. An ANN with the correct architecture will learn the task presented by the training set but also assimilate rules that are general enough to correctly predict outputs for unseen test set examples. More importantly for intelligent data analysis, the weights of a smaller ANN will be easier to interpret and the smallest net gives the most general rules for interpreting the data.

There are many possible networks that will model any given data. The process of finding the minimal network establishes a model that is a good approximation to the canonical model that would properly represent an accurate analysis of all available data.

Sietsma and Dow [19] describe an interactive pruning method that uses several heuristics to identify units that fail to contribute to the solution and therefore can be removed with no degradation in performance. This approach removes units with constant outputs over all the training patterns as these are not participating in the solution. Also, units with identical or opposite activations for all patterns can be combined. The approach to merging hidden units detailed in Sietsma and Dow's paper is useful, however it only covers perfectly correlated, binary activations.

The method presented here generalises correlated activity pruning to all real valued positively and negatively, highly correlated activation sets from hidden neurons.

3.1   CAPing Equations

Each hidden unit within a three layer neural network produces an activation when a set of inputs is presented to it. If two hidden units produce output signals that are 100 percent correlated for the entire training set then the activations for these two hidden units can be merged, with no loss of performance. Two hidden units with correlated activities can be simplified into one hidden unit by recalculating the weight vector from the remaining hidden unit to the output node using equations 4 or 5. A brief derivation of these equations is also shown. Where $Signal'$ is the sum of activation profiles from 2 hidden nodes, X and Y.

$$If Signal_X = Signal_Y \quad then \quad Signal' = Signal_y + Signal_y \quad (1)$$

If Signal X and Signal Y correlate with a correlation coefficient (ρ) of 1 then

$$Signal' = Signal_y + \frac{\sigma_x}{\sigma_y}(Signal_y - \mu_y) + \mu_x \quad (2)$$

$$= Signal_y(1 + \frac{\sigma_x}{\sigma_y}) + (\mu_x - (\frac{\sigma_x}{\sigma_y} * \mu_y)) \quad (3)$$

$$therefore, W' = W_y + (\frac{\sigma_x}{\sigma_y} * W_y) and Bias' = Bias + (\mu_x - (\frac{\sigma_x}{\sigma_y} * \mu_y)) \quad (4)$$

If Signal X and Signal Y correlate with (ρ) of -1 then

$$W' = W_y - (\frac{\sigma_x}{\sigma_y} * W_y) and Bias' = Bias + (\mu_x + (\frac{\sigma_x}{\sigma_y} * \mu_y)) \quad (5)$$

Bias' = new bias weight. W' = new hidden unit to output weight. W_x, μ_x and σ_x = output weight, mean and standard deviation of activations for hidden unit with lowest weighting

W_y, μ_y and σ_y = output weight, mean and standard deviation of activations for hidden unit with highest weighting

The equation for the correlation coefficient used is: $\rho_{xy} = \frac{Cov(x,y)}{\sigma_x, \sigma_y}$

Where $-1 \leq \rho_{xy} \leq 1$ and $Cov(x,y) = \frac{1}{n}\sum_{i-1}^{n}(x_i - \mu_x)(y_i - \mu_y)$

Correlation coefficients of 1 or -1 are not common for activations from any two hidden units of an ANN, but high correlations do exist in networks that use more hidden units than are required to learn a problem. Error can be introduced into the networks performance if the correlation coefficient is not equal to 1 or -1, but the closer the correlation the smaller the error.

Closely correlated activations can occur for several reasons:

1. Similar weights from input units to hidden units.
2. Each of the two hidden units detecting a different, co-dependent feature. For example, if high values always occur for input X when there are high (or low) values for input Y.
3. If two factors have the same effect. For example, if high values for input X have the same effect on the output as high (or low) values for input Y.

3.2   Benefits of CAPing

Analysis of Correlated units. The weights of the input layer to hidden unit connections for merged units are equivalent in terms of their influence on the network's output. They therefore represent eqivalent features. Identifying these equivalencies allows for a greater understanding of related data events.

Network Optimisation. Trial and error is a common approach to network optimisation. CAPing of a network can reach a near optimum structure in one training session.

Speed Up. An ANN can be trained in fewer epochs if more hidden units are used than are required. Parallel processing allows the addition of more hidden units without an accompanying increase in training time per epoch. Therefore an ANN can be trained quickly using more hidden units and then capped to a near optimum architecture.

3.3   CAPing in Practice : A Curve Fitting Example

Training an ANN to learn a complex, multiple variable equation is an accepted method of analysing network performances. This is performed as a curve fitting exercise (eg [21]) or as boolean equation solving (eg. [20]).

In this case, a 4 variable equation is used and random numbers are used as variables to generate an output, x. The equation is:

$$x = \frac{a - (b * c) + (\frac{b}{(d+1)}) + K}{J}$$ (6)

In equation 6 a,b,c,d are variables in the range [0,1]. J and K are set so that x then varies in the range [0,1].

A three-layer back propagation network is then trained using 500 data sets, each consisting of 4 inputs (A,B,C,D) and an output X(a,b,c,d). The level of accuracy to be achieved is set to an RMS of less than 0.03 and a maximum error of less than 0.1. The minimum number of hidden units needed to learn this can be found, by exhaustive trial, to be five. A network can then be trained using 20 hidden units, many more than are required. This takes fewer epochs to reach the required accuracy and, because of the parallel architecture used (a 64 processor SIMD machine), each epoch takes a similar training time as a 4 hidden unit network. The CAPing equations can then be used to recursively merge hidden units, with the most highly correlated units merged first. The activations of two such units and the activations of the combined unit are shown in figure 1.

Hidden unit 11 was merged into hidden unit 15 to produce a new weight from hidden unit 15 and a new bias into the output unit. The effect of changing the weight from hidden unit 15 and the bias is shown in figure 1.

Very little error is introduced until units with correlation coefficients of less than 0.8 are used. The correlation coefficients are given above the error in figure 2. Merging to even fewer hidden units is possible if some post-CAPing retraining is carried out, this retraining needs approximately 10 percent of the epochs of training from random weights.

Analysis of the weights from the input units of the merged hidden units shows some sets of weights to be highly correlated but other sets to be non-correlated (Table 2). This illustrates how CAPing removes two types of redundancy. Firstly when weight vectors represent similar features, and secondly differing features that have an equivalent effect on the network's output.

Table 2. Input to hidden unit weights for pairs of weight vectors with highly correlated hidden unit activation profiles.

Weight from:	to hidden unit 1	to hidden unit 2	to hidden unit 3	to hidden unit 4
Variable A	0.2756	0.04419	-1.4898	-1.4746
Variable B	-0.573	-0.7546	-0.6228	-1.6257
Variable C	-0.2666	-0.4243	-0.8889	-1.1646
Variable D	-0.9353	-1.0815	-1.0183	-1.2954
Comment	*Correlation coefficient for weights to 1 and 2 = 0.9991*		*Correlation coefficient for weights to 3 and 4 = -0.1099*	

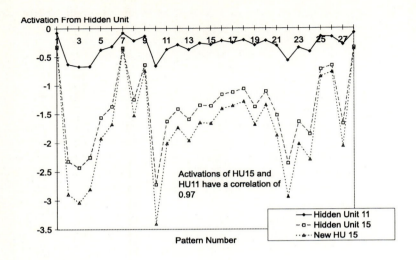

Fig. 1. The effects on activations of merging two correlated hidden units.

3.4   Real CAPing example: modelling pollution effects

ANNs have been applied to the problem of predicting leaf damage to crop plants
([1, 2]). ANNs with more hidden units than were required could be CAPed down
to architectures with a near optimal number of hidden units. An examination
of the weights of hidden units with correlated activations showed that similar
weight redundancy' was the main form of redundancy removed, but there were
also some cases of non-correlated weights giving correlated activations.

For example, in one model involving weekly pollutant and climate variables,
the following equations were synthesised from two hidden units and were found
to be correlated:

**F[7hr mean light + Max light - (2*AOT40)]   and
F[7hr mean Ozone + Max Ozone - MaxLight]**

Both of these units detect the same situation in different ways ie. when
average ozone and light levels are fairly high (but not very high). Such situations
can be very damaging to plants since stomatal aperture is large and ozone can
easily enter the plant.

4   Conclusions

While ANN's are capable of processing complex data, they can only be used for
intelligent data analysis if their functionality can be explained. Relative weight

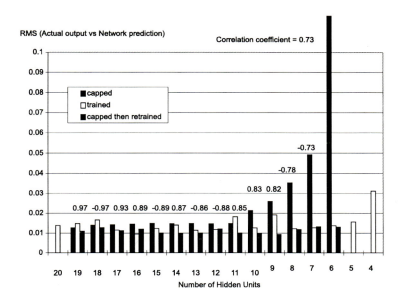

Fig. 2. The effect of merging hidden units with less correlated activations.

analysis and equation synthesis support this process. CAPing can be a valuable tool in minimising network architecture by detecting redundancy at the level of hidden unit activations. The comparison of hidden units with correlated activations also leads to a model of the relationships between causal agents.

5   Acknowledgements

We would like to thank the UK Department of the Environment for financial support of the work in this project (Project number PECD 7/12/145).

References

1. Benton J, Fuhrer J, Gimeno BS, Skarby L, Balls G, Palmer-Brown D, Roadknight C and Sanders G.1996. ICP-Crops and critical levels of ozone for injury development. Exceedences of Critical Loads and Levels. In Exceedences of Critical Loads and Levels. Eds. M. Knoflacher, J. Schneider and G. Soja. Umweltbundesamt (Federal Environment Agency) Wein, Austria. 97-112.
2. Benton J, Fuhrer J, Gimeno BS, Skarby L, Balls G, Roadknight C and Sanders-Mills G. 1996. The critical level of ozone for visible injury on crops and natural vegetation (ICP-Crops). In: Critical Levels of Ozone in Europe: Testing and Finalising the Concepts. UN ECE Workshop Report. Eds. L Karenlampi and L Skarby. University of Kuopio, Finland. 44 - 57.

3. Heck, WW, Taylor O. and Tingley DT. (eds) 1988. Assessment of Crop Loss from Air Pollutants. Elsevier Applied Science, New York.

4. Funahashi K. 1989. On the approximate realization of continuous mappings by neural networks. Neural Networks, 2, 183 - 192.

5. Burke HB, Hoang A and Rosen DB. 1995. Survival function estimates in cancer using artificial neural networks. Proceedings of WCNN. Vol II. p.748-749.

6. Orr RK. 1995. Use of probabilistic neural networks to predict mortality following cardiac surgery. Proceedings of WCNN. Vol II. p. 754-757.

7. Davalo E, Niam P. 1990. Neural Networks. Macmillan Press. p. 111-112.

8. Tan H, Prokhorov DV and Wunsch DC. 1995. Probabilistic and time-delay neural network techniques for conservative short-term stock trend prediction. Proceedings of World Congress on Neural Networks. Vol II. p. 44-47.

9. Fu L. 1994. Neural Networks in Computer Intelligence. McGraw-Hill International Editions. p.351-369.

10. Andrews R, Diederich J and Tickle AB. 1995. A survey and critique of techniques for extracting rules from trained Artificial Neural Networks. Knowledge Based Systems Vol 8 (6, December) p. 373-389.

11. Sestito S and Dillon T. 1994. Automatic knowledge acquisition. Prentice Hall.

12. Towell G and Shavlik J. 1993. The extraction of refined rules from knowledge based neural networks. Machine Learning Vol 131, p71-101.

13. Karin ED. 1990. A simple procedure for pruning back-propagation trained neural networks. I.E. Trans. Neural Networks, vol.1 no.2. p239-242.

14. Weigend AS, Rumelhart DE and Huberman BA. 1991. Generalization by weight elimination with applications to forecasting. In Advances in Neural Information Processing (3). Lippmann R, Moody J and Touretzky D. Eds. p. 875-882.

15. Balls GR, Palmer-Brown D, Cobb AH and Sanders GE. 1995. Towards unravelling the complex interactions between microclimate, ozone dose and ozone injury in clover. Journal of Water, Air and Soil Pollution. 85, 1467 - 1472.

16. Balls GR, Palmer-Brown D, and Sanders GE. 1996. Investigating microclimate influences on ozone injury in clover (Trifolium subterraneum) using artificial neural networks. New Phytologist, 132, 271 -280

17. Roadknight CM, Palmer-Brown D and Sanders GE. 1995. Learning the equations of data. Proceedings of 3rd annual SNN symposium on neural networks (eds. Kappen B and Gielen S) Springer-Verlag. 253-257.

18. Roadknight CM, Balls GR, Sanders GE and Palmer-Brown D. Modelling complex environmental data. IEEE Transactions on Neural Networks: Special edition on everyday applications. (In Press July 1997.)

19. Sietsma J and Dow RJF. 1988. Neural net pruning - Why and how. Prc. IEEE Int. Conf. Neural Networks. Vol 1. p. 325-333.

20. Wiersma FR, Poel M and Oudshoff AM. 1995. The BB neural network rule extraction method. Proceedings of 3rd annual SNN symposium on neural networks (eds. Kappen B and Gielen S) Springer-Verlag. 69-73.

21. Ripley BD. 1995. Statistical ideas for selecting network architectures. Proceedings of 3rd annual SNN symposium on neural networks (eds. Kappen B and Gielen S) Springer-Verlag. 183-190.

Simulation Data Analysis Using Fuzzy Graphs

Klaus–Peter Huber and Michael R. Berthold

Institute for Computer Design and Fault Tolerance (Prof. D. Schmid)
University of Karlsruhe – P.O. Box 6980 – 76128 Karlsruhe – Germany
eMail: [kphuber,berthold]@Informatik.Uni-Karlsruhe.de

Abstract

Analysis of simulation models has gained considerable interest in the past. However, their complexity still remains a considerable drawback in practical applications. A promising concept is to analyze the data from simulation experiments. Existing approaches are either restricted to simple models or are hard to interpret. We present an efficient algorithm that constructs a fuzzy graph model from simulation data and we show that the resulting system approximates also complex model functions with an adjustable precision. In addition the Fuzzy Graph allows the analyst to directly access easy to interpret if–then–rules. These rules help to understand the original simulation model, which is shown with a real world token bus model.

1 Introduction

The use of modeling and simulation techniques has gained considerable influence for the development or the optimization of large systems. Unfortunately the complexity of the corresponding simulation model increases with the complexity of the real systems. This leads to several drawbacks: simulation becomes a highly time consuming task which makes it impossible to perform interactive simulations and in addition the analysis of the resulting simulation model is extremely complicated.

A common solution is to build an auxiliary model, the so–called *metamodel* which is less complicated and therefore easier to interpret. Depending on the task of analysis there may exist different kinds of metamodels for example to analyze the parameter sensitivity [5], or to optimize models [7]. In all cases the generation of a metamodel helps to reduce the complexity of the model that is being analyzed.

Of particular interest are approaches that build the metamodel solely through observations of the original model, i.e. through analysing data from simulation experiments (Figure 1). One way is to analyze the data with statistical methods to find some unknown dependencies but in high–dimensional feature space the mathematical solution is not trivial and the interpretation is not straightforward.

X. Liu, P. Cohen, M. Berthold (Eds.): "Advances in Intelligent Data Analysis" (IDA–97)
LNCS 1280, pp. 347–358, 1997.   © Springer–Verlag Berlin Heidelberg 1997

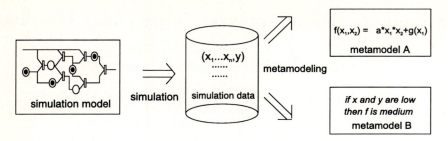

Fig. 1. Data based Metamodeling Process

Other approaches use methods from the machine learning area and find a set of rules or a decision tree.

In this paper we present an approach that analyses data resulting from simulation experiments and constructs a so–called Fuzzy Graph. This allows an easy interpretation of the input–output behavior since the metamodel can be represented with fuzzy rules. Analysis of the rules helps to uncover the underlying dependencies between factors and one output parameter and the rule base can also be used as a fast simulator.

2 Building a Metamodel from Data

Here, we concentrate on models where the behavior can be described by a function

$$f_{model}(\mathbf{x}) = f_{model}(x_1, \cdots, x_n) = y \tag{1}$$

where n factors $x_1, \cdots, x_n$ are considered and y is the output parameter of interest. In practice this function can not be extracted explicitly from the model description and is therefore unknown. To build a suitable metamodel means to perform m experiments with the model resulting in m data points $(x_1^i, \cdots, x_n^i, y^i)$ ($i \in [1, m]$) and to build a metamodel representing a function $f_{meta}(x_1, \cdots, x_n) = \tilde{y}$ with $\tilde{y}$ being the approximated value of y. If the data is generated by a stochastic simulation process, y is a stochastic variable. There may exist several different values of y for the same input parameters, e.g. by simulation with different random number streams. This should be taken into account when building a metamodel. To validate the quality of the metamodel the difference between the observed y values and the approximated values $\tilde{y}$ can be evaluated.

Metamodeling approaches can be categorized into *statistical approaches* that use the data to adapt a special kind of function with statistical means and *Machine Learning Algorithms* that use the data to generate a rule set.

2.1 Statistical Approaches for Metamodeling

Ad hoc methods use handfitted curves or graphical approaches, other methods use linear regression models with a least square approach to define the parameters [6]. Regression functions are very popular since the resulting models are easy to handle and interpret, and statistical methods like t-tests and F-tests can be used to validate the quality of the model [2]. To deal with nonlinear behavior metamodeling regression functions are often defined as:

$$f_{meta}(\mathbf{x}) = \beta_0 + \sum_{j=1}^{n} \beta_j \cdot x_j + \sum_{i,l} \beta_{i,l} \cdot x_i \cdot x_l + \epsilon \qquad (2)$$

with ϵ representing an error term and $\beta_{i,l}$ representing some user–defined in pair dependencies. To find a "best fitting" metamodel the regression parameters β_0, β_j and $\beta_{i,l}$ can be obtained by minimizing the mean square error in respect to the example data. The resulting function can be used for approximation or for analysis, e.g. to obtain some information about the sensitivity of each factor. But in practice often the behavior of a simulation model can not be described with such simple first–order models (see for example [2]).

2.2   Machine Learning Approaches for Metamodeling

To avoid the need for background knowledge nowadays several methodologies have been proposed that make use of rule learning algorithms from the machine learning area. In [7] preprocessed (i.e. cleaned and digitized) data from the simulation model was used to directly extract decision rules and in [5] a similar approach was used for sensitivity analysis. Both concepts are based on methods that build a classifier so they are restricted to applications where the output parameter is not continuous.

   Another kind of learning techniques that are especially well suited to deal with "noisy" or "stochastic" data originate from the Soft Computing or Fuzzy Systems area [11]. They offer an easy way to model soft data points, for example values with a corresponding confidence interval. Unfortunately, most known methods that build fuzzy systems from data have severe limitations in this context. Some require an a priori defined set of rules that is just fine tuned during training [9, 10], others construct the ruleset but the result depends heavily on the order of training examples [8]. A more sophisticated algorithm [3] tries to divide individual attributes step by step using an increasing number of membership functions but tends to split the feature space into too many tiles. For the purpose of data analysis and metamodeling it is of much more interest, however, to find only few rules that cover a large portion of the feature space.

   In the following a new approach is proposed that allows the usage of *Fuzzy Graphs* [11] to represent the discovered knowledge.

3   Data Driven Construction of Fuzzy Graphs

The definition and interpretation of a *Fuzzy Graph* is manifold. Here, we use it to represent a function with means of Fuzzy Logic instead of mathematical equations, analogous to [11]. The domains of the input parameters and the domain of the output parameter are described with so–called linguistic variables, represented through a set of individual membership functions. The mapping from

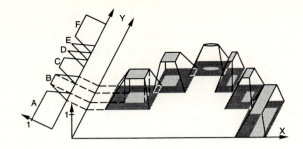

Fig. 2. A One–dimensional Fuzzy Graph

input to output is defined by a collection of Fuzzy Points. Well–known techniques like the center–of–gravity calculation allow to approximate real valued functions based on this Fuzzy Graph. Here, a Fuzzy Graph consists of a collection of Fuzzy Points that can be represented as if–then–rules. Other ways to represent Fuzzy Graphs are described in [11]. The main advantage of the Fuzzy Graph concept is the very compact and easy to understand representation of a function.

The algorithm presented in this section automatically constructs a Fuzzy Graph based on a set of data examples. Fuzzy Graphs built by our approach use a fixed soft granularization on the output variable which means that the user defines the membership functions of y a–priori. This is helpful to focus automatic generation of the Fuzzy Graph on specific regions of the output value and to weaken constraints (and therefore the evolving number of rules) on regions with a low focus of attention. Figure 2 shows an example where the output variable y was already partitioned into 6 soft regions with user defined membership functions. The partitioning of the input variables is determined from the data examples.

Each Fuzzy Point corresponds to one region of the output parameter and is described by its *core*–region and a larger *support*–region. The core represents the smallest area where data examples of the corresponding output–region were found and therefore is given a membership value of 1. The larger support–area contains no examples of other regions, and towards its boundaries the membership value declines linearly to 0. This leads to the trapezoidal membership functions illustrated in Figure 3.

Each such Fuzzy Point represents exactly one if–then–rule that can be described as:

IF x_1 $\in [b_1, c_1] \subset (a_1, d_1)$
AND $\cdots$
AND x_n $\in [b_n, c_n] \subset (a_n, d_n)$
THEN y is of k
(weight: w)

The input parameters are restricted through core $[b_i, c_i]$ and support–regions (a_i, d_i) with $a_i < b_i \leq c_i < d_i$, and each rule corresponds to one region k of

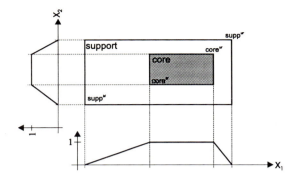

Fig. 3. A Two–dimensional Fuzzy Point

the output parameter. Additionally, the rule weight shows the number of data examples that are covered by the core of the rule and is therefore an indication for the reliability of the rule.

3.1 Automatic Construction of the Fuzzy Graph

Construction of the Fuzzy Graph is done using a set of data examples. Each of these examples consists of values for the input parameters with a corresponding output. The output value (or *target*) is given as a soft value, using an individual membership function μ_T. For practical applications these soft targets can originate from stochastic simulation experiments. For sharp targets a singleton can be used as the corresponding output membership function. From this soft target membership function μ_T the membership values μ_k for each of the pre–defined output–regions k $(1 \leq k \leq c)$are computed using a fuzzy and–operator (min) between the target μ_T and the k–th region membership set μ_k. This leads to example data consisting of an input vector $\mathbf{x} = (x_1, \cdots, x_n)$ with the corresponding $\boldsymbol{\mu} = (\mu_1, \cdots, \mu_c)$ $(0 \leq \mu_i \leq 1)$ where c denotes the number of predefined soft regions y is divided into.

The method presented in this paper makes sure that each data example is covered by a rule (Fuzzy Point) of the region with the highest membership value and that rules of regions with membership values $= 0$ do not cover the example. This is useful to tolerate moderately noisy patterns or small oscillations along region boundaries. The basic algorithm for one epoch, i.e. one presentation of all data examples, is presented in figure 4. The algorithm is based on three steps that introduce new rules when necessary and adjust the core– and support–regions of existing rules:

– **covered**: if a new data example lies inside the support–region (defined through supp^{bl} and supp^{ur}) of an already existing rule belonging to the region with the maximum μ_k, the core–region (core$^{bl/ur}$) of this rule is extended to cover the new example, which is, even in high–dimensional space, an easy task.

```
algo for one epoch
  //reset weights:
  forall 1 ≤ k ≤ c do
    forall 1 ≤ i ≤ m_k do p_i^k.w = 0
    endfor
  endfor
  //one complete epoch
  forall example data (x, μ) do
    //region with highest degree of membership
    k = argmax{μ_k : 1 ≤ k ≤ c}
    //find rule with highest activation
    l = argmax{p_l^k(x) : 1 ≤ l ≤ m_k}
    if p_l^k(x) > 0 then
      // "covered"
      if p_l^k.w = 0 then p_l^k.core^{bl/ur} = x
      else adjust p_l^k.core^{bl/ur} to cover x
      p_l^k.w = p_l^k.w + 1
    else
      // "commit": introduce new rule
      add new rule p_{m_k+1}^k with:
      p_{m_k+1}^k.core^{bl/ur} = x
      p_{m_k+1}^k.supp^{bl} = (-∞, ···, -∞)
      p_{m_k+1}^k.supp^{ur} = (∞, ···, ∞)
      p_{m_k+1}^k.w = 1
      m_k = m_k + 1
    endif
    // "shrink": adjust conflicting rules
    forall 1 ≤ k ≤ c ∧ μ_k = 0 do
      forall 1 ≤ i ≤ m_k do
        if p_i^k(x) > 0 then
          shrink p_i^k.supp^{bl/ur}
      endfor
    endfor
  endfor
end algo
```

$x = (x_1, \cdots, x_n)$: input vector,
$\mu = (\mu_1, \cdots, \mu_c)$ $(0 \leq \mu_k \leq 1)$ corresponding region membership values,
p_i^k: one rule (or Fuzzy Point) of region k, index i $(1 \leq i \leq m_k)$,
$p_i^k.w$: weight of this rule,
$p_i^k.core^{bl}$, $p_i^k.core^{ur}$: bottom left and upper right corner of core–region,
$p_i^k.supp^{bl}$, $p_i^k.supp^{ur}$: bottom left and upper right corner of support–region,
$p_i^k(x)$: degree of membership of x for this rule.

Fig. 4. The Fuzzy Graph Algorithm

— **commit**: if a new example is not covered, a new rule with the region corresponding to the maximum μ_k will be introduced. The core–region is defined by this new example and the support–region is set infinite.

— **shrink**: if a new example is incorrectly covered by an already existing rule of a conflicting region ($\mu_k = 0$), this rule's support–area will be reduced (e.g. shrunk) so that the conflict is solved. Here, a heuristic is used that tries to maximize the volume of the remaining support area. See [4] for details.

The whole generation process usually takes only about 4–5 epochs until the structure of the Fuzzy Graph automatically stops to change. Two conditions will hold for all example data (x, μ) after generation of the Fuzzy Graph: There is at least one rule of region k with the highest membership value μ_k which has x inside its core. And for all regions k with $\mu_k = 0$, x lies outside of their support area. This leads to a rule base where each example data is covered by an appropriate rule and not covered by those of conflicting regions. This resulting Fuzzy Graph can now be used for approximation.

3.2   Fuzzy Graph Function Approximation

To use the constructed Fuzzy Graph for function approximation for a given input parameter $\mathbf{x}$ the membership degree for each region of the output variable is typically computed using the maximum–operator:

$$\mu_k(\mathbf{x}) = \max_i \{p_i^k(\mathbf{x}) : 1 \le i \le m_k\}.$$

These values are then used to determine the resulting output membership function. If a crisp value is required, a defuzzification technique can be used, for example Center of Gravity. The combination of the respective parts of the membership functions μ_k leads to the soft value defined through μ_y:

$$\mu_y(\mathbf{x}) = \max_k \{\mu_k(\mathbf{x})\}.$$

Computing the center of this function leads to the crisp approximation y.

Different experiments with a variety of data sets show that the resulting Fuzzy Graph approximates the original function well, with a specific degree of accuracy (see [1] for experiments with different data sets). In regions containing "noise", the Fuzzy Graph ignores the oscillations and tends to produce plateaus. The degree of noise tolerance depends mainly on the width of the membership functions for the output parameter. Thus the amount of smoothing can be controlled by the output fuzzification. Using more and finer membership functions results in higher precision but forcing a system to follow the data points very closely, will result in a large number of rules that model the noise.

In the following an experiment with a complex model shows the usefulness of the presented approach for the analysis of simulation data.

4   Example of a Token Bus

To demonstrate how the presented approach can be used to find rules in data resulting from experiments, a real world simulation model was chosen. The used Token Bus system belongs to the class of field bus systems, i.e. a special type of communications systems, designed to connect machines and computers in a manufacturing environment. Important requirements in this area are "real time" facility, high flexibility, and low costs. In this section the analysis will mainly focus on the real time facility of the model, that is its capability to respond to each request within a limited time. To guarantee this property for the given simulation model a metamodel was built using the presented method and its behavior depending on different parameter settings was explored.

The modeled token bus system corresponds to the seven level architecture of the ISO/OSI communication standard. Figure 5 shows the structure of the system. Many details like different message priorities and the token handling had to be taken into account when modeling the system with a queuing network model. The model was then implemented with a commercially available simulation environment. To illustrate the complexity of the underlying queuing

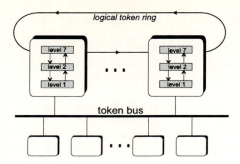

Fig. 5. The Model

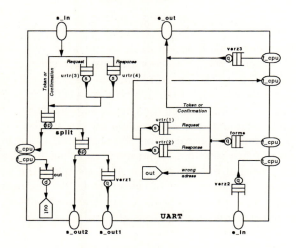

Fig. 6. The Internal Structure of one Module

network the internal structure of one module with its interfaces is illustrated on
Figure 6. Since each station is modeled by four different modules the whole model
consists of more than two hundred different queues and several hundreds connections. Due to the complexity of the internal structure a conventional analysis
of this model is extremely time consuming and complicated.

4.1   Data Generation and Analysis

Due to the large number of parameters (20 input and 10 output parameters) of
the complete model the example analysis presented in this section will focus on
the response time between two master stations in dependence of a selection of
parameters of interest. It is desirable that this response time always stays below
an upper bound to guarantee that the reaction of the system is always in time.
Four input parameters were chosen while the other parameters remained fixed:

- *average time for execution (cpu1)*: describes the performance of the CPU module of station 1, i.e., the average time required to execute one command. This value is varied within 0.1 (fast) — 3.4 (slow).
- *workload rate (workload)*: describes the average idle time between two requests, this value is varied within 0.02 (low idle time, high workload) — 1.0 (low workload).
- *maximum target-rotation-time (trt)*: maximum allowed time to process the token. This parameter controls the time each station has to send messages, values were set within $[0.01, 0.4]$
- *number of additional stations (stations)* represents the background workload on the network. Many additional stations communicating over the network will increase the traffic on the network: $[1, 15]$

Since the construction of the metamodel only depends on the example data these examples have to be representative. For this the planning of simulation experiments must be done carefully. In our application a full factorial design is not possible, therefore we used randomized settings for the input parameters. 350 simulation experiments were performed where the input parameter values were varied randomly within the given intervals. The averaged response time (rt) was measured within $(0.088, 9.75)$. Each simulation experiment was repeated five times with a different random number stream of the simulation tool. From these five values the minimum, the maximum, and the average were taken and a triangular target membership function was generated (Figure 7). With the Fuzzy

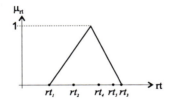

Fig. 7. Generation of a Soft Target

Graph approach these four–dimensional data–vectors with their corresponding target membership function were used for training. Since the main focus of attention were fast responses (i.e. low values of rt) the membership functions for low values are defined finer than those for bigger values (Figure 8). Three series of experiments were performed with two, five and ten membership functions. Fuzzy Graph construction required about 10 seconds on a SUN Sparc10 workstation. No training parameters besides the a priori definition of the output membership functions had to be considered or tuned. While a simulation run takes about 200 seconds the propagation of a new parameter set through the Fuzzy Graph is completed within fractions of a second, resulting in an increase in speed of two orders of magnitude. As expected the metamodel can be used for much faster simulation.

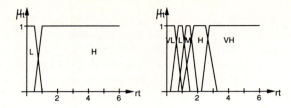

Fig. 8. Two Used Types of Membership Functions

4.2   Results

To judge the reliability of the complete rulebase the quality can be analyzed by computing the mean relative error of the approximation of the metamodel. For this analysis an independent dataset that was not used for building the Fuzzy Graph, the so called cross–validation set, was used. The dataset of 350 vectors was split into one tenth for testing and nine tenth for building the graph and using each tenth once for testing ten cross–validation runs were performed. The average error on the corresponding testdata was 4.4% ± 1.0% (avg. 32 rules) with two, 4.1% ± 1.2% (avg. 59 rules) with five, and 3.3% ± 0.9% (avg. 68 rules) with ten membership functions for the output. This approximation quality is sufficient because the primary goal of the presented approach is the extraction of few understandable rules instead of achieving minimal approximation errors.

One of the resulting rule bases from an experiment with two output regions was used for further analysis. In this case the regions are labeled *low* and *high*. Since the main focus of analysis were parameter settings which result in a low response time, rules of region L = low were investigated. From 29 rules 16 belong to this region and according to the rule weight the most important rule was:

$$
\begin{array}{llll}
\text{IF} & cpu1 & \in [0.11, 1.69] & \subset (-\infty, 1.70) \\
\text{and} & workload & \in [0.03, 0.99] & \subset (-\infty, +\infty) \\
\text{and} & trt & \in [0.01, 0.39] & \subset (-\infty, +\infty) \\
\text{and} & stations & \in [5, 15] & \subset (4, +\infty) \\
\text{THEN} & \multicolumn{3}{l}{low\ (rt \in [0.0, 0.5] \subset (-\infty, 1.0))} \\
\end{array}
$$
(weight: 116)

This rule demonstrates how the core always covers a confident subset of the support–region. Here, for two parameters, namely *workload rate* and *target–rotation–time*, the core covers the whole range of these parameters. Additionally it is only limited into one direction on the other two parameters, indicated by a support region having finite boundaries. The *performing time of CPU1* has to be below 1.70 and the *number of additional stations* above 4. This indicates that a certain amount of computation power together with some background stations guarantees fast responses no matter what settings are chosen for *workload rate* and *performing time*. In addition the weight of this rule can be used to judge its reliability. The weight indicates the number of examples that are covered by this rule. In the rule shown above 116 examples fall inside its core region.

This means that about 36% of all examples are covered by this rule, indicating a high reliability. Rules with low weight on the other hand might be indicators for outliers, irregularities in the dataset or regions of high sensitivity, i.e. regions where small changes of the attributes result in large variations of the output. Another question of interest is the influence of some parameters considering the output. In this example the rule above indicates that the *target–rotation–time* has no influence on the response time if the CPU is fast (below 1.7) and at least 5 background stations exist because the support of *trt* is not restricted and the core covers nearly the whole domain of this parameter.

It can also be of interest to find "bad" examples, i.e. regions where the response time is very high. These indicate parameter settings that should be avoided. For example, the rule with the highest weight for *response time = high* was:

IF   *cpu1* $\in [2.71, 3.39]$ $\subset (2.70, +\infty)$
and   *workload* $\in [0.09, 0.98]$ $\subset (-\infty, +\infty)$
and   *trt* $\in [0.14, 0.39]$ $\subset (0.13, +\infty)$
and   *stations* $\in [1, 15]$ $\subset (-\infty, +\infty)$
THEN   high ($rt \in [0.5, 1.0] \subset (1.0, +\infty)$)
(weight: 38)

This rule indicates that if the CPU is very slow and the *target–rotation–time* is above a certain value the response time is high no matter what workload is considered (represented by background stations and the time between requests). Therefore if the system includes a slow CPU module the *target–rotation–time* should be set carefully. Since only 85 examples are of region *high* the weight of 38 is an indication for a high reliability also of this rule.

These results illustrate the applicability of the presented approach for meta-modeling tasks. The approximation error indicates the reliability of the meta-model and the Fuzzy Graph can be used for new simulation experiments. The example rules deliver helpful information about dependencies between factors and the output of interest.

5   Conclusions

In this paper a new approach to handle complex simulation models through an analysis of data from experiments has been presented. The analysis is based on Fuzzy Graphs which makes it possible to use not only real valued but also soft or noisy values. This is especially well suited for the analysis of simulation data due to their stochastic character. It was demonstrated that rules from the Fuzzy Graph deliver meaningful information about the relation between input parameters and the output, a very helpful information when analyzing complex models like the used token bus. Since the presented method is easy to use and no parameters are needed, the proposed methodology allows to handle complex simulation models.

6   Acknowledgments

We thank Prof. D. Schmid for his support and the opportunity to work on this interesting project. Thanks also to Markus Weihrauch, who helped with the actual implementation, and to Raffaele Carluccio, who built the queuing network of the token bus.

References

1. Michael R. Berthold and Klaus-Peter Huber. Building fuzzy graphs from examples. In *IEEE International Conference on Fuzzy Systems*, 1, pages 608–613, September 1996.
2. Linda W. Friedman and Israel Pressman. The metamodel in simulation analysis: Can it be trusted? *Journal of the Operational Research Society*, 39(10):939–948, 1988.
3. Charles M. Higgins and Rodney M. Goodman. Learning fuzzy rule-based neural networks for control. In *Advances in Neural Information Processing Systems*, 5, pages 350–357, California, 1993. Morgan Kaufmann.
4. Klaus-Peter Huber and Michael R. Berthold. Building precise classifiers with automatic rule extraction. In *IEEE International Conference on Neural Networks*, 3, pages 1263–1268, 1995.
5. Klaus-Peter Huber and Helena Szczerbicka. Sensitivity analysis of simulation models with decision tree algorithms. In *Proceedings of the European Simulation Symposium ESS'94*, volume 1, pages 43–47, 1994.
6. J.P.C. Kleijnen. Regression metamodels for generalizing simulation results. *IEEE Transactions on Systems, Man and Cybernetics*, 9(2):93–96, 1979.
7. Henri Pierreval. Rule-based simulation metamodels. *European Journal of Operational Research*, 61:6–17, 1992.
8. Patrick K. Simpson. Fuzzy min-max neural networks – part 2: Clustering. *IEEE Transactions on Fuzzy Systems*, 1(1):32–45, january 1993.
9. Volkmar Uebele, Shigeo Abe, and Ming-Shong Lan. A neural-network-based fuzzy classifier. *IEEE Transactions on Systems, Man, and Cybernetics*, 25(2), february 1995.
10. Li-Xin Wang and Jerry M. Mendel. Generating rules by learning from examples. In *International Symposium on Intelligent Control*, pages 263–268. IEEE, 1991.
11. Lotfi A. Zadeh. Soft computing and fuzzy logic. *IEEE Software*, pages 48–56, november 1994.

Mathematical Analysis of Fuzzy Classifiers

Frank Klawonn[1] and Erich–Peter Klement[2]

[1] Fachbereich Elektrotechnik und Informatik
Fachhochschule Ostfriesland
Constantiaplatz 4, D-26723 Emden, Germany
[2] Fuzzy Logic Laboratorium Linz, Johannes Kepler University
A–4040 Linz, Austria

Abstract. We examine the principle capabilities and limits of fuzzy classifiers that are based on a finite set of fuzzy if–then rules like they are used for fuzzy controllers, except that the conclusion of a rule specifies a discrete class instead of a (fuzzy) real output value. Our results show that in the two–dimensional case, for classification problems whose solutions can only be solved approximately by crisp classification rules, very simple fuzzy rules provide an exact solution. However, in the multi–dimensional case, even for linear separable problems, max–min rules are not sufficient.

1 Introduction

Fuzzy controllers are well examined as function approximators. Piecewise monotone functions of one variable can be exactly reproduced by a fuzzy controller [1, 9] and for the multi–dimensional case fuzzy controllers are known to be universal approximators [2, 6, 12]. Although a lot of approaches for automatically learning fuzzy classifiers are proposed in the literature (see for instance [3, 4, 5, 10, 11, 13]), they are usually evaluated only on an experimental basis. A theoretical analysis of the principal capabilities of fuzzy classifiers aiming at the assignment of discrete classes to input vectors, is still lacking. A natural question concerning fuzzy classification rules is, whether they have any advantage over crisp classification rules when in the end for an input vector a unique assignment to one class has to be made. We will provide a positive answer to this question in the sense that already in the two–dimensional case fuzzy classification rules can solve problems for which only approximate solutions can be constructed on the basis of crisp classification rules. This paper is devoted to the question what kind of classification problems are solvable in principle by fuzzy classifiers using if-then rules. We do not discuss techniques for actually constructing suitable if-then rules from data.

The paper is organized as follows. Section 2 provides the formal definition of the type of fuzzy classifiers we are examining. In Section 3 we demonstrate that in the two–dimensional case quite general classification problems can be solved, whereas for higher dimensional problems simple max–min rules must fail. As shown in Section 4 this can be amended by using other operations than max or min.

X. Liu, P. Cohen, M. Berthold (Eds.): "Advances in Intelligent Data Analysis" (IDA–97)
LNCS 1280, pp. 359–370, 1997. © Springer–Verlag Berlin Heidelberg 1997

2   Formal Framework

Let us briefly introduce the formal framework we are considering. We consider fuzzy classification problems of the following form. There are p real variables $x_1, \ldots, x_p$ with underlying domains $X_i = [a_i, b_i]$, $a_i < b_i$. There is a finite set $\mathcal{C}$ of classes and a partial mapping

$$\text{class} : X_1 \times \ldots \times X_p \longrightarrow \mathcal{C}$$

that assigns classes to some, but not necessarily to all vectors $(x_1, \ldots, x_p) \in X_1 \times \ldots \times X_p$.

The aim is to find a fuzzy classifier that solves the classification problem. The fuzzy classifier is based on a finite set $\mathcal{R}$ of rules of the form $R \in \mathcal{R}$:

$$R: \text{If } x_1 \text{ is } \mu_R^{(1)} \text{ and } \ldots \text{ and } x_p \text{ is } \mu_R^{(p)} \text{ then class is } C_R.$$

$C_R \in \mathcal{C}$ is one of the classes. The $\mu_R^{(i)}$ are assumed to be fuzzy sets on X_i, i.e. $\mu_R^{(i)} : X_i \longrightarrow [0, 1]$. In order to keep the notation simple, we incorporate the fuzzy sets $\mu_R^{(i)}$ directly in the rules. In real systems one would replace them by suitable linguistic values like *positive big*, *approximately zero*, etc. and associate the linguistic value with the corresponding fuzzy set.

In Section 3, where we present our main results, we restrict ourselves to max–min rules, i.e., we evaluate the conjunction in the rules by the minimum and aggregate the results of the rules by the maximum. Therefore, we define

$$\mu_R(x_1, \ldots, x_p) = \min_{i \in \{1, \ldots, p\}} \left\{ \mu_R^{(i)}(x_i) \right\} \tag{1}$$

as the degree to which the premise of rule R is satisfied.

$$\mu_C^{(\mathcal{R})}(x_1, \ldots, x_p) = \max \left\{ \mu_R(x_1, \ldots, x_p) \mid C_R = C \right\} \tag{2}$$

is the degree to which the vector $(x_1, \ldots, x_p)$ is assigned to class $C \in \mathcal{C}$. The defuzzification – the final assignment of a unique class to a given vector $(x_1, \ldots, x_p)$ – is carried out by the mapping

$$\mathcal{R}(x_1, \ldots, x_p) = \begin{cases} C & \text{if } \mu_C^{(\mathcal{R})}(x_1, \ldots, x_p) > \mu_D^{(\mathcal{R})}(x_1, \ldots, x_p) \\ & \text{for all } D \in \mathcal{C}, D \neq C \\ unknown \notin \mathcal{C} & \text{otherwise.} \end{cases}$$

This means that we finally assign the class C to the vector $(x_1, \ldots, x_p)$ if the fuzzy rules assign the highest degree to class C for vector $(x_1, \ldots, x_p)$. If there are two or more classes that are assigned the maximal degree by the rules, then we refrain from a classification and indicate it by the symbol *unknown*. Note that we use the same letter $\mathcal{R}$ for the rule base and the induced classification mapping.

Finally,

$$\mathcal{R}^{-1}(C) = \{(x_1, \ldots, x_p) \mid \mathcal{R}(x_1, \ldots, x_p) = C\}$$

denotes the set of vectors that are assigned to class C by the rules (after defuzzification).

3 Max–Min Rules

Let us first take a look at crisp classification rules in the sense that the fuzzy sets $\mu_R^{(i)}$ are assumed to be characteristic functions of crisp sets, say intervals. Then it is obvious that in the two–dimensional case each rule assigns those inputs to the class appearing in the conclusion of the rule that are in the rectangle that is induced by the two intervals appearing as characteristic functions in the premise of the rule.

A classification problem with two classes that are separated by a hyperplane, i.e. a line in the two–dimensional case, is called linear separable. Obviously, a linear separable classification problem can be solved only approximately by crisp classification rules by approximating the separating line by a step function (see Figure 1).

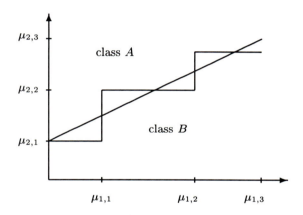

Fig. 1. Approximate solution of a linear separable classification problem by crisp classification rules

For fuzzy classification rules the situation is much better. The following lemma and its corollary show that in the two–dimensional case classification problems with two classes that are separated by a piecewise monotone function can be solved exactly using fuzzy classification rules.

Lemma 1. *Let* $f : [a_1, b_1] \longrightarrow [a_2, b_2]$ *($a_i < b_i$) be a monotone function. Then there is a finite set* $\mathcal{R}$ *of classification rules to classes* P *and* N *such that*

$$\mathcal{R}^{-1}(P) = \{(x, y) \in [a_1, b_1] \times [a_2, b_2] \mid f(x) > y\},$$
$$\mathcal{R}^{-1}(N) = \{(x, y) \in [a_1, b_1] \times [a_2, b_2] \mid f(x) < y\}.$$

Proof. Let us abbreviate $X = [a_1, b_1]$, $Y = [a_2, b_2]$.

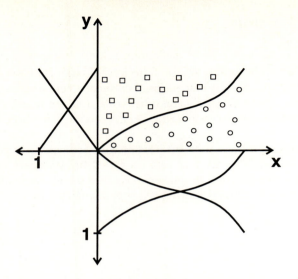

Fig. 2. The fuzzy sets for the classification rules

Define the fuzzy sets

$$\mu_1 : X \longrightarrow [0,1], \qquad x \mapsto \frac{b_2 - f(x)}{b_2 - a_2},$$

$$\mu_2 : X \longrightarrow [0,1], \qquad x \mapsto \frac{f(x) - a_2}{b_2 - a_2} = 1 - \mu_1(x),$$

$$\nu_1 : Y \longrightarrow [0,1], \qquad y \mapsto \frac{y - a_2}{b_2 - a_2},$$

$$\nu_2 : Y \longrightarrow [0,1], \qquad y \mapsto \frac{b_2 - y}{b_2 - a_2} = 1 - \nu_1(y).$$

The fuzzy sets are illustrated in Figure 2. The rule base consists of the two rules:

R_1: If x is μ_1 and y is ν_1 then class is N.

R_2: If x is μ_2 and y is ν_2 then class is P.

It is easy to verify that these rules solve the classification problem. □

Note that the proof is based on a very similar technique as the proof for constructing a fuzzy controller for rebuilding a function with one argument [1]. It is obvious that we can extend the result of this lemma to piecewise monotone functions, simply by defining corresponding fuzzy sets on the intervals where the class separating function is monotone (see Figure 3) and defining corresponding rules for each of these intervals so that we have the following corollary.

Corollary 2. *Let $f : [a_1, b_1] \longrightarrow [a_2, b_2]$ $(a_i < b_i)$ be a piecewise monotone function. Then there is a finite set $\mathcal{R}$ of classification rules to classes P and N*

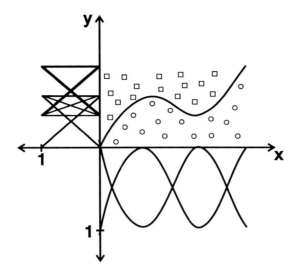

Fig. 3. The fuzzy sets for the classification rules of a piecewise monotone function

such that

$$\mathcal{R}^{-1}(P) = \{(x,y) \in [a_1,b_1] \times [a_2,b_2] \mid f(x) > y\},$$
$$\mathcal{R}^{-1}(N) = \{(x,y) \in [a_1,b_1] \times [a_2,b_2] \mid f(x) < y\}.$$

A direct consequence of Lemma 1 and its proof is that we can solve two–dimensional linear separable classification problems with only two fuzzy classification rules incorporating simple triangular membership functions so that we are in a much better situation than in the case of crisp classification rules. However, the result cannot be extended to more than two dimensions since the following theorem shows that even three–dimensional linear separable classification problems cannot be solved (evaluating the rules by the max–min schema).

Theorem 3. *Let $f : \mathbb{R}^3 \longrightarrow \mathbb{R}$, $(x_1, x_2, x_3) \mapsto x_1 + x_2 + x_3$. For fixed $X_i = [a_i, b_i]$, $a_i < b_i$, $(i = 1, 2, 3)$, denote*

$$P = \{(x_1, x_2, x_3) \in X_1 \times X_2 \times X_3 \mid f(x_1, x_2, x_3) > 0\},$$

$$N = \{(x_1, x_2, x_3) \in X_1 \times X_2 \times X_3 \mid f(x_1, x_2, x_3) < 0\}.$$

For any choice of the intervals X_1, X_2, X_3 and any finite set of classification rules $\mathcal{R}$ with classes CP and CN, at least one of the following conditions is not satisfied:

(i) $P \neq \emptyset$ and $N \neq \emptyset$
(ii) The sets $\mu_R^{(i)}$ ($i \in \{1, 2, 3\}$, $R \in \mathcal{R}$) are piecewise monotone and continuous.
(iii) $\mathcal{R}(CP) = P$ and $\mathcal{R}(CN) = N$.

Proof. Assume the conditions (i), (ii), and (iii) could be satisfied simultaneously. Then we can choose X_1, X_2, X_3 and $\mathcal{R}$ in such a way that the cardinality of $\mathcal{R}$ is minimal, i.e., no matter how we choose X_1', X_2', X_3' (satisfying (i)), a rule base $\mathcal{R}'$ that guarantees for (ii) and (iii) will contain at least as many rules as $\mathcal{R}$.

The continuity of the fuzzy sets enforces the continuity of $\mu_{CP}^{(\mathcal{R})}$ and $\mu_{CN}^{(\mathcal{R})}$. Therefore, we have for all $(x_1, x_2, x_3) \in X_1 \times X_2 \times X_3$ satisfying $f(x_1, x_2, x_3) = 0$ that $\mu_{CP}^{(\mathcal{R})}(x_1, x_2, x_3) = \mu_{CN}^{(\mathcal{R})}(x_1, x_2, x_3)$ holds.

The set

$$Z = \{(x_1, x_2, x_3) \in X_1 \times X_2 \times X_3 \mid f(x_1, x_2, x_3) = 0\}$$

– the boundary between the classes P and N – is the intersection of the plane $x_1 + x_2 + x_3 = 0$ and the cube $X_1 \times X_2 \times X_3$. According to condition (i), part of the plane lies in the interior of the cube.

Assume, there is a point $(x_1, x_2, x_3) \in Z$ in the interior of $X_1 \times X_2 \times X_3$ for which the rules do not fire to the same degree, i.e., there is a rule $S \in \mathcal{R}$ s.t.

$$\mu_S(x_1, x_2, x_3) < \max_{R \in \mathcal{R}} \{\mu_R(x_1, x_2, x_3)\}.$$

Then there is an entourage $X_1' \times X_2' \times X_3' \subseteq X_1 \times X_2 \times X_3$ of (x_1, x_2, x_3) s.t. for all $(z_1, z_2, z_3) \in X_1' \times X_2' \times X_3'$

$$\mu_S(z_1, z_2, z_3) < \max_{R \in \mathcal{R}} \{\mu_R(z_1, z_2, z_3)\}$$

holds. But this means that the rule base $\mathcal{R} \backslash \{S\}$ satisfies (i), (ii), and (iii) on $X_1' \times X_2' \times X_3'$, which is a contradiction to the minimality of $\mathcal{R}$. Thus we have for all $(x_1, x_2, x_3) \in Z$

$$\mu_R(x_1, x_2, x_3) = \mu_S(x_1, x_2, x_3) \tag{3}$$

for all $R, S \in \mathcal{R}$.

Therefore, we can define the function

$$g : Z \longrightarrow [0, 1], \qquad (x_1, x_2, x_3) \mapsto \mu_R(x_1, x_2, x_3)$$

independent of the choice of $R \in \mathcal{R}$.

The fuzzy sets $\mu_R^{(i)}$ are piecewise monotone, i.e., for each $\mu_R^{(i)}$ there are only finitely many points which have no entourage in which $\mu_R^{(i)}$ is monotone. Let z be such a point. For each z and $\mu_i^{(R)}$ we obtain a plane

$$H_{z, \mu_R^{(i)}} = \{(x_1, x_2, x_3) \in \mathbb{R}^3 \mid x_i = z\}.$$

Since $H_{z, \mu_R^{(i)}}$ is a plane parallel to two axes, whereas

$$\hat{Z} = \{(x_1, x_2, x_3) \in \mathbb{R}^3 \mid x_1 + x_2 + x_3 = 0\}$$

is not axes–parallel, their intersection is a line. Thus, if we consider the set

$$\hat{Z} \setminus \left(\bigcup_{R,i,z} H_{z,\mu_R^{(i)}} \right),$$

we simply cut out a finite set of lines from the plane $\hat{Z}$. Therefore, and since $\hat{Z}$ has a non–empty intersection with the interior of the cube $X_1 \times X_2 \times X_3$,

$$\left(\hat{Z} \cap (X_1 \times X_2 \times X_3) \right) \setminus \left(\bigcup_{R,i,z} H_{z,\mu_R^{(i)}} \right)$$

is non–empty. Choose an element (z_1, z_2, z_3) from this set which lies in the interior of $X_1 \times X_2 \times X_3$ and an entourage

$$X_1' \times X_2' \times X_3' \;=\; [z_1 - \delta, z_1 + \delta] \times [z_2 - \delta, z_2 + \delta] \times [z_3 - \delta, z_3 + \delta] \subseteq X_1 \times X_2 \times X_3$$

that does not contain any of the points of

$$\bigcup_{R,i,z} H_{z,\mu_R^{(i)}}.$$

We now prove that g is constant on $Z \cap (X_1' \times X_2' \times X_3')$. We can write any point

$$(x_1, x_2, x_3) \in Z \cap (X_1' \times X_2' \times X_3')$$

in the form $(z_1 + \varepsilon_1, z_2 + \varepsilon_2, z_3 + \varepsilon_3)$ with $\varepsilon_1 + \varepsilon_2 + \varepsilon_3 = 0$ and $|\varepsilon_i| \le \delta$.

Let us consider the point

$$(z_1 + \varepsilon_1, z_2 + \varepsilon_2, z_3 + \varepsilon_3) \in Z \cap (X_1' \times X_2' \times X_3').$$

Since a variation of the value z_3 in (z_1, z_2, z_3) leads to a change in the classification (from *unknown* to CP or CN), there has to be a rule $R \in \mathcal{R}$ with

$$\mu_R(z_1, z_2, z_3) \;=\; \mu_R^{(3)}(z_3).$$

Otherwise a (sufficiently small) variation of z_3 would neither change the value $\mu_{CP}(z_1, z_2, z_3)$ nor $\mu_{CN}(z_1, z_2, z_3)$, resulting in the wrong classification *unknown*. This implies

$$\begin{aligned}
g(z_1 + \varepsilon_1, z_2 - \varepsilon_1, z_3) &= \mu_R(z_1 + \varepsilon_1, z_2 - \varepsilon_1, z_3) \\
&\le \mu_R^{(3)}(z_3) \\
&= \mu_R(z_1, z_2, z_3) \\
&= g(z_1, z_2, z_3).
\end{aligned}$$

The same argument as above guarantees the existence of a rule $S \in \mathcal{R}$ with

$$\mu_S(z_1 + \varepsilon_1, z_2 - \varepsilon_1, z_3) \;=\; \mu_S^{(1)}(z_1 + \varepsilon_1).$$

Thus we obtain

$$
\begin{aligned}
g(z_1 + \varepsilon_1, z_2 + \varepsilon_2, z_3 + \varepsilon_3) &= \mu_S(z_1 + \varepsilon_1, z_2 + \varepsilon_2, z_3 + \varepsilon_3) \\
&\le \mu_S^{(1)}(z_1 + \varepsilon_1) \\
&= \mu_S(z_1 + \varepsilon_1, z_2 - \varepsilon_1, z_3) \\
&= \mu_R(z_1 + \varepsilon_1, z_2 - \varepsilon_1, z_3) \\
&\le g(z_1, z_2, z_3).
\end{aligned}
$$

By exchanging the roles of $(z_1 + \varepsilon_1, z_2 - \varepsilon_1, z_3)$ and (z_1, z_2, z_3), we can prove that

$$
g(z_1, z_2, z_3) \le g(z_1 + \varepsilon_1, z_2 + \varepsilon_2, z_3 + \varepsilon_3)
$$

also holds so that g has to be constant on $Z \cap (X_1' \times X_2' \times X_3')$, say $g(x_1, x_2, x_3) = \alpha$ for all $(x_1, x_2, x_3) \in Z \cap (X_1' \times X_2' \times X_3')$.

Assume there exist $i \in \{1, 2, 3\}$, $R \in \mathcal{R}$, and $|\varepsilon| < \delta$ s.t.

$$
\mu_R^{(i)}(z_i + \varepsilon) < \alpha.
$$

Without loss of generality let $i = 1$. This leads to the contradiction

$$
\alpha = \mu_R(z_1 + \varepsilon, z_2 - \varepsilon, z_3) \le \mu_R^{(1)}(z_1 + \varepsilon) < \alpha.
$$

Thus we have for all $i \in \{1, 2, 3\}$, for all $x \in X_i'$, and for all $R \in \mathcal{R}$

$$
\mu_R^{(i)}(x) \ge \alpha. \tag{4}
$$

Since a variation of the value z_1 in (z_1, z_2, z_3) leads to a change in classification, there must be an $\varepsilon > 0$ and a rule $R \in \mathcal{R}$ s.t.

$$
\mu_R(z_1 + \varepsilon, z_2, z_3) \ne \mu_R(z_1, z_2, z_3) = \alpha.
$$

By inequality (4) we obtain

$$
\begin{aligned}
\min\{\mu_R^{(1)}(z_1 + \varepsilon), \mu_R^{(2)}(z_2), \mu_R^{(3)}(z_3)\} &= \mu_R(z_1 + \varepsilon, z_2, z_3) \\
&> \alpha \\
&= \mu_R(z_1, z_2, z_3) \\
&= \min\{\mu_R^{(1)}(z_1), \mu_R^{(2)}(z_2), \mu_R^{(3)}(z_3)\}.
\end{aligned}
$$

Thus we have

$$
\mu_R(z_1, z_2, z_3) = \mu_R^{(1)}(z_1) < \min\{\mu_R^{(2)}(z_2), \mu_R^{(3)}(z_3)\}.
$$

Taking the monotonicity of $\mu_R^{(1)}$ into account, we derive that $\mu_R^{(1)}$ has to be increasing. The continuity of the fuzzy sets $\mu_R^{(i)}$ guarantees the existence of $\tilde{\varepsilon} > 0$ s.t.

$$
\begin{aligned}
\alpha &= \mu_R^{(1)}(z_1) \\
&< \mu_R^{(1)}(z_1 + \tilde{\varepsilon}) \\
&\le \min\{\mu_R^{(2)}(z_2 - \tilde{\varepsilon}, \mu_R^{(3)}(z_3)\}
\end{aligned}
$$

which leads to the final contradiction

$$\alpha = \mu_R(z_1 + \tilde{\varepsilon}, z_2 - \tilde{\varepsilon}, z_3)$$
$$= \mu_R^{(1)}(z_1 + \tilde{\varepsilon})$$
$$> \alpha.$$

$\square$

4 Other t–Norms and t–Conorms

The fact that linear separable higher dimensional classification problems cannot be solved with fuzzy classification rules can be amended by replacing the maximum by another t–conorm (an associative, commutative, monotone increasing binary operation with unit 0 on the unit interval, see for instance [7]), namely the bounded sum, or by replacing the minimum by another t–norm (an associative, commutative, monotone increasing binary operation with unit 1 on the unit interval), namely the Łukasiewicz–t–norm. The following two theorems show that it is sufficient to replace either the minimum or the maximum by a suitable t–norm, respectively t–conorm. The function f appearing in these theorems describes an arbitrary hyperplane that separates the two classes to be distinguished by the classifiers.

Theorem 4. *Let* $f : [a_1, b_1] \times \ldots \times [a_p, b_p] \longrightarrow \mathbb{R}$, $(x_1, \ldots, x_p) \mapsto c + \sum_{i=1}^{p} c_i x_i$ *($a_i < b_i$). Then there is a finite set $\mathcal{R}$ of classification rules to classes P and N such that*

$$\mathcal{R}^{-1}(P) = \{(x_1, \ldots, x_p) \in [a_1, b_1] \times \ldots \times [a_p, b_p] \mid f(x_1, \ldots, x_p) > 0\}, \quad (5)$$

$$\mathcal{R}^{-1}(N) = \{(x_1, \ldots, x_p) \in [a_1, b_1] \times \ldots \times [a_p, b_p] \mid f(x_1, \ldots, x_p) < 0\} \quad (6)$$

when the minimum in (1) is replaced by an arbitrary t–norm and the maximum in (2) is replaced by the bounded sum, i.e.

$$\mu_C^{(\mathcal{R})}(x_1, \ldots, x_p) = \min\left\{ \sum_{R \in \mathcal{R} : C_R = C} \mu_R(x_1, \ldots, x_p), 1 \right\}.$$

Proof. Without loss of generality let $c \geq 0$. (Otherwise consider the function $-f$ and exchange the rules for the classes P and N.) Without loss of generality, let

$$\alpha = \max\left\{ c, \max_{i \in \{1, \ldots, p\}} \left\{ \sup_{x \in [a_i, b_i]} \{|c_i x|\} \right\} \right\} < \frac{1}{2p}.$$

Otherwise (5) and (6) could be defined equivalently by

$$\mathcal{R}^{-1}(P) = \{(x_1, \ldots, x_p) \in [a_1, b_1] \times \ldots \times [a_p, b_p] \mid \frac{1}{2p\alpha} f(x_1, \ldots, x_p) > 0\}$$

$$\mathcal{R}^{-1}(N) = \{(x_1, \ldots, x_p) \in [a_1, b_1] \times \ldots \times [a_p, b_p] \mid \frac{1}{2p\alpha} f(x_1, \ldots, x_p) < 0\}.$$

Define $\mathcal{R} = \{R, R_1, \ldots, R_p\}$ where

$$\mu_R^{(i)} = \tfrac{1}{2} + c \qquad\qquad (i = 1, \ldots, p)$$

$$\mu_{R_i}^{(i)} = c_i x_i + \tfrac{1}{2p}$$

$$\mu_{R_i}^{(j)} = 1 \qquad\qquad \text{for } j \neq i$$

$$C_R = N$$

$$C_{R_i} = P \qquad\qquad (i = 1, \ldots, p).$$

It is easy to verify that these rules solve the classification problem. □

Theorem 5. *Let* $f : [a_1, b_1] \times \ldots \times [a_p, b_p] \longrightarrow \mathbb{R}$, $(x_1, \ldots, x_p) \mapsto c + \sum_{i=1}^p c_i x_i$ *($a_i < b_i$). Then there is a finite set $\mathcal{R}$ of classification rules to classes P and N such that*

$$\mathcal{R}^{-1}(P) = \{(x_1, \ldots, x_p) \in [a_1, b_1] \times \ldots \times [a_p, b_p] \mid f(x_1, \ldots, x_p) > 0\},$$

$$\mathcal{R}^{-1}(N) = \{(x_1, \ldots, x_p) \in [a_1, b_1] \times \ldots \times [a_p, b_p] \mid f(x_1, \ldots, x_p) < 0\}$$

when the minimum in (1) is replaced by the Łukasiewicz t–norm, i.e.

$$\mu_R(x_1, \ldots, x_p) = \max\left\{1 - p + \sum_{i=1}^p \mu_R^{(i)}(x_i), 0\right\} = \bigotimes_{i=1}^p \mu_R^{(i)}(x_i) \qquad (7)$$

and the maximum in (2) is replaced by an arbitrary t–conorm.

Proof. With the same argument as in the proof of Theorem 4 we may assume without loss of generality

$$\max\left\{|c|, \max_{i \in \{1, \ldots, p\}}\left\{\sup_{x \in [a_i, b_i]}\{|c_i x|\}\right\}\right\} < \varepsilon = \frac{1}{8p}.$$

Define

$$\delta = \frac{p - \tfrac{1}{4}}{p}.$$

Define $\mathcal{R} = \{R_P, R_N\}$ where

$$\mu_{R_P}^{(i)}(x) = \delta + c_i(x - a_i)$$

$$\mu_{R_N}^{(1)}(x) = -c + \frac{3}{4} - \sum_{i=1}^{p} c_i a_i$$

$$\mu_{R_N}^{(i)}(x) = 1 \qquad\qquad \text{for } (i \neq 1)$$

$$C_{R_P} = P$$

$$C_{R_N} = N$$

It is easy to verify that these rules solve the classification problem.   □

5   Conclusions

Our analysis of fuzzy if–then classification rules shows that their principal capabilities are superior to simple crisp classification rules. Thus it is worthwhile to design efficient learning algorithms for such systems. However, one has to take into account the limitations of such classifiers, as they are described in Theorem 3. One possibility is to use other operations than simply max and min. An alternative is the design of hierarchical fuzzy classifiers on the basis of max–min rules. Since we have shown that quite general classification problems can be solved with max–min rules in the two–dimensional case, a hierarchical fuzzy classifier consisting of cascaded rules where each single rule is restricted to two variables might be a promising approach.

References

1. Bauer, P., Klement, E.P., Leikermoser, A., Moser, B.: Interpolation and approximation of real input–output functions using fuzzy rule bases. In: [8], 245–254
2. Castro, J.L., Trillas, E., Cubillo, S.: On consequence in approximate reasoning. Journal of Applied Non-Classical Logics 4 (1994), 91–103
3. Genther, H., Glesner, M.: Automatic generation of a fuzzy classification system using fuzzy clustering methods. Proc. ACM Symposium on Applied Computing (SAC'94), Phoenix (1994), 180–183
4. Ishibuchi, H.: A fuzzy classifier system that generates linguistic rules for pattern classification problems. In: Fuzzy Logic, Neural Networks, and Evolutionary Computation, Springer, Berlin (1996), 35–54
5. Klawonn, F., Kruse, R.: Derivation of fuzzy classification rules from multidimensional data. In: Lasker, G.E., Liu, X. (eds.): Advances in Intelligent Data Analysis. The International Institute for Advanced Studies in Systems Research and Cybernetics, Windsor, Ontario (1995), 90–94

6. Kosko, B.: Fuzzy systems as universal approximators. Proc. IEEE International Conference on Fuzzy Systems 1992, San Diego (1992), 1153–1162

7. Kruse, R., Gebhardt, J., Klawonn, F.: Foundations of fuzzy systems. Wiley, Chichester (1994)

8. Kruse, R., Gebhardt, J., Palm, R. (eds.): Fuzzy systems in computer science. Vieweg, Braunschweig (1994)

9. Lee, J., Chae, S.: Analysis on function duplicating capabilities of fuzzy controllers. Fuzzy Sets and Systems **56** (1993), 127–143

10. Meyer Gramann, K.D.: Fuzzy classification: An overview. In: [8], 277–294

11. Nauck D., Kruse, R.: NEFCLASS – A neuro–fuzzy approach for the classification of data. In: George, K.M, Carrol, J.H., Deaton, E., Oppenheim, D., Hightower, J. (eds.): Applied Computing 1995: Proc. of the 1995 ACM Symposium on Applied Computing. ACM Press, New York (1995), 461–465

12. Wang, L.X.: Fuzzy systems are universal approximators. Proc. IEEE International Conference on Fuzzy Systems 1992, San Diego (1992), 1163–1169

13. Weber, R.: Fuzzy–ID3: A class of methods for automatic knowledge acquisition. In: Proc. 2nd International Conference on Fuzzy Logic and Neural Networks, Iizuka (1992), 265–268

Neuro–Fuzzy Diagnosis System with a Rated Diagnosis Reliability and Visual Data Analysis

A. Lapp and H.–G. Kranz

Bergische Universität GH Wuppertal
Laboratorium für Hochspannungstechnik
Fuhlrottstr. 10, 42097 Wuppertal, Germany
E–Mail: lapp@uni-wuppertal.de, kranz@uni-wuppertal.de

Abstract. This paper introduces an automated diagnosis system with an increased and rated diagnosis reliability in a case where a modelling of a technical diagnosis task is impossible so far. To achieve this a redundant diagnosis concept which features three independent feature extraction methods and three independent classifier which are based on Distance Classification Methods (DCM), Fuzzy Sets (FS) and Neural Networks (NN) were developed. Furthermore the data can be visualised in different plots and also rated by the user.

1 Introduction

This paper deals with the soft computing and pattern recognition potentials to solve technical diagnosis problems. Diagnosis methods and achievable diagnosis reliability are introduced giving an example in the field of insulation techniques where a modelling of the complete system is not possible due to the complex interaction of mechanical and physical measuring and operating influences. Therefore the state of the system can be diagnosed only by symptoms.

Webster's [13] defines the term diagnosis as "*The art or act of identifying a disease from its signs and symptoms*".

A sufficient reliable technical diagnosis can be achieved when a convincing modelling of a system or process is available. Usually the diagnosis has to answer two questions:

1. What defect type is apparent / in which state is the process?
2. Where has the defect occurred?

Methods of artificial intelligence or pattern recognition deliver a probability answer to these questions. A powerful classification method can always determine, whether the current state is the desired one, when a good process model is applied [7].

Much more complex is the diagnosis task when the engineer is in the same situation as a physician who is forced to come to a vital decision which is based

X. Liu, P. Cohen, M. Berthold (Eds.): "Advances in Intelligent Data Analysis" (IDA–97)
LNCS 1280, pp. 371–382, 1997.   © Springer–Verlag Berlin Heidelberg 1997

on limited information of an insufficiently described system/process. This decision can have far-reaching economical consequences. Therefore an automated computer based diagnosis of complex technical systems can only succeed if the following conditions are given:

- no misclassifications occur
- a rated diagnosis reliability is offered including a warning if the reference data base is insufficient
- a convincing performance in the case of the "unknown defect type" is guaranteed which is especially crucial for technical applications

The main field of application of Neuro-Fuzzy systems in diagnosis tasks is the pattern recognition of fast replicating signals. The human brain is not capable to resolve these patterns if they have a repetition frequency which is higher than some patterns per second. As known from experience even primitive neural networks with less than 1000 neurones are superior to the human brain in cases of fast replicating signals. Therefore e.g. a computer based diagnosis system is able to avoid catastrophes by switching of a power system even in critical situations like the example (partial discharge diagnosis) described here.

Pattern recognition methods are well advanced and powerful classifiers, e.g. to determine the distances in a N–dimensional space, have been developed early. Groß [2] showed that conventional and Neuro-Fuzzy classification algorithms can be described by a uniform mathematical description as long as the decision is based on a feedforward algorithm. Different types of neural networks (NNs) can be distinguished whether they are *recurrent* or *feedforward*. When using feedforward NNs the feedback of the output is only necessary during training (weights evaluation). Whereas a recurrent NN evaluates the decision vector by feeding back the output to the input vector. Therefore these two NN topologies learn differently by supervised respectively non supervised learning. In principle one could assume that the non supervised learning would be more suitable for the diagnosis task discussed here. But the authors have experienced that with every new entry the knowledge base acquired so far is lost completely. In this case even an expert is overtaxed with the task to set up a reference data base. For a technical diagnosis system it is better to apply the non supervised learning to identify the already known defect type correctly and reliable as long as it is possible to avoid misclassifications in the case of the *unknown* pattern.

Therefore this work combines the advantages of different classification algorithms to a so called redundant diagnosis system. A certain amount of redundancy is apparent if different input algorithms are used. These requirements are meet when conventional distance classifiers, neural networks and Fuzzy sets are used due to their different system theoretical basis.

For technical diagnosis tasks a second basic principle is valid: "*Even the most powerful classifiers are not able to differentiate if the input features are not significant*". The features extracted from the process signals play therefore an deciding part. Only if the features are sufficiently significant a correct diagnosis result is possible. In technical applications the acquired measurements are faulty and noisy which accounts for the fact that the significant patterns are lost in increasing noise. Again, it is perceptible that in fast changing situations a computer based diagnosis is superior to a human being and that a computer is capable in certain cases to substitute the ability of the human brain to generalise by performing pattern or signal recognition algorithms repeatedly.

Therefore a reliable technical diagnosis should not only answer the question *"Which kind of defect type is apparent and where has this defect occurred?"* but also should tell an expert how reliable the diagnosis decision is, to enable the expert to think about the consequences of this diagnosis result and not to force him or her to think about the diagnosis decision itself.

2   The Redundant Diagnosis Concept

The Redundant Diagnosis System (RDS) introduced here is employed to evaluate the insulation state of a high voltage (HV) component, apparatus etc. Defects in HV equipment cause so called partial discharge (PD) which are transient signals in the ns range and a magnitude in the pico Coulomb (pC) range.

Using a suitable computer file format the information can be processed automatically in such a way that it is suitable for an automated computer based defect identification as well as a visual evaluation of the insulation state by an expert.

In the upper part of Fig. 1 a so called discharge pattern is displayed. This is a phase resolved plot of an electrical PD measurement of an insulation defect where each pixel represents one single electrical PD pulse. The position of the pulses is determined by their appearance within a cycle of an AC test voltage period (50 Hz). The grey value determines the magnitude of this pulse. Different system states (defect types) cause different patterns which are discernible by these plots.

These plots enable the user to get an idea of the characteristics of different defect types and to trace the diagnosis results of the RDS. But it should be taken into account that a computer has the ability to manage a huge amount of data whereas the abilities of most human beings are limited. An operator has to learn numerous patterns to reach only a part of the diagnosis potential of the RDS.

Figure 1 shows the basic principle of pattern recognition. Characteristic features are extracted from a pattern. These features are combined into a feature vector which is located in a N-dimensional feature space. A classifier compares this feature vector with reference vectors and determines the correspondence with different reference classes in a library. These references determine which scope a certain class (which represents e.g. a certain defect type or state of a process) inhabits in feature space. This scope is not necessarily congruent to the true scope where feature vectors of a certain class are located. It depends on the choice and the number of references used. Therefore the use of a single classifier can cause misclassifications due to the discrepancy between the scope in feature space determined by references and the true scope. To avoid this, one can apply several classifier simultaneously for a mutual check of the results [4, 8, 10]. To

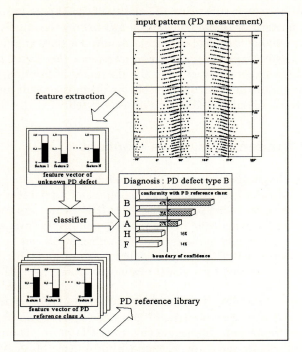

Fig. 1. The basic principle of pattern recognition

make a redundant diagnosis concept work the implemented feature extraction methods and classifiers have to be independent to inhabit a different scope in feature space which should only be congruent at the true scope where feature vectors are located [8].

E.g. the following feature extraction methods can be employed [4, 5, 12]:

- statistical methods
- orthogonal transforms
- neural networks which learn non supervised (e.g. Kohonen Feature Map)

The two first methods can be described uniformly by (1). The feature $m_1 \ldots m_N$ are extracted from a discrete time signal $q(n\tau)$ by a function $\mathbf{f}_v$.

$$\mathbf{m}_v = \begin{pmatrix} m_1 \\ \vdots \\ m_N \end{pmatrix} = \mathbf{f}_v \left(q\left(n\tau \right) \right) \tag{1}$$

$$\mathbf{m}_v = \text{feature vector extracted by feature extraction method } v$$
$$N = \text{dimension of feature vectors}$$
where
$$m_i = i\text{--th feature of feature vector } \mathbf{m}$$
$$q\,(n\tau) = \text{discrete time signal which has to be classified}$$
$$\mathbf{f}_v\,(q) = \text{feature extraction method } v$$

When there is a linear relation between feature vectors (2) which are extracted by different methods these vectors inhabit the same scope in feature space when they are normalised (3).

$$\mathbf{m}_{v1} = \mathbf{f}_{v1}\,(q(n\tau)) = a \cdot \mathbf{m}_{v2} = a \cdot \mathbf{f}_{v2}\,(q(n\tau)); \qquad a \in \mathbb{R} \qquad (2)$$

$$\frac{m_{v1,i}}{m_{v1,max}} = \frac{a \cdot m_{v2,i}}{a \cdot m_{v2,max}} = \frac{m_{v2,i}}{m_{v2,max}} \qquad (3)$$

Linear independent features (4) do not inhabit necessarily the same scope in feature space. Investigations of practical applications [3, 6] have proved that a linear independence is sufficient for a reliable diagnosis result of a redundant diagnosis system (RDS).

$$\mathbf{m}_{v1} = \mathbf{f}_{v1}\,(q(n\tau)) \neq a \cdot \mathbf{m}_{v2} = a \cdot \mathbf{f}_{v2}\,(q(n\tau)) \qquad (4)$$

The redundant diagnosis concept can only work if all different classifier employ a uniformly structured reference data base i.e. employ the same reference classes. Otherwise a comparison of the classification results is not possible. Therefore a non supervised NN is not suitable for the RDS since in this case a uniform reference data base is not guaranteed. This accounts for the fact that the RDS employs a feedforward NN which learns supervised using the back error propagation (BEP) algorithm. Additionally this NN utilises a reference class which contains random values. This additional class improves the classification performance of the NN in the case of an unknown pattern [4]. Therefore it is called modified NN in Fig. 2.

An example of a redundant diagnosis concept which has been verified in practical industrial applications is shown in Fig. 2. The features are extracted by Fourier Analysis Walsh-Hadamard and Haar transforms which are all orthogonal transforms [12].

For example Eqn. 5 to 8 show the feature extraction by Fourier Analysis [5]. For practical applications 6 spectral components ($k = 1\ldots6$) are sufficient to separate different reference classes but still provide the ability to generalise. So a feature vector has a dimension of $N = 24$.

Where $q_p(\varphi)$ is the apparent charge of a PD pulse at phase position φ within test voltage cycle p and P_0 if the number of considered cycles.

Equation (8) takes the changes of the input pattern with respect to time into consideration.

$$\mathbf{m} = (m_1 \ldots m_{24})^T = (\bar{a}_1 \ldots \bar{a}_6, \ \bar{b}_1 \ldots \bar{b}_6, \ |\overline{\Delta a_1}| \ldots |\overline{\Delta a_6}|, \ |\overline{\Delta b_1}| \ldots |\overline{\Delta b_6}|)^T \quad (5)$$

$$\bar{a}_k = \frac{1}{P_0} \sum_{p=1}^{P_0} a_k\,(p) \qquad \text{with} \qquad a_k\,(p) = \frac{\int_{\varphi=0}^{2\pi} q_p(\varphi) \cdot \cos(k\varphi)\, d\varphi}{\sqrt{\int_{\varphi=0}^{2\pi} q_p^2(\varphi)\, d\varphi}} \qquad (6)$$

$$\bar{b}_k = \frac{1}{P_0} \sum_{p=1}^{P_0} b_k\,(p) \qquad \text{with} \qquad b_k\,(p) = \frac{\int_{\varphi=0}^{2\pi} q_p(\varphi) \cdot \sin(k\varphi)\,d\varphi}{\sqrt{\int_{\varphi=0}^{2\pi} q_p^2(\varphi)\,d\varphi}} \qquad (7)$$

$$|\overline{\Delta a_k}| = \frac{1}{P_0} \sum_{p=1}^{P_0} |a_k\,(p) - a_k\,(p-1)|; \qquad |\overline{\Delta b_k}| = \frac{1}{P_0} \sum_{p=1}^{P_0} |b_k\,(p) - b_k\,(p-1)| \quad (8)$$

The PD signal changes with respect to the magnitude of the applied test voltage. Therefore it is possible to describe one input pattern with up to 30 feature vectors depending of the changes within the pattern.

In the case of the Haar and Walsh-Hadamard transform the sinusoidal functions are replaced by orthogonal Haar functions which are only locally defined respectively by Walsh-Hadamard functions which are discrete functions of values 1 and -1. [5, 12]

A L_2–distance classifier (L2) a Fuzzy System (FS) and a NN are used as classifiers. The choice of the feature extraction methods and classification algorithms is motivated in [5].

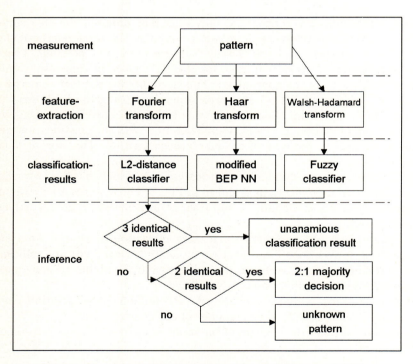

Fig. 2. Concept of the Redundant Diagnosis System (RDS)

The L2 is based on the normalised Euklidean distance (9). Despite it's simple algorithm the classification results have a good physical traceability [5] (see also section 3).

$$d_j = \sqrt{\frac{1}{N} \sum_{i=1}^{N} (m_i - m_{\text{Ref},ji})^2}$$ (9)

$$
\begin{aligned}
\text{whereas} \quad & d_j = L_2\text{–distance of a pattern to reference class } j \\
& m_i = i\text{–th component of feature vector } \mathbf{m} \\
& m_{\text{Ref},ji} = i\text{–th component of feature vector } \mathbf{m}_{\text{Ref},j} \\
& \qquad\qquad \text{of reference class } j \\
& N = \text{dimension of feature vectors}
\end{aligned}
$$

Furthermore a FS is implemented which is based on statistically evaluated membership functions which describe the scope a reference class inhabits. The inference of the membership values is evaluated by a γ–operator. Figure 3 gives an example of the membership function and Eqn. 10 of the geometric-arithmetic averaging γ–inference operator $M_{ga}(\gamma)$ for a feature m_i implemented in the RDS with $\mu_i = \mu_i(m_i)$.

$$M_{ga}(\gamma) = \gamma \cdot \sqrt[N]{\prod_{i=1}^{N} \mu_i} + (1-\gamma) \cdot \frac{1}{N} \sum_{i=1}^{N} \mu_i, \qquad \gamma \in \{\mathrm{I\!R} \mid 0 \le \gamma \le 1\}$$ (10)

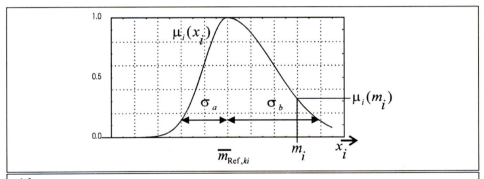

with
$\mu(x_i) = f(\bar{m}_{\text{Ref},ki}, \sigma_a, \sigma_b)$
$\bar{m}_{\text{Ref},ki} = $ mean value of the i–th component of reference class k
$\sigma_a = $ variance for $x_i < \bar{m}_{\text{Ref},ki}$ from variance and skewness of component ki
$\sigma_b = $ variance for $x_i > \bar{m}_{\text{Ref},ki}$ from variance and skewness of component ki

Fig. 3. Fuzzy membership function $\mu_i(x_i)$: non-symmetrical normal distribution

More detailed information and other membership functions which were investigated by the authors can be found in [1, 4]. It is also possible to set up a rule based Fuzzy classifier in which the knowledge of experts can be implemented [11]. In this case rules can also be set up to formulate exclusion criteria e.g. to determine areas where no signals should occur for a certain reference class.

The uniform mathematical description of the different classification algorithms in Tab. 1 shows that the elements a_i of an activity vector $\mathbf{a}$, which determines the membership of a feature vector $\mathbf{m}$ to a reference class i, are generally linear independent.

Table 1. Mathematical description for distance classifier neural network and Fuzzy system

Classifier	Evaluation of the membership value to a reference class
L_2–distance classifier	$\mathbf{a} = \begin{pmatrix} a_1 \\ \vdots \\ a_K \end{pmatrix} = \begin{pmatrix} \|\mathbf{m}_{\mathrm{Ref},1} - \mathbf{m}\| \\ \vdots \\ \|\mathbf{m}_{\mathrm{Ref},K} - \mathbf{m}\| \end{pmatrix}$
Back–Error–Propagation (BEP) neural network (NN)	$\mathbf{a} = \begin{pmatrix} a_1 \\ \vdots \\ a_K \end{pmatrix} = \begin{pmatrix} f\left(\mathbf{w}_{21}^T \cdot \mathbf{f}\left(W_1 \cdot \mathbf{m} + \mathbf{b}_1\right) + b_{21}\right) \\ \vdots \\ f\left(\mathbf{w}_{2K}^T \cdot \mathbf{f}\left(W_1 \cdot \mathbf{m} + \mathbf{b}_1\right) + b_{2K}\right) \end{pmatrix}$
Fuzzy classification system (FS)	$\mathbf{a} = \begin{pmatrix} a_1 \\ \vdots \\ a_K \end{pmatrix} = \begin{pmatrix} Op_\gamma\left(\boldsymbol{\mu}_{\mathbf{m}_{\mathrm{Ref},1}}(\mathbf{m})^T\right) \\ \vdots \\ Op_\gamma\left(\boldsymbol{\mu}_{\mathbf{m}_{\mathrm{Ref},K}}(\mathbf{m})^T\right) \end{pmatrix}$

with $\mathbf{a}$ = activity vector
$\mathbf{m}$ = feature vector of an unknown pattern
$\mathbf{m}_{\mathrm{Ref},j}$ = feature vector of reference class j
K = number of reference classes
W_i = weight matrix of the i–th layer
$\mathbf{w}_{ij}$ = j–th weight vector of the i–th layer
$\mathbf{b}_i$ = bias vector of the i–th layer
b_{ij} = j component of bias vector i
$f(x)$ = functional evaluation of a scalar
$\mathbf{f}(\mathbf{x}) = (f(x_1)\ldots f(x_N))^T$ = functional evaluation of the components of $\mathbf{x}$
$Op_\gamma(\mathbf{x}^T)$ = γ–operator with $\gamma \in \{\mathbb{R} \mid 0 \leq \gamma \leq 1\}$
$\boldsymbol{\mu}_{\mathbf{m}_{\mathrm{Ref},K}}(\mathbf{x})$ = membership function of the K–th reference class

So we can conclude that all classifier are feed by linear independent features and gain independently an originally classification result. These single results are compared by the RDS and three cases can be distinguished (see also Fig. 2):

1. All three single classification results are identical. In this case it is very likely that the correct defect type/process state is found: unanimous classification result $\Rightarrow$ **reliable identification**
2. Two of the three classifier are reaching the same result. Also in this case it is likely that the correct defect type/process state was identified: majority

decision ⇒ **sufficient reliable identification**

3. All three classification results are different. This suggests the fact that the pattern belongs to a defect type/process state which is not represented in the employed reference data base. Therefore the user is told that the RDS is not able to reach a decision: an unknown class is found which has to be documented and the RDS has to be trained with this pattern which is a new reference for the library.

To increase the performance of the RDS further a boundary of confidence was introduced. This boundary states whether a pattern has a greater likeness with all references than with a single reference class. If this boundary is crossed the classification result is not used for the RDS due to a lack of confidence.

Because of all these measures it was possible to compile an automated computer based diagnosis system which is able to perform a reliable and rated diagnosis decision.

3   Reliability of a Diagnosis System

To determine the quality of feature extraction methods and/or classifiers a so called verification matrix can be employed. The columns in such a matrix (Fig. 4) contain the different references classes which are in use whereas the rows contain repetition measurements of these classes which are not references themselves. The average conditional probability in percent by which the repetition measurements are placed in the different reference classes is displayed in the matrix elements. Ideally when using a perfect library the main diagonal contains the value 100% whereas the other elements equal zero. Fig. 4 shows a good classifier which is able to identify different patterns reliable.

Beside this mathematically formulated quality criteria like Average Decision Distance (ADD) and Logarithmic Performance Rate (LPR) can be employed (Tab. 2) which enable a more objective rating of the potential of a diagnosis system [4]. In the case of a technical diagnosis application it can be feasible to mark physical similar patterns by corresponding average conditional probabilities (physical similar patterns are framed in Fig. 4) instead of employing a classifier which is as ideal as possible. This furnishes the diagnosis system with ability to generalise up to a certain extend and it enables the user to trace the diagnosis results. It shows that it is necessary to use all quality criteria parallel, the numerical ones as well as the verification matrix to get a complete picture of the capability of a feature extraction method or a classifier.

4   Verification of the RDS

The components of the RDS are in practical operation for some years at an industrial HV test field. This application verifies that the RDS is a reliable tools to evaluate different insulation states as far as the reference data base is sufficiently significant [3, 10, 9].

Repetition measurements / Reference defects	Detached Electrode on Ground Potential in Air	Tip on High Potential in Air	Dull Tip on High Potential in Air	Tip on Ground Potential in Air	Internal Voids in Air	Fixed Conducting Particle on Dielectrics in Air	Surface Discharge in Air	Background Noise	Boundary of Confidence
DeElGrPo	74	12	11	11	20	15	27	16	30
TiHiPo	12	61	30	9	10	14	12	16	30
DuTiHiPo	11	30	92	10	10	13	12	16	30
TiGrPo	11	9	10	52	11	17	11	16	30
IntVoi	20	11	10	11	81	14	17	16	30
FiConPar	15	14	13	18	14	79	22	17	30
SurfDis	28	12	12	11	17	21	77	16	30
Noise	16	16	16	16	15	16	16	79	39

Fig. 4. Physical traceable verification matrix by L_2–distance classifier

In addition the RDS is up to a certain degree resistant against stochastic disturbances. This is especially important for on-site measurements where no shielding exists. Table 3 shows the classification result of the RDS when several thousand stochastic noise pulse are added to the input pattern. The maximal amplitude of the disturbances was increased from $10\,\mathrm{pC}$ up to $10000\,\mathrm{pC}$.

Up to a maximal noise level of $100\,\mathrm{pC}$ all input patterns were classed with the original reference class and few patterns were disturbed so much by the noise pulses that they resemble no class within the reference library sufficiently. Therefore they were classified as *unknown* by the RDS.

If the noise level is increased further several measurements were classified as reference classes which represent the background noise respectively a reference class which has an input pattern of stochastic pulses.

But the fact that a PD pattern which is superimposed with a large scale noise is classified as background noise can't be considered as a real misclassification. In contrast this performance is correct if the PD pulses from the defect are covered by noise. These classification results first take place with PD signals with a very low average PD magnitude(about 10 pC). This fact justifies the classification of the RDS. Still more than 50PD measurements were classified correctly. Taking

Table 2. Objective performance criteria for feature extraction methods and diagnosis systems

$$ADD = \frac{1}{N \cdot (N-1)} \sum_{k=1}^{N} \frac{1}{N_k} \sum_{j=1}^{N_k} \sum_{\substack{i=1 \\ i \neq k}}^{N} \left(P\left(\mathbf{m}_{\text{Ref},k} | \mathbf{m}_{k_j}\right) - P\left(\mathbf{m}_{\text{Ref},i} | \mathbf{m}_{k_j}\right) \right)$$

$$LPR = \frac{1}{2 \cdot N \cdot (N-1)} \sum_{k=1}^{N} \frac{1}{N_k} \sum_{j=1}^{N_k} \sum_{\substack{i=1 \\ i \neq k}}^{N} \log \left(\frac{P\left(\mathbf{m}_{\text{Ref},k} | \mathbf{m}_{k_j}\right)}{P\left(\mathbf{m}_{\text{Ref},i} | \mathbf{m}_{k_j}\right)} \right)$$

with $\quad P\left(\mathbf{m}_{\text{Ref},i} | \mathbf{m}_{k_j}\right) \neq 0$

where
$\quad\quad\quad N$ = number of reference classes
$\quad\quad\quad N_k$ = number of repetition measurements of the k–th reference class
$\quad\quad\quad \mathbf{m}_{k_j}$ = feature vector of measurement no. j of reference class k
$\quad\quad \mathbf{m}_{\text{Ref},i}$ = feature vector of reference class i
$P\left(\mathbf{m}_{\text{Ref},i} | \mathbf{m}_{k_j}\right)$ = conditional probability that feature vector $\mathbf{m}_{k_j}$
$\quad\quad\quad\quad\quad\quad$ is of reference class i

into account that the magnitude of most of this "useful pulses" is about 5 to 10 times smaller than the added noise, this is a convincing result and one can conclude that the RDS is tolerant against stochastic disturbances [4, 6, 8].

Table 3. Classification result of the RDS with noisy input patterns

magnitude of noise pulses	0 pC	10 pC	20 pC	50 pC	100 pC	500 pC	1000 pC	10000 pC
no. of input patterns classed with the original reference class	78	68	67	66	64	48	35	11
no. of input patterns unknown to the RDS	0	10	11	12	14	15	27	28
no. of input patterns classed with noise related references	0	0	0	0	0	15	16	39

5   Conclusion

The Redundant Diagnosis System (RDS) introduced here has the capability to perform a reliable diagnosis even for a technical application in the field of electrical insulations where no modelling is possible so far. Besides the user is informed about the reliability of the diagnosis result and can visualise the patterns in several plots to trace the diagnosis results. By these means misclassification can be avoided especially in the case when a pattern is unknown i.e. when it is not represented by a reference library.

Further investigations have shown that the RDS is tolerant to stochastic disturbances as long as the pattern is not completely unrecognisable.

Besides the diagnosis potential of a diagnosis system depends strongly on the quality of the reference data base in use [10]. In this field further research is necessary to achieve an automated set up of the reference data base. This can avoid a reduction of the diagnosis abilities of the RDS due to incorrectly defined reference classes by the user (expert).

References

1. Groß, A., Hücker, T., Kranz, H.-G.: "Rechnergestützte Auswertung von TE-Meergebnissen - Fuzzy-Logic", ETG-Fachbericht Nr. 56, pp. 79–88, Esslingen 1995
2. Groß, A., "Realzeitfähige Unterdrückung impulsförmiger Störsignale bei Teilentladungsmessungen mit einem neuronalen Signalprozessorsystem", Doctor Thesis, Bergische Universität, Wuppertal, 1996
3. Hücker, T., Kranz, H.-G., Krump, R., Haberecht, P.: "Computergestützte Teilentladungsdiagnostik in der industriellen Prfpraxis", ETG-Fachbericht Nr. 56, pp. 153–158, Esslingen 1995
4. Hücker, T., "Computergestützte Teilentladungsdiagnostik unter Berücksichtigung praxisrelevanter Randbedingungen", Doctor Thesis, Bergische Universität, Wuppertal, 1995
5. Hücker, T., Kranz, H.-G., "Requirements of Automated PD Diagnosis Systems for Fault Identification in Noisy Conditions", IEEE Transactions on Dielectrics and Electrical Insulation, Vol. 2 No. 4, pp. 544 ff., 1995
6. Hücker, T., Kranz, H.-G., Lapp, A., "A Partial Discharge Defect Identification System with Increased Diagnosis Reliability", 9th ISH, pp. 5612, Graz, Austria, 1995
7. Isermann, R., Ayoubi, M., "Fault Detection and Diagnosis with Neuro-Fuzzy-Systems", EUFIT '96, pp. 1479-1491, Aachen, Deutschland, 1996
8. Kranz, H.-G., Lapp, A., "Neuro-Fuzzy-Diagnosesystem mit bewerteter Diagnosezuverlässigkeit und hoher Rauschtoleranz", Symposium Anwendungen von Fuzzy Technologien und neuronalen Netzen, Berlin, Germany, 1996
9. Kranz, H.-G., Lapp, A., "The Influence of the Quality of the Reference Data Base on PD Pattern Recognition Results", 10th ISH, Montréal, Canada, 1997 (in print)
10. Lapp, A., Kranz, H.-G., "Discussion of the Quality of the Reference Data Base influencing PD Pattern Recognition Results", 8. MVK-TVN, pp. 99, Stará Lesná, Slowakei, 1996
11. Mayer, A., Mechler, B., Schlindwein, A., Wolke, R., "Fuzzy Logic - Einführung und Leitfaden zur praktischen Anwendung", Addison-Wesley, Advanced Book Program, 1974
12. Rao, K.R., Ahmed, N., "Orthogonal Transforms for Digital Signal Processing", Springer Verlag, Berlin, Heidelberg, New York, 1975
13. "Webster's New Encyclopedic Dictionary", Black Dog & Leventhal Publishers Inc., Revised Edition 1995

Genetic Fuzzy Clustering by Means of Discovering Membership Functions

Meltem Turhan

Department of Computer Engineering
Middle East Technical University
06531 Ankara, Turkey

Abstract. It has been observed that in the previous *Genetic Algorithms* (GA) based *Fuzzy Clustering* (FC) works only some of the parameters of an FC system are developed. Here, a new approach is proposed to develop directly the membership functions for the clusters using GA. This new technique is implemented and tested on common test data. A comparative study of the results against the quotations in literature reveals that the standard c-means FC technique is outperformed by the proposed technique in the count of misclassifications aspect.

1 Introduction

Cluster analysis is a powerful tool for data analysis, image segmentation and pattern recognition. Its aim is to partition distinguisable objects into subsets, called clusters. Crisp clustering requires the clusters to be disjoint, that is, one object is assigned to one and only one cluster. This restriction makes crisp clustering unsuccessful when clusters are intersecting. The use of fuzzy sets in clustering was first proposed in [1] and several classification schemes were developed by [2]. The first fuzzy clustering algorithm was developed in 1969 by Ruspini [3]. Following this, Dunn [4] developed the first fuzzy least square approach to clustering.

In this approach an entity is not forced to be a member of a single class/set only. With some degree of membership (*Membership Value)*, an entity, may be an element of any set. The idea is to form cluster sets and determining the MV values for each of the elements of some subject data with respect to these cluster sets. Various works using this technique is presented.

In the following section Fuzzy Clustering Algorithm will be presented. After this study a new technique for Fuzzy Clustering is presented. The technique is implemented as a C program and the results obtained are evaluated and compared to the literature.

1.1 Fuzzy-C-Means Clustering

The most popular fuzzy clustering algorithm is developed by J. Bezdek as the Fuzzy-C-Means (FCM) Algorithm.

X. Liu, P. Cohen, M. Berthold (Eds.): "Advances in Intelligent Data Analysis" (IDA–97)
LNCS 1280, pp. 383–393, 1997.   © Springer–Verlag Berlin Heidelberg 1997

The Fuzzy-C-Means algorithm tries to find a prototype for each cluster, and assigns suitable membership degrees to each object with respect to each cluster. This method uses an Euclidean distance function to calculate the distance between an object and the prototype of the cluster. The algorithm aims to minimize the function:

$$J_m(X, U, V) = \sum_{i=1}^{c} \sum_{k=1}^{n} (u_{ik})^m d_{ik}^2 (v_i, x_k)$$

where

$$\sum_{k=1}^{n} u_{ik} > 0 \qquad \text{for} \quad 1 \leq i \leq c$$

i. e., the membership degree should be greater than zero for each cluster and

$$\sum_{i=1}^{c} u_{ik} = 1 \qquad \text{for} \quad 1 \leq k \leq n$$

which means that the sum of all membership degrees of an object should be 1. Furthermore

- $X = x_1, \ldots, x_n$ is the data set,
- c is the number of clusters,
- $u_{ik} \in [0, 1]$ is the membership degree of x_k to cluster i,
- v_i is the prototype for cluster i,
- $d_{ik}(v_i, x_k)$ is the distance between prototype v_i and object x_k.
- m is a parameter that is called the *fuzziness index*.

1.2   Genetic Fuzzy Clustering

Clustering algorithms that use calculus based methods may set stuck in local minima. Genetic Algorithm can avoid local minima because of its random seach mechanism.

Many researhces uses Genetic Algorithms in FC just to evolve some parameters of Fuzz-C-Means (FCM) algorithm [5, 6, 7, 8]

Yuan et. al [9] again uses FCM algorithm but in their work GA is used to determine the best distance functions (among Minkowski, Euclidean, Hamming, Maximum, Mahalonabis). Not every GA-based FC assumes the number of clusters to be known a priori [10]. In this work GA evolves the number of clusters and also the shape of each fuzzy cluster.

2   Proposed Approach

In the above mentioned approaches, the idea is to find several members of the feature vectors which will become the centers of various clusters. The motive of these approaches is then to minimize the distances of the feature vectors to these cluster centers. While doing this, each feature vector is assigned a membership value relative to each cluster. These values add up to 1.0. The importance of a distance value is determined simply by multiplying the distance of a feature vector to that center with a positive power of the membership value to that cluster. The whole task of various algorithms is to

- determine the center feature vectors for each cluster
- and the membership values of each individual feature vector for each cluster.

Our approach differs from the literature. Instead of finding centers we will find directly the membership functions for each of the N clusters. If the feature vectors are members of $\mathcal{R}^d$ then this means that for each cluster we will have a vector of dimension d that has functions as components each of which is of the form

$$F : \mathcal{R} \mapsto [0, 1]$$

So if c marks a cluster then

$$[\mathbf{f}^c]_i : \mathcal{R} \mapsto [0, 1]$$

for notational convenience we will denote $[\mathbf{f}^c]_i$ by f_i^c.

The overall goal of clustering is converted into finding such functions that maximize the square sum of the membership values under the constraint that the sum of all average membership values of a feature vector is 1.0. Here the *average membership value* is defined as the average of each component's membership value (calculated by plugging it into the relevant f_i^c). Formally speaking,

$$[\mathbf{f}^c(\mathbf{x})]_i \stackrel{\triangle}{=} f_i^c(x_i)$$

$$S = \sum_{\substack{c \in \text{clusters} \\ \mathbf{x} \in \text{feature vectors}}} \mathbf{f}^c(\mathbf{x}) \cdot \mathbf{f}^c(\mathbf{x})$$

$$1.0 = \sum_{c \in \text{clusters}} \sum_{i=1}^{d} \frac{f_i^c(x_i)}{d}$$

The goal is to determine all $\mathbf{f}^c$ such that S is maximal under the constraint above.

In this work the functions $f_i^c(x)$ are restricted to triangular functions (we assume that the x values are normalized to 1.0):

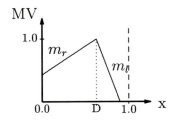

where the slopes m_l, m_r of the left and right lines, respectively, are restricted as:

$$0 \leq m_l < \infty \qquad -\infty < m_r \leq 0$$

So any function $f_i^c(x)$ is defined as $\tilde{f}_i^c(x)/\kappa(\mathbf{x})$ where:

$$\tilde{f}_i^c(x) = \begin{cases} \max(0, m_l(x-D)+1) & \text{if} \quad 0 \leq x \leq D \\ \max(0, m_r(x-D)+1) & \text{if} \quad D < x \leq 1 \\ undefined & \text{otherwise} \end{cases}$$

and κ is a normalization scalar calculated for each feature vector that serves the fulfillment of (1), which says that the sum of the cluster membership values for any feature vector shall be unity. So, it is defined as

$$\kappa(\mathbf{x}) = \sum_{c \in \text{clusters}} \sum_{i=1}^{d} \frac{f_i^c(x_i)}{d}$$

With this definition, any function $f_i^c(x)$ is uniquely represented by a 3-tuple $\langle m_l, m_r, D \rangle_i^c$.

As will be explained in the next sections, a Genetic Algorithms approach will be used to determine these 3-tuples such that the goal of clustering is satisfied.

2.1   GA Representation of the New Approach

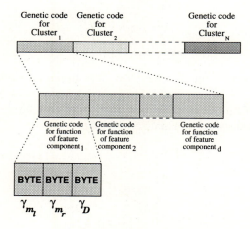

Fig. 1. Chromosome Structure

The mapping between the allele values and the $\langle m_l, m_r, D \rangle_i^c$ information is as follows:

$$\langle m_l, m_r, D \rangle_i^c = \langle \xi \gamma_{m_l}, -\xi \gamma_{m_l}, \gamma_D / 255 \rangle_i^c$$
$$\langle \gamma_{m_l}, \gamma_{m_r}, \gamma_D \rangle_i^c = \langle \gamma_P, \gamma_{P+1}, \gamma_{P+2} \rangle \quad \text{where} \quad P = 3(d(c-1) + i - 1)$$

Here d is the dimension of the feature vectors. ξ is a multiplicative factor which is determined by the maximal slope value that is admissable. The very nature of the implementation (as all computer implementations do) disables the use of a ∞ concept. So we have to decide for a maximal slope value. In the implementation a ξ value of 0.2 has been observed to be quite reasonable. This sets a maximal slope value about 50.

Any chromosome Γ constitutes $N \times d \times 3$ adjacent genes that we will label as

$$\gamma_0, \gamma_1, \ldots, \gamma_{3dN-1}$$

A gene occupies 1 byte. Therefore for each gene we have 256 different allele values.

A GA usually has three genetic operators that act on the chromosomes. In the following subsections these three operators are reviewed.

Reproduction The heart of this operation is the objective function or the so called 'fitness function'. The fitness function is a mapping from the chromosome domain into **R**. This function evaluates a chromosome, a candidate set of cluster defining membership functions, grading it on the following aspects and returns a weighted sum of them. These aspects are:

- The success in maximizing the sum defined in equation (1). So this is a component of the evaluation criteria which directly the square-sum value defined in 1. Since a maximization is aimed, the bigger this number, the more successful is that chromosome.
- Avoiding super-cluster formation. It is observed that local minima exists around the solutions where one cluster takes all feature vectors and leave the others with nothing. This is certainly not a good solution because we seek N clusters and not one and putting all in a bag is the trivial solution. Therefore, in order to enhance the convergence process we introduce a *penalty* for letting clusters remain 'underdeveloped'. A chromosome which leaves a cluster without a minimal (set as a parameter of the GA) number of feature members will pay this penalty.

The reproduction operator first evaluates all the chromosomes of a new generation and then replaces a certain proportion of the worst with the bests of the previous generation (provided that the replacing ones are better than the replaced ones).

Crossover The mating is exhaustive, i.e. there remains no individual chromosome that does not enter the crossover process in a generation. The mating is binary, without sex and absolutely stochastic. After couples are formed an n-point crossover is performed. The position of the n crossover points are totally determined randomly among gene boundaries. The value of n is among the dynamic parameters of the GA, set by the user before the run. Each couple will produce two off-springs. All of these off-springs form the new generation on which the reproduction operator will act. Hence the size of the population is invariant under the crossover operation.

If Γ and Γ' are two chromosomes that are mated we formally define the crossover operator which will produce the two offsprings $\tilde{\Gamma}$ and $\tilde{\Gamma}'$ as

$$\tilde{\gamma}_i = \begin{cases} \gamma_i \& \bar{\eta}_i \mid \gamma_i' \& \eta_i \text{ if } i \in I \\ \gamma_i \quad\quad\quad\quad \text{otherwise} \end{cases} \quad\quad \tilde{\gamma}'_i = \begin{cases} \gamma_i' \& \bar{\eta}_i \mid \gamma_i \& \eta_i \text{ if } i \in I \\ \gamma_i' \quad\quad\quad\quad \text{otherwise} \end{cases}$$

where

$$\Gamma = \langle\, \gamma_i \mid i = 1, 2, \ldots, M \,\rangle \quad\quad \Gamma' = \langle\, \gamma_i' \mid i = 1, 2, \ldots, M \,\rangle$$
$$I_i = \{\, k \mid k_{i-1} \le k < k_i, \ k \in \mathbf{N} \,\} \quad \text{where } k_1 = 1, \ k_i = \mathrm{Rnd}(\{2, 3, \ldots, M\}) \ni$$
$$k_{i-1} < k_i, \ i = 2, 3, \ldots, n$$
$$I = \bigcup_{even(i)} I_i$$
$$\eta_i = \mathrm{Rnd}(\{x \mid 0 \le x \le 255, \ x \text{ has } R \text{ bits set}\})$$

Here R is a GA engine parameter, set by the user. M is $3dN$, namely the count of genes in a chromosome. Furthermore $\&$ and $\mid$ denote bitwise-and and bitwise-or operations, respectively.

Mutation Mutation is a probabilistic choice of k chromosomes of the pool, and performing a totally random alternation of the genes at m points. The value of k and m are among the dynamics of the GA and is fixed by the user before the run. A mutation operator that will act on an individual chromosome Γ which is chosen for mutation and produce the off-spring $\tilde{\Gamma}$ is defined as

$$\tilde{\gamma}_i = \begin{cases} \mathrm{Rnd}(\alpha_i) \text{ if } i \in J \\ \gamma_i \quad\quad\quad \text{otherwise} \end{cases}$$

Where

$$\alpha_i \equiv \text{Allele of gene position}(i)$$
$$J = \{\, k_i \mid k_i = \mathrm{Rnd}(\{1, 2, \ldots, k\}), \ i = 1, 2, \ldots, m \,\}$$

$\tilde{\Gamma}$ replaces Γ in the population.

2.2   Genetic Engine

The GA engine has the standard approach.An outline of the engine and some typical parameter values are as follows:

- Take in the user defined configuration parameters from command line.
- Read in the feature vectors
  (≈ 150 − 250 depending on the test case)
- Generate a random population of chromosomes and evaluate it.
  (Pool size = 100 chromosomes)
  Chromosome length: M ≈ 25 − 125 (genes))

Repeat :
- Mutate.
  (once per ≈ 200 gene exchange (normal case) or if fitness ≤ 0.5 (strong case))
- Crossover.
  (n = 10 crossover points, random position, R = 4)
- Reproduction.
  (Keep ratio= 6%)
- Display/Record performance result.
- If it was not the last generation the user demanded,
  goto Repeat.

Two sets of World wide known data have been used as the test bed of the proposed method. All the data were floating point numbers and were mapped to the $[0.0, 1.0]$ range by a linear transformation in the read in phase. Engine wise the data specifications are tabulated in Table 1.

Table 1. Settings of GA Dynamics for Test Data

TEST DATA	FV DIM. d	FV COUNT P	CLUSTER COUNT N	CHROM.LEN. M	MIN CLUSTER/ PENALTY
iris	4	150	3	24	40/0.4
wine	13	178	3	117	45/0.4

3   Time and Space Complexity

It is known that the GA have $\mathcal{O}(n)$ time complexity with respect to chromosome length. But there are more than one basic factors that contribute to the complexity. In Table 2 a tabulated analysis of the time complexity of the implementation is made in terms of the various dynamics of the program.

d : Dimension of the feature vectors.
N : Count of distinct clusters (to be formed).
T : Count of feature vectors.
P : Count of chromosomes in the pool.
n : Count of crossover points per chromosome (n-point crossover).
m : Count of mutation points per chromosome.
k : Count of chromosomes picked for mutation in a generation cycle.

Table 2. Time Complexity

GA PHASE	COMPLEXITY
initialization	$\mathcal{O}(d \times N \times P + T)$
mutation (normal)	$\mathcal{O}(k \times m)$
mutation (strong)	$\mathcal{O}(d \times N \times P)$
crossover	$\mathcal{O}(n \times P)$
evaluation	$\mathcal{O}(d \times N \times P \times T)$
reproduction	$\mathcal{O}(P \times \lg P)$

The overall time complexity for each generation cycle is the sum of all those in the table. It is aslo worth to mention that teh constants which will multiply the $\mathcal{O}()$ values are considerably large and may vary also row wise in the table above (this is so due to hardware and implementation environment differences).

As far as the space complexity is concerned the answer is simple. The phases of the implemented genetic algorithm does not exhibit changes in space requirements. It is not of any dynamic nature. The space complexity for all the program is $\mathcal{O}(d \times N \times P + T)$.

4   Results & Discussions

A front–end that is developed as part of the implementation displays rowwise the cluster membership functions. The ticks on the axises mark 1.0. The left-most graph displays the membership function discovered that will act on the first component of a feature vector. The next in the row is a similar membership function that is devised to act on the second component of a feature vector. All membership functions in a row contribute to the decision of being in the 'same cluster X' in an additive manner. Below (Figure 2) is the best solution generated by the GA engine for the `iris` data.

This solution has a 0.98 fitness (out of 1.00) and clustered the data 94% correctly. This correctness is measured according to the cluster information provided with the data.This information has never been used in any phase of the solution finding process.

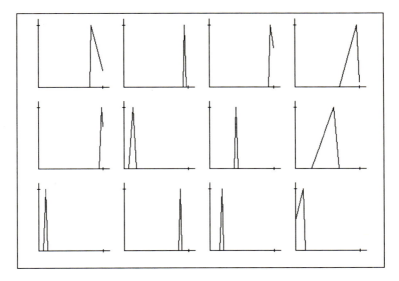

Fig. 2. Best solution for the `iris` data

The usual convergence behavior of the evolution of a GA solution is exactly observed. Figure 3 displays the fitness value versus the generation graph. The curve with higher values is the fitness of the best in the population. Below this is the average fitness value of the population plotted.

4.1   Discussions of the Results

The first data, namely `iris` was chosen because of its widely usage as a test bed. In [11] an Integrated Adaptive Fuzzy Clustering (IAFC) algorithm is presented. This work claims fewer misclassifications then other fuzzy and neuro-fuzzy clustering algorithms (such as FCM (15 misclassifications) and FKCN (10-13 misclassifications)) for `iris`. The best solution among various trials of this algorithm claims 7 misclassifications. The enhancement is presented in [12] and claimed that it outperforms FCMA on a set of data (which includes `iris` as well) does not report any improvement on misclassification ratio, but claims it reaches same convergence FCMA a does, in 65% CPU-time. In [9], an extensive study of FCMA is made. The authors enhance the algorithm by making use of various distance measures (Euclidean, Minkowski, Hamming, Maximum, Mahalanobis, etc.) Their claim is that, for `iris`, FCMA (with Euclidean distance) yields 15 misclassifications where their enhanced method yields 6 misclassifications.

The proposed approach is original in finding a vector of membership functions and using GA for this. For `iris` it results in 9 misclassifications. With this result it outperforms standard FCMA. It is very likely that through some fine tunings

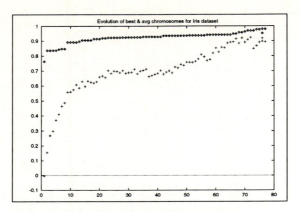

Fig. 3. Fitness value versus generation graph.

a further performance increase will be observed. We have plans to extend this work in this direction.

5   Conclusion

In this area of application, the combination of two methods inspired from nature, namely Fuzziness and Evolution is possible. Some attempts towards this idea have been made in the literature. Though the results are promising, the subject is far from being thoroughly explored. It appears that much work can and needs to be done as far as *Genetic Fuzzy Clustering* (GFCL) is concerned.

The work has proved itself to be a good and new solution for GFCL. The results were promising and quantitatively have outperformed *Fuzzy C-means*, the most common method in the field.

As for the future work Genetic Engine may be improved to avoid local minima solutions, much broader spectrum of membership functions can be implemented.

References

1. R.E. Bellman, R.A. Kalaba, and L.A. Zadeh.: Abstraction and pattern classification. *Journal of Math. Anal. Appl.*, 13:1–7, 1966.
2. I. Gitman and M. Levine.: An algorithm for detecting unimodal fuzzy sets and its application as a clustering technique. *IEEE Trans. Computers*, 19:917–923, 1970.
3. E.H. Ruspini.: A new approach to clustering. *Information and Control*, 15:22–32, 1969.
4. J. Dunn.: A fuzzy relative of the isodata process and its use in detecting compact well-seperated clusters. *Journal of Cybernetics*, 3:32–57, 1974.
5. J. Bezdek and R. Hathaway.: Optimization of fuzzy clustering criteria using genetic algorithms. In *Proc. of First IEEE Conf. on Evolutionary Computation*, number 589-594, Orlando, 1994.

6. L.O. Hall, J.C. Bezdek, S. Boggavarpu, and A. Bensaid.: Genetic fuzzy clustering. In *Int.Proc. North American Fuzzy Information Processing Society Biannual Conference (NAFIPS'94)*, pages 411–415, San Antonio, 1994.

7. J. Liu and W. Xie.: A genetic-based approach to fuzzy clustering. In *Proc. of Fourth IEEE Int. Conf. on Fuzzy Systems*, pages 2233–2237, Yokohama, 1995.

8. T.V. Le.: Evolutionary fuzzy clustering. In *Second IEEE Conf. on Evolutionary Computation*, volume 2, Perth, 1995.

9. B. Yuan, G.J. Klir, and J.F. Swan-Stone.: Evolutionary fuzzy C-Means clustering algorithm. In *Proc. Fourth IEEE Int. Conf. on Fuzzy Systems*, pages 2221–2226, 1995.

10. R. Srikanth, R. George, N. Warsi, D. Prabhu, F.E. Petry, and B.P. Buckles.: A variable-length genetic algorithm for clustering and classification. *Pattern Recognition Letters*, 16:789–800, 1995.

11. Y.S. Kim and S. Mitra.: Integrated adaptive fuzzy clustering algorithm. In *Proc. of Int. Conf. on Fuzzy Systems*, pages 1264–1268, San Francisco, March 1993.

12. M.S. Kamel and S.Z. Selim.: A relaxation approach to the fuzzy clustering problem. *Fuzzy Sets and Systems*, 61:177–188, 1994.

Section V:

Knowledge Discovery and Data Mining

A Strategy for Increasing the Efficiency of Rule Discovery in Data Mining

David McSherry

School of Information and Software Engineering, University of Ulster,
Coleraine BT52 1SA, Northern Ireland

Abstract. Increasing the efficiency of rule discovery is currently a major focus of research interest in data mining. Strategies available to the data miner include data sampling, knowledge-guided discovery, attribute reduction, parallelisation of the discovery process, and focusing on the discovery of a restricted class of rules, or those which appear most promising according to some measure of rule interest. This paper presents a new approach which combines the strategies of focusing on rules which appear most interesting, exploiting structural features of the data set when possible, and decomposition of the discovery process into sub-tasks which can be executed independently on parallel processsors.

1   Introduction

Often in data mining the objective is the discovery of classification rules, for example for assessing the creditworthiness of customers, predicting their loyalty to a certain product, or increasing the likelihood of their retention [9, 10, 6]. A typical classification rule has one or more conditions on the left-hand side (LHS), and a single outcome class on the right-hand side (RHS). A probabilistic rule states the probability of an outcome class given its conditions, while an exact rule holds for all instances in the data set. Given the enormous size of many business and financial databases, it is not surprising that increasing the efficiency of rule discovery is currently a major focus of research interest in data mining.

Strategies available to the data miner include data sampling [2], knowledge-guided discovery [6], attribute reduction [13], parallelisation of the discovery process [12], and focusing on the discovery of a restricted class of rules, or those which appear most promising according to some measure of rule interest [7, 11]. For example, KID3 focuses on the discovery of exact, or almost exact, rules [7], while in ITRULE, the data miner can limit the number of conditions in a discovered rule [11]. Measures of rule interest include the information theoretic J-measure used in ITRULE and the Piatetsky-Shapiro (PS) measure:

$$PS(E, H) = p(E)(p(H|E) - p(H)),$$

where E and H are the LHS and RHS of a discovered rule [7].

X. Liu, P. Cohen, M. Berthold (Eds.): "Advances in Intelligent Data Analysis" (IDA–97)
LNCS 1280, pp. 397–408, 1997.   © Springer–Verlag Berlin Heidelberg 1997

This paper presents a new approach to increasing the efficiency of rule discovery which combines the strategies of focusing on rules which appear most interesting, exploiting structural features of the data set when possible, and decomposition of the discovery process into sub-tasks which can be executed independently on parallel processsors. Section 2 describes an algorithm for rule discovery from data sets in which each outcome class is represented by a single instance in the data set [3]. In Section 3, the rule discovery process is generalised to produce a new algorithm for rule discovery from any data set. The results of experiments with two well-known data sets are reported in Section 4.

2   A Special Case of Data Mining

A special case of data mining arises when each outcome class is a unique individual or object, and is represented by a single instance in the data set. In this case, the objective may be to discover rules for identifying individuals or objects from their database descriptions, or extract features from the data which characterise a selected object or individual. Potential applications include characterising products, investments or designs (successful or otherwise), biological classification, and image recognition.

A data set in which each object is represented by a single instance, like the example data set in Table 1, will be called *heterogeneous*.

Table 1. A heterogeneous data set [3]

Body Cover	Eats Meat	Gives Milk	Lays Eggs	Flies	Animal
fur	yes	yes	no	no	tiger
hair	no	yes	no	no	horse
fur	no	yes	no	no	rabbit
feathers	yes	no	yes	no	penguin
fur	yes	yes	no	yes	bat
feathers	no	no	yes	yes	pigeon
fur	yes	yes	yes	no	platypus
feathers	yes	no	yes	yes	eagle

Although existing algorithms can be applied to heterogeneous data sets, the efficiency of rule discovery can be greatly increased by a discovery process which takes account of their unusual structure. As the following theorem shows, the PS measure of rule interest takes a simpler form, and all exact rules $E \rightarrow H$ are of equal interest in a heterogeneous data set [3].

Theorem 1. *For any feature E of an object H in a heterogeneous data set, $PS(E, H) = (n - f(E))/n^2$, where n is the size of the data set and $f(E)$ is the frequency of E in the data set. The maximum possible value of $PS(E, H)$ is $(n - 1)/n^2$, and occurs if and only if E is unique to H in the data set.*

Proof. Since E is a feature of H and the data set is heterogeneous,

$$
\begin{aligned}
PS(E, H) &= p(E)(p(H|E) - p(H)) \\
&= \frac{f(E)}{n} \left(\frac{1}{f(E)} - \frac{1}{n} \right) \\
&= \frac{1}{n} - \frac{f(E)}{n^2} \\
&= \frac{n - f(E)}{n^2}.
\end{aligned}
$$

The maximum possible value occurs when $f(E) = 1$, a condition which is satisfied if and only if E is unique to H in the data set.   □

It follows from Theorem 1 that the features E of a given object H which maximise $PS(E, H)$ are those which are *distinctive* features of H according to the following definition.

Definition 2. In a heterogeneous data set, a feature E of an object H will be called a distinctive feature of H if no other feature of H is shared by fewer objects in the data set.

From Table 1, for example, (body cover = fur) and (lays eggs = yes) are both distinctive features of *platypus*, while (flies = yes) is the only distinctive feature of *bat*. A goal-driven algorithm for rule discovery from a heterogeneous data set, called Aurum-1, is outlined in Fig. 1. Given a target object H, the algorithm's goal is the discovery of exact rules with one or more conditions on the LHS and H on the RHS. Starting with any distinctive feature of H as the first condition of a candidate rule, it generates a sequence of nested subsets of the data set in search of additional conditions, each of which must be a distinctive feature of H in the current subset. All the discovered rules are therefore *characteristic* in the sense of the following definition.

Definition 3. In a heterogeneous data set, an exact rule $E_1, E_2, ..., E_n \rightarrow H$ will be called a characteristic rule if E_1 is a distinctive feature of H in the data set and for $2 \leq k \leq n$, E_k is a distinctive feature of H in the subset of the data set consisting of all objects with features $E_1, E_2, ..., E_{k-1}$.

Some of the characteristic rules discovered by Aurum-1 from the example data set with *penguin*, *bat* and *platypus* as the target objects were:

Rule 1.   if (body cover = feathers) and (flies = no) then penguin
Rule 2.   if (flies = yes) and (gives milk = yes) then bat
Rule 3.   if (lays eggs = yes) and (gives milk = yes) then platypus
Rule 4.   if (body cover = fur) and (lays eggs = yes) then platypus

To discover characteristic rules for a target object H in a heterogeneous data set X :

for each distinctive feature of H in X **do**
begin
      form a subset S of X consisting of all objects with this feature
      if H is the only object in S
          then features on the path from X to S are conditions of a characteristic rule
          else if there are no more attributes
                then discard the current path
                else recursively apply this procedure to S
end

Fig. 1. Algorithm for the discovery of characteristic rules from a heterogeneous data set

Although Rule 1 may not be valid in a larger data set, Rule 2 accurately characterises the bat, as it is the only flying mammal in the animal kingdom. Similarly, the platypus is accurately characterised by Rule 3 as it is the only mammal known to lay eggs. Although Aurum-1 discovered all valid rules in the example data set, the existence of an interesting rule does not guarantee its discovery. For example, Rule 3 may not be discovered in a larger data set in which (lays eggs = yes) is no longer a distinctive feature of *platypus*.

3   Generalising the Rule Discovery Process

In this section, the discovery process in Aurum-1 is generalised and extended to produce a new algorithm called Aurum-2 for rule discovery from any data set. By Theorem 1, a distinctive feature of an object H in a heterogeneous data set is one for which the PS measure with respect to H is maximum. The concept can therefore be generalised as in the following definition, thus extending the definition of a characteristic rule to an arbitrary data set.

Definition 4. In any data set, an attribute value E which maximises $PS(E, H)$ for a given outcome class H will be called a distinctive feature of H.

An important difference is that while $PS(E, H)$ is maximum for any feature E which uniquely identifies an object H in a heterogeneous data set, it is not necessarily maximum for a feature E which is unique to an outcome class H in an arbitrary data set. Thus an attribute value which is unique to an outcome class H in an arbitrary data set is not necessarily a distinctive feature of H. Nevertheless, such attribute values, and the corresponding rules, may be of considerable *qualitative* interest [4]. To ensure they are not overlooked in Aurum-2, the scope of the discovery process must be extended to include the discovery of rules other than characteristic rules.

Aurum-2, like its predecessor, is goal driven. It can be applied with an outcome class selected by the data miner as the target outcome class, or with each outcome class taking its turn as the target outcome class. Given a target outcome class H, it selects an attribute value which maximises the PS measure with respect to H. Such an attribute value is, by definition, a distinctive feature of H. If more than one distinctive feature of H is found, one of them is arbitrarily selected. For any other attribute value which is unique to H in the data set, a corresponding rule is added to the list, initially empty, of discovered rules.

A subset of the data set is now formed consisting of all instances with the selected feature and the discovery process is recursively applied to the subset. In this way, a sequence of nested subsets of the data set is constructed. Before each new subset is formed, any attribute value which is unique to the target outcome class in the current subset is identified, and a corresponding rule is added to the list of discovered rules. The process continues until a subset containing only the target outcome class is reached or no further attributes remain. In the former case, features on the path from the original data set to the subset containing only the target outcome class provide the conditions of a rule which is characteristic for H.

All attributes involved in the rules discovered from the first sequence of subsets are now eliminated and a new sequence of subsets is constructed, starting with the subset generated by the most interesting of the remaining attribute values, which need not be a distinctive feature of the target outcome class. On completion of the second sequence, attributes involved in any additional rules discovered are also eliminated and a third sequence is generated if possible. This cycle of subset generation, rule discovery, and attribute elimination is repeated until no further attributes remain.

Aurum-2 is not strictly a generalisation of Aurum-1 and may not produce the same results when applied to a heterogeneous data set. Unlike Aurum-1, it does not focus exclusively on the discovery of characteristic rules and is not guaranteed to discover all such rules. When applied to the data set in Table 1, for example, it discovered only 11 of the 17 rules discovered by Aurum-1. Another advantage of Aurum-1 is that less computational effort is required to identify distinctive features as defined for a heterogeneous data set. Thus by exploiting the unusual structure of a heterogeneous data set, Aurum-1 increases the efficiency of rule discovery.

While Aurum-1 remains the more attractive of the two algorithms when the data set is heterogeneous, only Aurum-2 can be applied in the absence of heterogeneity. The two algorithms have been incorporated as alternative strategies in a single algorithm for rule discovery called Aurum. Heterogeneity in the data set is automatically detected by Aurum, and determines its choice between the alternative strategies as follows:

if data set is heterogeneous **then** apply Aurum-1 **else** apply Aurum-2

In both Aurum-1 and Aurum-2, the search for rules with each outcome class as the target outcome class is an independent sub-task of the discovery process.

Their goal-driven approach therefore provides a natural basis for decomposition of the discovery process into sub-tasks, one for each outcome class, which can be executed independently on parallel processors if required. Even without parallelisation, both algorithms are computationally inexpensive compared, for example, with algorithms like ID3 [8] which recursively *partition* the data set, creating at each node a subset corresponding to every value of a selected attribute.

4   Experimental Results

In this section, two well-known data sets from the UCI repository of machine learning data sets [5], neither of which is heterogeneous, are used to illustrate the discovery process in Aurum-2.

4.1   The Contact Lens Data

The contact lens data set is based on a simplified version of the optician's real-world problem of selecting a suitable type of contact lenses, if any, for an adult spectacle wearer [1]. Outcome classes in the data set are no contact lenses, soft contact lenses, and hard contact lenses. The attributes are age of patient, astigmatism, tear production rate, and spectacle prescription. Table 2 shows the frequencies of the outcome classes and attribute values in the data set.

When Aurum-2 is applied to the contact lens data, the discovery process begins with no contact lenses, the likeliest outcome class, as the target outcome class. Initially, the only attribute values in the data set which can increase the probability of the target outcome class are (age = presbyopic), (tear production rate = reduced), (astigmatism = present), and (spectacle prescription = hypermetrope). One of these attribute values must be the one which maximises the PS measure with respect to no contact lenses. For example,

$$PS(\text{age} = \text{presbyopic}, \text{no contact lenses}) = \frac{8}{24}\left(\frac{6}{8} - \frac{15}{24}\right) = 0.04.$$

Similarly, the PS values for (tear production rate = reduced), (astigmatism = present), and (spectacle prescription = hypermetrope), with respect to no contact lenses, are 0.19, 0.02, and 0.02. The only distinctive feature of no contact lenses is therefore (tear production rate = reduced). Since no other attribute value is unique to the target outcome class, no rules can be discovered before the subset corresponding to the selected feature is generated. As no contact lenses is the only surviving outcome class in this subset, a characteristic rule for no contact lenses is discovered:

> Rule 1. if tear production rate = reduced
>   then no contact lenses

As the only attribute involved in the discovered rule, tear production rate is now eliminated and the discovery process is repeated for the remaining attributes. With no contact lenses still as the target outcome class, the attribute

Table 2. Frequencies of outcome classes and attribute values in the contact lens data

	None (15)	Soft (5)	Hard (4)		None (15)	Soft (5)	Hard (4)
Age of patient				*Astigmatism*			
young	4	2	2	present	8	0	4
pre-presbyopic	5	2	1	absent	7	5	0
presbyopic	6	1	1				
Tear production rate				*Spectacle prescription*			
normal	3	5	4	myope	7	2	3
reduced	12	0	0	hypermetrope	8	3	1

Heading above columns: Contact lens type

value which now maximises the PS measure is (age = presbyopic). This attribute value, though not a distinctive feature of no contact lenses, is used to generate the first subset in a new sequence of nested subsets.

Table 3. Frequencies for astigmatism and spectacle prescription in the subset of the contact lens data [1] with (age = presbyopic)

	None	Soft	Hard		None	Soft	Hard
Astigmatism				*Spectacle prescription*			
present	3	1	0	myope	3	1	0
absent	3	0	1	hypermetrope	3	0	1

Heading above columns: Contact lens type

Table 3 shows the frequencies for astigmatism and spectacle prescription in the new subset. As it happens, none of the attribute values in the subset can change the probability (currently 0.75) of the target outcome class, so all are equally uninteresting according to the PS measure. Nor is any attribute value unique to no contact lenses in the current subset.

Table 4. Frequencies for astigmatism in the subset of the contact lens data [1] with (age = presbyopic) and (spectacle prescription = myope)

Contact lens type:	None	Soft	Hard
Astigmatism			
present	1	1	0
absent	2	0	0

Table 4 shows the resulting frequencies for astigmatism when (spectacle prescription = myope) is arbitrarily selected to generate the next subset in the nested sequence. As the only remaining attribute value for which the PS measure is positive with respect to no contact lenses, (astigmatism = absent) is now selected to generate the final subset. Since no contact lenses is the only surviving outcome class in this subset, another rule is discovered:

> Rule 2. if age of patient = presbyopic
> and spectacle prescription = myope
> and astigmatism = absent
> then no contact lenses

When the attributes involved in Rule 2 are eliminated, no further attributes remain, so the search for rules with no contact lenses as the target outcome class terminates. The discovery process now continues with soft contact lenses as the target outcome class and with all attributes in the data set once again available. Initially, (tear production rate = normal) and (astigmatism = absent) are equally interesting with respect to soft contact lenses, and more interesting than any other attribute value. Fig. 2 shows the sequence of nested subsets generated by Aurum-2 when (tear production rate = normal) is arbitrarily selected from the two distinctive features of soft contact lenses. The surviving outcome classes in each subset, and their frequencies, are also shown.

Only two outcome classes survive in the subset with (tear production rate = normal) and (astigmatism = absent). Although the attribute value selected to generate the next subset in the nested sequence is (spectacle prescription = hypermetrope), two values of age are unique to soft contact lenses in the current subset. As indicated by the dashed lines in Fig. 2, the following rules are therefore discovered:

> Rule 3. if tear production rate = normal
> and astigmatism = absent
> and age = young
> then soft contact lenses

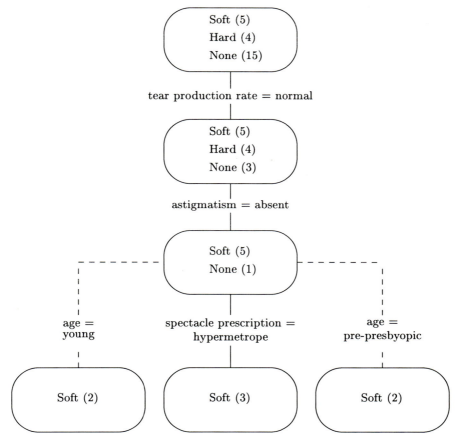

Fig. 2. In Aurum-2, a single sequence of nested subsets is enough to discover all the rules for soft contact lenses from the contact lens data [1]

> Rule 4. if tear production rate = normal
> and astigmatism = absent
> and age = pre-presbyopic
> then soft contact lenses

As soft contact lenses is the only outcome class surviving in the subset generated by (spectacle prescription = hypermetrope), a characteristic rule for soft contact lenses is now discovered:

> Rule 5. if tear production rate = normal
> and astigmatism = absent
> and spectacle prescription = hypermetrope
> then soft contact lenses

Since no further attributes remain when those involved in Rules 3, 4 and 5 are eliminated, the discovery process now continues with hard contact lenses as the target outcome class. The process terminates with the discovery of two rules for hard contact lenses, the second of which is a characteristic rule:

> Rule 6. if tear production rate = normal
> and astigmatism = present
> and age = young
> then hard contact lenses

> Rule 7. if tear production rate = normal
> and astigmatism = present
> and spectacle prescription = myope
> then hard contact lenses

The rules discovered by Aurum-2 from the contact lens data include all the valid rules for soft contact lenses and hard contact lenses. The only rules it failed to discover were two of the valid rules for no contact lenses [1].

4.2   The Voting Records Data

The congressional voting records data [5] collected by Jeff Schlimmer from a 1984 session of the United States Congress, records the party affiliations of congressmen and their votes on 16 budget issues such as education spending, freezing of physician fees, and synthetic fuel funding. When Aurum-2 was applied to the data set with Democrat as the target outcome class, 30 rules were discovered including the characteristic rule:

> if physician fee freeze vote = n
> and budget resolution vote = y
> then Democrat

This is in fact the most strongly supported of all exact rules with one or two conditions in the data set. All the Democrat rules had (physician fee freeze vote = n) as their first condition and were discovered from a single sequence of nested subsets. The most interesting *probabilistic* rule for Democrat with one or two conditions discovered by ITRULE had the same attribute value as its only condition [11].

With Republican as the target outcome class, Aurum-2 discovered 22 rules, the most strongly supported of which was a characteristic rule. Its first two conditions were the same as the conditions in the most interesting probabilistic rule for Republican with one or two conditions discovered by ITRULE. Like the Democrat rules, all the Republican rules were discovered from a single sequence of nested subsets.

5 Conclusions

An efficient algorithm for rule discovery called Aurum-2 has been presented. Its strategy of exploring a sequence of nested subsets in search of a characteristic rule for a target outcome class, combined with opportunistic discovery of additional rules by identifying features which are unique to the target outcome class in each subset, appears to produce a good yield for relatively small computational effort. When applied to the voting records data, in which the number of possible rules is of the order of 8.59×10^9, it had only to explore two short sequences of nested subsets to discover 52 rules.

While Aurum-2 can be applied to any data set, its predecessor Aurum-1 remains the more attractive of the two algorithms when the data set is heterogeneous. As well as guaranteeing the discovery of all characteristic rules, it exploits the unusual structure of a heterogeneous data set to increase the efficiency of rule discovery. The two algorithms have been incorporated as alternative strategies in a single algorithm for rule discovery called Aurum. The goal-driven approach used in both algorithms provides a natural basis for decomposition of the discovery process into sub-tasks which can be executed independently on separate processors if required.

References

1. Cendrowska, J.: PRISM: an algorithm for inducing modular rules. International Journal of Man-Machine Studies **27** (1987) 349–370
2. Frawley, W.J., Piatetsky-Shapiro, G., Matheus, C.J.: Knowledge Discovery in Databases: an Overview. In Piatetsky-Shapiro, G., Frawley, W.J. (eds.) Knowledge Discovery in Databases (AAAI Press, Menlo Park, CA, 1991) 1–27
3. McSherry, D.: An algorithm for the discovery of characteristic rules. Digest No. 96/198 (Institution of Electrical Engineers, London, 1996) 4/1–3
4. McSherry, D.: Qualitative assessment of rule interest in data mining. Proceedings of the Sixteenth Annual Technical Conference of the BCS Specialist Group on Expert Systems, Cambridge, December 1996, 204–215
5. Murphy, P.M., Aha, D.W.: UCI Repository of Machine Learning Databases. http://www.ics.uci.edu/~mlearn/MLRepository.html (1995)
6. Nelson, C.: Improving customer retention with knowledge guided data mining. BCS Specialist Group on Expert Systems Newsletter, No. 33 (1995) 15–20
7. Piatetsky-Shapiro, G.: Discovery, analysis and presentation of strong rules. In Piatetsky-Shapiro, G., Frawley, W.J. (eds.) Knowledge Discovery in Databases (AAAI Press, Menlo Park, CA, 1991) 229–248
8. Quinlan, J.R.: Induction of decision trees. Machine Learning **1** (1986) 81–106
9. Shortland, R.J., Scarfe, R.T.: Data mining applications in BT. BT Technology Journal **12** (1994) 17–22
10. Simoudis, E., John, G., Kerber, R., Livezey, B., Miller, P.: Developing customer vulnerability models using data mining techniques. Proceedings of IDA-95, Baden-Baden, August 1995, 181–185
11. Smyth, P., Goodman, R.M.: Rule induction using information theory. In Piatetsky-Shapiro, G., Frawley, W.J. (eds.) Knowledge Discovery in Databases (AAAI Press, Menlo Park, CA, 1991) 159–176

12. Thompson, S., Bramer, M.A.: Parallel knowledge discovery: a review of existing techniques. Digest No. 96/198 (Institution of Electrical Engineers, London, 1996) 5/1–5

13. Ziarko, W.: Discovery, analysis, and representation of data dependencies in databases. In Piatetsky-Shapiro, G., Frawley, W.J. (eds.) Knowledge Discovery in Databases (AAAI Press, Menlo Park, CA, 1991) 195–209

Intelligent Text Analysis for Dynamically Maintaining and Updating Domain Knowledge Bases

Klemens Schnattinger and Udo Hahn

Computational Linguistics Lab – Text Knowledge Engineering Group
Freiburg University, Werthmannplatz, D-79085 Freiburg, Germany
{schnattinger,hahn}@coling.uni-freiburg.de
http://www.coling.uni-freiburg.de/

Abstract. We propose a knowledge-intensive text analysis approach which deals with the continuous assimilation of new concepts into domain knowledge bases. Text understanding and knowledge acquisition proceed in tandem on the basis of terminological reasoning. Concept learning is considered an evidence-based choice problem the solution of which balances the "quality" of various clues from the linguistic structure of the texts and conceptual structures in the knowledge bases.

1 Introduction

Considering the increasing number of textual databases and their exploding growth rates, the natural language processing (NLP) community is beginning to respond to these new terms of trade. Lexical, statistical approaches to text analysis are inherently restricted to simple document referral services which provide references to relevant documents only [5]. Accounting for factual knowledge, relevant assertions, etc. in these texts requires more sophisticated NLP technology. Corresponding content-oriented efforts at text analysis, however, must somehow cope with the problem of providing sufficiently rich linguistic knowledge sources (grammars, lexicons) and domain knowledge sources (ontologies) in order to cope with the myriad of language phenomena and concepts any NLP system is going to encounter in real-world texts. As hand-coding of complex grammar and domain knowledge is obviously precluded, a growing number of NLP researchers are turning to machine learning methodologies to automatically augment these knowledge sources. Actually, there has been significant progress with regard to learning linguistic knowledge from large corpora of texts and data sets (e.g., [13, 15]). Approaches dealing with the acquisition of (non-toy) domain knowledge bases are much more rare [18, 19, 12, 8]. By this, we mean not only the automatic extraction of knowledge from some source repository (data bases, texts, etc.) but also its coherent integration in a corresponding knowledge base framework. The kind of update and maintenance procedures we aim at include the knowledge items' proper positioning in the concept hierarchy and their

X. Liu, P. Cohen, M. Berthold (Eds.): "Advances in Intelligent Data Analysis" (IDA–97)
LNCS 1280, pp. 409–422, 1997.   © Springer–Verlag Berlin Heidelberg 1997

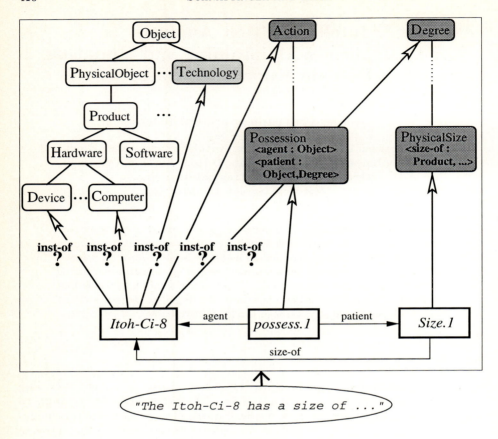

Fig. 1. A Sample Learning Scenario: Initial Setting

linkage with proper attributes, attribute values and value restrictions (integrity constraints).

2   Learning Scenario

In order to illustrate the problem we face, consider the following learning scenario. Suppose, your knowledge of the information technology domain tells you nothing about an *Itoh-Ci-8*. Imagine, one day your favorite technology magazine features an article starting with *"The Itoh-Ci-8 has a size of ..."*. Has your knowledge increased? If so, what did you learn already from just this phrase?

The problem is restated in conceptual terms in Fig. 1. The learning process starts upon the reading of the unknown lexical item *"Itoh-Ci-8"*. In this initial step, the corresponding hypothesis space incorporates all the top level concepts available in the ontology for the new lexical item *"Itoh-Ci-8"*. So, the concept ITOH-CI-8 may be a kind of an OBJECT, an ACTION, a DEGREE, etc. As a consequence of processing the noun phrase *"The Itoh-Ci-8 ... "* as the grammatical subject of the verb *"has"*, ITOH-CI-8 is related via the AGENT role to the ACTION concept POSSESSION. Since POSSESSION, the concept denoted by *"has"*, requires its AGENT to be an OBJECT, ACTION and DEGREE are no longer admitted as valid concept hypotheses. Hence, an already significant reduction of the huge initial hypothesis space is achieved (pruning of hypotheses by strict conceptual constraints is marked by darkly shaded boxes). The learner then aggressively specializes the remaining single hypothesis to the immediate subordinates of OBJECT, *viz.* PHYSICALOBJECT and TECHNOLOGY, in order to test more restricted hypotheses which – according to more specific constraints – are more easily falsifiable.

Up to now, primarily conceptual constraints have been considered. The verb *"has"*, however, relates *"Itoh-Ci-8"* and *"size"* in terms of a POSSESSION relationship based on a semantic interpretation process. The underlying linguistic constraints not only require the grammatical *subject* to be interpreted in terms of the conceptual AGENT role, but also indicate that the grammatical *direct object* relation is to be interpreted in terms of a conceptual PATIENT role. This rule also reflects the conceptual constraint that the AGENT of the POSSESSION action must be an OBJECT (as already discussed), and also requires the PATIENT to be either another OBJECT or a DEGREE. In a second semantic interpretation step, possible conceptual relations between the AGENT and PATIENT are tried. Exploiting the conceptual roles attached to PHYSICALSIZE, the system recognizes that (all specializations of) PRODUCT can be related to the concept PHYSICALSIZE (via the role SIZE-OF), while for TECHNOLOGY no such relation can be established. Based on this observation, we *prefer* the conceptual reading of ITOH-CI-8 as a kind of a PRODUCT over the TECHNOLOGY hypothesis. Since the origin of this computation step lies in heuristic considerations, it is considered a less reliable constraint than a purely conceptual one (pruning of hypotheses motivated by heuristics is marked by lightly shaded boxes). With this decision being taken, and following the aggressive hypothesizing strategy, the immediate subordinates of PRODUCT are selected for further consideration, *viz.* HARDWARE or SOFTWARE.

The next series of learning steps is discussed considering the processing of the phrase *"The DIP switch of the Itoh-Ci-8 ... "* (cf. Fig. 2). Being left with the HARDWARE and SOFTWARE hypothesis for ITOH-CI-8, a semantically unconstrained linking between ITOH-CI-8 and DIP-SWITCH via the relation SWITCH-OF is established, since the genitive construction does not allow more rigid constraints. (Note that any other conceptual role of DIP-SWITCH which is conceptually compatible with the current set of hypotheses for ITOH-CI-8 is linked to ITOH-CI-8, too.) From a conceptual point of view, the only constraint required for DIP-SWITCH is that the conceptual item it can be related to must be a kind of an OUTPUTDEVICE, STORAGEDEVICE, COMPUTER or CENTRALUNIT. Accordingly, the concepts SOFTWARE and DIP-SWITCH (as a transitive subordinate of HARDWARE) can be ruled out. In addition, a conceptual evaluation heuristic

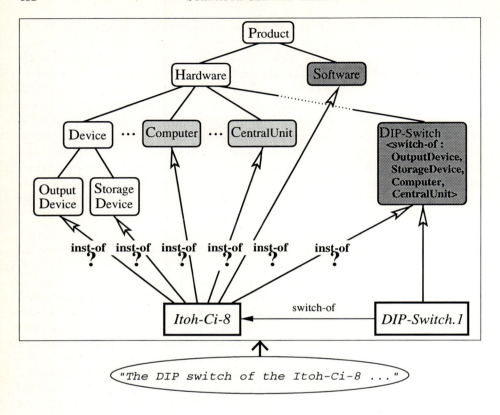

Fig. 2. A Sample Learning Scenario: Ongoing Hypothesis Refinement

indicates that clustered and cohesive areas of the knowledge base (such as DE-
VICE, OUTPUTDEVICE and STORAGEDEVICE) are preferred over non-clustered,
more or less distinct classes (like COMPUTER or CENTRALUNIT). Again, these
preference-based pruning decisions (as opposed to the violation of strict conceptual constraints) are marked by lightly shaded boxes. At this level, we have
narrowed down the set of hypotheses to the concept subhierarchy denoted by
OUTPUTDEVICE and STORAGEDEVICE. We will continue to elaborate on this
example in more technical terms in Section 4.

3   Quality-Based Learning

Generalizing from this scenario, we propose a concept learning methodology that,
based on some initially supplied high-level ontology, uses various linguistic as well
as conceptual clues (e.g., the occurrence as a subject, direct object or as a genitive
vs. the occurrence in a concept cluster) to generate, refine, evaluate and, finally,

select plausible concept hypotheses on the basis of a continuous stream of textual input. The reasoning mechanism in control of the hypothesis management is supplied with knowledge about natural language, knowledge about the domain of discourse, and metaknowledge about how these two knowledge sources can be exploited for making plausible guesses about new concepts while reading a text.

The work reported in this paper is part of a large-scale project aimed at the development of SYNDIKATE, a German-language text knowledge assimilation system. Two real-world application domains are currently under active investigation: test reports on information technology products, mainly chosen for pretest purposes (101 documents with 10^5 words) [8]; and, the major application, medical reports (approximately 120,000 documents with 10^7 words) [10].

In order to break the enormous complexity of our problem we have taken considerable care to design and implement a natural language parser [9, 17] that is *inherently robust* and has various strategies to get nearly optimal results out of deficient, i.e., underspecified knowledge sources in terms of *partial, limited-depth parsing*. The price we pay for this approach is uncertainty (in terms of alternative hypotheses) associated with the kinds of knowledge items we generate as a result of text analysis. To cope with this problem, we build on expressively rich knowledge representation models in the underlying domain [22]. We provide a start-up core ontology (such as the Penman Upper Model [3]) in the format of terminological assertions (for a survey, cf. [24]). Concept hypotheses relating to this domain knowledge emerge on the basis of two types of evidence, *viz.* the type of *linguistic* construction in which an unknown lexical item occurs in a text; and, *conceptually* motivated annotations of concept hypotheses reflecting structural patterns of consistency, mutual justification, analogy, etc. in the text knowledge base. This initial evidence is represented by a set of *quality labels*.

By way of standard terminological reasoning, information as supplied by the current text and the domain knowledge base is used to generate and further refine the concept hypotheses under consideration. By way of terminological *meta*reasoning on quality labels, the credibility and plausibility of these concept hypotheses is continuously evaluated and, finally, the most significant ones are selected. A first selection round for the assessment of hypothesis spaces is based on threshold levels which take mostly linguistic clues into account (cf. the discussion related to the AGENT and PATIENT role of the verb *"has"* in Fig. 1, which illustrates the emergence of the quality label CASEFRAME that yields quite a positive weight for the associated conceptual representation structure). In Section 4 this selection level will be referred to as **TH**. Those hypothesis spaces that have fulfilled the threshold criterion **TH** will then be classified relative to different credibility levels which take only conceptual quality labels into account (cf. the discussion related to DEVICE in Fig. 2, which illustrates the emergence of the label MULTIPLYDEDUCED, one that yields a lot of credit for the hypothesis it is related to). In Section 4 this selection level will be referred to as **CB**.

Terminological reasoning and terminological metareasoning are formally combined in the so-called *qualification calculus* [20]. It contains a reification schema for terminological assertions, a translation schema for mapping first-order terminological assertions to second-order ones, and *vice versa*; and, in particular, it contains a terminological classifier extended by an evaluation metric for quality-based selection criteria, itself fully embedded in a terminological language [21], which determines the most credible concept hypotheses.

4 Evaluation

In this section, we present some data from an empirical evaluation of the quality-based concept aquisition system. We focus here on the issues of learning accuracy and the learning rate. Due to the given learning environment, the measures we apply deviate from those commonly used in the machine learning community. In concept learning algorithms like IBL [1] there is no hierarchy of concepts. Hence, any prediction of the class membership of a new instance is either true or false. However, as such hierarchies naturally emerge in terminological frameworks, a prediction can be more or less precise, i.e., it may approximate the goal concept at different levels of specificity. This is captured by our measure of *learning accuracy* which takes into account the conceptual distance of a hypothesis to the goal concept of an instance,[1] rather than simply relating the number of correct and false predictions, as in IBL.

In our approach, learning is achieved by the refinement of *multiple* hypotheses about the class membership of an instance. Thus, the measure of *learning rate* we propose is concerned with the reduction of hypotheses as more and more *information* becomes available about one particular new instance. In contrast, IBL-style algorithms consider only one concept hypothesis per learning cycle and their notion of *learning rate* relates to the increase of correct predictions as more and more *instances* are being processed.

We considered a total of 101 texts taken from a corpus of information technology magazines. For each of them 5 to 15 learning steps were considered. A learning step is operationalized here by the representation structure that results from the semantic interpretation of an utterance which contains the unknown lexical item. In order to clarify the input data available for the learning system, consider Table 1. It consists of nine single learning steps for the unknown item *"Itoh-Ci-8"* already discussed in Section 1. Each learning step is associated with a particular natural language phrase in which the unknown lexical item occurs[2] and the corresponding semantic interpretation in the text knowledge base (the interpretation data also contain an annotation of the type of syntactic construction in which the unknown item occurred).

4.1 Learning Accuracy

In a first series of experiments, we investigated the *learning accuracy* of the system, i.e., the degree to which the system correctly predicts the concept class

---

[1] [11] argue for the adequacy of conceptual distance measures for network-based knowledge representation structures under certain ontological engineering assumptions.

[2] Note that our text database consists of German language data. The translations we provide give only rough English correspondences.

Step	Phrase	Semantic Interpretation
1.	*Ithoh-Ci-8 has*	(CaseFrame,possess.1,agent,Itoh-Ci-8)
2.	a *size* of ..	(CaseFrame,possess.1,patient,Size.1)
3.	*DIP switch* of the *Itoh-Ci-8* ..	(GenitiveNP,Itoh-Ci-8,has-switch,DIP-Switch.1)
4.	*case* from the *Itoh-Ci-8* ..	(PP-Attach,Itoh-Ci-8,has-case,Case.1)
5.	*Itoh-Ci-8* with a *main memory* ..	(PP-Attach,Itoh-Ci-8,has-memory,MainMemory.1)
6.	*Itoh-Ci-8's LED lines* ..	(GenitiveNP,Itoh-Ci-8,has-part,LED-Line.1)
7.	*Itoh-Ci-8's toner supply* ..	(GenitiveNP,Itoh-Ci-8,has-part,TonerSupply.1)
8.	*paper cassette* of the *Itoh-Ci-8* ..	(GenitiveNP,Itoh-Ci-8,has-part,PaperSupply.1)
9.	*Itoh-Ci-8* with a *resolution rate* ..	(PP-Attach,Itoh-Ci-8,has-resolution,Resolution.1)

Table 1. Parse Fragments of a Text Featuring *Itoh-Ci-8*

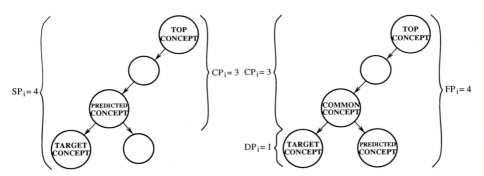

Fig. 3. LA Configuration for an Under-specified Concept Hypothesis

Fig. 4. LA Configuration for a Slightly Incorrect Concept Hypothesis

which subsumes the target concept under consideration (the *target* being the new item to be learned). Learning accuracy (LA) is here defined as (n being the number of concept hypotheses for a single target):

$$LA := \sum_{i \in \{1 \ldots n\}} \frac{LA_i}{n} \quad \text{with} \quad LA_i := \begin{cases} \dfrac{CP_i}{SP_i} & \text{if } FP_i = 0 \\ \dfrac{CP_i}{FP_i + DP_i} & \text{else} \end{cases}$$

SP_i specifies the length of the *shortest path* (in terms of the number of nodes being traversed) from the TOP node of the concept hierarchy to the maximally specific concept subsuming the instance to be learned in hypothesis i; CP_i specifies the length of the path from the TOP node to that concept node in hypothesis i which is *common* both for the shortest path (as defined above) and the actual

Learning Hypotheses	LA −	LA TH	LA CB	Learning Hypotheses	LA −	LA TH	LA CB
PHYS.OBJ.(176)	0.30	0.30	0.30	PRODUCT(136)	0.50	0.50	0.50
MENT.OBJ.(0)	0.16	0.16	0.16	MENT.OBJ.(0)	0.16		
INF.OBJ.(5)	0.16	0.16	0.16	INF.OBJ.(5)	0.16		
MASSOBJ.(0)	0.16	0.16	0.16	MASSOBJ.(0)	0.16		
NORM(3)	0.16	0.16	0.16	NORM(3)	0.16		
TECHNOLOGY(1)	0.16	0.16	0.16	TECHNOLOGY(1)	0.16		
MODE(5)	0.16	0.16	0.16	MODE(5)	0.16		
FEATURE(0)	0.16	0.16	0.16	FEATURE(0)	0.16		
	ϕ:0.18	ϕ:0.18	ϕ:0.18		ϕ:0.21	ϕ:0.50	ϕ:0.50
Learning step 1				**Learning step 2**			
COMPUTER(5)	0.50	0.50		NOTEBOOK(0)	0.43	0.43	
				PORTABLE(0)	0.43	0.43	
				PC(0)	0.43	0.43	
				WORKSTATION(0)	0.43	0.43	
				DESKTOP(0)	0.43	0.43	
OUTPUTDEV.(9)	0.80	0.80	0.80	PRINTER(3)	0.90	0.90	0.90
				VISUALDEV.(2)	0.66	0.66	0.66
				LOUDSPEAKER(0)	0.66	0.66	0.66
				PLOTTER(0)	0.66	0.66	0.66
STORAGEDEV.(5)	0.58	0.58	0.58	RW-STORE(2)	0.50	0.50	0.50
				RO-STORE(1)	0.50	0.50	0.50
SCANNER(0)	0.50	0.50	0.50	SCANNER(0)	0.50	0.50	
CENTRALU.(0)	0.58	0.58	0.58	CENTRALU.(0)	0.58	0.58	
	ϕ:0.59	ϕ:0.59	ϕ:0.62		ϕ:0.55	ϕ:0.55	ϕ:0.65
Learning step 3				**Learning step 4**			

Table 2. Some Concept Learning Results for a Text Featuring *Itoh-Ci-8*

path to the predicted concept (whether correct or not); FP_i specifies the length of the path from the TOP node to the predicted (in this case *false*) concept and DP_i denotes the node *distance* between the predicted node and the most specific concept correctly subsuming the target in hypothesis i, respectively. Figures 3 and 4 depict sample configurations for concrete LA values involving these parameters. Fig. 3 illustrates a correct, yet too general prediction with $LA_i = .75$, while Fig. 4 contains an incorrect concept hypothesis with $LA_i = .6$.

Given the measure for learning accuracy, Table 2 and Table 3 illustrate how the various learning hypotheses for ITOH-CI-8 develop in accuracy from one step to the other. The numbers in brackets in the column **Learning Hypotheses** indicate for each hypothesized concept the number of concepts subsumed by it in the underlying knowledge base; **LA CB** gives the accuracy rate for the full qualification calculus including threshold and credibility criteria, **LA TH** for threshold criteria only, while **LA −** depicts the accuracy values produced by the terminological reasoning component without incorporating the qualification calculus. As can be seen from Table 2 and 3, the full qualification calculus pro-

Learning Hypotheses	LA –	LA TH	LA CB	Learning Hypotheses	LA –	LA TH	LA CB
NOTEBOOK(0)	0.43	0.43		NOTEBOOK(0)	0.43	0.43	
PORTABLE(0)	0.43	0.43		PORTABLE(0)	0.43	0.43	
PC(0)	0.43	0.43		PC(0)	0.43	0.43	
WORKSTATION(0)	0.43	0.43		WORKSTATION(0)	0.43	0.43	
DESKTOP(0)	0.43	0.43		DESKTOP(0)	0.43	0.43	
LASERPRINT.(0)	1.00	1.00	1.00	LASERPRINT.(0)	1.00	1.00	1.00
INKJETPRINT.(0)	0.75	0.75	0.75				
NEEDLEPRINT.(0)	0.75	0.75	0.75				
CENTRALU.(0)	0.58	0.58					
	ϕ:0.58	ϕ:0.58	ϕ:0.83		ϕ:0.52	ϕ:0.52	ϕ:1.00
Learning step 5				**Learning step 6**			
NOTEBOOK(0)	0.43	0.43		NOTEBOOK(0)	0.43	0.43	
PORTABLE(0)	0.43	0.43		PORTABLE(0)	0.43	0.43	
PC(0)	0.43	0.43		PC(0)	0.43	0.43	
WORKSTATION(0)	0.43	0.43		WORKSTATION(0)	0.43	0.43	
DESKTOP(0)	0.43	0.43		DESKTOP(0)	0.43	0.43	
LASERPRINT.(0)	1.00	1.00	1.00	LASERPRINT.(0)	1.00	1.00	1.00
	ϕ:0.52	ϕ:0.52	ϕ:1.00		ϕ:0.52	ϕ:0.52	ϕ:1.00
Learning step 7				**Learning step 8**			
LASERPRINT.(0)	1.00	1.00	1.00				
	ϕ:1.00	ϕ:1.0	ϕ:1.00				
Learning step 9							

Table 3. Some Concept Learning Results for a Text Featuring *Itoh-Ci-8* (continued)

duces either the same or even more accurate results, the same or fewer hypothesis spaces (indicated by the number of rows), and derives the correct prediction more rapidly (in step 6) than the less knowledgeable variants (in step 9).

The data also illustrate the continuous specialization of concept hypotheses achieved by the terminological classifier, e.g., from PHYSICALOBJECT in step 1 via PRODUCT in step 2 to COMPUTER and NOTEBOOK in step 3 and 4, respectively. The overall learning accuracy – due to the learner's aggressive specialization strategy – may even temporarily decrease in the course of hypothesizing (e.g., from step 3 to 4 or step 5 to 6 for **LA** – and **LA TH**), but the learning accuracy value for the full qualification calculus (**LA CB**) always increases.

Generalizing from the learning behavior for a single concept like ITOH-CI-8, Fig. 5 depicts the learning accuracy curve for the entire data set (101 texts). We also have included the graph depicting the growth behavior of hypothesis spaces (Fig. 6; **NH** indicates the number of hypothesis spaces being generated per learning step distinguishing between the three criteria –, **TH, CB**). In Fig. 5 the evaluation starts at LA values in the interval between 48% to 54% for **LA** –, **LA TH** and **LA CB**, respectively, in the first learning step, whereas the **NH** values range between 6.2 and 4.5 (Fig. 6). In the final step, LA rises up to 79%,

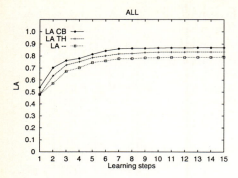

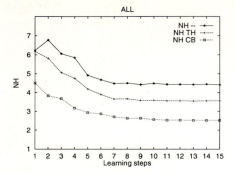

Fig. 5. Learning Accuracy (LA) for the Entire Data Set

Fig. 6. Number of Hypotheses (NH) for the Entire Data Set

83% and 87% for **LA −**, **LA TH** and **LA CB**, respectively, and the **NH** values reduce to 4.4, 3.6 and 2.5 for each of these three criteria.

The pure terminological reasoning machinery which does not incorporate the qualification calculus always achieves an inferior level of learning accuracy and generates more hypothesis spaces than the learner equipped with the qualification calculus. Furthermore, the inclusion of conceptual criteria (**CB**) supplementing the linguistic criteria (**TH**) helps a lot to focus on the relevant hypothesis spaces and to further discriminate the valid hypotheses (on the range of 4% of precision). Note that an already significant plateau of accuracy is usually reached after the third step (*viz.* 67%, 73%, and 76% for **LA −**, **LA TH**, and **LA CB**, respectively, in Fig. 5; the corresponding numbers of hypothesis spaces being 6.1, 5.1, and 3.7 for **NH −**, **NH TH**, and **NH CB**, respectively, in Fig. 6). This indicates that our approach finds the most relevant distinctions in a very early phase of the learning process, i.e., it requires only a *few* examples.

In a knowledge acquisition application operating on real-world texts, it should be fair to ask what level of precision one is willing to accept as a satisfactory result. We may discuss this issue in terms of degrees of learning accuracy. Under ideal conditions, one might require a 100% learning accuracy. Fig. 7 gives the number of texts being processed under this constraint and the associated number of learning steps, given the three types of criteria, *viz.* **LA -**, **LA TH**, and **LA CB**. Under the rigid condition that the most specific concept specializing the unknown item is to be learned, the three criteria require almost the same number of learning steps on the average. This is simply due to the fact that the knowledge base we supply has not a full coverage of any domain segment,

	# texts	steps
LA −	26	4,68
LA TH	31	4,16
LA CB	39	4,15

	# texts	steps
LA −	58	4,69
LA TH	74	4,36
LA CB	85	3,71

Fig. 7. Learning Steps for LA = 1.0

Fig. 8. Learning Steps for LA = .8

even in a limited domain such as information technology. (The knowledge base currently comprises 325 concept definitions and 447 conceptual relations.) The picture changes remarkably (cf. Fig. 8), if we only require a level of precision that does not fall below a learning accuracy of 80%. This also means that more general or slightly incorrect concept descriptions are accepted as a proper learning result, though more specific and entirely correct concept descriptions might have been worked out, at least in principle. About double the number of texts are being processed, but the number of learning steps are decreasing only for the full qualification calculus (*viz.* applying criterion **LA CB**). We may thus conclude that, granted an LA level of 80%, at least for the full qualification calculus only 15% of the learned concepts will be erroneous still. (Note the contrast to the LA values depicted in Fig. 5, which are derived from *averaging* single LA values.)

Summarizing this discussion, we may conclude that lowering the requirements on acceptable precision rates of learning results – within reasonable limits, of course – produces still valid learning hypotheses that are built up in less than four learning steps on the average.

4.2 Learning Rate

The learning accuracy focuses on the predictive power of the learning procedure. By considering the *learning rate* (LR), we supply data from the step-wise reduction of alternatives of the learning process. Fig. 9 depicts the mean number of transitively included concepts for all considered hypothesis spaces per learning step (each concept hypothesis denotes a concept which transitively subsumes various subconcepts). Note that the most general concept hypothesis in our example denotes OBJECT which currently includes 196 concepts. In general, we observed a strong negative slope of the curve for the learning rate.

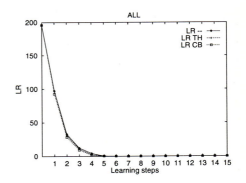

After the first step, slightly less than 50% of the included concepts are pruned (with 93, 94 and 97 remaining concepts for **LR CB**, **LR TH** and **LR** −, respectively). Again, learning step 3 is a crucial

Fig. 9. Learning Rate (LR) for the Entire Data Set

point for the reduction of the number of included concepts (ranging from 9 to 12 concepts). Summarizing this evaluation experiment, the quality-based learning system yields competitive accuracy rates (a mean of 87%), while at the same time it exhibits significant and valid reductions of the predicted concepts (up to two, on the average).

5   Related Work

Our approach bears a close relationship to the work of [7], [16], [18], [6], [19], [23], and [12], who aim at the automated learning of word meanings from context using a knowledge-intensive approach. But our work differs from theirs in that the need to cope with *several competing* concept hypotheses and to aim at a *reason-based selection* is not an issue in these studies. Learning from real-world textual input usually provides the learner with only sparse, highly fragmentary clues, such that multiple hypotheses are likely to be derived from that input. So we stress the need for a hypothesis generation and evaluation component as an integral part of large-scale real-world text understanders [10] operating in tandem with concept learning devices.

Note that this requirement also distinguishes our approach from the currently active field of information extraction (IE) [2] and [14]. The IE task is defined in terms of a *fixed* set of *a priori* given templates which have to be instantiated (i.e., filled with factual knowledge items) in the course of text analysis. In particular, no *new* templates have to be created. This step would correspond to the procedure we described in this contribution. Some work close to ours has been carried out by [18]. As in our approach, several concept hypotheses are generated from linguistic and conceptual data. Unlike our approach, the selection of hypotheses depends only on an ongoing discrimination process based on the availability of these data but does not incorporate an inferencing scheme for reasoned hypothesis selection. The difference in learning performance – in the light of our evaluation study in Section 4, at least – amounts to 8%, considering the difference between **LA** − (plain terminological reasoning) and **LA CB** values (terminological metareasoning based on the qualification calculus).

As a long-term goal we also subscribe to the perspectives underlying the work of [25], [18], [4], and [26] whose research is directed at the automatic generation of text knowledge bases from real-world texts. Unlike IE tasks, these studies are not limited to an *a priori* fixed set of templates that the analytic machinery concentrates on, though, usually (with the exception of [18]), no learning steps are required.

6   Conclusion

We have introduced a knowledge-based approach to the analysis of real-world natural language texts, which addresses the particular needs of dealing with new knowledge items. In order to incrementally expand the underlying domain knowledge base, we exploit qualitative knowledge about linguistic constructions in natural language texts and that about structural patterns in the emerging knowledge base. These clues were used to generate concept hypotheses, rank them according to plausibility, and select the most credible ones for assimilation into the conceptual knowledge base. We have demonstrated the feasibility of our approach and discussed the results of an empirical evaluation in terms of concept learning rates and learning accuracy.

The knowledge acquisition mechanism we propose is fully integrated in the text understanding mode. No specialized learning algorithm is needed, since concept formation is actually turned into a (meta)reasoning task carried out by the classifier of a terminological reasoning system. Quality labels can be chosen from any knowledge source that seems convenient. This will allow for the easy adaptation of our approach via the selection of quality labels coming from sources other than linguistic and conceptual knowledge (e.g., a vision system for the interpretation of photography). In our evaluation experiment the quality labels have turned out as a representation device that can be reasoned about in an intuitively perspicacious and natural way. These labels also have the necessary discriminative power for distinguishing between different concept hypotheses and they achieve a high degree of pruning of the search space for hypotheses in very early phases of the learning cycle. This is of major importance in considering our approach a viable contribution to a robust text analysis technology.

Acknowledgments. We would like to thank our colleagues in the CLIF group for fruitful discussions and instant support, in particular Joe Bush who polished the text as a native speaker. K. Schnattinger is supported by a grant from DFG (Ha 2097/3-1).

References

1. D. Aha, D. Kibler, and M. Albert. Instance-based learning algorithms. *Machine Learning*, 6:37–66, 1991.
2. D. Appelt, J. Hobbs, J. Bear, D. Israel, and M. Tyson. FASTUS: A finite-state processor for information extraction from real-world text. In *IJCAI'93 - Proc. 13th Intl. Joint Conf. on Artificial Intelligence*, pages 1172–1178, 1993.
3. J. Bateman, R. Kasper, J. Moore, and R. Whitney. A general organization of knowledge for natural language processing: The PENMAN upper model. Technical report, USC/ISI, 1990.
4. F. Ciravegna, R. Tarditi, P. Campia, and A. Colognese. Syntax and semantics in a text interpretation system. In *RIAO'91 - Proc. 3rd Conf. on Intelligent Text and Image Handling*, pages 684–694, 1991.
5. S. Dumais. Text information retrieval. In M. Helander, editor, *Handbook of Human-Computer Interaction*, pages 673–700. Amsterdam: North-Holland, 1990.
6. F. Gomez and C. Segami. Knowledge acquisition from natural language for expert systems based on classification problem-solving methods. *Knowledge Acquisition*, 2:107–128, 1990.
7. R. Granger. FOUL-UP: A program that figures out meanings of words from context. In *IJCAI'77 - Proc. 5th Intl. Joint Conf. on Artificial Intelligence*, pages 172–178, 1977.
8. U. Hahn, M. Klenner, and K. Schnattinger. Learning from texts: A terminological metareasoning perspective. In S. Wermter, E. Riloff, and G. Scheler, editors, *Connectionist, Statistical and Symbolic Approaches to Learning for Natural Language Processing*, pages 453–468. Berlin: Springer, 1996.
9. U. Hahn, S. Schacht, and N. Bröker. Concurrent, object-oriented dependency parsing: The PARSETALK model. *International Journal of Human-Computer Studies*, 41:179–222, 1994.

10. U. Hahn, K. Schnattinger, and M. Romacker. Automatic knowledge acquisition from medical texts. In *AMIA'96 - Proc. AMIA Annual Fall Symp. Beyond the Superhighway: Exploiting the Internet with Med. Informatics*, pages 383–387, 1996.

11. U. Hahn and M. Strube. PARSETALK about functional anaphora. In *AI'96 - Proc. 11th Biennial Conf. of the Canadian Society for Computational Studies of Intelligence*, pages 133–145. Berlin: Springer, 1996.

12. P. Hastings. Implications of an automatic lexical acquisition system. In S. Wermter, E. Riloff, and G. Scheler, editors, *Connectionist, Statistical and Symbolic Approaches to Learning in Natural Language Processing*, pages 261–274. Berlin: Springer, 1996.

13. M. Hearst. Automatic acquisition of hyponyms from large text corpora. In *COLING'92 - Proc. 15th Intl. Conf. on Computational Linguistics*, pages 539–545, 1992.

14. T. Kitani, Y. Eriguchi, and M. Hara. Pattern matching in the TEXTRACT information extration system. In *COLING '94 - Proc. 15th Intl. Conf. on Computational Linguistics*, pages 1064–1070, 1994.

15. C. Manning. Automatic acquisition of large subcategorization dictionary from corpora. In *Proc. 31st Meeting Assoc. for Comp. Linguistics*, pages 235–242, 1993.

16. R. Mooney. Integrated learning of words and their underlying concepts. In *CogSci'87 - Proc. 9th Conf. of the Cognitive Science Society*, pages 974–978, 1987.

17. P. Neuhaus and U. Hahn. Trading off completeness for efficiency: The PARSETALK performance grammar approach to real-world text parsing. In *FLAIRS'96 - Proc. 9th Florida Artificial Intelligence Research Symposium*, pages 60–65, 1996.

18. L. Rau, P. Jacobs, and U. Zernik. Information extraction and text summarization using linguistic knowledge acquisition. *Information Processing & Management*, 25(4):419–428, 1989.

19. U. Reimer. Automatic acquisition of terminological knowledge from texts. In *ECAI'90 - Proc. 9th European Conf. on Artificial Intelligence*, pages 547–549, 1990.

20. K. Schnattinger and U. Hahn. A terminological qualification calculus for preferential reasoning under uncertainty. In *KI'96 - Proc. 20th Annual German Conf. on Artificial Intelligence*, pages 349–362. Berlin: Springer, 1996.

21. K. Schnattinger and U. Hahn. Plausible learning from heterogeneous evidence in a text understanding system. In *KI'97 - Proc. 21st Annual German Conf. on Artificial Intelligence*. Berlin: Springer, 1997.

22. K. Schnattinger, U. Hahn, and M. Klenner. Terminological meta-reasoning by reification and multiple contexts. In *EPIA'95 - Proc. 7th Portuguese Conf. on Artificial Intelligence*, pages 1–16, 1995.

23. S. Soderland, D. Fisher, J. Aseltine, and W. Lehnert. CRYSTAL: Inducing a conceptual dictionary. In *IJCAI'95 - Proc. 14th Intl. Joint Conf. on Artificial Intelligence*, pages 1314–1319, 1995.

24. W. Woods and J. Schmolze. The KL-ONE family. *Computers & Mathematics with Applications*, 23:133–177, 1992.

25. G. Zarri. Knowledge acquisition for large knowledge bases using natural language analysis techniques. *Expert Systems for Information Management*, 1(2):85–109, 1988.

26. P. Zweigenbaum and M. Cavazza. Extracting implicit information from free text technical reports. In *RIAO'91 - Proc. 3rd Conf. on Intelligent Text and Image Handling*, pages 695–706, 1991.

Knowledge Discovery
in Endgame Databases

Michael Schlosser

Dept. of Electrical Engineering, Fachhochschule Koblenz,
Finkenherd 4, D-56075 Koblenz, Germany,
e-mail: schlosser@fh-koblenz.de

Abstract. In many application domains we have seen an explosive growth in the capabilities to both generate and collect data. Representative examples are business, medical, and scientific databases. In the game of chess (and similar games), we have a similar situation. Moreover, in chess we have gigantic databases with perfect information available. An endgame database is a very rare case of an information source with complete knowledge. However, this information, although complete, is not in a form which is particularly useful to human beings. Therefore, the raw data inside an endgame database have to be transformed into knowledge in understandable form. This task concerning chess endgame databases is, in principle, the same as in KDD in general.

This paper summarizes a few results from literature where knowledge (especially patterns) has been discovered for simple chess endgames - in most cases this has been done empirically, by intuition, and interacting with the computer, but not automatically. Up to now, there are almost no satisfying results for practical endgames to extract knowledge in an automatic way.

The main contribution of this paper is to present a new chess endgame database from which we have extracted knowledge automatically using decision trees - a method well-known from Machine Learning. Additionally, a mechanism for automatic feature discovery is introduced.

1 Introduction

Knowledge Discovery in Databases (KDD) is defined as "the non-trivial process of identifying valid, novel, potentially useful, and ultimately understandable patterns in data" [FPSS96]. A pattern, describing facts, is simpler (in some sense) than the enumeration of all facts. Extracting patterns is a first and necessary step on the way from raw data to knowledge.

The notion of pattern is relevant to Artificial Intelligence in general and for the evaluation of game positions in particular (concerning chess see [Kur77, Hee84]). A pattern is defined to be any property (relation) concerning some pieces on the board which may be present or absent. In chess, patterns are geometrically defined. Certain patterns may occur in a game more than once

X. Liu, P. Cohen, M. Berthold (Eds.): "Advances in Intelligent Data Analysis" (IDA–97)
LNCS 1280, pp. 423–435, 1997.   © Springer–Verlag Berlin Heidelberg 1997

depending on the distance to a goal. Examples from the literature are simple chess endgames, such as KRK [Mic77, Sei86], KPK [BC80], or KBNK [Kor84][1]. In most cases, however, the patterns have been found by hand. In the future, this has to be done automatically, because the most work lays in feature discovery.

The paper introduces a method to find patterns automatically from a so-called endgame database which contains complete knowledge of a certain chess endgame. Moreover, we give an alternative method to derive knowledge from an endgame database using a hierarchical sequence of decision trees.

The paper is organized as follows. Section 2 presents patterns for the simple endgames KRK and KPK which have been found by hand. Section 3 discusses the construction of endgame databases, containing complete knowledge for a certain endgame. Moreover, we illustrate the method by a small endgame database of our own. Section 4 shows two different ways of automatic knowledge discovery from the endgame database constructed in section 3, both by using decision trees. Finally, we summarize the paper and show that knowledge discovery from endgame databases and KDD are quite similar.

2 Patterns in Simple Chess Endgames

2.1 Patterns in KRK

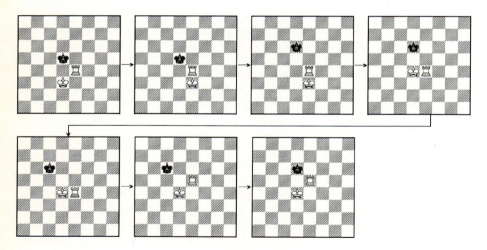

Fig. 1. 'North cycle' of mating the black King

The probably first attempt to discover patterns in chess endgames is given in [Mic77]. MICHIE investigated the elementary chess endgame KRK, in idealized

---

[1] KRK - King and Rook against King; KPK - King and Pawn against King; KBNK - King, Bishop, and Knight against King.

form, namely on an infinite board with two edges meeting at a corner (assume the north-west corner). On such a board mate can be forced. To mate the black King, it is necessary to drive it to one edge of the board; consequently, the white Rook and King must be on the "far" side of the black King. In [Mic77], an optimal strategy of this problem is given. In Figure 1, we give only a small part of this strategy, namely the so-called 'north cycle', consisting of six steps, that forces black to yield one rank per cycle.

By repetition of the sequence of six steps in Figure 1, the black King is forced to the north edge of the board. The positions of this sequence can be considered as patterns in KRK. Afterwards, a 'west cycle' is used. Finally, the black King will be checkmated in the corner.

Another way, the so-called deductive approach, is characterized by systematic isolation of necessary and sufficient conditions. Again we consider KRK, now on a normal board. Assume the black King stays in the center of the board, i. e. three steps from the edges. Since the mate can take place only with the black King on the edge, a mating sequence must necessarily contain a situation where the black King goes to a square of distance 2 to an edge, then distance 1, until the black King finally enters the edge. This reasoning yields the *ring-structure* of the board, i. e. a partition of the squares according to their distance to the nearest edge [Sei86] (see Figure 2).

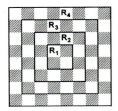

Fig. 2. Ring-structure for KRK

The strategy for KRK works as follows:
The black King is not naturally interested in going outside and must be compelled by White to do so. Assume the black King is on ring R_i ($i < 4$). The necessary and sufficient condition for being forced to proceed to R_{i+1} is that he can neither move to the next inner ring R_{i-1} nor in R_i itself. For R_4 such a situation is identical with mate or stalemate.
By this way we have a basic theorem:

Black King compelled to go to R_{i+1}.

is equivalent to

There exists a mate or a stalemate pattern on R_i.

Therefore, in any mating sequence the following stages must occur: The black King, say on R_1 is "mated" or "stalemated" on R_1, that means a mate or stalemate pattern is forced by White; the same for R_2 and R_3, and finally the black King will be mated on R_4.

Using this strategy we obtain a sequence of positions representing one pattern on the rings R_1 to R_4 (see Figure 3).

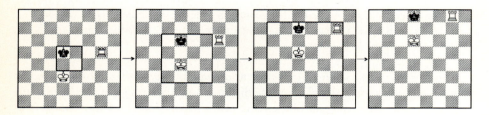

Fig. 3. Using the ring-structure to mate in KRK

2.2   Patterns in KPK

BEAL and CLARKE [BC80] have demonstrated that it is possible to create an efficient algorithm for the endgame KPK that uses only simple, easily defined patterns. If the side with the Pawn (suppose it to be White) is to win, we must have a method of repeatedly advancing the Pawn safely until it reaches the 8th rank. Figure 4 shows a move sequence often to occur in this endgame. The first position is a key pattern in KPK. Black has no possible move to stop White to advance the Pawn safely.

In the above cases, the patterns have been found empirically, by intuition. More difficult problems will need means of automatic pattern generation. The reasons are [Mic77]:

- Constructing patterns from the user's head is burdensome, and in a difficult domain prohibitive. Further, the problem of guarding against human error becomes unmanageable.
- Looking at the patterns above, no proof, only plausible argument, has been given that the strategy for using patterns is optimal. For formal proof to be tractable the patterns need to be the product of a fully defined procedure. The guarantee of optimality can then be built into the generating algorithm itself.

These two reasons led us to a method to overcome the difficulties. In the following section we show how to obtain complete information for a certain endgame, i. e. we sketch the construction of so-called endgame databases. However, the information in an endgame database, although complete, is not easily

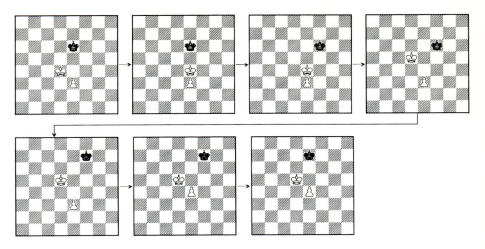

Fig. 4. Move sequence often to occur in KPK

usable by humans. That is the reason why it is very important to use KDD methods of "finding special patterns that may be interpreted as useful or interesting
knowledge" [FPSS96].

3 Endgame Databases

3.1 Optimal Strategy

It has been well-known since ZERMELO (1912) [Zer12] that for finite games,
such as chess or checkers, an optimal strategy can be constructed on the basis
of set-theoretic considerations. We present a rough overview of the notions and
notations to be used.

The set M of all possible legal positions[2] p can be partitioned into three
disjoint subsets

$M_W = \{$p: White has a forced win after n plies, n $\geq$ 0$\}$,
$M_B = \{$p: Black has a forced win after n plies, n $\geq$ 0$\}$,
$M_D = M \setminus (M_W \cup M_B) = \{$p: p is drawn $\}$
with $M = M_W \cup M_B \cup M_D$, $M_W \cap M_B = \emptyset$, $M_W \cap M_D = \emptyset$, $M_D \cap M_B = \emptyset$.

For the classes M_W and M_B, further partitions are possible resulting in
sequences $M_W^0, M_W^1, M_W^2, \ldots$, and $M_B^0, M_B^1, M_B^2, \ldots$, with the following meaning:

M_W^0: Black to move, Black has been checkmated;
M_W^1: White to move, checkmate after one ply;
M_W^2: Black to move, checkmate after two plies;

$\ldots$

---

[2] A position p is legal if it can be obtained from the starting position by legal moves.

The actions on the board are forced in the following sense:

if $p \in M_W^{2i}, i = 1, 2, \ldots$, then *each* black move leads to $p' \in M_W^k$ ($k < 2i$, k odd);

if $p \in M_W^{2i+1}, i = 0, 1, 2, \ldots$, then *at least one* white move exists leading to $p' \in M_W^{2i}$.

An optimal sequence of moves has its corresponding positions in M_W^i, M_W^{i-1}, $M_W^{i-2}, \ldots, M_W^0$, successively. Analogous considerations apply to M_B. Therefore, we have: $M_W = M_W^0 \cup M_W^1 \cup \ldots \cup M_W^{nW}$; $M_B = M_B^0 \cup M_B^1 \cup \ldots \cup M_B^{nB}$.[3]

This existence of an optimal strategy can be used to construct the subsets of M subsequently backwards, starting at a goal. These ideas resulted in complete endgame databases for a large number of chess endgames, starting with results in [Str70].

3.2   The Construction of an Endgame Database

It is not in the scope of this paper to explain, how a computer can construct an endgame database (EDB), but in short, the method is to work backwards from certain end, or terminal positions, calculating the result of more and more positions until eventually every legal position has been evaluated. The terminal positions may be simply the mating positions, but in complicated endgames more often they are the positions which arise when the material balance has been changed.

The end result of an EDB construction is a collection (a table, a list) of all (theoretically) possible positions (i. e. of all combinations of squares for each piece) of a certain endgame. Usually, for each position, an EDB contains one or more optimal moves and the number of moves to win. For further details see e. g. [Sch92].

3.3   An Example: the Endgame Database QK

Some years ago, the author created a new EDB to get new results in another field of chess [Sch88]. The generated EDB differs from usually considered EDBs in one point only, namely that there is a fixed number of immobile pieces on the board (Figure 5). This somewhat artificial position of black pieces avoids moves of the white King, but it is not a restrictive condition for the EDB construction.

We constructed an EDB in which all positions contain white Queen (wQ) and black King (bK) in addition to the pieces in Figure 5. All positions of the EDB differ from each other only in the placing of wQ and bK. These two chessmen are alone active during the play. The main results of this EDB construction are as follows (cf. [Sch88, Sch92]):

---

[3] nW, nB are the maximum lengths for a forced winning variation of White or Black, respectively. All sets appearing in this way are finite.

Fig. 5. Four immobile pieces in the QK database

- The mating position Qh5/Kh3 is only attainable with the black King inside a region consisting of 8 squares, namely h3, h4, h5, h6, g4, g5, f3, and f4. Outside this region, checkmating the black King on h3 cannot be forced. Other ways to win, e. g. checkmating with King and Queen against King, after capturing the three black pieces need more than 9 moves. They are not considered here.
- If the mating position Qh5/Kh3 is attainable at all, then in at most 9 moves (assuming optimal play).

This small EDB (2 mobile pieces only; one of them, the black King, is restricted to 8 squares) can be written on one page (see Table 1).

For each position of wQ and bK in addition to the pieces of Figure 5, a table entry contains

$$\left\{ \begin{array}{l} \text{number-square[/square ...] if } \textit{there is a win};\\ \qquad\qquad\quad \text{the elements denote}\\ \qquad\qquad\quad \text{number: number of moves to win (mate),}\\ \qquad\qquad\quad \text{square: target square(s) of a wQ move, allowing for equi-optimal moves.}\\ \\ \quad \text{draw if } \textit{there is no win}\\ \\ \quad \text{ill if } \textit{the position is illegal.} \end{array} \right.$$

The wQ/bK EDB can be used to play this artificial endgame perfectly in the following way:

1. Select a White to move position (wQ and bK in addition to Figure 5).
2. Play the move contained in the corresponding table entry.

wQ pos.	bK position								
	h3	h4	h5	h6	g4	g5	f3	f4	any
a1	3-g7	5-e5	3-g7	draw	4-f6	draw	5-e5	draw	ill or draw
a2	3-e6/g8	4-g8	6-e6	4-g8	7-d5	5-f7	draw	6-e6	ill or draw
a3	5-a6/d6/e7/f8	5-e7	draw	draw	6-f8	draw	ill	draw	ill or draw
a4	5-a6/c6	ill	7-e8	draw	ill	draw	7-e8	ill	ill or draw
a5	1-h5	5-e5	ill	draw	7-d5	ill	5-e5	draw	ill or draw
a6	3-e6	2-g6	6-e6	ill	4-f6	draw	draw	6-e6	ill or draw
a7	3-g7	5-e7	3-g7	draw	8-d7	5-f7	7-e7	draw	ill or draw
a8	3-g8	4-g8	7-e8	4-g8	6-f8	draw	ill		ill or draw
b1	3-f5	2-g6	9-h7	draw	8-g6	draw	9-e1	draw	ill or draw
b2	3-g7	5-e5	3-g7	draw	4-f6	draw	5-e5	draw	ill or draw
b3	3-e6/g8	4-g8	6-e6	4-g8	7-d5	5-f7	ill	6-e6	ill or draw
b4	5-b6/d6/e7/f8	ill	draw	draw	ill	draw	7-e7	ill	ill or draw
b5	1-h5	5-e5	ill	draw	7-d5	ill	5-e5	draw	ill or draw
b6	3-e6	2-g6	6-e6	ill	4-f6	draw	draw	6-e6	ill or draw
b7	3-g7	5-e7	3-g7	draw	7-d5	5-f7	ill	draw	ill or draw
b8	3-g8	4-g8	7-e8	4-g8	6-f8	draw	5-e5	ill	ill or draw
c1	5-c6/c8/h6	5-h6	draw	ill	draw	ill	9-e1	ill	ill or draw
c2	3-f5	2-g6	9-h7	draw	8-g6	draw	draw	draw	ill or draw
c3	3-g7	5-e5	3-g7	draw	4-f6	draw	ill	draw	ill or draw
c4	3-e6/g8	ill	ill	4-g8	ill	5-f7	draw	ill	ill or draw
c5	1-h5	5-e5/e7	ill	draw	6-f8	ill	5-e5	draw	ill or draw
c6	3-e6	2-g6	6-e6	ill	4-f6	draw	ill	6-e6	ill or draw
c7	3-g7	5-e5/e7	3-g7	draw	8-d7/e5	5-f7	5-e5	ill	ill or draw
c8	ill	4-g8	6-e6	4-g8	ill	draw	7-e8	6-e6	ill or draw
d1	1-h5	6-d7/f3	ill	draw	ill	draw	ill	draw	ill or draw
d2	5-d6/d8/h6	5-h6	draw	ill	7-d5	ill	9-e1	ill	ill or draw
d3	3-f5	2-g6	9-h7	draw	7-d5	draw	ill	draw	ill or draw
d4	3-g7	ill	3-g7	draw	ill	draw	5-e5	ill	ill or draw
d5	1-h5	4-g8	ill	4-g8	8-d7/e5	ill	ill	6-e6	ill or draw
d6	3-e6	2-g6	6-e6	ill	4-f6	draw	ill	6-e6	ill or draw
d7	ill	5-e7	3-g7	draw	ill	5-f7	7-e7/e8	6-e6	ill or draw
d8	3-g8	ill	7-e8/f6	4-g8	4-f6	ill	7-e7/e8	draw	ill or draw
e1	3-e6	5-e5/e7	6-e6	draw	8-e5	draw	5-e5	6-e6	ill or draw
e2	1-h5	5-e5/e7	ill	draw	ill	draw	ill	6-e6	ill or draw
e3	3-e6	5-e5/e7/h6	6-e6	ill	8-e5	ill	ill	ill	ill or draw
e4	3-e6/f5	ill	6-e6	draw	ill	draw	ill	ill	ill or draw
e5	1-h5	5-e7	ill	draw	4-f6	ill	7-e7/e8	ill	ill or draw
e6	ill	2-g6	7-e8/f6	ill	ill	5-f7	5-e5	draw	ill or draw
e7	3-e6/g7	ill	3-g7	draw	4-f6	ill	5-e5	6-e6	ill or draw
e8	1-h5	2-g6	ill	4-g8	6-f8	5-f7	5-e5	6-e6	ill or draw
f1	3-f5	6-f3/f8	7-f6	draw	4-f6	5-f7	ill	ill	ill or draw
f2	3-f5	6-f3/f8	7-f6	draw	4-f6	5-f7	ill	ill	ill or draw
f3	1-h5	6-f8	ill	draw	ill	5-f7	ill	ill	ill or draw
f4	3-f5	ill	7-f6	ill	ill	ill	ill	ill	ill or draw
f5	ill	2-g6	ill	draw	ill	ill	ill	ill	ill or draw
f6	3-e6/f5/g7	ill	3-g7	ill	6-f8	ill	ill	ill	ill or draw
f7	1-h5	2-g6	ill	4-g8	4-f6	draw	ill	ill	ill or draw
f8	3-f5/g7/g8	4-g8	3-g7	ill	4-f6	5-f7	ill	ill	ill or draw
g1	ill	ill	ill	ill	ill	ill	ill	ill	ill or draw
g2	ill	ill	ill	ill	ill	ill	ill	ill	ill or draw
g3	ill	ill	ill	ill	ill	ill	ill	ill	ill or draw
g4	ill	ill	ill	4-g8	ill	ill	ill	ill	ill or draw
g5	1-h5	ill	ill	ill	ill	ill	5-e5	ill	ill or draw
g6	1-h5	4-g8	ill	ill	ill	ill	7-e8	6-e6	ill or draw
g7	3-g8	2-g6	7-f6	ill	ill	ill	5-e5	draw	ill or draw
g8	3-e6/g7	2-g6	3-g7	draw	ill	ill	7-e8	6-e6	ill or draw
h1	ill	ill	ill	ill	ill	ill	ill	ill	ill or draw
h2	ill	ill	ill	ill	draw	draw	draw	draw	ill or draw
h3	ill	ill	ill	ill	ill	draw	draw	6-e6	ill or draw
h4	ill	ill	ill	ill	ill	ill	7-e7	ill	ill or draw
h5	ill	ill	ill	ill	ill	ill	ill	draw	ill or draw
h6	ill	ill	ill	ill	4-f6	ill	draw	ill	ill or draw
h7	ill	ill	ill	ill	8-d7/g6	5-f7	7-e7	draw	ill or draw
h8	ill	ill	ill	ill	4-f6	draw	5-e5	draw	ill or draw

Table 1. wQ/bK EDB

4   Knowledge Discovery from EDBs

The title of this section and of the whole paper is slightly paradoxical, because an EDB already contains total information about the endgame in question. However, this information, although complete, is not in an explicit form, like rules or patterns. Although databases have been constructed for all four-man endgames and almost all five-man endgames (for a survey see e. g. [Tho86]), they have had little effect on tournament chess. The quantity of information in such an EDB is daunting: nobody can memorize even a fraction of the material in a database, and faced with what seems to be a hopeless task, most players have not even made the attempt.

There are only a handful of chess experts who tried to learn from EDBs. Let us give one noteworthy example. The British International Grandmaster JOHN NUNN has been experimenting with EDBs for a few years. As a result of his tremendous work, he has published three books on four- and five-piece endgames in which he presents important discoveries from the databases and, moreover, he gives practical tips to chess players [Nun92] [Nun94] [Nun95][4]. NUNN was able to acquire high-level mastery of the play in a lot of endgames. However, he used intuition and interaction with the computer. We are concerned here with automating this process.

In the next two subsections, we use decision trees (see e. g. [Qui83]) - a method well-known from Machine Learning - to automatically extract knowledge from an EDB. For this purpose, we use an EDB in a restricted field of chess (see section 3). This EDB is, in comparison with practical EDBs constructed earlier, small enough to get some results in the field of automatic discovery. For further details on knowledge discovery from this EDB see [PS95].

4.1 Conceptualization of Knowledge Using Patterns

One typical recurring pattern in QK is exhibited in Figure 6. wQ and bK have a distance of a Knight's move. This pattern was discovered by a decision tree using the relative positions of King and Queen as a new attribute. This finding is very important for playing the endgame under consideration in an optimal way. An evaluation of the QK patterns found can be given by showing optimal play in the endgame. The QK patterns are subgoals on the way to the ultimate goal, i. e. to the mating position Qh5/Kh3 (please follow the arrows in Figure 6, all positions with Black to move; positions with White to move are omitted).

4.2 Hierarchical Classification

The complete wQ/bK EDB may be classified according to the outcome of the game (win and draw in our example). Using a hierarchical sequence of decision trees, we will finally find a small region for the black King in which mate may be

---

[4] We quote from the introduction to [Nun92]: "... anyone who has used a large database will know the problem involved in extracting anything useful from megabyte after megabyte of raw data. The human author had to perform two main functions. The first was to select the important, useful or entertaining information from the huge mass of uninteresting material. The second, more significant, function was to act as an interpreter. All the computer can do is say which moves win and how long the win will take. It cannot explain why some moves win while other, apparently similar, moves do not. It is also unable to derive characteristic themes which recur time and time again. These are the ideas which over-the-board players should know ..."

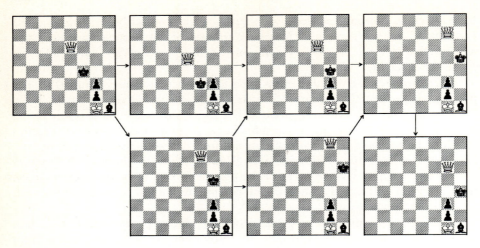

Fig. 6. One typical recurring pattern in QK

forced. The successive steps, which may be thought of as being in hierarchical sequence, are as follows:

1. Initial situation: the black King is allowed in any legal position on the board.
2. A first decision tree shows that no win is possible with the black King on the files a to e inclusive, i. e. the leftmost five files of the board are a drawing zone for the black King. Thus, a new binary attribute (file number of bK < 6) that is relevant to the problem was found (see Figure 7a).
3. Considering the remaining positions, another decision tree shows that the 7th and 8th ranks also belong to the drawing zone (see Figure 7b).
4. Finally, a decision tree identifies three more squares as belonging to the drawing zone (see Figure 7c).

It follows that only the residual squares of the black King (i. e. inside the 8-square region) are relevant to the wins of this endgame. We note that the same result was already given in section 3.3, immediately received from the EDB. Using decision trees, we have found new binary attributes which are relevant for the considered endgame.

We are able to derive simple heuristic rules from the decision trees in Figure 7. By comparison with the original database, heuristic rules are greatly simplified by disregarding exceptions, also known as special cases. While simplified rules have the advantage of comprehensibility, they risk being misleading in special cases. This is an instance of the classical conflict between completeness and simplicity in any instance of applied problem solving.

One way of finding heuristic rules is the pruning of exact decision trees. We present an example. From the decision tree implied by Figure 7a, we can derive a simple heuristic rule for White:

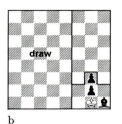
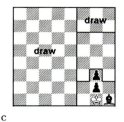
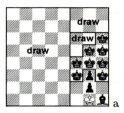

b   c

Decision trees (parts only)[5]:

```
K1                      K2                      K1
=1 --> 0 (313)          =8 --> 0 (118)          =6
=2 --> 0 (303)          =7 --> 0 (114)             and K2
=3 --> 0 (295)          =6 ...                            =6 --> 0 (34)
=4 --> 0 (291)          =5 ...                            =5 --> 0 (34)
=5 --> 0 (293)          =4 ...                            ...
=6 ...                  =3 ...                    =5
=7 ...                  =2 ...                       and K2
=8 ...                  =1 ...                            =6 --> 0 (39)
                                                          ...

                                                  ...
```

New rules (using new binary attributes):

```
IF K1<6                 IF K2>6                 IF (((K1=6) AND
                                                        (K2 in [5,6]))
                                                   OR ((K1=7)AND(K2=6)))
THEN class=0            THEN class=0            THEN class=0
ELSE ...                ELSE ...                ELSE ...
```

Fig. 7. Hierarchical decision trees, the corresponding diagrams, and the new rules

The black King must not enter the drawing zone (files a to e inclusive).

This heuristic translates to practical play as follows:

If the black King is on the f-file, the Queen must guard the relevant squares on the e-file,

or, much more simply (but not fully equivalently):

The Queen must occupy a safe square on the e-file.

The heuristic above fails in some cases. For instance, if the black King is on f4, the Queen moving to, say, e1 cannot prevent the black King from proceeding to f5 or g5 (which form part of the drawing zone, though not included in the simple definition of the drawing zone in the heuristic above).

5   Conclusions and Future Work

There are huge databases in almost all scientific and business domains as well as in the game of chess. Such a chess endgame database is a very rare case of an information source with complete knowledge. However, this information, although complete, is not in a form which is particularly useful to chess players. Therefore, the data inside an endgame database have to be transformed into knowledge. From that, a lot of questions arise, e. g.:

- How to make the knowledge more comprehensive to human beings?
- How to introduce concepts into the knowledge to make it manageable?
- How to reduce the amount of knowledge?
- How to deduce new rules from the database?
- How to extract a structure of knowledge?
- How to learn data dependencies?

These questions and certainly a few ones more concerning chess endgame databases are, in principle, the same as in KDD in general.

This paper summarized a few results from the literature, where knowledge (especially patterns) has been discovered for (simple) chess endgames. Up to now there are almost no acceptable results for practical endgames to answer the questions above in an automatic way. In contrast to earlier papers, we focussed on automatic ways to achieve our results.

This paper presented a new EDB in a restricted field of chess. This EDB is, in comparison with practical EDBs constructed earlier, small enough to get some results. We have extracted knowledge from this EDB automatically using decision trees - a method well-known from Machine Learning.

It seems promising to use this small EDB as a test-bed for Artificial Intelligence investigations in general. Results obtained so far might be generalized to much more complicated endgames as well as to other fields of Artificial Intelligence and KDD, e. g. to problem solving and diagnosis.

Acknowledgements

The author would like to thank Andreas Ittner, Christian Posthoff, Rainer Staudte, and the anonymous reviewers for helpful comments and suggestions. Special thanks to Rolf Rossius for his help to clarify the problems w.r.t. Latex.

References

[BC80]   D. F. Beal and M. R. B. Clarke. The Construction of Economical and Correct Algorithms for King and Pawn against King. In M. R. B. Clarke, editor, *Advances in Computer Chess 2*, pages 1–30, Edinburgh, Scotland, 1980. Edinburgh Univ. Press.

[FPSS96] U. M. Fayyad, G. Piatetsky-Shapiro, and P. Smyth, editors. *Advances in Knowledge Discovery and Data Mining*. AAAI/MIT Press, Boston, 1996.

[Hee84]   A. Heeffer.   Automated Acquisition of Concepts for the Description of
          Middle-Game Positions in Chess. Technical Report No. TIRM-84-005, The
          Turing Institute, 1984.

[Kor84]   J. Korst. Het genereren van regels voor schaak eindspelen ofwel eindspelen,
          moeilijker dan je denkt! (in Dutch). Master's thesis, Technische Universität
          Delft, Delft, 1984.

[Kur77]   R. Kurz. *Musterverarbeitung bei der Schachprogrammierung (in German)*.
          PhD thesis, Universität Stuttgart, Stuttgart, 1977.

[Mic77]   D. Michie. King and Rook against King: Historical Background and a Prob-
          lem on the Infinite Board. In M. R. B. Clarke, editor, *Advances in Computer
          Chess 1*, pages 30–59, Edinburgh, Scotland, 1977. Edinburgh Univ. Press.

[Nun92]   J. Nunn. *Secrets of Rook Endings*. B. T. Batsford Ltd., London, 1992.

[Nun94]   J. Nunn. *Secrets of Pawnless Endings*. B. T. Batsford Ltd., London, 1994.

[Nun95]   J. Nunn. *Secrets of Minor Piece Endings*. B. T. Batsford Ltd., London, 1995.

[PS95]   Chr. Posthoff and M. Schlosser. Optimal Strategies - Learning from Exam-
          ples - Boolean Equations. In K. P. Jantke and S. Lange, editors, *Proc. Work-
          shop "Algorithmic Learning for Knowledge-Based Systems"*, pages 363–390,
          Springer-Verlag, Berlin, 1995.

[Qui83]   J. R. Quinlan. Learning Efficient Classification Procedures and their Appli-
          cation to Chess End Games. In R. S. Michalski, J. G. Carbonell, and T. M.
          Mitchell, editors, *Machine Learning - An Artificial Intelligence Approach*,
          pages 463–482, Morgan Kaufmann, Los Altos, Cal., 1983.

[Sch88]   M. Schlosser. Computers and Chess Problem Composition. *ICCA Journal*,
          11(4):51–55, 1988.

[Sch92]   M. Schlosser. A Test-Bed for Investigations in Machine Learning. Gosler-
          Report No. 18, TH Leipzig, October 1992.

[Sei86]   R. Seidel. Deriving Correct Pattern Descriptions and Rules for the KRK
          Endgame by Deductive Methods. In D. F. Beal, editor, *Advances in Com-
          puter Chess 4*, pages 19–36, Oxford, UK, 1986. Pergamon Press.

[Str70]   T. Ströhlein. *Untersuchungen über kombinatorische Spiele (in German)*.
          PhD thesis, TU München, München, 1970.

[Tho86]   K. Thompson. Retrograde Analysis in Certain Endgames. *ICCA Journal*,
          9(3):131–139, 1986.

[Zer12]   E. Zermelo. Über eine Anwendung der Mengenlehre auf die Theorie des
          Schachspiels (in German). In *5. Int. Mathematikerkongreß*, volume 2, pages
          501–504, Cambridge, 1912.

Parallel Induction Algorithms for Data Mining

John Darlington, Yi-ke Guo, Janjao Sutiwaraphun, and Hing Wing To

Department of Computing,
Imperial College, London SW7 2BZ, U.K.
E-mail: {jd, yg, js11, hwt}@doc.ic.ac.uk

Abstract. In the last decade, there has been an explosive growth in the generation and collection of data. Nonetheless, the quality of information inferred from this voluminous data has not been proportional to its size. One of the reasons for this is that the computational complexities of the algorithms used to extract information from the data are normally proportional to the number of input data items resulting in prohibitive execution time on large data sets. Parallelism is one solution to this problem. In this paper we present preliminary results on experiments in parallelising C4.5, a classification-rule learning system using decision-trees as a model representation, which has been used as a base model for investigating methods for parallelising induction algorithms. The experiments assess the potential for improving the execution time by exploiting parallelism in the algorithm.

1   Introduction

In the last decade there has been an explosive growth in the generation and collection of data. Nonetheless, the use of information made out of these mountains of stored data does not seem to be proportional to their sizes. This has lead to the growth in research in methods for automated discovery of knowledge from databases. One such method is *classification data mining*, which finds a function that classifies the data in a database into predefined sets. One technique in classification data mining is the *inductive decision tree*, which is widely used because of the accuracy of its classification.

One of the greatest problems with the inductive decision tree technique is its prohibitive computation time on large data sets. Han et al. [2] show that the computational complexity of finding an optimal classification decision tree is NP-hard. Existing algorithms use local heuristics to decrease the computational complexity. For example, C4.5 [5], CDP [2] and SLIQ [2] have computational complexities ranging from $O(ANlogN)$ to $O(AN(logN)^2)$, where A is the number of attributes or fields and N is the number of cases in the training set. Thus the computational complexities of these algorithms increase with the number of attributes and training cases. This leads to prohibitive execution times when the discovery is conducted on realistic training sets.

Parallelism is one possible solution to this problem. Three approaches to parallelising the induction process have been proposed by Hedberg [3]:

X. Liu, P. Cohen, M. Berthold (Eds.): "Advances in Intelligent Data Analysis" (IDA–97)
LNCS 1280, pp. 437–445, 1997.   © Springer–Verlag Berlin Heidelberg 1997

1. To parallelise a single induction algorithm and execute on multiple processors.
2. To only parallelise the primitive database operations, such as select or join, in the data mining process.
3. To apply multiple analysis programs over partitioned databases and then to integrate the results together in some fashion.

This paper focuses on the first approach. An induction algorithm, C4.5, has been modified to execute in parallel on a Fujitsu AP1000. In Section 2 the sequential algorithm is outlined. Three different parallel implementations of C4.5 are then described in Section 3. Experimental result from applying these implementations to a large training set are given in Section 4.

2   Sequential C4.5

C4.5 is a classification-rule learning system which uses decision-trees as a model representation. Thus the aim of C4.5 is to find a function which maps the data items from a database into a set of predefined classes. The function, in this case a decision tree, is derived from a training set, which is usually some subset of the whole database. The training set is in the form of a flat file which contains a list of items or cases. Each case is a list of attributes, which are either continuous or discrete, and the class to which the item belongs. An example of a continuous attribute is *temperature*, whilst *gender* is an example of a discrete attribute. The output of the algorithm is a decision tree which is built from a training set.

At the core of the C4.5 algorithm is the tree construction process which aims to find the simplest decision tree that can describe the structure of the domain. The tree construction algorithm proceeds recursively. The main steps of the algorithm are:

1. Calculate the *information gain* of each attribute.
2. Select the attribute which yields the most information gain and use it as the node at the current point of the tree.
3. If the selected attribute is discrete, branch the node with all possible values. In the case of continuous attributes, the algorithm has to select a cut point that yields the most information gain and the node will be branched into those value less than and those values greater than the cut point.
4. Rearrange the data items into the corresponding branches.
5. Repeat all the steps in each branch of the tree.

The algorithm stops when *all* the data items at a particular node are of the same class.

The precise details of the tree construction algorithm vary between different implementations. Two features of the C4.5 implementation which are of particular interest here are the scheme used for storing the data items and the specific technique for handling continuous attributes. In the sequential implementation of C4.5, data items are referenced through an array of pointers. To improve efficiency C4.5 reorders the array of pointers such that items which are related to the current working branch appear contiguously in the array.

For a given branch, the cut point or threshold for a continuous attribute is usually found by first sorting the data items on the attribute. The set of candidate thresholds are then chosen to be the mid-points between the values of the continuous attribute of adjacent data items. The resulting threshold is the one which yields the highest information gain. C4.5 differs from this scheme by using as the threshold the highest value of the attribute from the *entire* training set which does not exceed the midpoint. This ensures that all threshold values appearing in the decision tree occur in the data set.

3 Parallel Implementation of C4.5

In this section we describe three parallel implementations of C4.5. All three implementations exploit parallelism by executing the recursive calls, in step 5 of the tree construction algorithm, in parallel. The three implementations mainly differ in the method used for partitioning the data. This component affects the communication overheads of an implementation.

3.1 The Scheme 1

In this scheme, the entire training set is initially duplicated onto all the processor. The master processor (usually processor 0) then begins to construct the tree depth first until there are as many leaves in the intermediate tree as there are processors. Each processor is then allocated one of these leaves by the master. The only communication required for allocating a leaf to a processor is the start and end positions of the portion of the training set forming the leaf. This is possible as each processor has a copy of the entire training set. Each processor then completes the tree construction for its leaf.

Additional communication is required when the intermediate tree is being constructed by the master. As described in Section 2 the C4.5 implementation rearranges the array of pointers to data items as it constructs the tree. The copies of the training sets on the other processors must be kept consistent to ensure that the correct tasks are allocated. Thus each time the training set is rearranged by the master the reordering must be broadcast to all the other processors.

3.2 The Scheme 2.1 and 2.2

The next two implementations differ from the first scheme by partitioning the training set across the processors, rather than duplicating it onto all the processors. Again the master processor begins by constructing the tree depth first until there are as many leaves in the intermediate tree as there are processors. Each processor is then allocated one of these leaves by the master. It is at this point that the two schemes differ from the first. To allocate a leaf to a processor the training cases which make up the leaf must be communicated from the master to the processor. Each processor then completes the tree construction for its leaf.

Schemes 2.1 and 2.2 differ in their method for handling continuous attributes. As described in Section 2, C4.5 has a special technique for handling continuous attribute values. In the algorithm each branch of the tree must use *all* the continuous values rather than just those allocated to that branch. To enable every processor to share the global view of continuous attribute values, two different techniques have been used leading to scheme 2.1 and 2.2 respectively. In scheme 2.1 all continuous attribute values are duplicated onto every processor. In scheme 2.2 communication is used to share the local values of each processor.

3.3 Analysis of the Different Schemes

The three schemes have quite different communication and memory costs. In particular scheme 1 duplicates the entire training set onto each processor. In contrast schemes 2.1 and 2.2 partition the data and send a partition to each processor. Thus scheme 1 should have the least communication costs. However, as the size of the training set increases this scheme will no longer be viable as the memory on each processor is limited.

There is a similar trade off between scheme 2.1 and 2.2. Scheme 2.1 reduces communication costs by duplicating the entire set of values of continuous attributes onto all the processors. In contrast scheme 2.2 saves on memory cost by partitioning the values, but has to perform more communication. The experiments detailed in the next section assess these differences.

4 Experimental Results

The performance of the three different parallel schemes were compared by applying their implementations to three training sets obtained from the UCI Repository of Machine Learning Databases [4]. The training sets differ by their proportion of discrete to continuous attributes.

4.1 The Training Sets

The first training set PEOPLE contains 27,722 cases of 14 attributes. The predefined classes of interest are people who have an income of less than $50,000 and people who have an income greater than $50,000 a year. Five out of the 14 attributes are represented by continuous attributes.

The second training set is LETTER-RECOGNITION. It contains 20,000 cases of 16 numerical attributes representing black-and-white rectangular pixels displayed as one of 26 capital letters in English alphabet. All of the attributes are continuous.

The third training set, CONNECT-4, contains all legal positions in games of Connect-4 in which neither player has yet won and in which the next move is not forced. Forty two attributes each of which corresponds to one Connect-4 square

are used to represent a stage in the game which would then be classified into either the game ending in Win, Loss or Draw. All are discrete attributes.

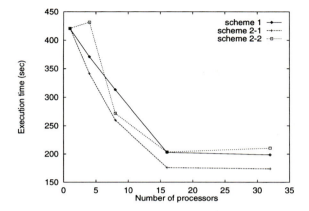

Fig. 1. Execution time of the different schemes on the PEOPLE training set.

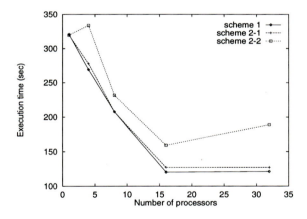

Fig. 2. Execution time of the different schemes on the LETTER-RECOGNITION training set.

4.2   Analysis of Results

The execution times of the different experiments are shown in Figures 1, 2 and 3, respectively. The time for executing the tree construction algorithm sequentially on one processor is the first point on each graph as this was used as the

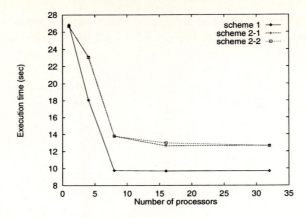

Fig. 3. Execution time of the different schemes on the CONNECT-4 training set.

parallel execution time for the one processor case. The accuracy of the results remained unchanged as the computation performed by the algorithms remained unchanged. Only the order of computation was changed, and the reorderings respected the data dependencies in the original program.

The performance of all the schemes scale to about 16 processors. After 16 processors the performance remains either static or degrades. This problem can be attributed to poor load-balancing. In each case the tree was constructed sequentially until there were as many branches as there were processors. Each branch was then assigned to a processor. Unfortunately, the sizes of the subtrees allocated to the processors varied leading to an uneven distribution of work between the processors. This is shown in Figure 4. The results indicate that

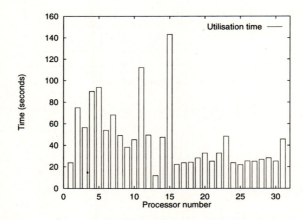

Fig. 4. Processor execution time for PEOPLE training set using scheme 2.2 with 32 processors.

although there were costs associated with the communication overheads, the main factor determining the performance was poor load-balancing. For more consistent scalability a load balancing scheme would have to be adopted. The performance of scheme 2.2 degrades after 16 processors. This can be attributed to the extra intercommunication costs incurred by the increase in the number of processors, coupled with no reduction in the largest task size allocated to a single processor.

It would be expected that scheme 1 would out-perform both the other schemes as the information used to allocate a task is significantly lower. Studying Figures 2 and 3, this would seem to be the case. Surprisingly, this is not the case for the PEOPLE training set, as shown in Figure 1. A closer examination of the tree construction phase shows that the the communication overheads involved in maintaining consistent copies of the data items, as described in Section 3.1, depends on the shape of the tree being generated. Further investigation is required to determine the characteristics of the PEOPLE training set which has led to the poorer performance of scheme 1.

In general, as shown in Figure 1, scheme 2.2 performs worse than 2.1. This is expected given the extra communication involved in scheme 2.2 for dealing with distributed continuous attributes. When there are no continuous attributes, as is the case for the CONNECT-4 training set, the performance of the two schemes is comparable. The full cost of distributing the continuous attributes can be seen in Figure 2 for the LETTER-RECOGNITION training set. With the prototype implementation the fully distributed scheme 2.2, was 26% slower than scheme 2.1. This indicates the potential for fully distributing the data sets which would be necessary as the size to the data sets grow and it is no longer possible to duplicate the entire set of continuous attributes on all processors. The details of further experiments on different training sets can be found in [6].

The initial results indicate that the performance of the different algorithms are extremely susceptible to the nature of the information in a training set. Future work will investigate methods for characterising different training set.

5 Further Work

From the experiments, it can be seen that some features of C4.5, for example the calculation of the thresholds on continuous attributes, prevent the exploitation of parallelism. Some of these problems can be avoided without affecting the accuracy of the algorithm, others will require solutions which may affect the accuracy of the algorithm. In the case of the threshold calculation, the problem can be avoided by delaying the calculation until the end of the tree construction process. The final thresholds are only required when the tree has been constructed and then later used for classification. Until then using the midpoints will still generate the same shape of tree. Therefore, the decision tree can be constructed in parallel without inter-processor communication and the cutpoint can then be updated after all subtrees have been gathered together at the master node.

Load-balancing is another important issue in parallelising C4.5. The heuristic used for determining the branching in this algorithm, information gain, is a single step look ahead technique. Thus the shape of the tree is not known in advance. With the current implementation, the allocation of work is static. Since it is not possible to determine the size of a task (shape of subtree) in advance this can lead to load-balancing problems. This can be overcome by allowing a processor to reallocate some of its work to idle processors. Such dynamic task allocation comes at the cost of higher communication and further research will be needed to determine its effectiveness. Some initial work in this area is described in [1]. Han et al. propose some interesting algorithms for balancing load, but as yet do not provide any performance results [2].

Sequential C4.5 has a feature known as *windowing*. With windowing, a decision tree is built from a subset of the training set. This decision tree is then used to classify the remaining data items. The mis-classified cases are then fed back as input to the tree construction process, which then constructs an improved tree from the enlarged training set. Quinlan suggests that the accuracy of the final decision tree can be improved by constructing different trees from different portions of the training set by using the windowing technique and then combining the different tress [5]. There are several different methods for combining multiple trees. The first is to select the one with the lowest predicted error rate. The second is generate the production rule for each tree and then construct the final rules from all the rules available. These techniques are of great interest here as they have a high potential for parallelisation. Future work intends to compare both the performance and accuracy of such approaches. There are other potential changes to the tree construction algorithm which will yield greater parallelism, but may alter the accuracy of the resulting tree. These are a fertile area for research.

6   Conclusions

In this paper we have investigated the potential for improving the computation time of data mining algorithms through the use of parallelism. C4.5, a classification-rule learning system using decision-trees as a model representation, has been used as the case study. Three different parallelisation schemes were implemented. The experimental results from applying the implementations to a sample training set were promising and showed that the execution time of C4.5 could be reduced through the use of parallelism. However, the results also highlighted the problems posed by poor load-balancing. Part of our future work includes the incorporation of load-balancing schemes into the parallel implementations.

Acknowledgements

The authors gratefully acknowledge support from the EPSRC funded project GR/K69988 and the Royal Thai Government. We would like to thank Fujitsu for providing the facilities at IFPC, which made this work possible.

References

1. Jaturon Chattratichat, John Darlington, Moustafa Ghanem, Yike Guo, Harald
   Hüning, Martin Köhler, Janjao Sutiwaraphun, Hing Wing To, and Dan Yang. Large
   scale data mining: The challenges and the solutions. In *Third International Conference on Knowledge Discovery and Data Mining, KDD-97*. American Association
   for Artificial Intelligence, 1997 (submitted).
2. E. Han, A. Srivastava, and V. Kumar. Parallel formulation of inductive classification learning algorithm. Technical Report 96-040, Department of Computer and
   Information Sciences, University of Minnesota, 1996.
3. S. R. Hedberg. Parallelism speeds data mining. *IEEE Parallel and Distributed
   Technology System and Applications*, 3(4):3–6, 1995.
4. C. J. Merz and P. M. Murphy. UCI repository of machine learning databases. University of California, Department of Information and Computer Science,
   http://www.ics.uci.edu/~mlearn/MLRepository.html, 1996.
5. J. R. Quinlan. *C4.5 Programs for Machine Learning*. Morgan Kaufmann Publishers, Inc, 1993.
6. Janjao Sutiwaraphun. Data mining on parallel machines. MSc thesis, Department
   of Computing, Imperial College, September 1996.

Data Analysis for Query Processing

J. Robinson and B.G.T. Lowden

Department of Computer Science, University of Essex, UK

Abstract. Data analysis is needed in connection with query processing, to produce data summary information in the form of rules or assertions that allow semantic query optimisation or direct query answering without consulting the data itself. The goal of an intelligent analyser in this context is to produce robust rules, stable in the presence of data changes, which allow easy rule maintenance as data changes, and provide rapid query reformulation, refutation or answering. It must also limit the rule set to rules useful for query processing.

1 Introduction

The purpose of the data analysis considered in this paper is to improve the speed of answering queries on the data. Analysis produces summary information that can either answer a query without consulting the data itself, or else modify the query to a form the data server will be able to process more quickly. This query-modification operation using knowledge of the data is known as Semantic Query Optimisation (SQO) [1, 3, 5, 7]. *Relational* data is discussed in this paper. This application (ie SQO) requires *continuous data analysis*, as the focus of query interest in the database changes with time. The analyser's activity is guided partly by information it discovers during examination of the data, and partly from query information such as query structure, history of data access, and the frequency of data change in certain areas of the database. Continuous background data analyser processes (which can utilise idle time on any available networked workstations) derive and shortlist useful rules (summary information) for the query reformulation module, and receive copies of queries and data updates. The query processor module can recognise any summary information affected by data changes, because of its simple format, and either revise it [8] or suspend it from use. Although fast data analysis is desirable there is no critical dependency: any current query will utilise information available at the time, but there is no delay waiting for analysis results. Because of this, the continuous analysis workload can be distributed across any available workstations in the local network, to run as background processes. However, the amount of spare computing capacity in a modern workstation is high even when the machine is 'in use'rather than 'screen saving', so analysis results tend to be available in time to be used for the next query. Previous work in connection with meta-data discovery for use in query processing has used a machine-learning approach where

X. Liu, P. Cohen, M. Berthold (Eds.): "Advances in Intelligent Data Analysis" (IDA–97)
LNCS 1280, pp. 447–458, 1997.

the result of a query is treated as a set of positive instances (a training set) [1, 3]. A disadvantage of this approach is that it can produce exactly the rules that are unlikely to be used by future queries, unless the same query is repeated in the near future. The reason for using query-triggering of rule induction was a fear of excesive numbers of discoverable rules. However, queries can be used by an intelligent data analyser to suggest beneficial directions for analysis without unnecessary restrictions. This paper examines the potential extent of discoverable knowledge and discusses ways to impose beneficial and appropriate restrictions. The products of data analysis are *Assertions and Inference Rules*.

The structure of paper is as follows: section 2 introduces the application requiring this form of data analysis, sections 3 and 4 discuss the rules produced by the analyser to support the application. This is followed by sections on rule maintenance and the structure of the system, integrating queries and analysis.

2   Query Reformulation

An example of a database query or subquery is:

"Obtain column6 values from rows of database table J, where

(column2 = "AB6") AND (column4 is BETWEEN 15 AND 46)."

The bracketed components are the query *Conditions*. Each is a Selection Condition, denoting a subset of the database table (ie the subset of tuples matching that condition). Conditions, or *Constraints*, have the form $(a\theta n)$ or $(n \leq a \leq m)$ where a is an attribute name and n, m are constants of the same type as attribute a. θ is an operator from the set $\{<, \leq, =, \geq, >\}$. *Condition matching* between queries and rules is needed, to reformulate the query. Matching is by *subrange containment*. This is also the mechanism for cascading rules to form the CD Graph (section 3.1). Containment is a many-to-many mapping, so rule conditions can be used without necessarily matching query conditions exactly - which would limit the usefulness of a rule. One measure of the usefulness, or *Utility*, of a rule is the number of different query conditions it can match. This *Containment Potential* of Conditions is one of the factors which an intelligent data analyser uses to guide its rule derivation operations. It is quantified as a pair of numbers: (i) condition range as a percentage of extreme value range for the attribute, and (ii) number of tuples selected by the condition as a percentage of the whole table. The data analyser uses this measure to judge potential use for new rules. In the case of Join rules, the percentage refers to the *joined table* from which the attribute-pair rules are extracted. A good rule, $A \Rightarrow B$, in general is one where the containment potential does not decrease much from A to B. The Query Rewrite Module uses information produced by data analysis to either answer or rewrite the query. The requirements of these operations dictate the form of information to be obtained by data analysis. It can either:

(i) answer a query immediately by

a) reporting that there are no values in the result set, or

b) returning the single-value of the query-specified attribute which is provided by the consequent of a rule, or

c) in specialised systems, providing an intensional answer rather than a set of data values [9].

or   *(ii) modify the query* before sending it on to the data server to process in the normal way (using conventional query optimisation to generate an execution plan). The modified query will provide a faster execution plan. *Query reformulation methods* include the following *seven operations.*

1. Query Condition Deletion

Either *(a)* because the deleted condition's truth is implied by another query condition, which will be tested during query processing in the data server, obviating the need to test both conditions. Any tuple that satisfies the first condition is known (by previous data analysis) to satisfy the second.

or *(b)* if the selection condition in the query selects all tuples in the database table. Its range contains both *extreme values* for the attribute. The condition is therefore true of all tuples in the table, so need not be tested during query processing.

2. Query Condition Substitution

Substituting an equivalent but faster-to-test condition. (Equivalent conditions denote the same set of rows in a database table).

3. Index Introduction

Adding one new condition to a query which contains no indexed attribute conditions. The new selection condition extracts a query-relevant subset of rows from the table, to avoid scanning the whole table, applying the other query conditions to each tuple. Therefore the intelligent data analyser must generate rules $A \Rightarrow B$ where B is a condition on an indexed attribute and selects a small superset of the tuples identified by condition A.

4. Query Refutation

Two ANDed query conditions A and B are incompatible if they denote different subsets of the data. Hence the query will produce no results. To support this, the data analyser must have produced rule $A \Rightarrow C$ where the attribute in conditions C and B is the same, but the two condition ranges are disjoint.

5. Query Answering

Query: "Get the values in the d column from all rows of a specified database table where the value in the e column is in the range (52..63)".

Rule: `e(25..90)` $\Rightarrow$ `(d = "secretary")`.

The rule says: If a tuple has an e value in the range (25..90) its d value is "secretary". So return the single result value, "secretary", to answer the query.

6. Join Elimination

From semi-join queries (where result values come from only one of the joined tables), the Join operation can be avoided if the data analyser has produced rules in advance from the appropriate joined table, which show the query condition on the second table is always true if one of the query conditions applied to the first table is true. The data analyser uses information from previous queries to decide *which tables to join, which join attributes to use,* and *which pair of attributes* in the joined table to use for antecedent and consequent conditions in its derived rule set. This restricts rule derivation to rules known to be relevant to queries.

7. Scan Reduction

In a Join query, if one of the tables has no selection condition then the whole table must be scanned repeatedly in comparing the join attribute values of two tables. However, a new selection condition added to the query can make the data server select a subset of the table before the scan phase. A condition can be added if it is implied by an existing query condition. This implication will have been noted, in the form of a rule, by appropriate data analysis for this purpose. The data analyser is aware that high selectivity is required in the consequent condition of rules it derives. It also knows that the consequent condition always selects a superset of the antecedent condition's set, so its systematic data analysis will terminate when either antecedent or consequent denote too large a percentage of the table. However, since tuple numbers rather than percentages are the basis of cost, it will also consider the size of the table when choosing the point at which to terminate analysis.

3   Data Analysis to Support Query Processing

Two types of information result from data analysis: *Assertions* and *Inference Rules*. An assertion is a statement about a table or attribute. *Domain Assertions* for an attribute include: the two *extreme values*; a set of *empty subrange* intervals on the number line between the two extreme values, each denoted by a pair of range limits; *percentile points*, useful during rule analysis to estimate the selectivity of range conditions, but not maintained afterwards. The *number of distinct values* in each column of the table, and the *set of distinct values* for category attributes, are examples of other domain assertions.

An *inference rule* is a pair of conditions, each on a different attribute in the table. For Join Rules the table is the one resulting from the Join operation, and each attribute in the rule is from a different base table. Attribute-pair rules, called 'Simple Rules' in [7], are particularly suitable for query processing. The single condition structure of antecedent and consequent optimises rule usability, (*all* conditions in a rule must be matched). The *robustness* [2] of a rule is the probability that the next data change will not affect it. Attribute-pair rules have only two attributes to be affected, so their robustness is correspondingly higher than rules with two antecedent conditions, for example. Other advantages of the 'simple' rule structure are easy rule extraction from data, fast access (for query processing and rule maintenance) and the ability to precompute all chains of inference that can occur with a query, avoiding delay during query processing. One of their main features is that each rule is an *edge* in a graph.

3.1   The Condition Dependency Graph

Rules can be cascaded. For example, in the three rules:

$$A \Rightarrow b(4..15) \qquad\qquad b(3..20) \Rightarrow C \qquad\qquad b(2..32) \Rightarrow D$$

where A, C, D and b(n .. m) are conditions. The consequent condition (assertion) b(4..15) satisfies antecedent condition b(3..20), so $A \Rightarrow C$. Similarly, by transitive inference, $A \Rightarrow D$ using the first and third rules. The *rule of inference* that allows the consequent of one rule to imply the antecedent of another rule is *subset containment* since a Condition denotes a set of tuples. In the case of range conditions, implication is also by *subrange containment.*

Several rules can be cascaded to form a chain or branched path. Rules are therefore edges in a Condition Dependency Graph, whose paths are chains of inference from a given Condition (such as a selection condition in a query) to another Condition which restricts possible values of a specific attribute. Semantic Query Optimisation can thus be seen as path discovery in the Condition Dependency Graph. Query conditions map to condition nodes in the graph, where paths reveal redundant query conditions, available conditions on indexed attributes that can be added to the query, lower cost Equivalent Conditions (denoting the same subset of data items) and in some cases paths to restriction conditions that immediately refute the query or provide a single-value answer to the query.

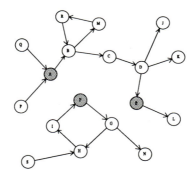

Fig. 1. A Condition Dependency Graph including query conditions A, F and E.

Fig 1 shows that condition E can be deleted from the query. The path from A to E means that testing condition A is a sufficient test for both conditions A and E. The rule $A \Rightarrow E$, denoted by this path, will have been derived by the data analyser in advance of query processing, so there is no graph traversal at query time, only table lookup in the condition-pair rules for pairs of conditions in the query. Rules which the query reformulator seeks initially, to see whether any conditions can be deleted from the query, are: $A \Rightarrow F, A \Rightarrow E, F \Rightarrow A, F \Rightarrow E, E \Rightarrow A, E \Rightarrow F$. Conditions F, G, H and I are *equivalent conditions.* They all denote the same set of tuples, so G, H or I could be substituted for F in the query. If rule $F \Rightarrow H$ gives best cost reduction, substitute H for F in the query. The data analyser has derived rules $F \Leftrightarrow G, F \Leftrightarrow H, F \Leftrightarrow I$, tagged as *equivalences* and labelled with degree of cost reduction between conditions measured as in [3]. If the query still lacks a condition on an indexed attribute and L is an indexed condition with suitable selectivity it can be added to the query because of the path (ie rule) $A \Rightarrow L$.

Fig 1 shows query reformulation is modification to *improve* the cost of the existing query. Relative cost of the two conditions in certain attribute-pair rules is examined by the data analyser. Operations 2 and 3 in section 2 involve relative cost, the rest do not. So only indexed-consequent rules and rules in a cycle of the CD graph need individual condition cost assessment by the data analyser.

3.2   Rule Discovery by Data Analysis

Rules relevant to queries must be derived. Standard SQL data types include numeric types for which ranges can be specified and comparisons other than equality can be used in Conditions. String types, in contrast, have only equality and Set conditions such as (job_title = "secretary") or (job_title IN { "doctor", "lawyer", "academic"}. Range or magnitude-comparison conditions are unlikely in queries on these attributes. String types are often used for categories, such as job titles, so the number of different values occurring in that column of the table is limited. The data analyser derives rules to operate on anticipated queries, so rules with string antecedent attributes have *equality* conditions; numeric conditions use *equality, magnitude comparison,* and *range membership.*

There are four forms of antecedent condition: $(a = n), (a \leq n), (a \geq n)$, and $(n \leq a \leq m)$. The fourth is very useful in query processing, but its rule set is largest. The number of possible rules with antecedent $(a \leq n)$ is N, where N is the number of different values that occur in the 'a' column of the table. Similarly there can be N rules with antecedent $(a \geq n)$ for that attribute. But the number of possible antecedents of type $(n \leq a \leq m)$ is the number of pairs of values that can be chosen from the N different values, to produce pairs of range limits. The number is $^{N}C_2$ so there are $N * (N - 1)/2$ different ranges. This is the maximum size of the rule set for this attribute. Eg: for N = 100 there could be up to 4950 rules. The value N is not necessarily related to the number of tuples in the table, but there is still a clear possibility that the rule set can be much larger than the database table it describes. (And this is only one ruleset for one antecedent attribute). The number of possible rules would obviously increase if rules with more than one antecedent condition were permitted. The numbers above refer to each attribute pair. A database table with M columns provides $^{M}P_2$ ordered pairs of attributes, each a potential rule set. The analyser therefore has to decide which sets of rules should be derived, and which should be retained as query requirements change with time. The size of each attribute pair rule set can be restricted by partitioning it into subsets or discarding rules on the basis of utility values for rules.

3.3   Utility Measures for Rules

Usefulness depends on proposed use. Section 2 identifies seven different uses for rules. Each attribute pair provides rules with recognisable uses. Every rule belongs to a specific Attribute-Pair Class (AP Class) according to the ordered pair of attributes it contains. The set of rules in an AP Class is called an AP Set. Each AP Class has specific Utility metric(s) for its rules. In general, the metric chosen for a particular AP Class depends on the types of the two attributes, the existence of indices, and the existence of rules for related attribute pairs.

For example, rules of the form:

(string attribute condition) ⇒ *(numeric attribute condition)*

can be used in operation 1 (query condition deletion) but cost and relative selectivity of query conditions may be important. If a query contains conditions A and B, rule $A \Rightarrow B$ allows deletion of B, but this may *increase* the cost of the query. A is a high cost condition which, in the absence of other query conditions, must be applied to all tuples in the table (assuming the attribute in A is not indexed). The numeric comparison in condition B is cheaper than the string comparison, so B could be used to extract a subset from the table for condition A to test. The total cost of condition B followed by A may be cheaper than condition A alone. It depends on the size of set(B) relative to the whole table, because condition A must be applied to *all tuples* in set(B). A value for selectivity(B) can be identified for an attribute pair rule set. Rules with selectivity(B) greater than this threshold indicate B conditions that can be deleted from queries, even if there are no other query conditions apart from A, because B does not restrict the set of tuples sufficiently. Condition A might as well be applied to the whole table as to the result set from B. Rules whose consequent condition B selects less than the threshold fraction of the table can be used for operation 3, if there is no AP rule set $A \Rightarrow C$ where C is an indexed attribute. An index can reduce the number of disk blocks accessed in query condition testing. This has a very significant effect on query cost.

Rules of the form: *(numeric attribute condition)* ⇒ *(string attribute condition)* will only be used for adding a condition to a query if the consequent is indexed. Otherwise cost of condition evaluation is not an issue for this sort of rule, nor is relative selectivity of antecedent and consequent. The rules can be used for operations 1, 4, 5 and 6 in section 2.

Other utility metrics for rules include the χ^2 test [4] and the *Search Ratio* [10].

4   Range Antecedent Rule Production

A scanning algorithm can be used in systematic data analysis, to discover rules with interval range antecedent conditions. If the antecedent attribute is sorted into ascending order, its distinct values become apparent and the set of possible ranges is readily extracted. Eg for sorted values: 1, 2, 2, 3, 5, 5, 5, 7, 9

1..2,	1..3,	1..5,	1..7,	1..9
	2..3,	2..5,	2..7,	2..9

$$3..5, \qquad 3..7, \qquad 3..9$$
$$5..7, \qquad 5..9$$
$$7..9$$

are the ranges identified by a systematic scanning algorithm. Each range corresponds to an antecedent condition in a rule. The corresponding range of data values for the consequent constraint is obtained from the tuples selected by the antecedent condition. The following example illustrates the process. The three attributes in the database table are named a, b and c. The table has been sorted using column 'a', because that is the required antecedent attribute at present. (Different workstations sort the table on different attributes, in order to simultaneously generate different rule sets). The scanning algorithm uses two pointers to generate range antecedents and corresponding attribute-pair rules:

	a	b	c
lower	1	6	12
	2	4	4
upper	2	18	12
	3	9	4
	5	3	12
	5	7	4
	5	1	4
	7	26	5
	9	28	12

$$a(1..2) \qquad b(4..18)$$

$$a(1..2) \qquad c(4..12)$$

Table 1.

Both pointers start at the first tuple. The upper-limit range pointer advances to find a data value different from that indicated by the lower limit pointer, then advances to the last occurrence of that data value in the sorted sequence. A range of tuples now lies between the two pointers. These tuples are the set selected by the range constraint a(1..2) at present. The analyser simply identifies the extreme values of b and c in this set to generate two rules shown to the right of the table. The upper limit pointer now advances to the final copy of the next value in column a *and notes the new tuples it passes*. In the example it advances to 3 and adds only one tuple to its *difference set*. Values in the difference set are examined to see whether they extend the range of the previous rule consequent. The value b = 9 is within the existing consequent range b(4..18), so the new rule is a(1..3) ⇒ b(4..18). This *subsumes* the previous rule: a(1..2) ⇒ b(4..18) which can therefore be deleted immediately from the rule set. The remaining rule is more useful than the deleted rule, and no information is lost by the deletion. It was a *logically redundant* rule. If retained it degrades system performance by slowing access to the rule set and increasing rule maintenance overhead.

Rule Subsumption Theorem

For two rules,   R1: $A \Rightarrow B$,   and   R2: $C \Rightarrow D$,

R1 subsumes R2 if range(A) contains range(C) AND range(D) contains range(B). The proof is as follows:

- If range(A) contains range(C), eg a(5..100) contains a(25..90), then C *implies* A. If C is true then A is true. Therefore $C \Rightarrow A$ and $A \Rightarrow B$ so $C \Rightarrow B$.
- now $C \Rightarrow B$ and $C \Rightarrow D$ but B is a more restrictive range condition than D. For example, compare the rules: $C \Rightarrow e(25..50)$ and $C \Rightarrow e(13..100)$. The first rule implies the second, by *subrange containment*:

A value in the range $(n..m)$ is also in any range $(\leq n.. \geq m)$.

The data-scanning algorithm continues on the example table above, advancing the range upper limit pointer and noting whether attribute values in tuples in the difference set extend the existing consequent range or not, and deleting the previous rule if not. For the c column, the only rule remaining at the end of this first scan is $a(1..9) \Rightarrow c(4..12)$. And this will be deleted because both conditions denote the whole table, so the rule has no information content. The data analyser uses the *extreme values* domain assertion for attributes, to recognise these redundant conditions and discard the corresponding rules.

The scanning algorithm now begins a new scan: the lower limit pointer advances to the *first instance* of the next distinct value of the sorted antecedent attribute and the upper limit pointer searches forward from this value to the *last instance* of the next distinct data value. This initialises the range for the next scan, and consequent conditions are computed from the initial set enclosed by the limit pointers. A useful feature of the scanning algorithm is the way its local *zone of interest* (the difference set) moves progressively downwards through the table during each scan. The zone is likely to fit in a memory page so the algorithm is well-suited to the virtual memory system of workstations. (Suitable for networked clusters of workstations). The maximum number of tuples examined in the algorithm (if equality antecedents are included) is $n * (n + 1)/2$, where n is the number of rows in the table. Duplicate values in the antecedent column reduce this number. Each repeat value eliminates one of the n scans (each containing $(n + 1)/2$ tuples, on average).

4.1   Path Extension Rule Production

The next phase of data analysis takes the set of rules produced by scanning and uses Antecedent and Consequent Triggering as a rule discovery algorithm. This extends CD Graph paths associated with rules (edges) known to be useful, and so adds value to those rules. Examples of rules known to be useful are indexed consequent rules (including Join rules) and Join rules with closely matching selectivity in antecedent and consequent conditions, so these are derived first by scanning. A consequent a(m..n) requires new rules with antecedent $a(\leq n.. \geq m)$

for path extension. Data examination reveals consequents in other attributes for these antecedent ranges. From a rule $A \Rightarrow B$ try first to produce a rule $B \Rightarrow A$ showing conditions A and B are *equivalent*, then generate a set of rules $B \Rightarrow C$, and hence infer rules $A \Rightarrow C$. This transitive deduction of attribute-pair rules from a CD graph path generates new rules that are independent of the rules used to reveal them. In a chain of inference corresponding to a path in the graph, intermediate rules can be eliminated by changes to data but the derived rule need not be affected. It is a rule linking values in two attributes, and only changes to those data values affect the rules.

Forward extension is simple. Eg from edge: $d(51..90) \Rightarrow f(16..36)$ use condition $f(16..36)$ to select tuples from the table. The range for each other attribute, in the set of selected tuples, gives a consequent C in a rule of the form: $f(16..36) \Rightarrow C$.

Backward path extension involves more work. Selection condition $d(51..90)$ derives positive instances. The negative set is selected by $((d < 51) OR (d > 90))$. Initialise each target antecedent's range condition, using the range denoted by the two extreme values for that attribute in the positive set of tuples. This condition may imply values of attribute **d** outside the range (51.. 90), so the antecedent range is progressively narrowed to eliminate all negative values. Hence antecedent ranges. The consequent range is correspondingly narrowed from $d(51..90)$ in the new rule.

5   Rule Maintenance

Generally there are two problem areas with regard to the maintenance of rules: updates to the rule set, and updates to the database. For previous authors, the first concerned the maintenance of a non-redundant and non-contradictory set of rules in the presence of changes and additions to the rule set [11, 12]. This problem is, however, associated only with rules which are imposed on the database, eg. integrity constraints, and not with rules which are derived by data analysis since, provided the latter are consistent with the current state of the database, then they must necessarily be consistent with each other. Adding a rule (an edge) to the C.D. graph allows new rules to be discovered by deduction, as paths in the graph. But rule deletion has no direct effect on other rules, whatever their method of discovery, since each rule depends only on the attribute pair it describes.

The problem of maintaining derived rules in the presence of updates to the database has been tackled by the authors elsewhere [8]. In that paper algorithms are proposed which detect rule falsification and initiate rule modification procedures where necessary. One or more AP Sets affected by new data may be temporarily withdrawn from use by the query reformulator while the rules are checked by background processes. (Checking is a lookup process in an AP set's data structure, rather than an examination of all rules in the set, so the rule set is soon back in use). Data *deletion* cannot make any rule invalid, so the modification process following this data change is not urgent. All rules in affected

AP sets continue to be used for query *reformulation*, but the use of consequent values from rules in the affected AP Classes as *answer values* to queries (instead of consulting the database) is temporarily disallowed, until a background process has updated the rule set. Rule maintenance is an incremental process. It is not necessary to re-analyse the data.

Many important data sets are static. Data Warehouses and Data Archives, for example, separate the active data from data for investigation (ie read-only access by queries). Any derived rule in these systems is permanent.

6   The Overall System

This paragraph gives a brief indication of the way the various components fit together. In a local network of computers with normal multiprocessing capabilities, one or more of the machines is a standard database server, and one machine (or process) provides a user interface to data. All changes to data (ie insert, delete, modify tuple) must occur through this interface. Any data read (query) wishing to benefit from SQO is done by using this interface as the SQL server. The interface module contains the query reformulator which matches query conditions with nodes in the CD graph and performs the operations described in section 2. If a query contains a pair of conditions for which no AP Class currently exists, then the task of data analysis to generate the required AP set of rules is delegated to a new process on any machine in the network. The interface maintains statistics on frequency of use of each AP set, in order to decide when to discard an AP set from the rulebase, and when to convert a partial AP set produced by path extension into a full set by scanning the table.

7   Discussion and Conclusions

Early work in SQO used Integrity Constraints as the rule base. This provided *immutable* rules, immune to data changes, but needed a person to specify the knowledge. Also integrity rules are not well suited to SQO requirements. More recent work has concentrated on automatic discovery of rules. The authors [3], and Hsu [1] used machine learning to generate rules that would have been useful if they had existed *before* the query that triggered their discovery. Siegel [7] used queries in a similar way, to suggest rules that would have been useful, but examines the data directly to test whether the rules are supported by the data. We now generalise the query-recommended rule to a pattern for a set of rules to be obtained by data analysis, which then examines and shortlists the most probably useful subset for future queries. Shekhar [5] used a classification grid to examine the distribution of data values and derive rules with a wide range of structures, whose applicability to a wide range of queries, robustness against data changes, ease of maintenance, and access speed is poor compared with attribute-pair rules. Our data analysis strategy is flexible in the amount of work it does and the number of rules it generates, but because of its more thorough examination of at least part of the data, it can choose to retain a more appropriate subset of the rulebase. It can also optimise range conditions in rules to maximise antecedents and minimise consequents by taking account of *empty subranges* (query ranges include an arbitrary amount of empty subrange which gets copied into rules by earlier query triggering systems). The ability to precompute the transitive closure of the CD graph eliminates much of the delay previously associated with query reformulation [6], where sequential application of rules was used to (in effect) build paths between query conditions, but with no guide to the choice or order of application of rules.

Our initial experimental results, using rules obtained by data analysis, are encouraging. Experiments carried out on large datasets, using a suitable query transformation algorithm, have shown significant savings in query execution time [10]. Performance improvement was directly related to the quality of the rules employed and ranged from several percent to nearly 100 %.

References

1. Hsu, C., Knoblock, C.A.: Rule Induction for Semantic Query Optimization. Proc. 11th International Conference on Machine Learning, 1994, pp 112-120.
2. Hsu, C.,Knoblock, C A.: Estimating the Robustness of Discovered Knowledge. Proc. 1st International Conf. on Knowledge Discovery and Data Mining, 1995.
3. Lowden, B G T., Robinson, J., Lim, K Y.: A Semantic Query Optimiser using Automatic Rule Derivation. Proc. WITS 95: 5th International Workshop on Information Technologies and Systems, Holland, 1995, pp 68-76.
4. Sayli, A.,Lowden, B G T.: The Use of Statistics in Semantic Query Optimisation. Proc. 13th. European Meeting on Cybernetics and Systems Research, pp 991-996, Vienna, 1996.
5. Shekhar, S., Hamidzadeh, B., Kohli, A., Coyle, M.: Learning Transformation Rules for Semantic Query Optimization: A Data-Driven Approach. IEEE Transactions on Data and Knowledge Engineering, 5(6), 1993, pp 950-964.
6. Shekhar, S., Srivastava, J., Dutta, S.: A Formal Model of Trade-off between Optimization and Execution Costs in Semantic Query Optimization. Proc. 14th International Conference on Very Large Databases, 1988, pp 457-467.
7. Siegel, M., Sciore, E., Salveter, S.: A Method for Automatic Rule Derivation to Support Semantic Query Optimization. ACM TODS 17(4) 563-600, 1992.
8. Sayli, A., Lowden, B G T.: Maintaining Derived Rules for Semantic Query Optimisation. Computer Science Memorandum 291, University of Essex, 1997.
9. Lowden, B G T., et al.: Modal Reasoning in Relational Systems. Journal of Database Technology (4) 4, pp 235-244, Pergamon Press, 1993.
10. Sayli, A., Lowden, B G T.: A Fast Transformation Method for Semantic Query Optimisation. Proc. IEEE International Database Engineering and Applications Symposium, Montreal, August 1997.
11. Yu, C., Sun, W.: Automatic Knowledge Aquisition and Maintenance for Semantic Query Optimisation. IEEE Transactions on Knowledge and Data Engineering, 1989.
12. Ishakbeyoglu, N., Ozsoyoglu, Z M.: On the Maintenance of Implication Integrity Constraints. DEXA '93: Proc. 14th Intl. Conf. on Database and Expert Systems Applications, 1993, pp 221-232. (LNCS 720)

Datum Discovery

Laurent Siklóssy and Marc Ayel

LIA - Université de Savoie
F-73376 Le Bourget du Lac - France
Email : {siklossy,mayel}@univ-savoie.fr

Abstract. In some contexts, it is more important to find a single datum than to find many data which satisfy some criteria of interest. In the domain of police analysis, the discovery of a single datum may make it possible to determine, for example, the structure of a criminal organisation from already known structures that appeared initially unrelated, or to discover the single identity of a criminal who was hiding behind several aliases.

To search for a potentially interesting datum, we suggest two approaches. The first approach makes use of our system to process incomplete, dynamic knowledge, contributed by several informants. In the second approach, we propose a single paradigm, the search in neighborhoods of a case, to search for and discover items of interest.

1 Introduction

In data-mining, often various statistical methods are applied to notice certain regularities (For some recent results in this area see [3], [4], [5], [7]). In datum discovery, we are interested in exploring the contents of the mine of data, but it is often for the purpose of finding a *single datum*. (We are also interested in finding certain regularities, but that aspect will be de-emphasized here.)

We are looking for isolated, single data which, for example, will allow us to connect structures of knowledge which have seemed independent until now (see [2] and [6]), or to realize that two seemingly different persons are in fact the same individual, etc.

We shall describe situations where datum discovery is relevant, and describe various approaches towards mining that useful, unique datum. Our ideas have had some initial implementations in the context of Interpol (section two), but much needs to be done to validate our approaches. Of course, the domain of Interpol is only one possible domain of application of our ideas, even if our description relies on the example of Interpol.

Datum discovery can be considered as a part of data mining, since the search for a single datum is not excluded in data mining, although it is not considered usually. However, the tools used in data mining, often statistical tools searching for regularities, and the general goal of data mining, finding many connected data, differ markedly from our approach towards the discovery of (often) a single

X. Liu, P. Cohen, M. Berthold (Eds.): "Advances in Intelligent Data Analysis" (IDA–97)
LNCS 1280, pp. 459–463, 1997.   © Springer–Verlag Berlin Heidelberg 1997

datum. We do not search for statistical regularities, since we are often interested in just one datum. We do not attempt to find many connected data, but rather single pieces of information which may, among others, connect so far disjoint structures of information, or more generally greatly fertilize already existing but apparently scattered information.

2   The Context of Interpol

Interpol, headquartered in Lyon, France, is an international police organisation. The members of Interpol include most countries in the world. [Our description of Interpol is necessarily sketchy and incomplete, but should suffice for our purposes here.]

A goal of Interpol is to help in the fight against criminality, in particular in certains domains, such as drugs trafficking, and the theft, transport and sales of art objects (e.g. paintings) or cars. Interpol receives data from many national police organisations, and keeps in its databases only data which have a high-level of certainty, and therefore can be *certified* correct. [Correct data need not reflect truth. For example, it may be certifiably correct that under certain circumstances a person entered France carrying a passport in the name of Siyel, but that does not necessarily imply that the person's name *is* Siyel.]

The analysts at Interpol analyze cases, for example a particular stolen car. Several cases can be united, if it appears that a gang of car thieves is at work, and the analyst will be interested in building structures, for example the structure describing the organisation of a gang. Considerable progress is made when very small amounts of information make it possible to connect cases which at first glance may appear totally disconnected.

This is exactly what happened recently in Belgium (seemingly, not a result of Interpol's work), where organisations as disparate as gangs of car thieves, organisations of pedophiles and high political spheres were found to be linked, and furthermore connected with murder cases. In the Belgian example, the various seemingly separate organisations became linked after one single datum of information here, and then another single datum of information there, made it possible to first conjecture links between the organisations and murder cases, awaiting subsequent proofs acceptable to the judiciary.

3   Our goal : Increase the Efficiency of Knowledge Workers

The *productivity* of an analyst at Interpol is from about one to a few cases per year. [A case is a set of similar activities which take place in similar temporal and geographical neighborhoods, for instance a series of art thefts in Southern France over a two months period.] The long time required by an analyst to *dig* into a case, while trying to solve it, is justified by the great amount of time it takes her to search vast amounts of data, almost all of which is not -or at least does not appear to be at first- relevant.

Our goal is to *industrialise* the empirical methods of the analysts of Interpol, and more generally of (certain types of) knowledge workers who often search, typically among large amounts of information, the datum which can significantly advance their work. We propose to develop an intelligent assistant who could greatly help our knowledge workers, and could therefore greatly increase the quality of their results (for example, by helping to build a more complete description of a criminal organisation), and by increasing the efficiency with which their results are produced.

4   Relevant Techniques

We assume that a knowledge worker, using various information, has built an initial description of a case; typically such a description will involve some individuals, some activities, some relations among the individuals, etc.

Presently, we hypothesize that a good paradigm for datum discovery is the exploration of neighborhoods of a case. Of course, we wish to discover data which advance significantly the existing discovery of a case. One approach, when it is relevant, is to first generate a variety of data, and then to test by simulation whether such data would significantly advance the comprehension of a case. The second approach explores the various neighborhoods of a case.

We shall describe now very briefly these two approaches.

4.1   Generation of a potentially very productive datum

At the Université de Savoie, we have developed a running system which processes incomplete dynamic data (see [1]).

More specifically, we assume that several informants provide independently correct but usually incomplete data to a central system; moreover, we assume that the informants are trustworthy, i.e. the information they provide is correct within its incompleteness.

For example, if an informant reports that between 23 and 27 individuals came to a criminal meeting (the informant could not provide better information since it was very dark...) then we assume that indeed between 23 and 27 individuals came to the meeting. Our system processes the contributions of the informants, together with a wide variety of domain constraints, to improve the precision of the available information. For example, using various other information and constraints, we could conclude that the number of persons attending the meeting was 26 or 27. Sometimes the constraint computation gives a precise result: the number of persons is 26, for example.

It now becomes easy -it would only cost much computer time- to find out which additional datum could significantly improve the existing database. [In a way, we are looking for information which gives us a lot of $bang-for-the-buck$!] Indeed, sometimes some seemingly small datum can have a propagating wave effect, which can reduce significantly the incomplete information contained in the system. Datum discovery now has become directed: try to find in the data mine (or possibly request from outside sources) that potentially very productive datum.

The potential productivity of a new datum could be measured by the amount of incomplete information which would be affected by this new datum. When many relations and constraints exist inside the set of incomplete information, the computation of the productivity of a datum is not obvious. Moreover, the semantic validity of that number decreases with the number of relations.

The selection of most productive data can lead the generation of warnings: when a hightly productive datum arrives, the database is modified and the Interpol analyst is alerted. At that point, the criminal analysis can be resumed by the analyst in a deeper way.

4.2   Exploration of neighborhoods

The neighborhood of a case could be defined as the set of elements (objects, persons, relations, structures, actions, etc.) which are closely connected to the case. To advance a case, we propose to search in the neighborhoods of a case again for that very productive datum (or two, or more...) which could significantly advance the case.

A case has many neighborhoods. Among some of them, we could mention:

-geographic neighborhoods Were there certain activities, certain persons present in the geographic vicinity of a case?
-time neighborhoods Did certain activities (e.g. telephone calls) take place during approximately the time interval of a case?
-spatio-temporal neighborhoods Did some potential suspects take a train or a plane some time before or some time after an incident?
-phonetic neighborhoods It has been noticed that some criminals use differently spelled, but phonetically identical or similar aliases.
-identification neighborhoods A criminal using many aliases must remember correctly her many identities.
Memory will be helped if some of the items to remember, e.g. place and/or date of birth, do not vary. In this way, we can build possible identification neighbors of a criminal.
-numeric neighborhoods The neighborhood depends on the transformations that could be applied to a number.
For example, when falsifying an automobile chassis number, it is relatively easy to transform a 3 into an 8, but not vice-versa. In this way, we can build potential neighbors of the chassis number of a stolen car.

At this point, we have experimented with some of these neighborhoods, in prototypical systems.

5   Conclusion

Perhaps as much as, and perhaps more than data mining, datum discovery could prove very useful to knowledge workers who try to build structures, connections and relations among diverse data. We have shown how datum discovery could prove particularly useful to increase the scope of the work and the efficiency of police analysts.

For datum discovery, we suggest two approaches, which have received a beginning of experimental confirmation and validation:

- simulation, to find potentially very powerful information, which could greatly advance a case. Our system to process incomplete information could be helpful in this context.
- the exploration of neighborhoods of a case, again searching for potentially very powerful information. There are many neighborhoods of a case, and we have described very briefly some of them.

References

1. Cimpan, S., Képes, I.: Incomplete information in databases. LIA Report. LIA - Université de Savoie. Chambéry, France. 1996.
2. Dumitrescu, A.: Incomplete information in frame-based systems. LIA Report. LIA - Université de Savoie. Chambéry, France. 1996.
3. Fayyad, U.: Data Mining and Knowledge Discovery: Making Sense Out of Data. IIIE Expert. Intelligent Systems and their Applications **17 (5)** (1996) 20-25.
4. Fayyad, U., Piatetsky-Shapiro, G., Smyth, P.: From Data Mining to Knowledge Discovery in Databases. AI Magazine **17 (3)** (1996) 37-54.
5. FinCEN The Global Fight Against Money Laundering FinCEN, Vienna, VA, USA
6. Ivask, E.: Incompleteness in rule-based systems. LIA Report. LIA - Université de Savoie. Chambéry, France. 1996.
7. Wasserman, S., Faust, K.: Social Network Analysis: Methods and Applications New York: Cambridge University Press (1996)

A Connectionist Approach to Extracting Knowledge from Databases

Zhou Yuanhui, Lu Yuchang, Shi Chunyi

Department of Computer Science
Tsinghua University
Beijing, 100084, P.R. of China

Abstract. Classification, which involves finding rules that partition a given data set into disjoint groups, is one class of data mining problems. Approaches proposed so far for mining classification rules for large databases are mainly decision tree based on symbolic learning methods. In this paper, we use artificial neural network to mine classification rules. We present a novel approach, called LBSB, composed of two phases to extract rules from artificial neural network and discover knowledge in databases. Some experiments have demonstrated that our method generates rules of better performance than the decision tree approach in noisy conditions.

1 Introduction

Although the stored data are a valuable asset of an organization, most organizations may face the problem of data rich but knowledge poor sooner or later. This situation aroused the recent surge of research interests in the area of data mining [1, 2]. One of the data mining problems is classification. Classification is the process of finding the common properties among different entities and classifying them into classes. The results are often expressed in the form of the classification rules. The classification problem is to obtain a set of rules by using a given training data set. By applying these rules to the testing set, the rules can be checked whether they generalize well.

In this paper, we apply the neural network to mine classification rules from databases with focus on articulating the classification rules represented by the trained neural network. The reason is that compared with symbolic inductive learning such as C4.5, artificial neural networks perform better than in noisy condition. But the major weakness of this approach is that the knowledge learned by a neural network is very difficult to interpret. It is becoming increasingly apparent that without some form of explanation capability, the full potential of trained artificial neural networks may not be realized. How to explain the behavior of the trained neural networks, how to understand the concept representation formed by the network and how to make this black box problem transparency have attract many researchers to work on it.

X. Liu, P. Cohen, M. Berthold (Eds.): "Advances in Intelligent Data Analysis" (IDA–97)
LNCS 1280, pp. 465–475, 1997.

We present a novel method including two phases to extract rules from the trained neural network. The first phase is to extract rules from hidden layer to output layer. In this phase, we use a learning method to identify the regions in the hidden activation space in order that all hidden vectors in them generate the activation of a given output larger than a certain threshold. The second phase is to extract rules between input layer and hidden layer in a searching method to select all rules so that all instances covered by them generate hidden activation vectors lying in these regions. Some experiments have been done to prove that the rules extracted in this method have many advantages on symbol learning method such as C4.5 [5].

2   Mining classification rules using NN

2.1   Neural network learning

A three layer neural network is trained in this step. The training phase aims to find the best set of weights for the network which allow the network to classify with a satisfactory level of accuracy. An initial set of weights is chosen randomly. Updating these weights is done by using information involving. The neural networks have fully-connected hidden units in a single layer. The number of hidden units used in each network is determined by cross-validation within the training set. For each training set, we reduce the number of the hidden units and use cross validation to pick the hidden units and use cross validation to pick the network that is to be trained on all of the data in the training set. This phase is terminated when the stopping criterion meet.

2.2   Rule extracted from trained network in LBSB

This process includes two algorithms: LB algorithm and SB algorithm. The LB algorithm is used to identify the regions of hidden activation space so that all hidden vectors within them belong to positive class. The SB algorithm is to search all feature combination to make the input instances covered by them lying in these regions.

LB algorithm This algorithm is used to extract a set of valid regions, called hypercubes whose dimensions are determined by the number of hidden nodes, in order that all input instances which cause their corresponding hidden activation vectors within these hypercubes belong to same class. Here, a hidden activation vector is composed of all hidden nodes' outputs for an input example. We assume that the activation function is the sigmod function

$$F(a) = \frac{1}{1 + e^{-\lambda a}} \tag{1}$$

where λ determines the steepness of the function and $\lambda > 0$. The activation of unit j (a hidden or output unit) is calculated by

$$O_j = F \left(\sum_i W_{ji} X_i - \Theta_j \right) \tag{2}$$

From (1), we can see that $F(a)$ is a monotonic function. The value of F will increase as a grows. From (2), we can see that the sensitivity pattern for input vector X which is directly connected with the output node O_j depends on the connected weight W_{ij}. If W_{ij} is greater than zero, then O_j will increase with W_{ij} increasing and the sensitivity pattern of X_i is '+'. Otherwise, will decrease and sensitivity pattern of is '−'. The sensitivity pattern do not change in the whole input space. This will give us a chance to easily extract a set of hypercubes which approximately represent the original valid space in which all examples belong to same class. The LB algorithm is briefly described below:

The LB(learning-based) algorithm
Step1: Initially, the examples used for extracting hypercubes, called failure positive examples, are trained positive examples.
Repeat
Step2: Using region clustering algorithm, according to the failure positive examples, continue to extract a set of valid hypercubes from the whole hidden activation space, in which all hidden activation vector generate a given output larger than a threshold.
Step3: Expand the size of these valid hypercubes, including upward expanding and downward expanding.
Step4: Randomly generate positive examples, which causes a given output larger than a threshold, calculate their corresponding hidden activation vectors and select the failure positive examples which do not lie in the valid hypercubes from the positive examples.
until stopping criterion meet.
Step5: Extract rules such as: If $(H \in D_1)$ or ... or $(H \in D_l)$,where H is the hidden vector, D_i represents ith valid hypercube.

We use a region clustering algorithm to extract a set of valid hypercubes according to the sensitivity pattern of hidden vector. We check each hypercube by finding the hidden vector which generates the smallest activation of the given output. If the smallest activation of output is larger than a given threshold, all vectors in this hypercube will cause the output exceeding to the threshold. So this hypercube is a valid region. Else, there must be some hidden vectors in it that do not satisfy the condition and the hypercube is invalid. Because the sensitivity pattern of a hidden vector doesn't change, this point must be the topmost of the hypercube. According to the sensitivity pattern, we can easily find this point. For example, a 3-dimensional hypercube with three valid intervals $[a_1, b_1]$, $[a_2, b_2]$, $[a_3, b_3]$ the sensitivity pattern of the hidden vector is +,+,−.

Then the hidden vector which produces the smallest output is (a_1, a_2, b_3). The region clustering algorithm is described as follow.

The region clustering algorithm
We start with $j_{max} = 1$
For all hidden activation vectors which are generated by the failure positive examples, do
begin
Step1: let x be the hidden vector currently under test. If it is in a valid hypercube D_j, continue this loop.
Step2: Set $j = 1$ and $flag ::= NotYetFound$
Step3: While $j \leq j_{max}$ and $flag ::= NotYetFound$, do
begin
Step3-1: Select the smallest hypercube D_j so that it can contain the valid hypercube D_j and x.
Step 3-2: Set $Status ::= OK$
Step 3-3: According to the sensitivity pattern, we can select the vertex of hypercube D_j, which causes the output smallest. If this output value is smaller than the given threshold, then set $Status ::= NotOK$.
Step3-4: If $Status ::= OK$, then set $flag = Found$ and exit from this level of loop. Otherwise, increment j as $j ::= j + 1$, and continue.
end.
Step4: If $flag = Found$ then $D_j = D_j$. If $flag = NotYetFound$, then create a new set containing only x, i.e.
begin
Step4-1: $j_{max} = j_{max} + 1$
Step4-2: Create $D_{jmax} = \{x\}$
end.
End.

The extracted hybercubes maybe cover a part of the positive hidden activation space. So we can expand these hybercubes to cover the space as more as possible to describe the network to an satisfactory degree of accuracy . This can be done according to the sensitivity pattern. The expanding proceeding includes upward expanding and downward expanding. Upward expanding is to enlarge the size of a hypercube along the directions which cause the output activation increasing. Downward expanding begins from the topmost of the hypercube, which generates the smallest output, to enlarge the hypercube along the direction that causes the output activation decreasing and exceeding the given threshold in the mean time. The enlarging direction can be decided according to the sensitivity pattern of hidden vector. For the two expanding methods, there is a same condition that all expanding hypercubes can not intersect. This can avoid generating redundant rules.

The stopping criterion can be done in different ways, such as the number of iteration, the ratio of the number of the examples that lies in the hypercubes to that of the whole positive examples, etc. If the stopping criterion meets, then stop. Otherwise continue extracting the hypercubes and expanding them in our clustering learning according to the failure examples.

The SB algorithm After LB algorithm has been used to extract a set of hypercubes, the SB algorithm heuristically searches through the rule space from input layer to hidden layer in terms of combinations of attributes so that all examples covered by the rules generate hidden activation vectors lying in the valid hypercubes. For a single hidden unit, there are several valid intervals extracted in the first phase. SB extracts rules for each hidden unit, which are expressed in terms of the units that feed into it so that all instances covered by them generate the activation of the hidden unit within the these valid intervals. SB is a searching algorithm which is similar to that of [6], but there are some differences. Different from other search-based algorithms, such as KT algorithm [4], this algorithm is based on a heuristic search which is conducted by combining the positive and negative weights. This algorithm first convert all negative weights to positive weights, then sort the weights in descend ordering and recursively extract all essential prime implicants(the maximum-general rules) by using the solution function and the bounding function in the depth-first backtracking tree search. The sorting of weights in descending order ensures that the search is done in an optimal order and non-prime implicants are not generated. The bounding function in the search process eliminates the possibility of generating any unnecessary subpath which can not be a part of any solution subsets. The bounding function is given as

$$\sum_{j=1}^{k} w_j x_j + \sum_{j=k+1}^{n} w_j \geq T \quad \text{with} \sum_{j=1}^{k-1} w_j x_j < T \tag{3}$$

where k is again the tree depth of a current subpath and is in $\{0,1\}$. The use of the solution function in the tree search ensures that no unnecessary superset of any solution subset is generated. The solution function is given as follow:

$$\sum_{j=1}^{k} w_j x_j \geq T \quad \text{with} \sum_{j=1}^{k-1} w_j x_j < T \tag{4}$$

where k is the tree depth of a current subpath and x_j is in $\{0,1\}$. The detail will see [6]. SB algorithm is different from backtrack tree algorithm on two points:
(a) SB may extract rules from a closed intervals, such as $a < \sum W_{ij} h_i < b$. When it generates all leaves, called positive leaves, which make $a < \sum W_{ij} h_i$, it will continue searching from these positive leaves and producing all leaves, called negative leaves that make $a \sum W_{ij} h_i < b$. Then extract rules from the final negative leaves according to the binary code from root to these leaves.
(b) SB will extract rules from a set of intervals for each hidden node. In order to reduce the complexity of running time, we use only one backtrack tree to finish this searching. First, sort the intervals bounds in a increasing order and divide them into two groups, one for lower bound $a_i \in A$ and one for higher

bound $b_i \in B$. Then search in this order to generate positive and negative leaves. When meeting the element of A, then generating all positive leaves that cause the $a_k < \sum W_{ij}h_j$ in the method [6] and continue searching from these leaves. When meeting the element of B, then generating all negative leaves that cause the $\sum W_{ij}h_j < b_k$ and extract rules according to the binary code from root to these negative leaves, and then continue searching from all positive leaves that cause the $b_k < \sum W_{ij}h_j$. This process continues until the largest bound has been searched.

The SB (searching-based) algorithm
For all hidden nodes, do
Step0: Covert the valid intervals of the currently hidden node into those of its input.
Step1: Sort the valid interval bounds in an ascending order and put them into the R.
Step2: For all elements in R, according to this order, do
begin
Step2-1: using backtracking tree algorithm to generate all positive leaves making $r_i < \sum W_{ij}x_i$ and in the mean time generate all negative leaves making $r_i > \sum W_{ij}x_i$.
Step2-2: If the element r_i is lower bound of an interval and there is no higher bound of the interval, extract rules from root to these positive leaves according to the binary code. If the element r_i is higher bound of an interval, extract rules from root to these negative leaves according to the binary code.
Step2-3: continue this loop from the positive leaves.
end
End

2.3   Rewriting the extracted rules

After finishing the LB and SB algorithm, a rewriting module is used to rewrite rules containing some symbols designating hidden units which do not correspond to predefined attributes or concepts so that these symbols are eliminated from the rules. This process contains two phrases: one is to eliminate the symbols representing the hidden nodes and the other is to combine the rules extracted from all hidden nodes. This module make it easy for us to understand the extracted rules.

3   Experiment and Results

In this section, we describe the databases and representations used in the experiments. We compare the rules extracted in LBSB with those extracted from decision tree to see which performs better in noisy condition.

3.1 The Databases and the Representations

We use two databases, which are Splice junction and Pima Indians diabetes, from the University of California Irvine data repository for machine learning and one real- world database from YuXi cigarette factory.

- Splice junction: Splice junctions are points on a DNA sequence at which 'superfluous' DNA removed during the process of protein creation in higher organism. The problem posed in this database is to recognize, given a sequence of DNA, boundaries between exon(the part of the DNA sequence retained after splicing) and intron (the part of the DNA sequence that are spliced out). This problem consists of two tasks: recognizing exon/intron boundaries(referred to as EI sites) and recognizing intron/extron boundaries(IE sites). The data set contains 3190 examples, in which 767 are for exon/intron boundaries(EI), 768 are for intron/exon boundares and 1655 for neither(N). Each example has 60 attributes which are 60 sequential DNA nucleotide position. Each attribute has four values A,G,C and T, coded as 1000,0100,0010 and 0001. We select 1500 examples for training and else for testing.
- Pima Indians Diabetes: This database was made by National Institute of Diabetes and Digestive and Kidney Diseases. Several constraints were placed on the selection of these instances from a larger databases. In particular, all patients here are females at least 21 years old of Pima Indian heritage. There are total 768 instances, 500 for class 0 and 268 for class 1. Each instance has 8 attributes, which are all numeric valued. They are sample code number, clump thickness, uniformity of cell size, uniformity of cell shape, marginal adhesion, single epithelial cell size, bare nuclei, bland chromatin, normal nucleoli and nitoses. We use half of them for training.
- A real-world database: It is a real-world database— Information management and Decision support system of the YuXi cigarette factory. We use the large database on market information of this factory to analysis the relationship of the economic parameters and extract decision knowledge from the database to offer decision supporting for managers. In this problem, we use 32 economic attributes that may have influence on the sales volume of several brands of cigarette made by YuXi cigarette factory, such as gross national product, gross fixed asset formation, general price index, total volume of retail sales, the number of population, standards of consumption, national income per capita, gross product of the tobacco, age structure of the people, education of the people, unemployment and employment, etc. We use 20 annual data of these attributes and the classes are the sales volume of the high-grade, the middle grade and the low grade cigarette of the YuXi factory, each class have three attributes: worse, better and general. By this, we hope to know which economic attributes have important influence on the sales volume and how they affect.

3.2   Experiment and the result

We compare the rules extracted from neural network in LBSB with those extracted from decision tree. The neural networks used for the three domains have fully-connected hidden units in a single layer. The number of hidden units used in each network is determined by cross-validation within the training set. For each training set, we reduce the number of the hidden units and use cross validation to pick the network that is to be trained on all of the data in the training set. Decision trees are also induced by these data and rules extracted from them. In testing the tolerance to noise, we added noise at different levels by changing the class labels of training instances randomly. For example, a 10% noise level means that we perturb 10% of the training instances by changing their class labels. We test it using the three databases. For example, in the Splice junction domain, we use 1500 trained examples for extracting. The rules extracted in LBSB achieved an error rate of 7.6% and C4.5 a rate of 9.3% in the absence of noise. At the 10% noisy level, the former gained an error rate of 9.1%. whereas the latter yield 13.9%. In the case of 20% noisy level, the former reached 11.7% error rate, whereas the latter got 19.4%.

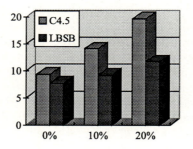

Fig. 1. The error rate comparison in the Splice Junction domain

In the Sales Volume domain, we use 360 positive examples for extracting. The rules extracted in LBSB achieved an error rate of 11.1% and C4.5 a rate of 14.7% in the absence of noise. At the 10% noisy level, the former gained an error rate of 13.7%,whereas the latter yield 17.2%. In the case of 20% noisy level, the former reached 9.7% error rate, whereas the latter get 13.8%.
From these experiments, we can see that in the noise condition, the error rate of the rules grows more quickly in C4.5 [5] than in LBSB. This demonstrates that in the noisy condition the rules extracted from neural networks in LBSB have stronger tolerance ability than those extracted from decision trees. The main reason is that the neural network performs better in noisy condition than C4.5. The results also show that using the neural network, high quality rules can be

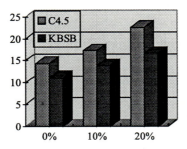

Fig. 2. The error rate comparison in the Sales Volume domain

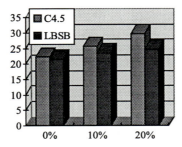

Fig. 3. The error rate comparison in the Pima Indians Diabetes domain

discovered from the databases.

Understandability is partly defined as being explicable in the sense that a prediction can be explained in terms of inputs. Being explicable is only one aspect of understandability. A rule with many conditions is harder to understand than a rule with fewer conditions. Too many rules also hinder humans understanding of the data under examination. The comparison is performed in fig 4 and 5.

4   Discussion and conclusion

In this paper, we apply the neural network to mine classification rules from databases with focus on articulating the rules represented by network. We present a novel approach to extract rules from the neural networks with distributed representation. Here we use the tolerance ability against noisy data to evaluate the quality of the rules extracted from the neural network.

The resistant against noisy data is very important because in the real-world problem, there may be some noisy data. From the experiment, we can see that

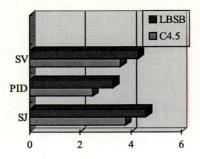

Fig. 4. Number of rules for the three databases

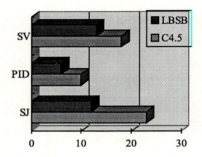

Fig. 5. Average number of a rule's conditions for the three databases

compared with the rules extracted from decision trees, the rules extracted from neural networks in LBSB can perform better in noisy condition. This demonstrates the advantage of neural network over decision tree in data mining.

There are also some future works to do: (a) how to efficiently represent the extracted rules. LBSB extracts only if-then rules. But for some problems, it will take a long time and generates many such kinds of rules to cover positive hidden activation space. The future work is to extract rule described by a more powerful representation to reduce the number of the rules.(b) how to reduce the number of rules conditions.

References

1. Ming-Syan Chen, Jiawei Han and Philip S.Yu: Data Mining: An Overview from Database Perspective. Technology Report, 1996.
2. Victor Ciensielski and Gregory Palstra: Using a Hybrid Neural / Expert System for Data Base Mining in Market Survey Data. In Proc. 1996 International Conference on Data Mining and Knowledge Discovery(KDD 96), Portland, Oregon, August,1996.

3. Mark W.Craven and Jude W.Shavlik: Using Sampling and Queries to Extract Rules from Trained Neural Networks. Proceeding of the Eleventh International conference in Machine Learning, Morgan Kaufmann, San Francisco, CA, 1994.
4. LiMin Fu: Rule Generation from Neural Networks. IEEE Transactions On Systems, Man and Cybernetics, Vol.24,NO.8, August 1994.
5. J.R.Quinlan: C4.5: Programs for Machine Learning. Morgan Kaufmann,1993
6. Ishwar K.Sethi and Jae H.Yoo: Symbolic approximation of feedforward neural network. In Pattern Recognition in Practice IV, pp 313-324, North-Holland,1994.
7. G.G.Towell and J.W.Shavlik: Interpretation of artificial neural networks: Mapping knowledge-based neural networks into rules. In NIPS 4, Denver, CO. Morgan Kaufmann.
8. G.G. Towell and J.W. shavlik , Refinement of approximately correct domain theories by knowledge-based neural networks, In proceeding of Eighth National Conference on Artificial Intelligence, pages 861-866, Boston, MA.
9. S.I.Gallant, Connectionist expert systems, Communications of the ACM, 31(2), pp.152-169,1988.
10. C.L.Giles and C.W. Omlin, Rule refinement with recurrent neural network., In Proccedings of the IEEE International Conference on Neural Network, 1993, IEEE Neural Network Council.
11. Y.Hayashi, A neural expert system with automated extraction of fuzzy if-then rules and its application to medical diagnosis., in Advances in Neural Information Processing Systems, Vol 3. San Mateo, CA:Morgan Kaufmann, 1990.
12. G.E.Hinton, Connectionist learning procedures., Artificial Intelligence, 40:185-234, 1989.
13. R.S.Michalski, A theory and methodology of inductive learning. Artificial Intelligence, 20:111-161.
14. Rudy Setiono and Huan Liu, Understanding Neural Networks via Rule Extraction, Proceedings of the Fourteenth International Joint Conference on Artificial Intelligence, Canada,1995.
15. Sebastian Thrun, Extracting rules form artificial neural networks with distributed representations, In Advances in Neural Information Processing Systems7, 1995.
16. L.G.Valiant, A theory of the learnable, Communications of the ACM, 27:1134-1142.
17. Zhang Zhaohui and Zhou Yuanhui, etc., Extracting rules from a GA-Pruned neural network, 1996 IEEE International Conference on Systems, Man and Cybernetics.
18. Zhou Yuanhui and Zhang Zhaohui, etc., Multistrategy learning using genetic algorithm and neural network for pattern classification, 1996 IEEE international Conference on Systems, Man and Cybernetics.

Section VI:

Estimation, Clustering

A Modulated Parzen–Windows Approach for Probability Density Estimation

G.C. van den Eijkel, J.C.A. van der Lubbe and E. Backer

Department of Electrical Engineering, Delft University of Technology
Mekelweg 4, 2628 CD Delft, The Netherlands

Abstract. The Parzen-window approach is a well-known technique for estimating probability density functions. This paper introduces a modulated Parzen-windows approach. This approach uses kernels at equidistant samples to obtain a probability density function more efficiently. Experiments on both artificial and real data show that the modulated Parzen-windows approach is more efficient in probability density function estimation, without costly preprocessing or severe loss of accuracy.

1   Introduction

The estimation of a probability density function (pdf) is often a first step in many (intelligent) data analysis applications. In most applications the pdf is used directly for classifier design [5] [1], whereas in some rule-based applications it is used for determining regions in the data space from which rule induction may start [6]. Since the pdf of the data is almost never known in advance, non-parametric techniques exist which estimate the pdf without assuming a particular density function. A well-known technique is the Parzen-windows approach [8] which is based on kernel functions. A disadvantage of the Parzen-windows approach is its computational cost. It may be argued that with todays computing power, computational cost is not an issue anymore. However, in the field of data analysis and data mining, the number of data to be analyzed is increasing rapidly, balancing the increase in computational power. Hence, the need for more efficient (and accurate) techniques remains. This paper introduces such a technique, a modulated Parzen-windows approach.

2   Parzen-Windows

Before introducing the modulated Parzen-windows approach we will first recall Parzen-windows (PW) [2]. In PW a density function (the kernel) $\phi(x)$ is used for which: $\phi(x) > 0$, $\int \phi(x)dx = V$. Having samples x_i, then the estimated pdf $\hat{p}(x)$ is obtained from:

$$\hat{p}(x) = \frac{1}{V_n n} \sum_{i=1}^{n} \phi_n(x - x_i) \tag{1}$$

X. Liu, P. Cohen, M. Berthold (Eds.): "Advances in Intelligent Data Analysis" (IDA–97)
LNCS 1280, pp. 479–489, 1997.   © Springer–Verlag Berlin Heidelberg 1997

Where n is the total number of data and V_n is the "kernel volume":

$$V_n = \int \phi_n(x)dx \qquad (2)$$

The kernel width is usually a function of $\frac{1}{\sqrt{(n)}}$ (i.e. the larger the number of data the smaller the kernel). An example of a PW estimate is given in Fig. 1.

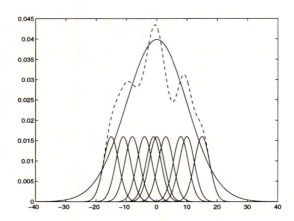

Fig. 1. Parzen-windows estimation of a Gaussian based on 10 observations

For an unlimited number of data it can be proven [2] that the sum of equation (1) converges to a convolution (convergence in mean square):

$$\lim_{n \to \infty} \frac{1}{V_n n} \sum_{i=1}^{n} \phi_n(x - x_i) = \frac{1}{V_n} \int \phi_n(x - x_i)p(x_i)dx_i \qquad (3)$$

If we now define:

$$\delta_n(x - x_i) = \frac{1}{V_n}\phi(x - x_i) \qquad (4)$$

then we may write:

$$\frac{1}{V_n} \int \phi_n(x - x_i)p(x_i)dx_i =$$
$$\int \delta_n(x - x_i)p(x_i)dx_i = p(x) \qquad (5)$$

Thus, the sum converges to the exact pdf if the number of data is unlimited and if the kernel approaches a Dirac delta function. It may also be observed that the best estimate for $p(x)$ is always a smoothed (filtered) version of the real pdf due to the convolution with the kernel ϕ_n, which should therefore be as small as possible. More details can be found in [8].

Although PW is a theoretically sound technique, in practice the number of data is limited and therefore the kernel can not be taken too small (otherwise "holes" and "spikes" appear in the estimated pdf). The optimal choice for the kernel width (often referred to as the smoothing-parameter) is still subject to research [9], however, in practice the Gaussian kernel is often used and a width is chosen which optimizes the classification performance. Another practical problem is the computational time and storage for estimating a pdf value. From equation (1) it is clear that for each pdf value x, all n kernels have to be evaluated and summed since each datum "carries" its own kernel. Several (costly) preprocessing approaches exist which reduce the load and storage by clustering the data [1] or by extracting a suitable subset from the data [5].

3 Modulated Parzen-Windows

The basic idea in the modulated Parzen-windows (MPW) approach is to estimate only a few points (samples) in the pdf using the data, and then obtain an estimated pdf from these samples (e.g. interpolation). This is reasonable if the underlying pdf of the data is a smooth curve, which is in general true for pdf's.

3.1 Analysis

We can write from equation (1) the estimate for a sample x_s:

$$\hat{p}(x_s) = \frac{1}{V_n n} \sum_{i=1}^{n} \phi_n(x_s - x_i) \tag{6}$$

This is the summation of kernels positioned at x_i for a value at x_s. If we require that the kernel is symmetric, i.e. $\phi(u) = \phi(-u)$, then the following holds:

$$\sum_{i=1}^{n} \phi_n(x_s - x_i) = \sum_{i=1}^{n} \phi_n(x_i - x_s) \tag{7}$$

This can be interpreted as a summation of kernel values of one kernel positioned at x_s instead of a summation of n kernels at position x_i (e.g. Fig. 2). Although the result for $\hat{p}(x_s)$ from the left and right-hand side of equation (7) is exactly the same, the interpretation is different.

If we have a total of S different samples x_s of which $\hat{p}(x_s)$ is known, then we may interpret this as a new dataset of S different "observations" x_s where each value is not only observed once but with an estimated frequency $\hat{f}_n(x_s) = V_n n \, \hat{p}_n(x_s)$. Hence, the total number of data in the new dataset is $N_S = \sum_{s=1}^{S} \hat{f}(x_s)$. An estimation for $\hat{p}(x)$ can be obtained from these N_S observations by applying equation (1) on the new dataset:

$$\hat{\hat{p}}_S(x) = \frac{1}{V_S N_S} \sum_{s=1}^{S} \phi_s(x - x_s) \hat{f}_n(x_s) \tag{8}$$

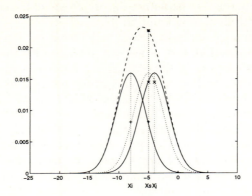

Fig. 2. One kernel at x_s (dotted-kernel), or two kernels at x_i and x_j (left and right) lead to the same summed estimate at x_s

This is interpreted as a summation of S "modulated" kernels, since each kernel has to be multiplied (modulated) by the number of times (frequency) that it is observed. The frequency is an estimation obtained from equation (6).

The user can now choose the kernel functions ϕ_n, ϕ_s and S sample points to obtain an estimation for the pdf. For future use of the data, e.g. for classification or reasoning, it will be sufficient to store only the values of $\hat{f}(x_s)$. However, instead of choosing one kernel function, a user is now burdened with specifying two kernel functions and a number of samples. We address this problem in the next section.

3.2   Full Modulated Parzen-Windows

Several interesting estimators can be obtained from equation (8). An interpolated estimate is obtained if we choose the kernel ϕ_s such that $\hat{p}_S(x_s) = \hat{p}(x_s)$ at all samples, which comes down to choosing interpolation kernels (e.g. block-functions, triangular-functions etc.). Estimators of this type are well known in the literature and are frequently referred to as binned-kernel estimators [3]).

Although interesting combinations of kernels for ϕ_n and ϕ_s can be obtained, we restrict our analysis in this section to cases where $\phi_n = \phi_s$. For these cases a useful relation can be derived between the PW kernel, the MPW kernel and the number of samples. We will refer to the MPW approach using this relation as *Full Modulated Parzen-Windows* (FMPW). Furthermore, we will use ϕ_m for the kernel used by the FMPW approach and ϕ_p for the kernel used by the PW approach.

From equations (8) and (6), it can be proven that if the number of samples is unlimited, the estimate $\hat{\hat{p}}_m(x)$ is a convolution with the PW estimate $\hat{p}_p(x)$:

$$\lim_{n,S\to\infty} \hat{\hat{p}}_m(x) = \frac{1}{V_S}\phi_m(x) * \hat{p}_p(x) \tag{9}$$

Using the Fourier convolution-theorem, we may write in the frequency domain:

$$\hat{\hat{P}}_m(\omega) = \Phi_m(\omega) \cdot \Phi_m(\omega) \cdot \sum_i^N e^{-j\omega x_i} \qquad (10)$$

Whereas the original Parzen-windows, using a kernel ϕ_p, in the case of unlimited data may be written as:

$$\hat{P}_p(\omega) = \Phi_p(\omega) \cdot \sum_i^N e^{-j\omega x_i} \qquad (11)$$

Hence, if we choose ϕ_m such that:

$$\Phi_m(\omega) = \sqrt{\Phi_p(\omega)} \qquad (12)$$

then the MPW estimate equals the PW estimate, given an unlimited number of data and samples.

The number of samples from which to obtain the PW estimate is now given by the Shannon sampling theorem, i.e. it is sufficient to sample (equidistantly) at twice the cut-off frequency of the PW kernel $\Phi_p(\omega)$ within the data range. The reader is referred to [7] for a treatise on sampling and interpolation. The advantage over straightforward interpolation from samples (as mentioned previously) is that by using equation (12) we make maximum use of the minimum number of samples. In interpolation often a higher sampling frequency is necessary to obtain a sufficiently smooth and accurate estimate.

3.3   Multi-dimensional extension

Many pattern recognition problems require a multi-dimensional approach. To extend the modulated Parzen-Windows approach to multi-dimensional problems we use a kernel function $\phi_D(\mathbf{u})$ to obtain the new data set and for estimating the continuous pdf. Without loss of generality, we require that the kernel is the product of one-dimensional kernels $\phi_d(u_d)$ (it is said that $\phi_D(\mathbf{u})$ is separable):

$$\phi_D(\mathbf{u}) = \prod_{d=1}^{N_D} \phi_d(u_d) \qquad (13)$$

Where N_D is the number of dimensions. This implies that we assume that, within the kernel width, the dimensions (features) are independent. This is a reasonable assumption since the kernel width (smoothing parameter) is often small with respect to the scale. However, the main reason for requiring separability at this stage is because of simplicity. If the multi-dimensional kernel is separable, then the multi-dimensional sample-grid is simply (hyper-)rectangular. Non-separable kernels are possible as well within the MPW framework, but require a sample-grid along the principal axes of the kernel for optimal sampling (see [7]). The multi-dimensional MPW approach is:

$$\hat{\hat{p}}_m(\mathbf{x}) = \frac{1}{V_S N_S} \sum_{s=1}^{S} \left[\phi_D(\mathbf{x} - \mathbf{x}_s) \sum_{i=1}^{n} \phi_D(\mathbf{x}_s - \mathbf{x}_i) \right] \tag{14}$$

In multi-dimensional problems the number of samples can be determined from the data-range in each dimension and the width of the ϕ_D kernel along each dimension (which is possible due to the separability of the kernel). However, the number of samples can become very large. If we have S_d samples for each dimension, then the total number of samples in the multi-dimensional data-range becomes $(S_d)^{N_D}$. To restrict this number of samples we truncate the summation over the samples by selecting only the "maximum-kernels" $\phi_m(\mathbf{x} - \mathbf{x}_m)$. A kernel $\phi_m(\mathbf{x} - \mathbf{x}_m)$ is a maximum-kernel if there exists at least one observation $\mathbf{x}_i$ for which

$$\phi_m(\mathbf{x}_i - \mathbf{x}_m) = \max_s [\phi_D(\mathbf{x}_i - \mathbf{x}_s)] \tag{15}$$

In this way, each example is represented by its nearest neighbor sample, which results in a maximum number of samples that is equal to the number of data. If the data is "dense" we obtain always less samples than the number of data. By truncating, we actually reject the samples that have a small, if any, contribution to the summation. We will refer to this use of the MPW as the *Reduced Modulated Parzen-Windows* (RMPW). Note that the second summation in equation (14) is *not* truncated.

Unlike the previous section, we now choose $\phi_m(\mathbf{u})$ such that:

$$\phi_m(\mathbf{u}) = \phi_p(\mathbf{u}) \tag{16}$$

where $\phi_m(\mathbf{u})$ is the PW kernel. The reason for this is that, by truncating, some variance is introduced in the estimator due to the variance in the original data set. Hence, we may not expect to obtain a very good estimate with a kernel width smaller than the PW kernel-width. Therefore the RMPW kernel is chosen to be the same as the PW kernel.

4 Experimental Results

In our experiments we used one-dimensional artificial data to compare the FMPW and PW approach for pdf estimation. Furthermore, we used the Iris dataset [4] to show the applicability of the RMPW approach.

4.1 PDF estimation in one dimension

We used a Gaussian kernel for the PW approach with a standard deviation (SD) of σ_p. From equation (12) we obtained a Gaussian kernel for the FMPW approach having a SD of $\frac{1}{\sqrt{2}}\sigma_p$, and a sampling frequency of σ_p (which is approximately twice the cut-off frequency of a Gaussian). The actual number of samples was always rounded to the nearest integer to obtain equidistant-sampling such that the first sample lied at the minimum value observed in the data, and the last sample lied at the maximum value observed in the data.

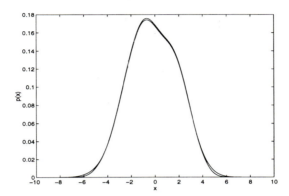

Fig. 3. Estimation of a Gaussian using Gaussian kernels from 80 draws for both the PW and FMPW approach (the FMPW result is obtained from 12 kernels).

Figure 3 shows the PW and the FMPW estimate for a Gaussian having a SD of 2, and a kernel for which $\sigma_p = 1$. It was obtained from 80 draws (the number of data the pdf was estimated from) and hence the PW approach needs 80 kernels, whereas in this case the FMPW approach only uses about 12 kernels (depending on the observed minimum and maximum value in the draws: the draw-range). The mean square error (obtained from 100 values of $p(x)$ in the draw-range) between the PW and FMPW estimates in this specific case is in the order of 1e-4. Since the mean value (within the draw-range) for the PW estimate is in the order of 1e-1, this is a very good result. We repeated this experiment for several numbers of draws (10, 40, 160 and 640 draws) and several kernel SD's (from 0.1 up to 2.5, step 0.1). We also used an exponential and a Poisson distribution (from 160 draws). Fig. 4 shows the obtained mean square error (mse) for all six experiments, the highest error curve corresponds to the Gaussian estimate using 10 draws. We observed that the error decreases with an increasing number of data. The reason for this effect may be that, by using finite sampling, the true convolution is estimated by a discrete convolution (equation (9) only holds for a true convolution). It appears that the error induced by the discrete convolution decreases for larger number of data. Still, the error between the FMPW and the PW estimate is very small (maximum mse over all six experiments is approximately 1e-3, minimum mse is 1e-6). In Fig. 5 we plotted the reduction ratio of the number of kernels (samples) used by the FMPW (N_s) and by the PW approach (n) as a function of SD for all the cases where we

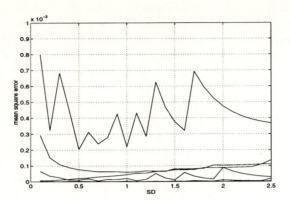

Fig. 4. Mean square error as a function of σ_p, the highest error curve corresponding to 10 draws, lowest error curve to 640 draws.

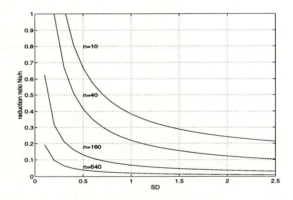

Fig. 5. Reduction ratio of the number of kernels used in the FMPW and PW approach as a function of σ_p. Plots showing results for several number of draws (n).

estimated the Gaussian pdf (similar results were obtained for the Poisson and exponential distribution). For large values of the SD the FMPW approach is most "efficient", we also observe that there is always a SD where the FMPW approach is less efficient than the PW approach, i.e. it uses more samples than the number of draws. The break-even point lying at $\sigma_p = \frac{x_{max} - x_{min}}{n}$, where n is the number of draws, and where x_{max} and x_{min} define the draw-range in which we sampled with frequency σ_p. However, beyond the break-even point both the PW and the FMPW estimate became very erratic and lost their smoothness. In practice larger SD values (relative to the "width" of the pdf to be estimated) are used, and hence a high computational reduction can be obtained. It is difficult to estimate the reduction ratio in general, since it depends on both the number of available data and the real pdf. However, for our experiments good estimates (mse < 1e-3) of the real pdf were obtained at $\sigma_p = 1$, corresponding to a reduction ratio of 0.4, 0.2, 0.07 and 0.02 for respectively 10, 40, 160 and 640 draws.

4.2 Multi-dimensional pdf estimation

The *Reduced Modulated Parzen-Windows* was applied to the Iris dataset. We used Gaussian kernels, with a diagonal covariance matrix $h^2 * \Sigma^2$, where h is a smoothing parameter and Σ^2 the empirical covariance matrix of which only the diagonal was used to keep the kernels separable and the sampling simple. Sampling was done equidistantly in each dimension using the sampling frequency of $1/(h * \Sigma)$, approximately the cut-off frequency of a multidimensional Gaussian with smoothing parameter h. We calculated the mse on the basis of approximately 6500 values for a range of smoothing parameters h. Results are shown in figure 6 and figure 7. The results show that, although we have severely truncated the discrete convolution over the available samples in the four dimensional space, the *Reduced Modulated Parzen-Windows* still gives very accurate results. However, we also observe that the reduction ratio is not very small and, hence, we do not gain a lot of efficiency. This is due to the relatively small size of the iris dataset. The increase of the mse by several orders of magnitude for very small kernel-widths is due to the fact that the RMPW and PW estimate become "spiky" for small kernel-widths (this same effect can somewhat be observed in the one-dimensional case and is a direct consequence of the increase in variance for small kernel-widths (see figure 4)). However, the mse still remains much smaller than the mean of the PW estimate (the mean is in the order of 1e-3). Overall, we may conclude that the multi-dimensional RMPW approach is very accurate when estimating smooth pdf's, however a large gain in efficiency can only be obtained for large datasets.

5 Conclusions and Further Research

In this paper the modulated Parzen-windows approach is introduced. This approach is a feasible and efficient method for estimating smooth probability density functions, which does not need costly preprocessing. It is theoretically proven that the *Full Modulated Parzen-Windows* estimate is equal to the Parzen-windows estimate of the pdf. However, in practice the *Reduced Modulated Parzen-Windows* should be used for an efficient and accurate estimation of the pdf. Experimental results on both artificial and real data (in one and more dimensions) show that nearly equal estimates are obtained more efficiently than by using the Parzen-Windows method.

We are currently studying the use of the Reduced Modulated Parzen Windows approach for classification in multi-dimensional data spaces. Furthermore, we study its applicability for rule-induction and rule-generalization by using relatively large kernel functions on a (hyper-)rectangular grid.

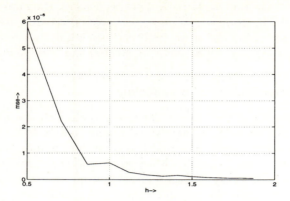

Fig. 6. The mean square error between the PW and RMPW estimate as a function of smoothing parameter h.

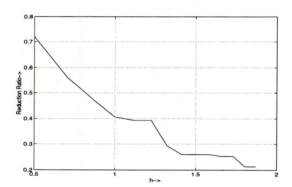

Fig. 7. Reduction ratio of the number of kernels used in the RMPW and PW approach as a function of smoothing parameter h.

References

1. G.A. Babich and O.I. Campus, Weighted Parzen Windows for Pattern Classification, *IEEE Trans. Pattern Analysis and Machine Intelligence*, vol. 18, pp. 567-570, 1996.
2. R.O. Duda and P.E. Hart, *Pattern Classification and Scene Analysis*, New York: John Wiley & Sons Inc., 1973.
3. J. Fan and J.S.Marron, Fast Implementations of Nonparametric Curve Estimators, *J. Computational and Graphical Statistics*, vol 3, pp. 35-56, 1994
4. R.A. Fisher, The use of multiple measurements in taxonomic problems, *Annals of Eugenics*, vol. 7, pp. 179-188, 1936

5. K. Fukunaga, *Statistical Pattern Recognition*, San Diego, Calif: Academic Press Inc., 1990
6. L.B. Gamage, R.G. Gosine and C.W. de Silva, Extraction of Rules from Natural Objects for Automated Mechanical Processing, *IEEE Trans. Syst., Man, Cybern.*, vol. 26, pp. 105-120, 1996.
7. R.J. Marks II, *Introduction to Shannon Sampling and Interpolation Theory*, New York: Springer-Verlag Inc., 1991
8. E. Parzen, On estimation of a probability density function and mode, *Ann. Math. Statistics*, vol. 33, pp. 1065-1076, 1962.
9. S.J. Raudys and A.K. Jain, Small Sample Size Effects in Statistical Pattern Recognition: Recommendations for Practitioners, *IEEE Trans. Pattern Analysis and Machine Intelligence* vol. 13, pp. 252-264, 1991.

Improvement on Estimating Quantiles in Finite Population Using Indirect Methods of Estimation

M. Rueda García[1], A. Arcos Cebrián[1], E. Artés Rodríguez[2]

[1] Department of Statistics and Operations Research
University of Granada, 18071 Granada, Spain
[2] Department of Statistics and Operations Research
University of Almería, 04120 Almería, Spain

Abstract. New methods for estimating confidence limits for quantiles in a finite population are proposed. These methods use auxiliary information through the ratio, difference and regression estimator of the population distribution function. They may be applied to any type of sampling. Simulation studies based of two real populations show that the methods proposed in this paper can be considerably more efficient than the customary classic method.
Key words: auxiliary information, finite population quantiles, ratio, difference and regression type estimator, confidence intervals.

1 Introduction

In survey practice, it is often of interest to study variables with a highly skewed distribution. In such situations, it is useful to make inferences about finite population quantiles. Sample medians have long been recognized as simple robust alternatives to sample means, for estimating location of heavy-tailed or markedly skewed populations from simple random samples. A large class of robust estimates of location, including the sample median, was investigated in the Princeton simulation study (D.F. Andrew et al (1972)). Although the sample median did not emerge as best estimate in many nonstandard populations simulated in the study, its robustness in small samples for medium and large deviations from normality was clearly demonstrated. Its simplicity relative to other robust estimates, indicated its choice. Unfortunately, while there is an extensive literature on the estimation of means and totals, relatively less research has been done to development of efficient methods for estimating finite population quantiles. Moreover, most of these methods in simple random sampling (*Gross* 1980, *Sedransk* and *Meyer* 1978, *Smith* and *Sedransk* 1983) does not make explicit use of auxiliary variable is available, it is natural to expect that the auxiliary information can be incorporated to construct an estimator more efficient than the direct estimator (sample quantile).

X. Liu, P. Cohen, M. Berthold (Eds.): "Advances in Intelligent Data Analysis" (IDA–97)
LNCS 1280, pp. 491–499, 1997.   © Springer–Verlag Berlin Heidelberg 1997

The use of indirect methods for estimating a finite population mean has been widely studied (see *Cochran* 1977), however, it is not immediately clear how these well–established techniques, such as the regression estimator, can be extended to the case of estimating the quantiles.

Increasingly, this need is being recognized, so point estimation of finite population quantiles that uses auxiliary information has received considerable attention (*Chambers* and *Dunstan*, 1986, and *Rao, Kovar* and *Mantel*, 1990), both suggested estimating quantiles by in inverting improved estimates of the distribution functions in presence of auxiliary information. Other references are *Kuk* and *Mak* (1989,94), *Mak* and *Kuk* (1993).

In this paper, we suggest alternative procedures for determining confidence intervals for a finite population median and other quantiles, under simple random sampling and using an auxiliary variable.

Let $y_1, y_2, \ldots, y_N$ denote the values of the population elements $U_1, U_2, \ldots, U_N$, for the variable of interest y. For any y ($-\infty < y < \infty$), as the population distribution $F_Y(y)$ is defined as the proportion of elements in the population that are less than or equal to y.

The finite population β quantile is defined as

$$Q_Y(\beta) = F_Y^{-1}(\beta),$$

where F_Y^{-1} is the inverse function of F_Y.

The general procedure to estimate the population quantile $Q_Y(\beta)$, using data y_k for $k \in s$, where s is a simple random sample can be summarized as follows: we first produce an estimated distribution function, $\widehat{F}_Y(y)$, and then, estimate $Q_Y(\beta) = F_Y^{-1}(\beta)$ as $\widehat{Q}_Y(\beta) = \widehat{F}_Y^{-1}(\beta)$, where the inverse $\widehat{F}_Y^{-1}$ is to be understood in the same way as F_Y^{-1} above. This method has been in use for a long time; the first published account is probably *Woodruff* (1952).

Woodruff (1952) describes a general method of obtaining confidence intervals for medians and others positions measures using a principle that has been applied to sample random sampling and extending it to any type of sampling. These confidence limits can be approximated for any sampling design where the variance of the percentage of items less than a stated value can be acceptably estimated (in general, where large samples are involved).

We present new methods to derive the confidence interval finite population quantiles in Section 2 and 3. Ratio, difference and regression estimators of the population distribution function based on an auxiliary variable is the key to these methods. We also compare the methods that we propose and Woodruff's method using simulation studies, in Section 4.

2   Confidence intervals for the quantiles using ratio, difference and regression estimators

Consider the simple random sampling design. Suppose that the population under study consists of N units, and attached to each of these units are the values of the survey variable y of interest and an auxiliary variable x. It's assumed that only the population βth quantile $Q_X(\beta)$ of x is known and that $Q_Y(\beta)$ is to be estimated on the basis of a simple random sample of size n. Let $(x_1, y_1), \ldots, (x_n, y_n)$ be the associated values of the variables x and y for the units in the sample.

Consider the ratio, difference and regression estimators

$$\widehat{F}_R(Q_Y(\beta)) = \frac{\widehat{F}_Y(Q_Y(\beta))}{\widehat{F}_X(Q_X(\beta))}\beta$$

$$\widehat{F}_D(Q_Y(\beta)) = \widehat{F}_Y(Q_Y(\beta)) + \left(\beta - \widehat{F}_X(Q_X(\beta))\right)$$

$$\widehat{F}_{Reg}(Q_Y(\beta)) = \widehat{F}_Y(Q_Y(\beta)) + b\left(\beta - \widehat{F}_X(Q_X(\beta))\right)$$

(b is a known constant) and the constants c_1^i and c_2^i, $i = 1, 2, 3$ such that

$$P\left\{c_1^1 \leq \widehat{F}_R(Q_Y(\beta)) \leq c_2^1\right\} = 1 - \alpha.$$

$$P\left\{c_1^2 \leq \widehat{F}_D(Q_Y(\beta)) \leq c_2^2\right\} = 1 - \alpha.$$

$$P\left\{c_1^3 \leq \widehat{F}_{Reg}(Q_Y(\beta)) \leq c_2^3\right\} = 1 - \alpha.$$

Thus, the approximated $100(1 - \alpha)\%$ confidence intervals for $Q_Y(\beta)$ will be

$$\left[\widehat{F}_Y^{-1}\left(c_1^1 \frac{\widehat{F}_X(Q_X(\beta))}{\beta}\right), \widehat{F}_Y^{-1}\left(c_2^1 \frac{\widehat{F}_X(Q_X(\beta))}{\beta}\right)\right]$$

$$\left[\widehat{F}_Y^{-1}\left(c_1^2 - \left(\beta - \widehat{F}_X(Q_X(\beta))\right)\right), \widehat{F}_Y^{-1}\left(c_2^2 - \left(\beta - \widehat{F}_X(Q_X(\beta))\right)\right)\right]$$

$$\left[\widehat{F}_Y^{-1}\left(c_1^3 - b\left(\beta - \widehat{F}_X(Q_X(\beta))\right)\right), \widehat{F}_Y^{-1}\left(c_2^3 - b\left(\beta - \widehat{F}_X(Q_X(\beta))\right)\right)\right].$$

For large samples, $\widehat{F}_Y(Q_Y(\beta))$, $\widehat{F}_X(Q_X(\beta))$ and $\widehat{F}_R(Q_Y(\beta))$ are approximately normally distributed (see *Kuk* and *Mak*, 1989). Then, the asymptotic distributions of the estimation $\widehat{F}_D(Q_Y(\beta))$ and $\widehat{F}_{Reg}(Q_Y(\beta))$ approach a normal distribution, and we would choose the smallest confidence interval as

$$c_1^1 = \beta - z_{\frac{\alpha}{2}}\left\{V\left(\widehat{F}_R(Q_Y(\beta))\right)\right\}^{\frac{1}{2}}, \qquad c_2^1 = \beta + z_{\frac{\alpha}{2}}\left\{V\left(\widehat{F}_R(Q_Y(\beta))\right)\right\}^{\frac{1}{2}},$$

$$c_1^2 = \beta - z_{\frac{\alpha}{2}}\left\{V\left(\widehat{F}_D(Q_Y(\beta))\right)\right\}^{\frac{1}{2}}, \qquad c_2^2 = \beta + z_{\frac{\alpha}{2}}\left\{V\left(\widehat{F}_D(Q_Y(\beta))\right)\right\}^{\frac{1}{2}},$$

and

$$c_1^3 = \beta - z_{\frac{\alpha}{2}} \left\{ V\left(\widehat{F}_{Reg}(Q_Y(\beta))\right)\right\}^{\frac{1}{2}} \quad , \quad c_2^3 = \beta + z_{\frac{\alpha}{2}} \left\{ V\left(\widehat{F}_{Reg}(Q_Y(\beta))\right)\right\}^{\frac{1}{2}}.$$

We don't know $Q_Y(\beta)$, then the problem of evaluating the last unknown variances is not so simple. For example, to evaluate the variance $V\left(\widehat{F}_R(Q_Y(\beta))\right)$, we make the variables

$$e_0 = \frac{\widehat{F}_Y(Q_Y(\beta)) - F_Y(Q_Y(\beta))}{F_Y(Q_Y(\beta))}, \quad e_1 = \frac{\widehat{F}_X(Q_X(\beta)) - F_X(Q_X(\beta))}{F_X(Q_X(\beta))}.$$

Then, Taylor's series expansion yields

$$V\left(\widehat{F}_R(Q_Y(\beta))\right) \simeq F_Y(Q_Y(\beta))^2 \left(E\left(e_0^2\right) + E\left(e_1^2\right) - 2E\left(e_1 e_0\right)\right) =$$

$$= \left(V\left(\widehat{F}_X(Q_X(\beta))\right) + V\left(\widehat{F}_Y(Q_Y(\beta))\right) - 2\mathrm{Cov}\left(\widehat{F}_X(Q_X(\beta)), \widehat{F}_Y(Q_Y(\beta))\right)\right) =$$

$$= 2\frac{1-f}{n}\beta(1-\beta) - 2\mathrm{Cov}\left(\widehat{F}_X(Q_X(\beta)), \widehat{F}_Y(Q_Y(\beta))\right). \tag{1}$$

We have to calculate the value of $\mathrm{Cov}\left(\widehat{F}_X(Q_X(\beta)), \widehat{F}_Y(Q_Y(\beta))\right)$, therefore we consider the two–way classification

	$x_k \le Q_X(\beta)$	$x_k > Q_X(\beta)$	
$y_k \le Q_Y(\beta)$	$n_{11} \setminus N_{11}$	$n_{12} \setminus N_{12}$	$N_{1.}$
$y_k > Q_Y(\beta)$	$n_{21} \setminus N_{21}$	$n_{22} \setminus N_{22}$	$N_{2.}$
	$N_{.1}$	$N_{.2}$	

where n_{11} denotes the number of units in the sample with $x \le Q_X(\beta)$ and $y \le Q_Y(\beta)$; and N_{11} is the number of units in the population with $x \le Q_X(\beta)$ and $y \le Q_Y(\beta)$. Thereby,

$$(n_{11}, n_{12}, n_{21}, n_{22}) \simeq HG(N, n, N_{11}, N_{12}, N_{21}),$$

$n\widehat{F}_Y(Q_Y(\beta)) = n_{11} + n_{12}$ and similarly $n\widehat{F}_X(Q_X(\beta)) = n_{11} + n_{21}$.

Besides, we can verify that

$$\mathrm{Cov}\left(n\widehat{F}_Y(Q_Y(\beta)), n\widehat{F}_X(Q_X(\beta))\right) = \frac{N-n}{N-1}\frac{n}{N^2}\left(N_{11}N_{22} - N_{12}N_{21}\right).$$

Substituting the last expression in (1) we have

$$V\left(\widehat{F}_R(Q_Y(\beta))\right) = \frac{1-f}{n}2\left(\beta(1-\beta) - \frac{N_{11}N_{22} - N_{12}N_{21}}{N^2}\right). \tag{2}$$

Denoting Cramer's V coefficient as

$$\phi_\beta = \frac{N_{11}N_{22} - N_{12}N_{21}}{\sqrt{N_1 . N_2 . N_{.1} N_{.2}}},$$

we can rewrite the following expression

$$V\left(\widehat{F}_R(Q_Y(\beta))\right) = \frac{1-f}{n} 2\beta(1-\beta)(1-\phi_\beta). \tag{3}$$

Analogously, we evaluate the variances of $\widehat{F}_D(Q_Y(\beta))$ and $\widehat{F}_{Reg}(Q_Y(\beta))$ which are given by (3) and

$$V\left(\widehat{F}_{Reg}(Q_Y(\beta))\right) = \frac{1-f}{n}\beta(1-\beta)(1+b^2 - 2b\phi_\beta). \tag{4}$$

In practice ϕ_β is unobservable since $Q_Y(\beta)$ is unknown and therefore has to be estimated from the sample. Substituting n_{ij} for $\tilde{n}_{ij}$, based on a similar cross–classification

$$
\begin{array}{ccc}
 & x_k \le \widehat{Q}_X(\beta) & x_k > \widehat{Q}_X(\beta) \\
y_k \le \widehat{Q}_Y(\beta) & \tilde{n}_{11} \backslash \tilde{N}_{11} & \tilde{n}_{12} \backslash \tilde{N}_{12} \\
y_k > \widehat{Q}_Y(\beta) & \tilde{n}_{21} \backslash \tilde{N}_{21} & \tilde{n}_{22} \backslash \tilde{N}_{22}
\end{array} ,
$$

and then, we would consider the following estimator for ϕ_β

$$\tilde{\phi}_\beta = \frac{\tilde{n}_{11}\tilde{n}_{22} - \tilde{n}_{12}\tilde{n}_{21}}{\sqrt{\tilde{n}_1 . \tilde{n}_2 . \tilde{n}_{.1} \tilde{n}_{.2}}}.$$

So, the intervals

$$\left[\widehat{F}_Y^{-1}\left(\tilde{c}_1^1 \frac{\widehat{F}_X(Q_X(\beta))}{\beta}\right), \widehat{F}_Y^{-1}\left(\tilde{c}_2^1 \frac{\widehat{F}_X(Q_X(\beta))}{\beta}\right)\right],$$

$$\left[\widehat{F}_Y^{-1}\left(\tilde{c}_1^2 - \left(\beta - \widehat{F}_X(Q_X(\beta))\right)\right), \widehat{F}_Y^{-1}\left(\tilde{c}_2^2 - \left(\beta - \widehat{F}_X(Q_X(\beta))\right)\right)\right] \text{ and}$$

$$\left[\widehat{F}_Y^{-1}\left(\tilde{c}_1^3 - b\left(\beta - \widehat{F}_X(Q_X(\beta))\right)\right), \widehat{F}_Y^{-1}\left(\tilde{c}_2^3 - b\left(\beta - \widehat{F}_X(Q_X(\beta))\right)\right)\right] \text{ where}$$

$$\tilde{c}_i^1(\beta) = \tilde{c}_i^2(\beta) = \beta + (-1)^i z_{\frac{\alpha}{2}} \left\{\frac{1-f}{n} 2\beta(1-\beta)(1-\tilde{\phi}_\beta)\right\}^{\frac{1}{2}} \quad i = 1, 2,$$

$$\tilde{c}_i^3(\beta) = \beta + (-1)^i z_{\frac{\alpha}{2}} \left\{\frac{1-f}{n}\beta(1-\beta)(1+b^2 - 2b\tilde{\phi}_\beta)\right\}^{\frac{1}{2}} \quad i = 1, 2,$$

are $100(1-\alpha)\%$ confidence intervals for $Q_Y(\beta)$. When the sample size is small, the method should be applied with caution, as this method relies on several approximations.

3   Confidence intervals for quantiles using the optimum regression estimator

The optimum regression estimator, $\widehat{F}_{Reg}^{opt}\left(Q_Y(\beta)\right)$, that is, the regression type estimator with the smallest variance, is obtained in this section. The variance (4) is minimum to $b = \phi_\beta$, and then we consider the regression type estimator for the population distribution function as follows:

$$\widehat{F}_{Reg}^{opt}\left(Q_Y(\beta)\right) = \widehat{F}_Y\left(Q_Y(\beta)\right) + \phi_\beta\left(\beta - \widehat{F}_X\left(Q_X(\beta)\right)\right),$$

and its variance is given by

$$V\left(\widehat{F}_{Reg}^{opt}\left(Q_Y(\beta)\right)\right) = \frac{1-f}{n}\beta(1-\beta)(1-\phi_\beta^2).$$

This regression type estimator is always more precise than the simple estimator $\widehat{F}_Y\left(Q_Y(\beta)\right)$ (except to $\phi_\beta = 0$), although it has the same difficulty that the previous estimators since ϕ_β is unobservable, because $Q_Y(\beta)$ is unknown. To resolve this difficulty, we take the coefficient estimation corresponding to ϕ_β, but with $Q_Y(\beta)$ and $Q_X(\beta)$ replaced by $\widehat{Q}_Y(\beta)$ and $\widehat{Q}_X(\beta)$, respectively, which gives

$$\widehat{F}_{Reg}^{*}\left(Q_Y(\beta)\right) = \widehat{F}_Y\left(Q_Y(\beta)\right) + \check{\phi}_\beta\left(\beta - \widehat{F}_X\left(Q_X(\beta)\right)\right).$$

Now, the asymptotic distribution of $\widehat{F}_{Reg}^{*}\left(Q_Y(\beta)\right)$ can be derived through the following reasoning: if p_{11} denotes the proportions of units in the sample with $x \leq \widehat{Q}_X(\beta)$ and $y \leq \widehat{Q}_Y(\beta)$, and P_{11} the proportions of units in the populations with $x \leq Q_X(\beta)$ and $y \leq Q_Y(\beta)$, it can be see that

$$\widehat{F}_{Reg}^{*}\left(Q_Y(\beta)\right) = \widehat{F}_Y\left(Q_Y(\beta)\right) + \frac{p_{11} - \beta^2}{\beta(1-\beta)}\left(\beta - \widehat{F}_X\left(Q_X(\beta)\right)\right).$$

Since $\widehat{F}_X\left(Q_Y(\beta)\right) \to \beta$ in probability and $p_{11} - P_{11}$ is of order $O_p\left(n^{-\frac{1}{2}}\right)$ (see *Kuk* and *Mak*, 1989), then

$$\widehat{F}_{Reg}^{*}\left(Q_Y(\beta)\right) = \widehat{F}_{Reg}^{opt}\left(Q_Y(\beta)\right) + O_p\left(n^{-\frac{1}{2}}\right),$$

and $\widehat{F}_{Reg}^{*}\left(Q_Y(\beta)\right)$ has the same asymptotic distribution of $\widehat{F}_{Reg}^{opt}\left(Q_Y(\beta)\right)$. Hence $\widehat{F}_{Reg}^{*}\left(Q_Y(\beta)\right)$ is asymptotically normal with mean β and variance

$$\frac{1-f}{n}\beta(1-\beta)(1-\phi_\beta^2).$$

Considering this new regression estimator we can derive a confidence interval for the βth quantile $Q_Y(\beta)$ as follows:

$$c_j^4 = \beta + (-1)^j z_{\frac{\alpha}{2}}\left\{\widehat{V}\left(\widehat{F}_{Reg}^{*}\left(Q_Y(\beta)\right)\right)\right\}^{\frac{1}{2}}, \quad j = 1, 2,$$

where

$$\widehat{V}\left(\widehat{F}^*_{Reg}\left(Q_Y(\beta)\right)\right) = \frac{1-f}{n}\beta(1-\beta)(1-\tilde{\phi}_\beta^2).$$

Then,

$$\left[\widehat{F}_Y^{-1}\left(c_1^4 - \tilde{\phi}_\beta\left(\beta - \widehat{F}_X(Q_X(\beta))\right)\right), \widehat{F}_Y^{-1}\left(c_2^4 - \tilde{\phi}_\beta\left(\beta - \widehat{F}_X(Q_X(\beta))\right)\right)\right]$$

is a confidence interval with confidence coefficient $1 - \alpha$ for $Q_Y(\beta)$.

4   Simulation study

To compare the efficiencies of the proposed methods and Woodfruff's method; we use simulation studies. Choose and fix a $1-\alpha$ level of confidence and a sample size n, consider 1000 samples of size n from the population and for each sample compute the length of the confidence intervals by several methods. The average length of 1000 samples yields information about the precision of each method. Furthermore, their variances yield information about the representatively of the means.

We carry out empirical studies using two finite populations, the first one being the block population (*Kish*, 1965). The data consist of 270 blocks, and Y and X in this example are respectively the number of rented houses and the number of houses in each block, respectively.

Table 1 shows the average length, $\bar{l}$, and the variance length, σ_l^2, of the confidence intervals built using Woodfruff's method (classical) and the ratio, difference and regression methods that we propose, for 1000 samples of size n, for $n = 30, 35, 40, 45, 50$ and 100 selected from the population for $Q_Y(0.5)$ and $100(1-\alpha)\% = 90\%, 95\%$ and 99%.

From table 1, we can see that for this population there is considerable improvement between the average length of the confidence intervals built using the methods proposed in this paper and the classical method, for any quantile and confidence coefficient. For example, if $100(1-\alpha)\%=95\%$ and $n = 50$, the average length of 1000 confidence intervals determined using the ratio, difference and regression methods are, respectively, 63, 59 and 49 percent of the average length of the respective confidence intervals constructed using the classical method. In this population, the variables Y and X are well correlated, and the concordance is high.

The second population (*Fernández* and *Mayor*, 1994) consist of 1500 households. In this example Y and X are the annual food costs and annual income, respectively. Table 2 shows the results of this second simulation study.

For two population we compute the proportion of intervals that contains the actual population quantile (Cove). We observe this variable doesn't differ a lot from the nominal coverage. Only in the case of the first population this difference is noteworthy for small samples with the regression method. As the sample size increases, the Cove variable is on the increase too and even surpasses the nominal coverage, as it happened in the first population, moreover the average length keeps being lower than the direct method.

In the two populations, we verify that the proposed methods determine more precise confidence intervals for finite population quantiles than Woodruff's method.

References

[1] Andrew, D.F. et al: Robust estimates of location-surveys and advances. Princeton University Press, (1972)

[2] Chambers, R. L., Dunstan, R.: Estimating distribution functions from survey data. Biometrika **73** (1986) 597–604

[3] Chambers, R. L., Dorfman, A. H. and Wehrly, T. E.: Bias robust estimation in finite population using nonparametric calibration. J. Amer. Statist. Assoc. **88** (1991) 268–277

[4] Cochran, A. H.: Sampling Techniques, Third Edition. Wiley, New York (1977)

[5] Fernández García, F. R., Mayor Gallego, J. A.: Muestreo en Poblaciones Finitas: Curso Básico. P.P.U., Barcelona (1994)

[6] Gross, S. T.: Median estimation in sample survey. Proc. Surv. Res. Meth. Sect. Amer. Statist. Ass. (1980) 181–184.

[7] Haskell, J., Sedransk, J.: Confidence interval for quantiles and tolerance intervals of populations. Unpublished Technical Report. SUNY at Albany Dept. of Mathematic Statistics (1980) Albany NY

[8] Kish, L.: Survey Sampling. John Wiley and Sons (1965) New York

[9] Kuk, A. Y. C., Mak, T. K.: Median estimation in presence of auxiliary information. J. R. Statist. Soc. B **51** (1989) 261–269

[10] Kuk, A. Y. C., Mak, T. K.: A functional approach to estimating finite population distribution functions. Commun. Statist. Theor. Meth. **23** (1994) 883–896

[11] Mak, T. K., Kuk, A. Y. C.: A new method for estimating finite–population quantiles using auxiliary information. The Canadian Journal of Statistics **21** (1993) 29–38

[12] Sedransk, J., Meyer, J.: Confidence intervals for the quantiles of a finite population: simple random and stratified simple random sampling. J. Amer. Statist. Ass. **76** (1978) 66–77

[13] Smith, P., Sedransk, J.: Lower bounds for confidence coefficients for confidence intervals for finite population quantiles. Commun. Statist. Theor. Meth. **12** (1983) 1329–1344

[14] Woodruff, R. S.: Confidence intervals for medians and other position measures. J. Amer. Statist. Assoc. **47** (1952) 635–646

Table 1. Block population. $Q_Y(0.5)$.

n	method	\multicolumn{3}{c}{$100(1-\alpha)\%=90\%$}			\multicolumn{3}{c}{$100(1-\alpha)\%=95\%$}			\multicolumn{3}{c}{$100(1-\alpha)\%=99\%$}		
		l	Cove	σ_l^2	l	Cove	σ_l^2	l	Cove	σ_l^2
30	classical:	13.08	.941	24.78	15.83	.974	32.21	19.01	.986	36.92
	ratio:	8.75	.911	35.97	10.36	.941	53.06	13.49	.958	76.16
	difference:	8.22	.910	21.57	9.81	.942	26.78	13.05	.958	52.27
	regression:	5.79	.791	15.22	6.98	.832	18.61	9.52	.856	29.72
35	classical:	12.12	.928	17.34	12.06	.932	18.94	17.79	.996	26.34
	ratio:	7.68	.932	23.59	9.41	.956	46.44	11.68	.972	45.41
	difference:	7.35	.927	13.62	8.83	.955	19.79	11.36	.972	25.51
	regression:	5.56	.850	10.51	6.62	.889	13.66	8.92	.909	20.38
40	classical:	9.86	.903	13.83	12.04	.957	15.51	16.48	.990	20.83
	ratio:	7.17	.939	17.02	8.17	.962	28.79	11.42	.979	45.53
	difference:	6.99	.948	10.47	7.89	.962	13.39	10.79	.980	19.77
	regression:	5.36	.885	7.87	6.31	.905	10.67	8.72	.945	13.61
45	classical:	9.47	.937	10.27	11.43	.945	13.66	15.71	.994	18.28
	ratio:	6.20	.950	10.35	7.53	.964	15.05	10.21	.979	30.54
	difference:	5.98	.953	7.16	7.23	.965	9.68	9.66	.981	15.37
	regression:	4.90	.913	5.40	5.95	.929	7.40	8.07	.958	10.68
50	classical:	9.51	.934	11.10	10.99	.965	11.47	14.53	.993	15.99
	ratio:	5.93	.953	8.87	6.94	.967	11.84	9.37	.987	22.68
	difference:	5.63	.947	5.52	6.48	.962	7.56	8.78	.986	11.34
	regression:	4.77	.912	4.38	5.41	.928	5.74	7.44	.967	7.24
100	classical:	5.39	.925	2.49	6.28	.958	2.81	8.97	.996	4.15
	ratio:	3.25	.947	1.32	3.98	.972	1.97	5.03	.994	3.00
	difference:	3.10	.937	1.05	3.79	.969	1.42	4.85	.995	2.02
	regression:	2.88	.940	0.88	3.47	.959	1.06	4.52	.990	1.35

Table 2. Household population. $Q_Y(0.5)$.

n	method	\multicolumn{3}{c}{$100(1-\alpha)\%=90\%$}			\multicolumn{3}{c}{$100(1-\alpha)\%=95\%$}			\multicolumn{3}{c}{$100(1-\alpha)\%=99\%$}		
		l	Cove	σ_l^2	l	Cove	σ_l^2	l	Cove	σ_l^2
30	classical:	902.42	.920	70012.73	1120.36	.961	78944.98	1343.77	.989	87891.63
	ratio:	732.42	.845	102011.39	932.74	.924	163754.75	1269.66	.970	250355.84
	difference:	735.94	.872	72255.67	913.56	.937	96244.71	1244.99	.980	135796.34
	regression:	647.72	.864	57635.63	775.95	.923	68952.73	1043.31	.972	90731.77
35	classical:	869.32	.910	61117.19	1037.03	.967	66873.03	1416.70	.996	81143.97
	ratio:	717.95	.899	74739.91	838.36	.924	106418.91	1122.44	.976	169528.73
	difference:	705.56	.904	60664.25	853.60	.936	76192.88	1132.52	.990	101151.03
	regression:	615.78	.890	47201.46	729.89	.933	56199.54	965.02	.983	71993.33
40	classical:	815.62	.907	48657.10	987.70	.963	53706.22	1316.73	.991	67421.45
	ratio:	651.66	.885	58891.99	765.09	.922	77151.24	1044.07	.974	138418.58
	difference:	637.23	.886	43446.93	771.50	.934	57501.28	1052.02	.987	78183.77
	regression:	554.84	.884	34808.63	670.20	.923	43843.52	901.39	.978	53760.44
45	classical:	659.01	.865	33460.33	801.92	.931	42223.61	1100.41	.986	55879.21
	ratio:	619.80	.892	54675.29	735.40	.946	72341.49	977.01	.979	99075.30
	difference:	603.06	.885	39712.07	737.81	.960	48640.51	984.62	.983	60614.36
	regression:	533.12	.886	29573.68	647.30	.948	35393.71	857.71	.983	47882.70
50	classical:	659.93	.905	29497.58	784.47	.942	37390.02	1033.61	.989	44294.93
	ratio:	569.48	.878	38292.25	678.98	.934	50601.72	906.85	.973	84822.34
	difference:	570.99	.892	30036.28	686.16	.947	39557.14	901.79	.979	52860.35
	regression:	498.01	.895	24763.54	600.00	.937	31051.13	797.53	.984	40655.98
100	classical:	442.95	.895	11055.36	573.00	.963	14610.82	761.55	.998	19171.91
	ratio:	386.65	.877	12057.26	470.89	.944	15444.78	617.80	.985	19659.33
	difference:	385.23	.887	10022.00	473.97	.948	13011.61	612.82	.987	15872.20
	regression:	345.74	.878	8792.42	419.40	.945	10947.29	553.51	.987	14143.49

Robustness of Clustering under Outliers

Yurij Kharin

Dept. of Mathematical Modelling&Data Analysis, Belarusian State University,
4 Fr. Skariny av., 220050 Minsk, Belarus

Abstract. The problem of clustering of multivariate random data is considered in presence of outliers. The hypothetical model of data is described by a mixture of regular m-parametric probability densities. Clustering of data is made by the often used in practice decision rule which is derived by substitution of ML–estimators (on the unclassified sample) of parameters for their unknown true values in Bayesian decision rule. Robustness of probability of classification error is evaluated. The new clustering algorithm with smoothing is presented. Illustration for the case of the Gaussian hypothetical model and for the Fisher's data under outliers is given.

1 Introduction

The traditional algorithms of data analysis are based on the extremely poor family of hypothetical data models. In constructing of intelligent data analysis systems we need to expand this family and to increase adequacy of the models of real data sets. One of the approaches in generating of adequate models and effective data processing algorithms is given by the theory of robust statistical data analysis [1–3]. Here we develop this approach for the cluster analysis problems under distortions of data.

Cluster analysis problem is in construction of a decision rule by an unclassified training sample $A = (x_1, \ldots, x_n) \subset \mathbb{R}^N$ which divides the sample A into L clusters and can classify "new" observations: $x_{n+1}, x_{n+2}, \ldots$. Each of existing cluster analysis algorithms uses not only the experimental data A but also some (exogenous) prior model assumptions, e.g.: Gaussian distribution and independence of sample elements, absence of outliers, homogeneity of clusters. A system of these model assumptions is called a hypothetical model of data.

In practice, however, the assumed hypothetical model of data is usually distorted [1 — 7]. The classical clustering algorithms often lose their performance and become unstable [6, 7] under distortions. Therefore, the problems of robustness analysis for classical clustering algorithms and synthesis of new robust decision rules are very topical. In [6] these problems were solved for three types of distortions: 1) small–sample effects; 2) presence of runs in the sample; 3) presence of Markov type dependence of class indexes.

---

[1] These investigations were supported by INTAS (project INTAS-93-725-ext).

X. Liu, P. Cohen, M. Berthold (Eds.): "Advances in Intelligent Data Analysis" (IDA–97)
LNCS 1280, pp. 501–511, 1997.   © Springer–Verlag Berlin Heidelberg 1997

This paper is devoted to robustness evaluation of traditional clustering algorithms for a often met in practice type of distortions — presence of outliers in the sample A.

2 Hypothetical model of data and its distortions

Let in a feature space $\mathbb{R}^N$ a regular m-parametric family of N-dimensional p.d.f.

$$Q = \{q(x; \theta'), x \in \mathbb{R}^N : \theta' \in \Theta \subseteq \mathbb{R}^m\}$$

be defined, vector of the parameters $\theta' = (\theta'_k) \in \Theta$ be identifiable, and Θ be a compact set. Let the random observations of the objects from $L \geq 2$ classes $\Omega_1^0, \ldots, \Omega_L^0$ be registered in $\mathbb{R}^N$ with probabilities $\pi_1, \ldots, \pi_L$ ($\pi_i > 0, \pi_1 + \ldots + \pi_L = 1$). An observation from the class Ω_i^0 is a random vector $X_i \in \mathbb{R}^N$ with p.d.f. $q(x; \theta_i^0)$, $i \in S = \{1, 2, \ldots, L\}$; the true parameter values $\{\theta_i^0\} \subset \Theta$ are different and unknown.

An unclassified training sample $A \subset \mathbb{R}^N$ really consisting of L independent subsamples $A_1, \ldots, A_L$ is observed:

$$A = \cup_{i \in S} A_i, \ A_i \cap A_j = \emptyset \ (i \neq j);$$

A_i is a random sample from probability distribution with Tukey–Huber p.d.f.

$$p_i^\varepsilon(x; \theta_i^0) = (1 - \varepsilon_i)q(x; \theta_i^0) + \varepsilon_i h_i(x), \ x \in \mathbb{R}^N, \tag{1}$$

where $h_i(x)$ is any p.d.f. of outliers in A_i:

$$h_i(x) \geq 0, \ \int_{\mathbb{R}^N} h_i(x)dx = 1, \tag{2}$$

ε_i is unknown probability of an outlier presence in A_i ($0 \leq \varepsilon_i \leq \varepsilon_{i+} < 1$).

The distortions (1), (2) have evident illustration: a random element from A_i is an observation from the class Ω_i^0 with probability $1 - \varepsilon_i$, and with probability ε_i it is a random outlier described by the contaminating p.d.f. $h_i(x)$; ε_{i+} is a fixed distortion level (admissible fraction of outliers in A_i), $i \in S$. Note, that some preliminary investigations of these distortions we made in [8].

If the aggregate Lm-vector of parameters $\theta^{0T} = \left(\theta_1^{0T} : \ldots : \theta_L^{0T}\right)$ is known a priori, and the outliers are absent ($\varepsilon_{i+} \equiv 0$), then the Bayesian decision rule

$$d = d(x; \theta^0) = \arg \max_{i \in S}(\pi_i q(x; \theta_i^0)), \ x \in \mathbb{R}^N, \ d \in S \tag{3}$$

classifies the sample A and "new" observations $\{x_{n+1}, x_{n+2}, \ldots\}$ with minimal error probability r_0. As θ^0 is uknown, the so-called "plug-in" decision rule is often used:

$$d = d(x; \hat{\theta}) = \arg \max_{i \in S}(\pi_i q(x; \hat{\theta}_i)), \ x \in \mathbb{R}^N, \ d \in S, \tag{4}$$

where $\hat{\theta} \in \mathbb{R}^{Lm}$ is the ML-estimator of θ^0 constructed by the sample A from the mixture $p^0(\cdot)$ of L p.d.f.:

$$\hat{\theta} = \arg \max_{\theta \in \Theta^L} l(\theta),$$

$$l(\theta) = \frac{1}{n} \sum_{t=1}^n \ln p^0(x_t; \theta), \quad p^0(x; \theta) = \sum_{i \in S} \pi_i q(x; \theta_i). \tag{5}$$

To avoid the known in cluster analysis "phenomenon of classes ambiguity" we will assume that the vectors $\theta_1^0, \ldots, \theta_L^0 \in \Theta$ are lexicographically ordered.

3 Properties of $\hat{\theta}$ under outliers

Let us investigate influence of outliers (1), (2) on the estimator (5) entering the decision rule (4). As it follows from Section 2, the sample A is the random sample from the mixture of L p.d.f. (1):

$$p^\varepsilon(x; \theta^0) = \sum_{i \in S} \pi_i p_i^\varepsilon(x; \theta_i^0). \tag{6}$$

Introduce the notations: $\varepsilon_+ = \max_{i \in S} \varepsilon_{i+}$ is maximal distortion level; ∇_θ^k is operator of k-th order differentiation w.r.t. θ; $\mathbf{0}_{p \times q}$ is $(p \times q)$-matrix with all zero elements; $\mathbf{1}_{p \times q}$ is $(p \times q)$-matrix, all elements of which are equal to 1;

$$H_0(\theta^0; \theta) = - \int_{\mathbb{R}^N} p^0(x; \theta^0) \ln p^0(x; \theta) \, dx,$$

$$A_i(\theta) = \int_{\mathbb{R}^N} (h_i(x) - q(x; \theta_i^0)) \ln p^0(x; \theta) dx, \quad A_i^{(k)}(\theta) = \nabla_\theta^k A_i(\theta) \ (i \in S), \tag{7}$$

$$J^0 = - \int_{\mathbb{R}^N} p^0(x; \theta^0) \nabla_{\theta^0}^2 \ln p^0(x; \theta^0) dx.$$

Note, that in (7) $H_0(\theta^0; \theta^0)$ is Shannon entropy for the hypothetical mixture $p^0(\cdot)$, and J^0 is $(Lm \times Lm)$-Fisher information matrix. Note also, that in regularity conditions

$$\nabla_\theta H_0(\theta^0; \theta) \Big|_{\theta = \theta^0} = \mathbf{0}_{Lm}, \quad \nabla_\theta^2 H_0(\theta^0; \theta) \Big|_{\theta = \theta^0} = J^0. \tag{8}$$

Theorem 1. *If for any $\theta^0, \theta \in \Theta^L$ the integrals $H_0(\theta^0; \theta)$, $A_i(\theta)$ $(i \in S)$ are finite, then under outliers (1), (2) at $n \to \infty$ the almost sure convergence takes place:*

$$-l(\theta) \xrightarrow{a.s.} H_\varepsilon(\theta^0; \theta) = H_0(\theta^0; \theta) - \sum_{i \in S} \varepsilon_i \pi_i A_i(\theta). \tag{9}$$

Proof. According to the strong law of large numbers we have from (5), (6):

$$-l(\theta) \xrightarrow{a.s.} \mathbf{E}\{-\ln p^0(x_t; \theta)\} = -\int_{\mathbb{R}^N} p^\varepsilon(x; \theta^0) \ln p^0(x; \theta) dx.$$

Using (1), (5) we find:

$$p^\varepsilon(x; \theta^0) = p^0(x; \theta^0) + \sum_{i \in S} \varepsilon_i \pi_i \left(h_i(x) - q(x; \theta_i^0) \right),$$

and we come to the convergence (9).   □

Theorem 2. *If the family of p.d.f. Q satisfies the Chibisov regularity conditions [9], the functions $H_0(\theta^0; \theta), \{A_i(\theta)\}$ are thrice differentiable w.r.t. $\theta \in \Theta$, and the point $\theta^\varepsilon = \arg\min_{\theta \in \Theta^L} H_\varepsilon(\theta^0; \theta)$ is unique, then the almost sure convergence of the estimator (5) under outliers (1), (2) takes place at $n \to \infty$:*

$$\hat{\theta} \xrightarrow{a.s.} \theta^\varepsilon, \tag{10}$$

and $\theta^\varepsilon \in \mathbb{R}^{Lm}$ satisfies the asymptotic expansion:

$$\theta^\varepsilon = \theta^0 + \sum_{i \in S} \varepsilon_i \pi_i (J^0)^{-1} A_i^{(1)}(\theta^0) + \mathcal{O}(\varepsilon_+^2) \mathbf{1}_{Lm}. \tag{11}$$

Proof. The convergence (10) is proved by Theorem 1 in the same way as strong consistency property of ML-estimator $\hat{\theta}$ without outliers is [9]. The asymptotic expansion (11) is constructed by using of linear Taylor formula to the left side of the equation

$$\nabla_\theta H_\varepsilon(\theta^0; \theta) \Big|_{\theta = \theta^\varepsilon} = \mathbf{0}_{Lm}$$

and then by using (9), (8).   □

It is seen from Theorem 2, that in presence of outliers in the sample the estimator $\hat{\theta}$ can become non-consistent:

$$\theta^\varepsilon - \theta^0 = \sum_{i \in S} \varepsilon_i \pi_i (J^0)^{-1} A_i^{(1)}(\theta^0) + \mathcal{O}(\varepsilon_+^2) \mathbf{1}_{Lm}. \tag{12}$$

Note, that according to (7) $|A_i^{(1)}(\theta^0)|$ depends on the contaminating density $h_i(\cdot)$ (satisfying only (2)) and may have sufficiently large value $(i \in S)$.

4   Asymptotic expansion of error probability

If the outliers are absent in the sample A $(\varepsilon_+ = 0)$, and $\theta^0 \in \Theta^L$ is a priori known, then the Bayesian decision rule (3) has the probability of classification error

$$r_0 = 1 - \sum_{j \in S} \pi_j \int_{\mathbb{R}^N} q(x; \theta_j^0) \mathbf{1}_{V_j^0}(x) dx, \tag{13}$$

where

$$\mathbf{1}_{V_j^0}(x) = \prod_{k \neq j} \mathbf{1}\left(f_{kj}^0(x)\right)$$

is the indicator function of the region of decision making in favour of Ω_j; $V_j^0 = \{x \in \mathbb{R}^N : d(x; \theta^0) = j\}$; $\mathbf{1}(z)$ is the unit function; $f_{kj}^0(x) = \pi_j q(x; \theta_j^0) - \pi_k q(x; \theta_k^0)$ is the hypothetical Bayesian discriminant function for the pair of classes $\{\Omega_j^0, \Omega_k^0\}$.

According to Theorem 1 and Theorem 2 the decision rule (4) in presence of outliers $(\varepsilon_+ > 0)$ at $n \to \infty$ converges to the decision rule different from (3):

$$d = d(x; \theta^\varepsilon) = \arg\max_{i \in S}\left(\pi_i q(x; \theta_i^\varepsilon)\right), \quad x \in \mathbb{R}^N. \tag{14}$$

Let us characterize robustness of the decision rule (4) under outliers by the probability of classification error for the limit decision rule (14):

$$r_\varepsilon = 1 - \sum_{j \in S} \pi_j \int_{\mathbb{R}^N} q(x; \theta_j^0) \mathbf{1}_{V_j^\varepsilon}(x) dx, \tag{15}$$

its supremum:

$$r_+ = \sup_{\{\varepsilon_i\}} r_\varepsilon,$$

and robustness factor $(r_0 > 0)$:

$$\kappa = (r_+ - r_0)/r_0,$$

where

$$\mathbf{1}_{V_j^\varepsilon}(x) = \prod_{k \neq j} \mathbf{1}\left(f_{kj}^\varepsilon(x)\right), \quad f_{kj}^\varepsilon(x) = \pi_j q(x; \theta_j^\varepsilon) - \pi_k q(x; \theta_k^\varepsilon).$$

Introduce the notations: $(J^0)^{-1}{}_{(i)}$ is $(Lm \times m)$-matrix — the i-th block–column of the matrix $(J^0)^{-1}$; $\overset{\circ}{\Gamma}_{ij} = \{x : f_{ij}^0(x) = 0\} \subset \mathbb{R}^N$ is the hypothetical Bayesian discriminant $(N - 1)$-hypersurface for two classes $\{\Omega_i, \Omega_j\}, i \neq j$; $\overset{\circ}{\Gamma}_{ijk} = \{x : f_{ij}^0(x) = f_{jk}^0(x) = 0\} \subset \mathbb{R}^N$ is the hypothetical Bayesian discriminant $(N - 2)$-hypersurface for three classes $\{\Omega_i, \Omega_j, \Omega_k\}, i \neq j \neq k$;

$$[a_1(x), a_2(x)] = \left(|\nabla a_1(x)|^2 |\nabla a_2(x)|^2 - \left((\nabla a_1(x))^T \nabla a_2(x)\right)^2\right)^{\frac{1}{2}};$$

$$u_{ij}(x) = \pi_i (J^0)^{-1}{}_{(i)} \nabla_{\theta_i^0} q(x; \theta_i^0) - \pi_j (J^0)^{-1}{}_{(j)} \nabla_{\theta_j^0} q(x; \theta_j^0)$$

is Lm-vector–column;

$$G_L = \sum_{j=1}^{L-1} \sum_{i=j+1}^{L} \int\limits_{\overset{\circ}{\Gamma}_{ij}{}'} u_{ij}(x) u_{ij}^T(x) |\nabla f_{ij}^0(x)|^{-1} ds_{N-1} -$$

$$-\sum_{j\in S} \sum_{i\neq j} \sum_{k\notin\{i,j\}} \pi_j \int\limits_{\overset{\circ}{\Gamma}_{ijk}{}'} q(x;\theta_j^0) u_{ij}(x) u_{kj}^T(x) [f_{ij}^0(x), f_{kj}^0(x)]^{-1} ds_{N-2} \qquad (16)$$

is $(Lm \times Lm)$-matrix, where $\overset{\circ}{\Gamma}_{ij}{}'$, $\overset{\circ}{\Gamma}_{ijk}{}'$ are that parts of the surfaces $\overset{\circ}{\Gamma}_{ij}$, $\overset{\circ}{\Gamma}_{ijk}$ respectively which are the borders of V_j^0.

Theorem 3. *If the conditions of Theorem 2 hold and the surface integrals in (16) are bounded, then the error probability r_ε defined by (15), (14) under outliers (1), (2) satisfies the asymptotic expansion:*

$$r_\varepsilon = r_0 + \sum_{l_1,l_2=1}^{L} \rho_{l_1,l_2} \varepsilon_{l_1} \varepsilon_{l_2} + \mathcal{O}(\varepsilon_+^3), \qquad (17)$$

where

$$\rho_{l_1,l_2} = \frac{\pi_{l_1} \pi_{l_2}}{2} A_{l_1}^T(\theta^0) G_L A_{l_2}(\theta^0). \qquad (18)$$

Proof. At first, using (10), (11), (16) we construct the asymptotic expansion of the distorted discriminant functions $(k, j \in S, k \neq j)$:

$$g_{kj}(x) = f_{kj}^\varepsilon(x) - f_{kj}^0(x) = -\Big(\sum_{i\in S} \varepsilon_i \pi_i A_i^{(1)}(\theta^0)\Big)^T u_{kj}(x) + \mathcal{O}(\varepsilon_+^2).$$

At second, we construct the asymptotic expansion of the integral entering (15)

$$B_j = \int\limits_{\mathbb{R}^N} q(x;\theta_j^0) \mathbf{1}_{V_j^\varepsilon}(x) dx$$

by the lemmas 5.1, 3.1, 3.3 [6].

Putting this expansion into (15), collecting the main expansion terms w.r.t. $\{\varepsilon_i\}$ and using (13), (16), we come to the asymptotic expansion (17), (18). $\quad\square$

Note, that the matrix G_L is nonnegatively definite, therefore, the quadratic form in (17) is nonnegative.

Corollary 4. *In the case of two classes $(L = 2)$ the error probability satisfies the asymptotic expansion:*

$$r_\varepsilon = r_0 + \varepsilon_1^2 \rho_{11} + \varepsilon_2^2 \rho_{22} + 2\varepsilon_1 \varepsilon_2 \rho_{12} + \mathcal{O}(\varepsilon_+^3),$$

$$G_2 = \int\limits_{\overset{\circ}{\Gamma}_{12}{}'} u_{12}(x) u_{12}^T(x) |\nabla f_{12}^0(x)|^{-1} ds_{N-1}.$$

Proof. If $L = 2$, then the surface integrals on $\overset{o}{\Gamma}_{ijk}$ disappear in (16). □

Corollary 5. *In the case of two $(L = 2)$ equidistorted classes $(\varepsilon_1 = \varepsilon_2 = \varepsilon \le \varepsilon_+)$ supremum of error probability satisfies the expansion:*

$$r_+ = r_0 + \rho \varepsilon_+^2 + \mathcal{O}(\varepsilon_+^3), \tag{19}$$

$$\rho = \rho(h^\pi) = \Big| \int_{\mathbb{R}^N} h^\pi(x) G_2^{\frac{1}{2}} \nabla_{\theta^0} \ln p^0(x; \theta^0) dx \Big|^2, \tag{20}$$

and $h^\pi(x) = \sum_{i \in S} \pi_i h_i(x)$ is the mixture of contaminating p.d.f.

Proof. Proof is based on Corollary 4 and expressions (8), (18). □

Corollary 6. *Under conditions of Corollary 5*

$$\kappa = \bar{\kappa} + \mathcal{O}(\varepsilon_+^3), \bar{\kappa} \le \tau(h^\pi) \varepsilon_+^2, \tag{21}$$

where

$$\tau(h^\pi) = \int_{\mathbb{R}^N} h^\pi(x) Q(x) dx, \quad Q(x) = \nabla_{\theta^0}^T \ln p^0(x; \theta^0) G \nabla_{\theta^0} \ln p^0(x; \theta^0) \ge 0.$$

Proof. Proof is based on (19) and on applying of Cauchy–Schwarz–Bunyakovskii inequality to (20). □

Corollary 7. *If $\mathcal{X} \subset \mathbb{R}^N$ is the compact set of admissible values of outliers, then under conditions of Corollary 5*

$$\bar{\kappa} \le \max_{x \in \mathcal{X}} Q(x) \cdot \varepsilon_+^2.$$

Proof. Maximizing $\tau(h^\pi)$ w.r.t $h^\pi(\cdot)$, we find: $h_*^\pi(x) = \delta(x - x^*)$ — the Dirac delta function, where $x^* = \arg\max_{x \in \mathcal{X}} Q(x)$. □

5   The case of Gaussian hypothetical model under outliers

Let us apply the results of Section 4 to the most popular hypothetical model in statistical clustering — the Gaussian model $(m = N)$:

$$q(x; \theta) = (2\pi)^{-N/2} |\Sigma|^{-1/2} \exp\big(-(x - \theta)^T \Sigma^{-1} (x - \theta)/2\big), \quad x \in \mathbb{R}^N,$$

where Σ is any fixed nonsingular covariance matrix. For the case of two $(L = 2)$ equiprobable $(\pi_1 = \pi_2 = \frac{1}{2})$, equidistored $\varepsilon_1 = \varepsilon_2 = \varepsilon$ classes let us introduce the notations: $\Delta = \sqrt{(\theta_2^0 - \theta_1^0)^T \Sigma^{-1} (\theta_2^0 - \theta_1^0)}$ is Mahalanobis interclass distance; $\Phi(z)$ is the standard normal distribution function; $\phi(z) = \Phi'(z) = (2\pi)^{-1/2} \exp(-z^2/2)$ is p.d.f. of this distribution. Note, that

$$r_0 = \Phi(-\Delta/2).$$

By Corollary 4 and the properties of multivariate Gaussian p.d.f. we find:

$$G_2 = \begin{pmatrix} G_{11} & \vdots & G_{12} \\ \cdots & \cdots & \cdots \\ G_{12} & \vdots & G_{22} \end{pmatrix},$$

where the $(N \times N)$-blocks of this matrix are determined by the formula:

$$G_{ij} = (-1)^{i+j} \frac{2}{\Delta} \phi(\frac{\Delta}{2}) \left(\Sigma + \left(\frac{(-1)^{i+j}}{2} - \frac{1}{\Delta^2} \right) (\theta_2^0 - \theta_1^0)(\theta_2^0 - \theta_1^0)^T \right) + o \cdot 1_{N \times N},$$

and o denotes the remaindor. Then (21) gives:

$$\bar\kappa \le \varepsilon_+^2 \frac{2\varepsilon_+^2 \phi(\frac{\Delta}{2})}{\Delta \Phi(-\frac{\Delta}{2})} \sum_{i=1}^{2} \int_{\mathbb{R}^N} (x - \theta_i^0)^T \left(\Sigma^{-1} + \left(\frac{1}{4} - \frac{1}{\Delta^2} \right) bb^T \right) (x - \theta_i^0) h^\pi(x) dx,$$

where $b = \Sigma^{-1}(\theta_2^0 - \theta_1^0)$. In particular, if the contaminating p.d.f. are Gaussian densities: $h_1(x) = h_2(x) = h^\pi(x) = q(x; \theta^+)$, where $\theta^+ \in \mathbb{R}^N$ is mathematical mean of the outliers, then

$$\bar\kappa \le \varepsilon_+^2 \frac{4\phi(\Delta/2)}{\Delta \cdot \Phi(-\Delta/2)} \left(N - 1 + \frac{\Delta^2}{4} + \frac{1}{2} \sum_{i=1}^{2} \left((\theta^+ - \theta_i^0)^T \Sigma^{-1} (\theta^+ - \theta_i^0) + \right. \right.$$

$$\left. \left. + (\frac{1}{4} - \frac{1}{\Delta^2})((\theta^+ - \theta_i^0)^T \Sigma^{-1}(\theta_2^0 - \theta_1^0))^2 \right) \right). \tag{22}$$

This estimate of robustness factor gives an opportunity to evaluate stability of the clustering algorithm under outliers and to investigate dependence of its stability w.r.t. proportion of outliers ε_+, dimensionality of feature space N, Mahalanobis interclass distance Δ and mathematical mean of outliers θ^+.

For illustration let us consider the special case of two-dimensional Gaussian data $(N = 2)$ with "outliers in mean":

$$\theta_1^0 = \begin{pmatrix} -\frac{\Delta}{2\sqrt 2} \\ -\frac{\Delta}{2\sqrt 2} \end{pmatrix}, \ \theta_2^0 = \begin{pmatrix} \frac{\Delta}{2\sqrt 2} \\ \frac{\Delta}{2\sqrt 2} \end{pmatrix}, \ \Sigma = \mathbf{I}_2, \ \theta^+ = \begin{pmatrix} -\frac{\lambda \Delta}{2\sqrt 2} \\ \frac{\lambda \Delta}{2\sqrt 2} \end{pmatrix},$$

where $\lambda \in \mathbb{R}$ is the parameter characterizing the value of distortions. Note that the distance between the "hypothetical mean" θ_i^0 and "outlier mean" θ^+ $(i = 1, 2)$

$$|\theta^+ - \theta_i^0| = \frac{\Delta}{2} \sqrt{\lambda^2 + 1}$$

increases if $|\lambda|$ increases.

By the formula (22) we find the upper bound for the robustness factor:

$$\kappa \le \kappa_+ + \mathcal{O}(\varepsilon_+^3),$$

where the main term κ_+ is defined by the expression:

$$\kappa_+ = \varepsilon_+^2 \cdot \frac{\varphi\left(\frac{\Delta}{2}\right)}{\Phi\left(-\frac{\Delta}{2}\right)} \left(\frac{\Delta^3}{4} + (\lambda^2 + 1)\Delta + \frac{4}{\Delta}\right). \tag{23}$$

Let us fix now any admissible level of robustness factor $\delta > 0$ and find the corresponding level of outlier probability $\varepsilon_+ = \varepsilon_+(\delta, \Delta)$ from the equation

$$\kappa_+ = \delta.$$

We get the expression of this critical value using (23):

$$\varepsilon_+(\delta, \Delta) = \min\left\{\sqrt{\frac{\Phi\left(-\frac{\Delta}{2}\right)}{\varphi\left(\frac{\Delta}{2}\right)} \cdot \frac{\delta}{\frac{\Delta^3}{4} + (\lambda^2 + 1)\Delta + \frac{4}{\Delta}}}, 1\right\}.$$

The critical values of outlier probability $\varepsilon_+(\delta, \Delta)$ are presented in Table 1 for $\delta = 1$, different values of Mahalanobis distance Δ and different values of the parameter λ.

Table 1. Critical values of outlier probability $\varepsilon_+(\delta, \Delta)$

λ	Δ				
	1	2	3	4	5
0	0.408	0.331	0.216	0.142	0.098
1	0.374	0.286	0.191	0.130	0.092
3	0.248	0.165	0.116	0.086	0.066
5	0.170	0.108	0.077	0.059	0.047
10	0.091	0.056	0.041	0.032	0.026
100	0.009	0.006	0.004	0.003	0.002

One can see from Table 1, that if, for example $\Delta = 3$ ($r_0 = 0.067$) and $\lambda \leq 10$, then the classification error probability will not exceed the value $r = (1 + \delta)r_0 = 0.134$ (100%-increment) for all situations where the probability of outlier presence in the sample will not exceed the critical value $\varepsilon_+ = 0.041$.

6 Clustering algorithm with smoothing

As it is seen from (12) and from constructed estimates of robustness factor, presence of outliers in the sample A can influence on $\theta^\varepsilon - \theta^0$ and κ very strongly.

To reduce this influence we propose the clustering algorithm with smoothing, which can be presented as the following sequence of steps.

1. Let us construct any nonsingular robust estimate $S = (s_{kl})$ $(k, l = \overline{1, N})$ of common covariance matrix Σ by the sample A (e.g. Huber estimate [1]).

2. Construct the matrix $B = (b_{ij})$ of Mahalanobis distances between all elements of the sample A:

$$b_{ij} = \sqrt{(x_i - x_j)^T S^{-1}(x_i - x_j)} \geq 0 \ (i, j = \overline{1, n}).$$

3. For fixed integer m (parameter of the algorithm) and $i = \overline{1, n}$ using B:

− find m nearest neighbours of the point x_i: $x_{j_{i1}}, x_{j_{i2}}, \ldots, x_{j_{im}} \in A$;
− make smoothing of the observation x_i, that is to replace it by the local mean point:

$$x_i^{(1)} = \frac{1}{m+1}\left(x_i + \sum_{k=1}^{m} x_{j_{ik}}\right).$$

This step results in the "smoothed" sample: $A^{(1)} = \{x_1^{(1)}, x_2^{(1)}, \ldots, x_n^{(1)}\}$.

4. If it is necessary, repeat steps 2, 3 for reiterated smoothing of sample $A^{(1)}$.

5. Apply the well known clustering algorithm "L-means" to the smoothed sample $A^{(1)}$ and get the vector of classification into L classes: $\hat{D} = (\hat{d}_1, \ldots, \hat{d}_n)$, where $\hat{d}_i \in S$ is the estimate of the class index for the i-th obsevation $(i = \overline{1, n})$.

To illustrate this algorithm we use the well known in cluster analysis Fisher's "Iris Data" [10].

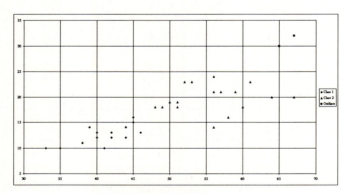

Fig. 1. Scatter diagram of A

The sample A consisted of $n = 40$ observations of $L = 2$ classes: twenty observations from Ω_1^0 ("Iris versicolor") and twenty observations from Ω_2^0 ("Iris virginica") in two-dimensional ($N = 2$) feature space (x_1 is length and x_2 is width of the petal). Two observations of the class Ω_2^0 were replaced by outliers ($\varepsilon_1 = 0$, $\varepsilon_2 = 0.1$). The scatter diagram of the sample A with outliers is presented in Figure 1. The classical algorithm "L-means" gives very bad clustering with the frequency of classification error $\hat{r} = 0.47$. After smoothing according to the described algorithm with $m = 7$ we got the sample $A^{(1)}$ scatter diagram of which is presented in Figure 2; clustering after smoothing gave the acceptable level of frequency of classification error $\tilde{r} = 0.10$.

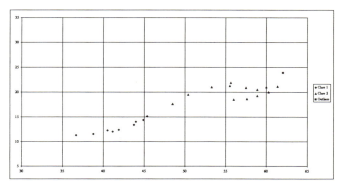

Fig. 2. Scatter diagram of $A^{(1)}$

Note in conclusion, that this clustering algorithm can be optimized w.r.t. the computational complexity by using the effective algorithms of sorting and preliminary processing of data.

References

1. Huber, P.: Robust statistics. John Wiley and Sons, N.Y., (1981)
2. Hampel, F. et al.: Robust statistics. John Wiley and Sons, N.Y., (1986)
3. Rieder, H. (Ed.): Robust statistics, data analysis and computer intensive metods. Lecture Notes in Statistics **109** (1996)
4. Bock, H.: Probabilistic aspects in cluster analysis. Proc. 13th Conf. of Classif. Society, Springer–Verlag, N.Y. (1989) 12 — 44
5. Aivazyan, S. et al.: Applied Statistics, v.3. Fin. i Stat., Moscow (1989)
6. Kharin, Yu.: Robustness in statistical pattern recognition, Kluwer Academic Publishers, Dordrecht (1996)
7. McLachlan, G. et al.: Mixture Models: inference and applications to clustering. Marcel Dekker, N.Y. (1988)
8. Kharin, Yu., Zhuk, E.: Asymptotic robustness in cluster analysis for Tukey–Huber distortions. Information and Classification. Springer–Verlag, N.Y. (1993) 31 — 39
9. Chibisov, D.: Asymptotic expansion for a class of estimators including ML-estimators. Prob. Theory and its Appl. **18** (1973) 303 — 311
10. Fisher, R.: The use of multiple measurements in taxonomic problems. Ann. Eugen. **7** (1936) 179 — 188

The BANG–Clustering System: Grid–Based Data Analysis

Erich Schikuta and Martin Erhart

Institute of Applied Computer Science and Information Systems
University of Vienna, Rathausstr. 19/4, A-1010 Vienna, AUSTRIA
Email: schiki@ifs.univie.ac.at

Abstract. For the analysis of large images the clustering of the data set is a common technique to identify correlation characteristics of the underlying value space. In this paper a new approach to hierarchical clustering of very large data sets is presented. The BANG-Clustering system presented in this paper is a novel approach to hierarchical data analysis. It is based on the BANG-Clustering method ([Sch96]) and uses a multi-dimensional grid data structure to organize the value space surrounding the pattern values. The patterns are grouped into blocks and clustered with respect to the blocks by a topological neighbor search algorithm.

1 Introduction

Clustering methods are extremely important for explorative data analysis, which is an important approach for the analysis of images. Previously presented algorithms can be divided into hierarchical algorithms, e.g. single-linkage, complete-linkage, etc. and partitional algorithms, e.g. K-MEANS, ISODATA, etc. (see [DJ80]). All of these methods suffer from specific drawbacks, when handling large numbers of patterns. The hierarchical methods provide structural information, as dendrograms, but are suitable only for a small number of patterns. With growing numbers the computational expense increases, due to the calculation of a dissimilarity matrix, where each pattern is compared to all others. The partitional methods are to some less resource consuming. However they lack methodical freedom because of the prerequisite of a "good guess" of structural information, e.g. the numbers and the positions of the initial cluster centers. If the choice of the initial clustering is not appropriate, the partitional methods also become very calculation intensive in computing new cluster centers.

A number of different algorithms have been proposed to overcome these problems (e.g. [Bru88, CCM92, CD91, Kur91, IK89, VR92, ZWB91]). Most of the algorithms compare the patterns to each other or to predefined cluster center. Via a calculated distance metric they organize the patterns by combining them into clusters. Another alternative is to cluster the patterns according to the structure of the embedding space, as first presented by Warnekar et al. [WK79]. They proposed a heuristic, hierarchical clustering algorithm based on detecting

X. Liu, P. Cohen, M. Berthold (Eds.): "Advances in Intelligent Data Analysis" (IDA–97)
LNCS 1280, pp. 513–524, 1997. © Springer–Verlag Berlin Heidelberg 1997

clusters by overlapping pattern cells. As Warnekar et al. pointed out, however, the algorithm suffered from the problem that all pattern cells were of the same size and so did not adapt to the real distribution of the patterns. This leads to a complex and quite costly algorithm, which decides how to combine the cells by an expensive distance calculation between possible neighbor cells. Broder [Bro90] solved this problem by means of using of an adaptable data structure, the kd-B-tree. His algorithm calculates the m nearest neighbors to a specified pattern in a k-dimensional value space. The algorithm is problematic for cluster analysis, because it needs a given cluster center as input, additionally the performance is low due to the necessary recursive traversal of the kd-B-tree index.

The hierarchical Grid-Clustering algorithm [Sch96] combines and refines both ideas. It introduces a density index to compare and evaluate the possible pattern cells to find cluster centers and to combine neighbor cells. Further, a new Grid-Structure to organize the value space surrounding the patterns was implemented, which overcomes the problems of the kd-B-tree. This algorithm achieves an appealing performance gain in comparison to all conventional algorithms.

2   BANG-Clustering

Conventional cluster algorithms calculate a distance based on a dissimilarity metric (e.g. Euclidean distance, etc.) between patterns or cluster centers. The patterns are clustered accordingly to the resulting dissimilarity index. The BANG-Clustering algorithm presented here uses the idea of Warnekar [WK79] to organize the value space containing the patterns. For that reason we use the BANG-Structure. The patterns are treated as points in a k-dimensional value space and are randomly inserted into the BANG-Structure. These points are stored accordingly to their pattern values, preserving the topological distribution. The BANG-Structure partitions the value space (shown in figure 1) and administrates the points by a set of surrounding rectangular shaped *blocks* (figure 2).

A block is a rectangular shaped cube containing up to a maximum of p_{max} patterns. $X = (x_1, x_2, ..., x_n)$ is a set of n patterns and x_i is a pattern consisting of a tuple of k describing features $(p_{i_1}, p_{i_2}, ..., p_{i_k})$, where k is the number of dimensions of the underlying value space of the data set.

The pictures shown in the figures are actual screen shots. In the example 50000 patterns, which are clustered into 3 groups with about 20 percent noise, were analyzed by the BANG-Clustering algorithm.

The BANG-Clustering algorithm uses the block information of the BANG-Structure and clusters the patterns accordingly to their surrounding blocks creating a respective dendrogram in turn (see figure 8).

2.1   The BANG-Structure

The BANG-Structure stores the patterns of the underlying value space by a grid structure, which is called *grid directory* (similarly to the Grid-File). This

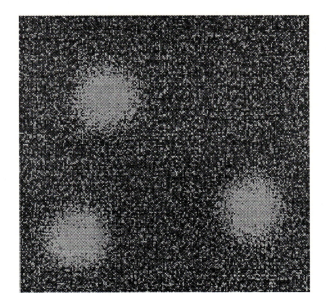

Fig. 1. 2-dimensional pattern set

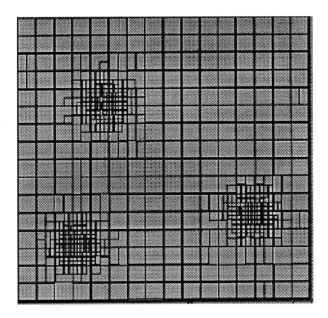

Fig. 2. BANG-Clustering

structure (see figure 3), which is administrated by *scales*, partitionates the k-dimensional value space into *grid regions* (rectangular shaped subspaces). Each scale represents one pattern attribute, and each scale entry resembles a (k-1) dimensional hyperspace partitioning the value space into two.

Each grid region is mapped to one *data block* containing the patterns, but a data block can be mapped by more than one grid region (1:m mapping). The union of these grid regions (mapping to the same data block) is called *block region*. The value space spanned by a block region is rectangular shaped (convex).

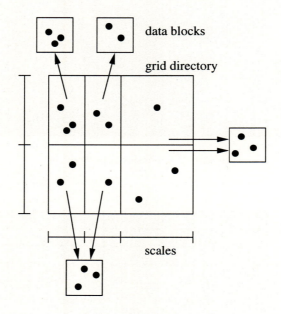

Fig. 3. BANG-Structure

The value space is partitioned into a *hierarchical* set of grid regions. Each region is uniquely identified by a pair of keys (r, l), where r is the region number and l is the level number. The partitioning is binary (a region is split into two equally shaped regions) in each dimension. The sequence of the split dimensions has to be uniquely defined. Region (0,0) comprises the whole value space and is partitioned according the defined scheme into subregions.

The structure of the block regions is defined by the following two axioms [Fre87]

— The union of all subregions into which the value space has been partitioned must span the whole value space.

  – If two subregions intersect, then one of these subregions completely encloses the other.

The second axiom allows nested regions, which is shown in figure 4. To reach compact structures algorithms are defined [Fre87], which guarantee a balance between the data blocks by redistribution. This proved extremely useful for clustered value sets.

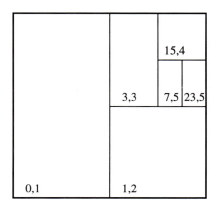

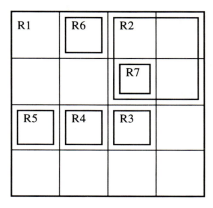

Fig. 4. BANG-Structure block regions   **Fig. 5.** Grid directory - Neighborhood

The BANG-Clustering algorithm uses the block information of the grid directory and clusters the patterns to their blocks accordingly.

2.2 Density Index

The algorithm calculates a *density index* of each block via the numbers of patterns and the spatial volume of the block (analogously to the Grid-Clustering algorithm [Sch96]). The spatial volume V_B of a block B is the Cartesian product of the extents e of block B in each dimension, i.e.

$$V_B = \prod e_{B_i},$$
$$i = 1, ..., k.$$

The density index D_B of block B is defined as the ratio of the actual number of patterns p_B contained in block B to the spatial volume V_B of B, i.e.

$$D_B = \frac{p_B}{V_B}$$

The blocks are sorted accordingly to their density indices. Blocks with the highest density index (obviously with highest pattern correlation) become clustering centers. The remaining blocks are then clustered iteratively in order of their density index, thereby building new cluster centers or merging with existing clusters. Only blocks adjacent to a cluster, i.e. *neighbors*, can be merged.

2.3   Neighbors

Two types of neighborhood can be distinguished in the BANG-Structure, *normal neighborhood*, i.e. neighbors respective block regions, and *refined neighborhood*, i.e. neighbors respective logical regions. Further a *neighbor degree* can be defined by the dimensionality of the "touching" area between 2 regions. Generally the dimensionality can vary between 0 (an point) and k-1 (a k-1 dimensional hyperplane). For the example shown in figure 5 (2-dimensional case) the level of dimensionality is 0 (a point) and 1 (an edge). A normal neighborhood exists e.g. between regions R2 and R1, R6, and R7, and a refined neighborhood between regions R2 and R1, R6, and R7.

Neighbors are found by comparison of the scale values of the grid directory. If regions are at the same level, the differences can be determined directly. If the levels are not at the same level, the lower level region has to be transformed to the higher level region and the comparison has to be done appropriately. In the example of figure 6 the regions and their identifiers are R1 = (0,0), R2 = (3,2), R3 = (9,4), R4 = (12,4), R5 = (8,4), R6 = (14,4), and R7 = (3,4). R6 and R2 are neighbors but on different region-levels (R6 on level 4, with x =1 and y = 3, and R2 on level 2, with x =1 and y = 1). Therefore we have to transform R2 to level 4, which yields in $x_{min} = 2$, $x_{max} = 3$, and $y_{min} = 2$, $y_{max} = 3$.

The region identifier are an ordered set of tuples. To find possible neighbor regions the algorithm accesses these tuples. To support this step efficiently we designed a novel administration structure for the region identifiers. Because of the numbering scheme of the grid regions we chose a binary tree for storing the grid structure. The basic scheme is shown in figure 6.

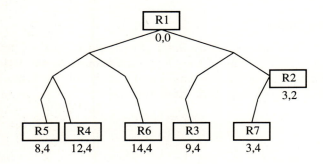

Fig. 6. Binary tree storing the grid directory of figure 5

The partitioning of a region is directly supported by the tree scheme, and a comprising region is simply found by backtracking the path to the root (representing the whole value space). The height of the tree is defined by the number of region levels.

2.4 Dendrogram

The dendrogram is calculated directly by the clustering algorithm, as depicted in figure 7. The density indices of all regions are calculated and sorted in decreasing order.

Starting with the first region (with the highest density index) all neighbor regions are determined and classified in decreasing order (step 1). The neighbor search is repeated for each processed region. The found regions are placed in the dendrogram to the right of the original regions (step 2), respective to the following rules,

- is R1 neighbor of R2 and R2 neighbor of R3 and R1 > R2 > R3, then build with R1, R2, and R3 a cluster (neighbor search starting from R3), and
- is R1 neighbor of R2 and R2 neighbor of R3 and R1 > R2 < R3, then build with R1, R2, and R3 a cluster (neighbor search starting from R2).

A calculated dendrogram based on the example depicted by figure 2 is shown in figure 8.

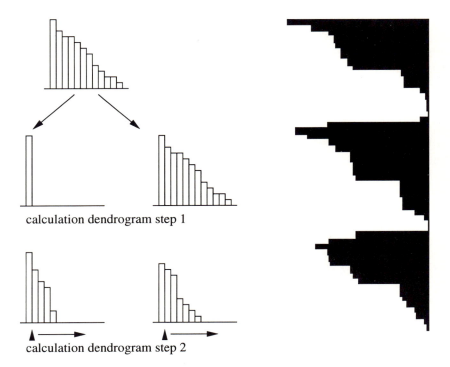

calculation dendrogram step 1

calculation dendrogram step 2

Fig. 7. Dendrogram creation

Fig. 8. Calculated dendrogram for example of figures 1 and 2

3   The BANG-Clustering System

The BANG-Clustering algorithm was implemented under the Unix operating
systems using X11 and the Motif libraries providing a user friendly environment
[Erh95]. Running versions are available for Linux, Sun and HP workstations.
Basically the system provides windows for

- the control of the program and the visual analysis of the data set,
- the graphical representation of the BANG structure and the state of the
  clustering process,
- the number and positions of detected cluster centers, and
- the layout of the dendrogram.

The system allows to control any parameter of the clustering process and to
follow iteratively each clustering step. The design of the system tries to maximize
the user's flexibility for the analysis of the data set. In figure 9 a screen shot of
the BANG-system is shown, with a sample of the most common windows. The
control, clustering, dendrogram, and data view window are easily identifiable in
the figure.

In figure 10 the iterative clustering approach is shown for a 2-dimensional
data set consisting of 10000 patterns. The user has the possibility to control
the clustering process at his will (he can advance or retrace in the process),
by positioning the scroll bar at the bottom of the main window showing the
pattern lay-out. For higher dimensional data sets the projection dimensions can
be chosen appropriately. In figure 10 three clustering states are depicted, for 0%,
50%, and 100% of the data set clustered. Interdependently the window showing
the BANG-file structure shadows the clustered block regions and blacks out the
last clustered region (the region with the lowest density index until now). Thus,
if the whole data set is clustered, all but one block regions are shadowed and the
last region is blacked out.

Accordingly to the clustering process also the dendrogram window is updated
showing the clustered data patterns in black and the remaining patterns in grey.
In figure 11 the situation is shown for 50% of the data set clustered.

4   Performance Analysis

For our analysis we compare the BANG-Clustering to several well known conventional clustering algorithms and to the Grid-Clustering [Sch96] as well. For
the conventional algorithms we used commercial statistical packages on workstations (WS) and, due to memory exhaustion and exceeding execution times
in the workstation environment, also on mainframes (MF). The times shown in
the charts are pure processing times for the completion of the algorithms (no
data set loading or display of information). The investigated algorithms in figure
12 were BANG-Clustering (workstation), GRID-Clustering (workstation), single
linkage (mainframe), and quick cluster (workstation).

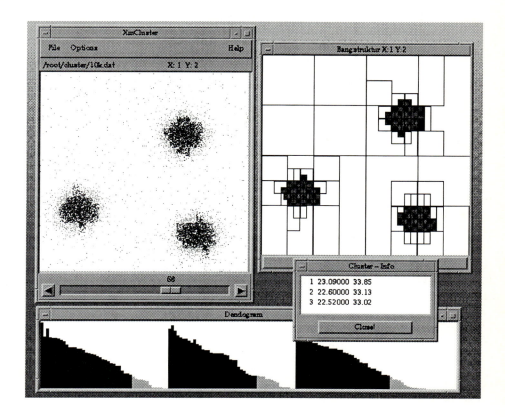

Fig. 9. BANG system screen layout

BANG-Clustering and Grid-Clustering outperformed the conventional methods by far. Both performed even better on workstations than any of the conventional methods on mainframes. More importantly they delivered results where the other algorithms failed because of exceeding runtime behavior or memory exhaustion.

Further we compared BANG-Clustering to Grid-Clustering directly. The block size (i.e. the number of patterns per region) is an influencing factor for grid structures, as shown in [Sch96]. Therefore we measured the execution times for constant block size (100 patterns) and for dynamic block sizes (10% of the data set). Due to the situation that both grid methods allow very large data sets, we present the execution times for data set sizes up to 100000 patterns (see figure 13).

For large data sets BANG-Clustering outperformed Grid-Clustering due to the linear growth rate of the BANG-structure size (one logical region per data block) compared to the over-linear growth of the Grid-structure size (often many

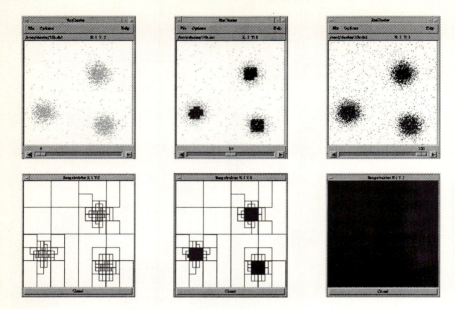

Fig. 10. Iterative data clustering process

Fig. 11. Dendrogram screen

regions per data block). BANG-Clustering is even capable to cluster data sets of up to 1 million patterns, which is beyond the possibilities of Grid-Clustering. The only restriction are the memory requirements for storing such huge data sets. Test showed that for 1 million 2-dimensional patterns (2 double values) a BANG-Structure of 48 MByte was created.

5   Summary

We presented a novel hierarchical clustering method, BANG-Clustering, which organizes the space comprising the pattern set. This method is an extension of the Grid-Clustering algorithm presented in [Sch96], and is capable to cluster even larger pattern sets more efficiently. It outperforms all conventional hierarchical and partitional algorithms and the Grid-Clustering method as well.

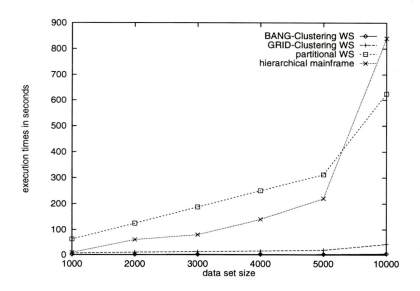

Fig. 12. Runtime comparison of different clustering algorithms

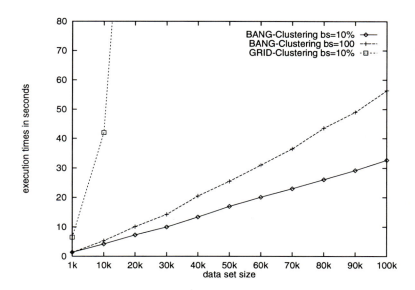

Fig. 13. Runtime comparison of BANG-Clustering for different block sizes (bs)

BANG-Clustering is therefore capable to analyze data sets, which were previously not tractable due to their size and/or dimensionality without any supplemental input information.

References

[Bro90] A.J. Broder. Strategies for efficient incremental nearest neighbor search. *Pattern Recognition*, 23:171–178, 1990.

[Bru88] M. Bruynooghe. A very efficient strategy for very large data sets clustering. In *Proc. 9th Int. Conf. on Pattern Recognition*, pages 623–627. IEEE Computer Society, 1988.

[CCM92] D. Chaudhuri, B.B. Chaudhuri, and C.A. Murthy. A new split-and-merge clustering technique. *Pattern Recognition Letters*, 3:399–409, 1992.

[CD91] Gowda K. Chidananda and E. Diday. Symbolic clustering using a new dissimilarity measure. *Pattern Recognition*, 24:567–578, 1991.

[DJ80] R. Dubes and A.K. Jain. *Clustering methodologies in exploratory data analysis*, volume 19, pages 113–228. Academia Press, 1980.

[Erh95] Martin Erhart. Entwurf und Implementation eines BANG-File-basierten Clusteranalyseverfahrens. Master's thesis, University of Vienna, September 1995.

[Fre87] M.W. Freestone. The bang file: A new kind of grid file. In *Proc. Special Interest Group on Management of Data*, pages 260–269. ACM, May 1987.

[IK89] M.A. Ismail and M.S. Kamel. Multidimensional data clustering utilizing hybrid search strategies. *Pattern Recognition*, 22:75–89, 1989.

[Kur91] T. Kurita. An efficient agglomerative clustering algorithm using a heap. *Pattern Recognition*, 24:205–209, 1991.

[Sch96] E. Schikuta. Grid clustering: An efficient hierarchical clustering method for very large data sets. In *Proc. 13th Int. Conf. on Pattern Recognition*, volume 2, pages 101–105. IEEE Computer Society, 1996.

[VR92] N.B. Venkateswarlu and P.S.V.S.K. Raju. Fast isodata clustering algorithms. *Pattern Recognition*, 25:335–342, 1992.

[WK79] C.S. Warnekar and G. Krishna. A heuristic clustering algorithm using union of overlapping pattern-cells. *Pattern Recognition*, 11:85–93, 1979.

[ZWB91] Q. Zhang, Q.R. Wang, and R. Boyle. A clustering algorithm for data-sets with a large number of classes. *Pattern Recognition*, 24:331–340, 1991.

Section VII:

Data Quality

Techniques for Dealing with Missing Values in Classification

W.Z. Liu[1], A.P. White[2], S.G. Thompson[1] and M.A. Bramer[1]

[1] Artificial Intelligence Research Group, Department of Information Science,
University of Portsmouth, Locksway Road, Milton, Hampshire PO4 8JF, U.K.
[2] School of Mathematics and Statistics, University of Birmingham, Edgbaston,
Birmingham B15 2TT, U.K.

Abstract. A brief overview of the history of the development of decision tree induction algorithms is followed by a review of techniques for dealing with missing attribute values in the operation of these methods. The technique of dynamic path generation is described in the context of tree-based classification methods. The waste of data which can result from casewise deletion of missing values in statistical algorithms is discussed and alternatives proposed.

Keywords: Missing values, Dynamic path generation, Intelligent data analysis, Inductive learning, Knowledge discovery, Data mining, Machine learning.

1   Introduction

In the information age, data is generated almost everywhere: satellites orbiting the moons of Jupiter; submarines in the deepest ocean trench; even electronic point of sale machines in the high street produce data. All of these systems generate millions of megabytes of data every day. Some of these data contain information that could lead to important discoveries in science; some data contain the knowledge that could predict a company's growth or collapse and other data contain knowledge that could mean the difference between life and death.

In order to analyse these important data and uncover hidden relationships and knowledge within the data, some sort of data mining approach is required. In the past, statistical methods such as logistic regression or discriminant analysis were the only tools available for such a task. Unfortunately, they are somewhat cumbersome in the sense that the *form* of the model needs to be specified beforehand, which is often not really feasible for an exploratory analysis involving a large number of variables. More recently, the massive increase of interest and research in this area has made a number of innovative techniques available, which have their origins in computer science, rather than mathematical statistics. These techniques include tree-based methods, neural networks, genetic algorithms, case-based reasoning and so on. They offer the possibility of automating the process of knowledge discovery to a greater degree than appears

X. Liu, P. Cohen, M. Berthold (Eds.): "Advances in Intelligent Data Analysis" (IDA–97)
LNCS 1280, pp. 527–536, 1997.   © Springer–Verlag Berlin Heidelberg 1997

possible with traditional statistical approaches. Because of their greater simplicity and transparency, tree-based classification techniques are of particular interest in this context.

Unlike some of the other newer techniques, tree-based classification methods have two points of academic origin. The first of these was the study of inductive learning. The influence of computer science in this field did not develop until the second half of this century.[3] Hunt was one of the early pioneers who modelled a theory of human concept learning using computer programs. He developed a series of algorithms called 'concept learning systems' (CLS-1 to CLS-9), described in Hunt (1962) and Hunt et al. (1966). Quinlan's well-known ID3 algorithm (Quinlan, 1979), was descended from these systems. Basically, ID3 was a procedure for discriminating between two classes in domains which were entirely free from uncertainty. (In fact, ID3 was developed initially for performing chess endgame analysis, discriminating between winning and non-winning positions).

The second point of origin lay in the discipline of mathematical statistics. It is interesting to note that, among mathematical statisticians, there has been a minority interest in these techniques for the last thirty years. However, this interest did not really come to the fore until the last twelve years or so, prompted by the work of Breiman et al. (1984) and the associated CART software for performing classification and regression using binary trees. In the last few years, tree-based classification and regression procedures have been incorporated into multi-purpose statistical software packages. For example, the statistical package 'S', recently developed for use by statisticians themselves and described by Clark & Pregibon (1992), includes a set of procedures for conducting classification and regression tasks by the use of binary trees. Similarly, the well-known package SPSS has recently become available with CHAID (CHi-squared Automatic Interaction Detector). This is based on earlier work by Kass (1980), which used multiple, rather than binary, branching.

In the statistical field, tree-based methods dealt with uncertainty from the beginning but, in computer science, the adaptation of this type of algorithm to deal with noisy domains took place much more recently. Quinlan (1986) extended the ID3 system, producing the C4.5 algorithm, to deal with the usual statistical situation in which the attributes (independent variables) provide probabilities of class membership, rather than definitive indications. Initially, computer scientists were unaware of the penalties of constructing over-large trees, which is actually equivalent to constructing models with more than an optimal number of parameters, which is understood by statisticians as 'overfitting'. However, Quinlan (1986) rediscovered the problem and dealt with it by incorporating a pruning phase into the algorithm. Thus, an over-large tree was grown to begin with and then cut back to protect against overfitting.

In retrospect, it is obvious that the applicability of the first generation of knowledge discovery systems of computer science ancestry, such as ID3, was very limited. In fact, they could be applied only in deterministic domains, such as

---

[3] A more detailed review of these methods from a computer science perspective is given in Liu & White (1991).

chess endgame analysis, in which there is no noise or uncertainty involved. Now, it is increasingly apparent that, in order for a knowledge discovery system to be able to deal with real-world applications, it must be able to handle noise. This is because noise is inevitable in most real-world applications. The data collected in the real-world are based either on measurements or subjective judgements. Both of these are subject to error. In order for the knowledge extracted from the data to be useful in helping future decision making, the knowledge obtained must be based on intrinsic relationship or structure in the data, rather than some *ad hoc* features of the data such as noise. A less obvious source of error lies in the relationship itself, which links the dependent variables with the attributes. In many real-world examples, the independent variables available provide only an incomplete indication of the value of the dependent variable, even when no errors are present in the dataset itself. Statistical models typically concatenate all these sources of error and express them as a single *error term* on the right-hand side of an equation specifying the model.

About ten years ago, Quinlan (1986) made some useful modifications to ID3 to deal with noise. He pointed out that two modifications to ID3 are necessary if it is to be able to operate with a noise-affected training set:

1. The algorithm must be able to deal with clashes (when two or more cases have identical values for each attribute but belong to different classes);
2. The algorithm must be able to decide when the testing of further attributes will not improve the predictive performance of the decision tree, i.e. to determine when to stop adding further branches to the decision tree.

The first goal is achieved by using probabilistic induction. When the branching process stops, if the cases at any given terminal node are not all of the same class, then probabilities for membership of the various classes are assigned instead. The conversion of these probabilities to predictions of class membership may then be done either by using the obvious strategy of selecting the most likely class at each terminal node or, if differential mis-classification costs are operating, either by some sort of cost minimisation procedure such as described by Breiman et al. (1984), or else by selecting an appropriate discrimination point on an ROC curve (or its equivalent), in the manner described elsewhere by Liu et al. (1994, 1996) and White & Liu (1997). There are two possible ways to achieve the second goal. What Quinlan suggested doing, is to use some kind of 'stopping rule' to prevent over-large decision trees being grown. The second solution to the problem is to grow an over-large tree to begin with, and to prune it back to the right size.[4]

The various techniques for dealing with uncertainty are very important and, in the past decade more and more research has been focused on problems in this area. However, comparatively little attention has been paid to methods of handling some special types of noise, such as missing values. Where data have been collected for a particular purpose, known beforehand, it is often possible

---

[4] A review of pruning techniques can be found in Mingers (1989).

to minimise, or even completely avoid, the occurrence of missing values for data items. On the other hand, where data are collected as a by-product of some other activity and subsequently subjected to some sort of data mining operation, missing values are much more likely to be present in substantial proportions. The intention of this paper is to review and summarise techniques for dealing with missing values that are used in tree-based classification methods and to discuss the possibility of adapting these techniques to other knowledge discovery approaches.

2   Decision-Tree Based Inductive Learning

The principle of tree-based inductive learning (Quinlan, 1986; Liu & White, 1991) is well-known. Basically, the idea is to build a learning algorithm to induce classification rules in the form of a decision tree, by operating on a training set. A *training set* usually consists of a set of past decision-making examples, each of which is comprised of a number of attributes (variables) and a class membership indicator. The decision tree obtained can then be used to classify future cases of unknown class membership.

The task of constructing a decision tree from a training set is typically handled by a recursive partitioning algorithm which, at each non-terminal node, branches on that attribute which discriminates best between the cases filtered down to that node. In order to decide which attribute to select to branch on, some suitable attribute selection measure is needed (Liu & White, 1994). There are many such measures which can be used for this purpose, such as transmitted information[5] and χ^2. Definitions for both these measures are given in White & Liu (1994, 1997). The importance of these criteria lies with their ability to measure the association between the class and the other independent variables. This enables the induced tree to reflect the classification structure of the original data.

In situations where there are no missing values in the training set, tree building can proceed in the expected manner. However, if missing values do exist in the training set, the way these missing values are dealt with will have some effect on the tree building process. In the next section, various techniques for handling missing values in such situations are reviewed.

After obtaining a classification tree, the next step is to use the tree to predict the class membership of test cases. Again, this is very simple if there are no missing values for the attributes of the case undergoing classification. However, if the value of a particular attribute is required in order to classify a particular case and that attribute has a missing value for that case, then simple classification immediately becomes impossible because we do not know which branch to take in order to classify the case. In order to carry out classifications under these

---

[5] Transmitted information (H_T) is actually algebraically equivalent to information gain, as described by Quinlan (1986). However, its formulation in the former terms is particularly useful because it represents information about class membership transmitted by the attribute concerned.

circumstances, other methods for handling missing values have to be used. Some of these techniques are described in Section 4.

3 Dealing with Missing Values in Training Cases

As mentioned earlier, at each non-terminal node of the decision tree, that attribute which gives the strongest association with class is selected to branch on. In situations when there are no missing values for an attribute, the calculation of association between class and that attribute is quite simple. It starts with cross-tabulating class against that particular attribute in the following way. Suppose that we are dealing with a problem with k classes and that an attribute, A, with l distinct values is under consideration at a particular node. The following contingency table (Table 1) can be constructed which represents the cross-tabulation of class and attribute values for A:

Table 1. A cross-tabulation of class and attribute values, for attribute A.

	a_1	a_2	...	a_l	
C_1	n_{11}	n_{12}	...	n_{1l}	n_{1*}
C_2	n_{21}	n_{22}	...	n_{2l}	n_{2*}
$\vdots$	$\vdots$	$\vdots$	...	$\vdots$	$\vdots$
C_k	n_{k1}	n_{k2}	...	n_{kl}	n_{k*}
	n_{*1}	n_{*2}	...	n_{*l}	n_{**}

where C_i ($i=1, 2, \ldots, k$) and a_j ($j=1, 2, \ldots, l$) represent class and attribute values respectively; n_{ij} ($i=1, 2, \ldots, k$; $j=1, 2, \ldots, l$) represent the frequency counts of cases with attribute value a_j and class C_i and:

$$n_{i*} = \sum_{j=1}^{l} n_{ij}$$

$$n_{*j} = \sum_{i=1}^{k} n_{ij}$$

$$n_{**} = \sum_{i=1}^{k} \sum_{j=1}^{l} n_{ij} = N$$

There are several ways to get around the problem of missing values of cases in the training set. Obviously, the simplest way to deal with unknown attribute values is just to ignore the cases containing them and base the calculation of association on the contingency table constructed from only those cases which have known values on this attribute. This is the method used in PREDICTOR (White, 1987).

The second type of technique in dealing with missing values is to try to determine these values using other information. For example, Kononenko et al.(1984) used class information to estimate missing attribute values. Let us assume that the case with missing value on attribute A is of class C. The idea is to assign the most probable value, a_i, of attribute A to the missing value, given the class membership of the case concerned. Another method suggested by Shapiro and described by Quinlan (1986) is to use a decision tree approach to decide the missing values of an attribute. It considers the subset S', of the training set S, which consists of those cases whose value of attribute A is known. In S', the original class is regarded as another attribute while the value of attribute A becomes the 'class' to be determined. Using S', a classification tree can be built for determining the value of attribute A from the other attributes and the class. Then, this tree can be used to classify each object in the set $S - S'$. Consequently, each missing value can be estimated. This is a very thorough technique and makes good use of all the information available from the class variables and all the other independent variables. However, it would appear that the technique is appropriate only for sparse concentrations of missing values. Difficulties arise if the same case has missing values on more than one attribute.

Quinlan (1986) proposed another two different methods. The first method is to treat 'unknown' as a new possible value for each attribute and deal with it in the same way as other values. However, this is appropriate only when the missing values are informative, e.g. values recorded as missing because they were too small or too large to be measured. Usually, missing values are missing at random and, in these circumstances, the value 'unknown' does not have the same status as a proper attribute value, i.e. whether or not a particular attribute has a known value for a particular case does not provide information about class membership of that case. Thus, this method cannot really be regarded as a *general* solution to the problem of missing values. The second method is based on the idea that cases with unknown values are distributed across the values of A in proportion to the relative frequency of these values in the training set. Consider a simple 2×2 contingency table with similar notation to that described earlier, with m_i ($i = 1, 2$) cases with missing value on attribute A, for each class respectively. Then each frequency count of the contingency table is adjusted as follows:

$$n'_{ij} = n_{ij} + m_i \frac{n_{*j}}{n_{**}}$$

where i, $j = 1, 2$. The attribute selection criterion is then calculated using the adjusted frequency counts. When an attribute has been chosen by the selection criterion, cases with unknown values of that attribute are discarded before going to the next step of branching. This method can be too conservative. The following example shows how it attenuates association between attribute and class. Consider the following 2×2 contingency table of those cases whose value on attribute A is known:

$$
\begin{array}{c|c|c|c}
 & a_1 & a_2 & \\
\hline
C_1 & 5 & 0 & 5 \\
C_2 & 0 & 5 & 5 \\
\hline
 & 5 & 5 & 10
\end{array}
$$

where C_i and a_i ($i = 1, 2$) represent class and attribute values respectively. Suppose there are another five cases of class 1 and five cases of class 2 with missing values on attribute A. Then, if we adjust the frequency counts according to the column proportions (as in the formula above), the following table can be derived:

$$
\begin{array}{c|c|c|c}
 & a_1 & a_2 & \\
\hline
C_1 & 7.5 & 2.5 & 10 \\
C_2 & 2.5 & 7.5 & 10 \\
\hline
 & 10 & 10 & 20
\end{array}
$$

The χ^2 and H_T of the first table are 10 and 1, while those of the second table are only 5.556 and 0.236 respectively. This is obviously undesirable and misleading.

The reason why this method can give such unsatisfactory estimates for missing values is revealed if we take a statistical view of the process involved. To put the matter simply, the procedure takes no account at all of the structure in the data set. Missing value estimates are assigned merely on the basis of prior attribute value probabilities. Kononenko's method is better, because the estimates are made conditional upon class membership. Thus, in the example just considered, the adjusted frequencies become:

$$
\begin{array}{c|c|c|c}
 & a_1 & a_2 & \\
\hline
C_1 & 10 & 0 & 10 \\
C_2 & 0 & 10 & 10 \\
\hline
 & 10 & 10 & 20
\end{array}
$$

This is clearly preferable.

4   Dealing with Missing Values in Test Cases

The other half of the story is how missing attribute values are dealt with during classification of test cases. When classifying a case, if the value of a particular attribute which was branched on in the tree is unknown, then classification immediately becomes impossible because we do not know which branch to take in order to classify this case. In order to carry out classifications under these circumstances, other methods for handling missing values have to be used.

The procedure implemented by Quinlan (1986) is to explore all branches (below the current node) and take into account that some are more probable than others. This seems to be very clumsy and unsatisfactory. The other method suggested by Breiman et al. (1984) is to use a *surrogate split* when a missing value is found in the attribute originally chosen. The surrogate attribute is the one which has the highest correlation with the original attribute. The efficacy of this method obviously depends on the magnitude of the correlation in the database between the original attribute and its surrogate.

There is another method, called *dynamic path generation*, proposed by White (1987), which can offer great flexibility in dealing with missing values of this type. Instead of generating the whole decision tree beforehand, the dynamic path generation method produces only the path (i.e. the rule) required to classify the case currently under consideration. This approach can deal with missing values very flexibly. Once a missing value is found to be present in an attribute of a new case, such an attribute is never branched on when classifying the case. In more detail, let us consider the process of building a classification rule (i.e. a path in a classification tree) to classify a new case O_1. At each step, the inductive algorithm chooses the most informative attribute on which to branch. However, if the value of the selected attribute is missing in case O_1, then this attribute cannot be branched on and the algorithm tries with the second most informative attribute. Thus, path generation is strictly *dynamic*. Of course, this approach is somewhat expensive in computational terms. However, if N-fold cross-validation is required, then the technique becomes much more economical, in comparative terms (Liu & White 1994). This is because, for N-fold cross-validation, a fresh model needs to be constructed for each case, whatever method is used. When combined with dynamic path generation, only a fresh *path* needs to be constructed for the classification of each case. In other words, with dynamic path generation, cross-validation imposes no extra cost penalty.

The approach of dealing with missing values in test cases in this way is also referred to as the *lazy decision tree* method (Friedman et al., 1996). The reason why this approach is called the lazy decision tree approach is because the creation of a single 'best' tree is deferred. Instead, it constructs the 'best' tree for each test instance. (In fact, only a classification path needs to be generated).

5   Discussion

In many real-world applications, missing values are often inevitable. Therefore, every intelligent data analysis tool should be equipped with facilities to deal with missing values. Unfortunately, many systems which have been built so far still have very limited power in dealing with missing values. For example, most orthodox statistical packages deal with missing values on a casewise deletion basis, for most statistical procedures. This means that if *any* of the available variables has a missing value for a particular case, then that case is omitted from the analysis. Clearly, this may cause a huge waste of data and, as a result, may not be satisfactory in some circumstances. For example, in the medical database reported by White et al. (1996), if the casewise deletion method is used, there are only 632 cases consisting entirely of non-missing values – fewer than a quarter of the 2692 cases available in the original database. By contrast, if the dynamic path generation method is used, missing values can be dealt with very simply.

In fact, some of the techniques for dealing with missing values in decision tree induction (reviewed in the previous sections) can be easily adapted to many other model-based methods of data analysis. Take some statistical classification method such as linear discriminant analysis or logistic regression as an example. It is clear that:

- If an overall model is required in the training phase, then it is always possible to estimate missing values by one of the techniques mentioned in Section 3.
- In order to deal with missing values in test cases, the 'lazy' approach could be easily adapted. Instead of producing a single set of linear discriminant functions (or a regression equation in the case of logistic regression) in the training phase, we could construct a set of discriminant functions or a regression equation for each test instance. This would ensure that variables with missing values on a particular test case did not occur in the model constructed to classify that case. In this way, missing values are handled very naturally.

To conclude, there is no fundamental reason why the lazy approach could not even be extended to other techniques, such as genetic algorithms, in order to prevent waste of information. Of course, the lazy approach can be expensive in computational terms. However, with modern computer technology this is becoming less and less of a problem.

References

Breiman, L., Friedman, J.H., Olshen, R.A. & Stone, C.J. (1984). *Classification and regression trees*. Belmont: Wadsworth.

Clark, L.A. & Pregibon, D. (1992). Tree-based models. In *Statistical Models in S*, edited by J.M. Chambers & T.J. Hastie, pp. 377–419. California: Wadsworth & Brooks/Cole.

Friedman, H.F., Kohavi, R. & Yun, Y. (1996). Lazy decision trees. in *Proceedings of the 13th National Conference on Artificial Intelligence*, pp. 717–724, AAAI Press/MIT Press.

Hunt, E.B. (1962). *Concept learning: an information processing problem*. New York: Wiley.

Hunt, E.B., Marin, J. & Stone, P.J. (1966). *Experiments in induction*. New York: Academic Press.

Kass, G.V. (1980). An exploratory technique for investigating large quantities of categorical data. *Applied Statistics*, **29**, 119–127.

Kononenko, I., Bratko, I. & Roskar, E. (1984). Experiments in automatic learning of medical diagnostic rules. *Technical Report*. Jozef Stefan Institute, Ljubjana, Yugoslavia.

Liu, W.Z. & White, A.P. (1991). A review of inductive learning. In *Research and Development in Expert Systems* VIII, edited by I.M. Graham and R.W. Milne, pp. 112–126. Cambridge: Cambridge University Press.

Liu, W.Z. & White, A.P. (1994). The importance of attribute selection measures in decision tree induction. *Machine Learning*, **15**, 25–41.

Liu, W.Z. White, A.P. & Hallissey, M.T. (1994). Early screening for gastric cancer using machine learning techniques. In *Machine Learning: ECML-94*, edited by F. Bergadano and L. De Raedt, pp. 391–394. Springer-Verlag, Berlin.

Liu, W.Z., White, A.P., Hallissey, M.T. & Fielding, J.W.L. (1996). Machine learning techniques in early screening for gastric and oesophageal cancer. *Artificial Intelligence in Medicine*, **8**, 327–341.

Mingers, J. (1989). An empirical comparison of pruning methods for decision tree induction. *Machine Learning*, **4**, 227–243.

Quinlan, J.R. (1979). Discovering rules by induction from large collections of examples. In *Expert Systems in the Micro-Electronic Age,* edited by D. Michie, pp. 168–201. Edinburgh: Edinburgh University Press.

Quinlan, J.R. (1986). Induction of decision trees. *Machine Learning*, **1**, 81–106.

White, A.P. (1987). Probabilistic induction by dynamic path generation in virtual trees. In *Research and Development in Expert Systems* III, edited by M.A. Bramer, pp. 35–46. Cambridge: Cambridge University Press.

White, A.P. & Liu, W.Z. (1994). Bias in information-based measures in decision tree induction. *Machine Learning*, **15**, 321–329.

White, A.P., Liu, W.Z., Hallissey, M.T. & Fielding, J.W.L. (1996). A comparison of two classification techniques in screening for gastro-oesophageal cancer. *Applications and Innovations in Expert Systems* IV, edited by A. Macintosh and C. Cooper, pp. 83–97. Cambridge: Cambridge University Press.

White, A.P. & Liu, W.Z. (1997). Statistical properties of tree-based approaches to classification. In *Machine Learning and Statistics: the Interface*, edited by R. Nakhaeizadeh and C. Taylor, pp. 23–44. ISBN 0-471-14890-3, John Wiley & Sons, Inc.

The Use of Exogenous Knowledge to Learn Bayesian Networks from Incomplete Databases

Marco Ramoni[1] and Paola Sebastiani[2]

[1] Knowledge Media Institute
The Open University
[2] Department of Actuarial Science and Statistics
City University

Abstract. Current methods to learn Bayesian Belief Networks (BBNs) from incomplete databases share the common assumption that the unreported data are missing at random. This paper describes a method — called *Bound* and *Collapse* (BC) — to learn BBNs from incomplete databases which allows the analyst to efficiently integrate information provided by the database and exogenous knowledge about the pattern of missing data. BC starts by *bounding* the set of estimates consistent with the information conveyed by the database and then *collapses* the resulting set to a point via a convex combination of the extreme points, with weights depending on the assumed pattern of missing data. Experiments comparing BC to Gibbs Sampling are provided.

1 Introduction

A Bayesian Belief Network (BBN) is defined by a set of *variables* $\mathcal{X} = \{X_1, \dots, X_I\}$ and a direct acyclic graph defining a model $\mathcal{M}$ of conditional dependencies among the elements of $\mathcal{X}$. A conditional dependency links a *child* variable X_i to a set of *parent* variables Π_i, and is defined by the conditional distributions of the child variable given the configurations of the parent variables. We shall consider discrete variables only. The structure $\mathcal{M}$ yields a factorization of the joint probability of a set of values $x_k = \{x_{1k}, \dots x_{Ik}\}$ of the variables in $\mathcal{X}$ as

$$p(\mathcal{X} = x_k) = \prod_{i=1}^{I} p(X_i = x_{ik}|\Pi_i = \pi_{ij}),$$

where π_{ij} denotes the state of Π_i in x_k. We will denote $X_i = x_{ik}$ by x_{ik}, and $\Pi_i = \pi_{ij}$ by π_{ij}. The number of states of X_i and Π_i will be denoted respectively by c_i and q_i.

Several efforts have been addressed to develop Bayesian methods to learn BBNs from databases [3, 2, 6]. Along this approach, we shall consider the conditional probabilities defining the BBN as being generated by parameters $\theta = \{\theta_{ijk}\}$, where $\theta_{ijk} = p(x_{ik}|\pi_{ij}, \theta)$. Thus the joint probability of a case x_k is

X. Liu, P. Cohen, M. Berthold (Eds.): "Advances in Intelligent Data Analysis" (IDA–97)
LNCS 1280, pp. 537–548, 1997.   © Springer–Verlag Berlin Heidelberg 1997

$$p(x_k|\theta) = \prod_{i=1}^{I} \theta_{ijk}.$$

The parameters θ_{ijk} are random variables, whose *prior* distribution represents the observer's belief about the conditional probabilities specifying the BBN.

Bayesian learning means to update the prior belief on the parameters from a database of n cases $\mathcal{D} = \{x_1, \ldots, x_n\}$ by using Bayes' theorem. The prior density $p(\theta)$ is updated in the *posterior* density

$$p(\theta|\mathcal{D}) = p(\theta)p(\mathcal{D}|\theta)/p(\mathcal{D}),$$

where $p(\mathcal{D}|\theta)$ is the *sampling model* and $p(\mathcal{D}|\theta) = \prod_k p(x_k|\theta)$, if the cases are independent given θ. Furthermore

$$p(\mathcal{D}) = \int p(\mathcal{D}|\theta)p(\theta)d\theta$$

is the *marginal probability* of the database. The standard Bayesian estimate of θ is then the *posterior expectation* $E(\theta|\mathcal{D})$. We assume that the parameters are mutually independent so that the joint prior density factorizes in $p(\theta) = \prod_i p(\theta_{ij})$, where the parameter vector $\theta_{ij} = \{\theta_{ij1}, \ldots, \theta_{ijc_i}\}$ is associated to the conditional probabilities $p(x_{ik}|\pi_{ij})$, $k = 1, \ldots, c_i$. Moreover, we assume that, for all i and j, θ_{ij} has a *Dirichlet* distribution with *hyper-parameters* $\{\alpha_{ij1}, \cdots, \alpha_{ijc_i}\}$, denoted by $\theta_{ij} \sim D(\alpha_{ij1}, \ldots, \alpha_{ijc_i})$, with $\alpha_{ijk} > 0$.

The prior hyper-parameters α_{ijk} can be regarded as frequencies of the imaginary cases needed to formulate the prior distribution. As a matter of fact, the probability of $x_{ik}|\pi_{ij}$ is α_{ijk}/α_{ij}, and $\alpha_{ij} = \sum_{k=1}^{c_i} \alpha_{ijk}$ is the *prior precision* on θ_{ij}. The situation of initial ignorance can be represented by assuming $\alpha_{ijk} = 1$ for all i, j, k, so that the probability of $x_{ik}|\pi_{ij}$ is constant as k varies. Then, as shown for instance in [10], the assumption of independence of the parameters and the particular form of the prior distribution imply that the posterior distribution is still a product of Dirichlet distributions and

$$\theta_{ij}|\mathcal{D} \sim D(\alpha_{ij1} + n(x_{i1}|\pi_{ij}), \ldots, \alpha_{ijc_i} + n(x_{ic_i}|\pi_{ij})),$$

where $n(x_{ik}|\pi_{ij})$ is the frequency of cases with $x_{ik}|\pi_{ij}$. The Bayes estimate of the conditional probability of $x_{ik}|\pi_{ij}$ is the posterior expectation of θ_{ijk}:

$$\frac{\alpha_{ijk} + n(x_{ik}|\pi_{ij})}{\sum_k \alpha_{ijk} + n(\pi_{ij})},$$

where $n(\pi_{ij}) = \sum_k n(x_{ik}|\pi_{ij})$.

Unfortunately, the situation is quite different when the database is incomplete, that is, some entries are reported as unknown. Suppose for instance that the case x_k is incomplete, with entries on X_i and/or its parents missing, so that $\mathcal{D}_i = \mathcal{D}_o \cup x_k$, where $\mathcal{D}_o$ denotes the part of the database with complete entries. Exact analysis would require the computation of the joint posterior distribution of the parameters by considering all possible completions of the incomplete case.

Assuming that independence is retained, exact updating can be performed locally, so that the posterior distribution of the parameters associated to the conditional distribution of $X_i | \pi_{ij}$ turns out to be the mixture

$$\sum_k D(\alpha_{ij1} + n(x_{i1}|\pi_{ij}) + \delta_1(k), \ldots, \alpha_{ijc_i} + n(x_{ic_i}|\pi_{ij}) + \delta_{c_i}(k)) p(x_{ik}, \pi_{ij}|\mathcal{D}_o)$$

$$+ D(\alpha_{ij1} + n(x_{i1}|\pi_{ij}), \ldots, \alpha_{ijc_i} + n(x_{ic_i}|\pi_{ij}))(1 - p(\pi_{ij}|\mathcal{D}_o))$$

where $\delta_i(k) = 1$ if $i = k$, $\delta_i(k) = 0$ otherwise, and the frequencies are based on $\mathcal{D}_o$. The first term in the mixture computes the possible completions of the child variable X_i conditioning on the fact that the configuration π_{ij} of the parents variables was observed. The last term is conditioned on completing the parent configuration to a state different from π_{ij}, so that the distribution of θ_{ij} is not updated. As the number of incomplete cases increases, exact local updating becomes apparently infeasible: its complexity is in fact exponential in the number of missing data [3], and approximate methods are therefore required.

Markov Chain Monte Carlo (MCMC) methods, such as Gibbs Sampling (see [5] for a recent review), provide estimates that do not rely on the independence of the parameters. When some of the entries in the database are missing, Gibbs Sampling treats the missing data as unknown parameters, so that in each iteration, for each missing entry, a value is sampled from the conditional distribution of the corresponding variable, given the available data and the values simulated for the other missing entries and the parameters. The algorithm is iterated to reach stability, and then a sample from the joint posterior distribution is taken which can be used to provide empirical estimates of the posterior mean [11]. The underlying assumption is that the unreported data are *Missing at Random* (MAR), so that the probability of an entry being missing does not depend on the state of the corresponding variable [7]. In other words, $\mathcal{D}_o$ is a representative sample of the complete but unknown $\mathcal{D}$. Experimental evidences [9] show that, when this assumption is violated, these methods are not robust with respect to the pattern of missing data. Furthermore, these methods are usually highly resource demanding, their convergence rate may be slow, and their execution time heavily depends on the number of missing data.

Ramoni and Sebastiani [9] introduced a deterministic method to learn conditional probabilities from incomplete databases which does not rely on this assumption. The method *bounds* the set of possible distributions consistent with the available information by computing the minimum and the maximum Bayes estimate that would be obtained from the possible completions of the database. This paper presents a technique to use prior knowledge about the pattern of missing data to identify a unique distribution within these bounds. This new method *collapses* the set of possible distributions into a single one via a convex combination of the extreme points with weights depending on the assumed pattern of missing data. An approximation of the variance of the estimates is also provided. Because of its strategy, we call this method *Bound* and *Collapse* (BC). The use of BC allows the encoding of prior knowledge on the pattern of missing data, such as the MAR assumption, without any need to guess the missing data. Experimental evaluations show that the estimates provided by BC are very closer to those provided by Gibbs Sampling, when data are MAR, and are more robust to departure from the true pattern of missing data. Furthermore, BC reduces the cost of estimating each conditional distribution of $X_i|\pi_{ij}$ to the cost of one exact Bayesian updating and one convex combination for each state of X_i, thus making the computational effort of stochastic methods vane.

2 Method

This section introduces a deterministic method, called BC, to estimate parameters from an incomplete database which does not guess the missing entries but *bounds* the possible estimates consistent with the available data, and then *collapses* the resulting interval to a point via a convex combination of the extreme estimates using information on the pattern of missing data.

2.1 Bound

In [9] it was shown that the missing entries in the database lead to bounds on the possible estimates that could be obtained from the complete database. Thus the *bound* step of the BC method computes, for each parameter, a probability interval whose extreme points are the minimum and the maximum Bayes estimate that would be inferred from all possible completions of the database.

Let X_i be a variable, and let $n^{\bullet}(x_{ik}|\pi_{ij})$ be the frequency of cases with $X_i = x_{ik}$, given the parent configuration π_{ij}, which have been obtained by completing incomplete cases. The incompleteness of a case can be due either to a missing observation in the parent configuration or to a missing observation of the variable child X_i. Denote by $n(?|\pi_{ij})$ the frequency of cases in which the entry on the child variable is missing and the configuration of the parent variable is known, by $n(x_{ik}|?)$ the frequency of cases in which the parent configuration is unknown and the child variable is known, and by $n(?|?)$ the frequency of cases in which the whole configuration parent-child is unknown, then

$$n^{\bullet}(x_{ik}|\pi_{ij}) = n(?|\pi_{ij}) + n(x_{ik}|?) + n(?|?).$$

Clearly $n^{\bullet}(x_{ik}|\pi_{ij}) = n_{ij}^{\bullet}$ for all k when either data are missing only on the child variable, or $n(x_{ik}|?)$ are all identical. Then, the Bayes estimate $E(\theta_{ijk}|\mathcal{D})$, that would be computed from the complete database, is bounded above by

$$p^{\bullet}(x_{ik}|\pi_{ij}, \mathcal{D}_i) = \frac{\alpha_{ijk} + n(x_{ik}|\pi_{ij}) + n^{\bullet}(x_{ik}|\pi_{ij})}{\alpha_{ij} + \sum_l n(x_{il}|\pi_{ij}) + n^{\bullet}(x_{ik}|\pi_{ij})} \tag{1}$$

and below by

$$p_{\bullet}(x_{ik}|\pi_{ij}, \mathcal{D}_i) = \frac{\alpha_{ijk} + n(x_{ik}|\pi_{ij})}{\alpha_{ij} + \sum_l n(x_{il}|\pi_{ij}) + \max_{h \neq k} n^{\bullet}(x_{ih}|\pi_{ij})} \tag{2}$$

Each maximum probability $p^{\bullet}(x_{ik}|\pi_{ij}, \mathcal{D}_i)$ is obtained from a Dirichlet distribution

$$D_k(\alpha_{ij1} + n(x_{i1}|\pi_{ij}), \ldots, \alpha_{ijk} + n(x_{ik}|\pi_{ij}) + n^{\bullet}(x_{ik}|\pi_{ij}), \ldots, \alpha_{ijc_i} + n(x_{ic_i}|\pi_{ij})$$

which identifies a unique probability for the other states of the variable X_i given π_{ij}:

$$p_{k\bullet}(x_{il}|\pi_{ij}, \mathcal{D}_i) = \frac{\alpha_{ijl} + n(x_{il}|\pi_{ij})}{\alpha_{ij} + \sum_h n(x_{ih}|\pi_{ij}) + n^{\bullet}(x_{ik}|\pi_{ij})}$$

and

$$p_{\bullet}(x_{ik}|\pi_{ij}, \mathcal{D}_i) = \min_l \{p_{l\bullet}(x_{ik}|\pi_{ij}, \mathcal{D}_i)\}.$$

These bounds depend only on the frequencies of observed entries in the database and the "artificial" frequencies of the completed cases, so that the whole data can be processed at once. This method is implemented using an algorithm based on discrimination trees to efficiently store information about the parameters [9].

The main feature of this method is its independence on the distribution of the missing data because it does not try to infer them, since an incomplete database can only give rise to constraints on the possible estimates that could be learned from the database. Furthermore, the bounds provides a new measure of information: the width of the interval accounts for the amount of information available in $\mathcal{D}_i$ about the parameter to be estimated, and represents a measure of the quality of the probabilistic information conveyed by the database about the parameter itself. In this way, intervals provide an explicit representation of the reliability of the estimates which can be taken into account when the extracted BBN is used to perform a particular task.

2.2   Collapse

The second step of BC collapses the interval estimated in the *bound* step into a point estimate using a convex combination of the distributions D_k induced by the maximum probabilities. This convex combination can be achieved either by the use of external information about the pattern of missing data or by a dynamic estimation of this pattern from the available information in the database.

Using Exogenous Knowledge Suppose that some external information is available on the pattern of missing data. The analyst can use this information to formulate a probability distribution describing, for each variable in the database, the probability of a completion as:

$$p(x_{ik}|\pi_{ij}, X_i =?) = \phi_{ijk} \tag{3}$$

where $k = 1, ..., c_i$ and $\sum_k \phi_{ijk} = 1$. This information can be used to identify a point estimate within the interval estimate via a convex combination of the extreme probabilities:

$$\hat{p}(x_{ik}|\pi_{ij}, \mathcal{D}_i, \phi_{ijk}) = \sum_{l \neq k} \phi_{ijl} p_{l\bullet}(x_{ik}|\pi_{ij}, \mathcal{D}_i) + \phi_{ijk} p^\bullet(x_{ik}|\pi_{ij}, \mathcal{D}_i). \qquad (4)$$

The intuition behind (4) is that the lower bound of the estimate of $p(x_{ik}|\pi_{ij})$ is obtained when all incomplete cases are assigned to $x_{ih}|\pi_{ik}$ and $h \neq k$ or $k \neq j$, while the upper bound is obtained when all incomplete cases are completed as $x_{ik}|\pi_{ij}$. Thus, if the user specifies that $p(x_{ik}|\pi_{ij}, X_i =?) = 1$ for a particular k, then (4) will return the upper bound of the interval probability as estimate of $p(x_{ik}|\pi_{ij})$, and $p_{k\bullet}(x_{ih}|\pi_{ij}, \mathcal{D}_i)$ as estimates of $p(x_{ih}|\pi_{ij})$, $h \neq k$. This case corresponds to the assumption that data are *systematically* missing about x_{ik}. On the other hand, when no information on the mechanism generating missing data is available, and therefore all patterns of missing data are equally likely, then $\phi_{ijk} = 1/c_i$.

As the number of missing entries decreases, $p^\bullet(x_{ik}|\pi_{ij}, \mathcal{D}_i)$ and $p_{l\bullet}(x_{ik}|\pi_{ij}, \mathcal{D}_i)$ approach $(\alpha_{ijk} + n(x_{ijk}|\pi_{ij}))/(\alpha_{ij} + n(\pi_{ij}))$ so that, when the database is complete, (4) returns the exact estimate $E(\theta_{ijk}|\mathcal{D})$. As the number of missing entries increases, then $p_{l\bullet}(x_{ik}|\pi_{ij}, \mathcal{D}_i) \to 0$ for all l, and $p^\bullet(x_{ik}|\pi_{ij}, \mathcal{D}_i) \to 1$, so that the estimate (4) approaches the prior probability ϕ_{ijk}, and coherently nothing is learned from an empty database.

The estimates so found define a probability distribution since

$$\sum_{k=1}^{c_i} \hat{p}(x_{ik}|\pi_{ij}, \mathcal{D}_i, \phi_{ijk}) =$$
$$\sum_{k=1}^{c_i} \phi_{ijk} p^\bullet(x_{ik}|\pi_{ij}, \mathcal{D}_i) + \sum_{k=1}^{c_i} \sum_{l \neq k} \phi_{ijl} p_{l\bullet}(x_{ik}|\pi_{ij}, \mathcal{D}_i) = \qquad (5)$$
$$\sum_{k=1}^{c_i} \phi_{ijk} \{p^\bullet(x_{ik}|\pi_{ij}, \mathcal{D}_i) + \sum_{l \neq k} p_{k\bullet}(x_{il}|\pi_{ij}, \mathcal{D}_i)\} =$$
$$\sum_{k=1}^{c_i} \phi_{ijk} = 1.$$

If $n^\bullet(x_{ik}|\pi_{ij}) = n_{ij}^\bullet$ then (4) simplifies to

$$\frac{\alpha_{ijk} + n(x_{ik}|\pi_{ij}) + n_{ij}^\bullet \phi_{ijk}}{\alpha_{ij} + \sum_k n(x_{ik}|\pi_{ij}) + n_{ij}^\bullet} \qquad (6)$$

so that the incomplete cases are distributed across the states of X_i according to the prior knowledge on the pattern of missing data. Note that (6) is the *expected posterior expectation*, conditional on the assumed pattern of missing data.

Using Available Information Suppose now that the user has sufficient information to believe that data are MAR, so that $\mathcal{D}_o$ is a *representative sample* of the complete but unknown database $\mathcal{D}$. Thus, the probability of a completion is $\phi_{ijk} = p(x_{ik}|\pi_{ij})$, and can be estimated from $\mathcal{D}_o$ as

$$\hat{\phi}_{ijk} = \frac{\alpha_{ijk} + n(x_{ik}|\pi_{ij})}{\alpha_{ij} + \sum_k n(x_{ik}|\pi_{ij})}.$$

Then $\hat{\phi}_{ijk}$ can be used to compute (4). As the number of missing entries increases, the estimate (4) still approaches the prior probability α_{ijk}/α_{ij}, so that again we have coherence of the estimation method and no updating is performed when data are totally missing.

In particular, if $n^{\bullet}(x_{ik}|\pi_{ij}) = n_{ij}^{\bullet}$ then (6) becomes

$$\hat{p}(x_{ik}|\pi_{ij}, \mathcal{D}_i) = \frac{\alpha_{ijk} + n(x_{ik}|\pi_{ij}) + n_{ij}^{\bullet}\hat{\phi}_{ijk}}{\alpha_{ij} + \sum_k n(x_{ik}|\pi_{ij}) + n_{ij}^{\bullet}} \tag{7}$$

which is a consistent estimate of (6). Consistency arises from the fact that $\hat{\phi}_{ijk}$ is the Generalized Maximum Likelihood Estimate of ϕ_{ijk} [1]. If $\alpha_{ijk} = 0$, then (6) is the classical Maximum Likelihood Estimate of θ_{ijk} [7].

Gibbs Sampling would complete the database by guessing the missing data from the current estimate of $p(x_{ik}|\pi_{ij}, \theta)$, use the completed database to learn the parameters and iterate the procedure until convergence is reached. Clearly, the estimates of the conditional probabilities returned in (4) with ϕ_{ijk} replaced by $\hat{\phi}_{ijk}$ are the expected estimates, and for large samples they will be the same. On the other hand, BC reduces the cost of estimating each conditional distribution of $X_i|\pi_{ij}$ to the cost of one exact Bayesian updating and one convex combination for each state of X_i.

2.3   Variance

The bounds (1) and (2) contain the estimates that could be obtained from all possible completed databases, and hence give an overall measure of the available information. The value in (4) is an estimate of the posterior expectation of θ_{ijk} that would be obtained from the complete database $\mathcal{D}$, before losing some of the entries. With a complete database, the exact posterior distribution would be $D(\alpha_{ij1} + n(x_{i1}|\pi_{ij}), \ldots, \alpha_{ijc_i} + n(x_{ic_i}|\pi_{ij}))$, so that the posterior variance of θ_{ijk} would be

$$V(\theta_{ijk}|\mathcal{D}) = \frac{(\alpha_{ijk} + n(x_{ik}|\pi_{ij}))(\alpha_{ij} - \alpha_{ijk} + n(\pi_{ij}) - n(x_{ik}|\pi_{ij}))}{(\alpha_{ij} + n(\pi_{ij}))^2(\alpha_{ij} + n(\pi_{ij}) + 1)}.$$

Note that

$$V(\theta_{ijk}|\mathcal{D}) = \frac{E(\theta_{ijk}|\mathcal{D})(1 - E(\theta_{ijk}|\mathcal{D}))}{\alpha_{ij} + n(\pi_{ij}) + 1}.$$

Since (4) is an estimate of $E(\theta_{ijk}|\mathcal{D})$, the posterior variance can be approximated as

$$\hat{V}(\theta_{ijk}|\mathcal{D}_i) = \frac{\hat{p}(x_{ik}|\pi_{ij}, \mathcal{D}_i, \phi_{ijk})(1 - \hat{p}(x_{ik}|\pi_{ij}, \mathcal{D}_i, \phi_{ijk}))}{\alpha_{ij} + n(\pi_{ij}) + 1} \qquad (8)$$

where $n(\pi_{ij})$ are the observed frequencies. Since $n(\pi_{ij})$ is smaller than or at most equal to the frequencies in $\mathcal{D}$, (8) will be an upper bound of $V(\theta_{ijk}|\mathcal{D})$, and approaches the exact posterior variance as the number of missing entries increases. If we want to approximate the marginal posterior distribution of θ_{ijk}, we can use a moment-matching approximation [4], such as

$$\theta_{ijk}|\mathcal{D}_i, \phi_{ijk} \sim D(\tilde{\alpha}_{ijk1}, \tilde{\alpha}_{ijk2}),$$

where $\tilde{\alpha}_{ijk1}$ and $\tilde{\alpha}_{ijk2}$ are such that

$$\hat{p}(x_{ik}|\pi_{ij}, \mathcal{D}_i, \phi_{ijk}) = \frac{\tilde{\alpha}_{ijk1}}{\tilde{\alpha}_{ijk1} + \tilde{\alpha}_{ijk2}}$$

$$\hat{V}(\theta_{ijk}|\mathcal{D}_i) = \frac{\tilde{\alpha}_{ijk1}\tilde{\alpha}_{ijk2}}{(\tilde{\alpha}_{ijk1} + \tilde{\alpha}_{ijk2})^2(\tilde{\alpha}_{ijk1} + \tilde{\alpha}_{ijk2} + 1)}.$$

An alternative approximation could be based on the use of BC to estimate the expected posterior precision, see [8] for details.

3   Experimental Evaluation

Gibbs Sampling is currently considered the most appropriate solution to the problem of learning BBNs from incomplete databases, although its limitations are well-known: its convergence rate is slow and resource consuming. The aim of these experiments is to compare the accuracy of the parameter estimates provided by Gibbs Sampling and BC as the available information in the database decreases. Therefore, in order to make the comparison with Gibbs Sampling meaningful, BC assumes that data are missing at random.

3.1   Materials and Methods

We compared an implementation of BC written in Common Lisp under CLISP version 1996/10/10 to the implementation of Gibbs Sampling provided by the BUGS version 0.5 [11] on a Sun Sparc 5 under SunOS 5.5. Figure 1 shows the graphical structure of the BBN — defined by three binary variables A, B, and C — used for this comparison. We generated a database of 1000 random cases from the probability distribution reported in Figure 1.

We ran two different experiments, one fulfilling and one violating the MAR assumption. In the first experiment, we started with a complete database, where all the parameters are independent and uniformly distributed, and ran both BC and Gibbs Sampling on it. Then, we proceed by randomly deleting 20% of the entries of the database and running the two methods until the database was

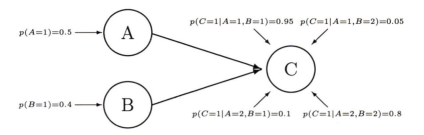

Fig. 1. The BBN used for the experimental evaluation.

empty. Each run of Gibbs Sampling is based on 2,000 iterations, which were sufficient to reach stability, and a final sample of 2,000 cases. The procedure followed during the second experiment is identical to the procedure used in the first one, except that, at each step, the 2% of the entries $B = 1$ were systematically deleted until no entry $B = 1$ was reported in the database.

3.2   Results

Figure 2 plots the parameter estimates given by the two systems against the percentage of entries still present in the database used in the first experiments. Dotted lines are the bounds computed by BC. Solid lines links the point estimates returned by BC under the MAR assumption, so that the weights used in the convex combination are computed from the observed entries in the database. Dashed lines display the 95% confidence intervals around the BC estimates, which were computed using the approximate posterior distributions for the parameters described at the end of Section 2. Stars represent the point estimates given by Gibbs Sampling and error bars reports the 95% confidence intervals for these estimates. Both the BC and Gibbs Sampling confidence intervals are based on the 2.5 % and 97.5 % quantiles. Not surprisingly BC under the MAR assumption returns the accurate estimates of Gibbs Sampling, with confidence intervals almost identical. The main difference between the performances of the two systems has been the execution time: in the worse case, Gibbs Sampling took over 3 minutes to run to completion, while BC ran to complete the same task in less than 20 milliseconds, independently of the number of missing entries. Figure 3 summarizes the results of the second experiment. As expected, the estimates returned by Gibbs Sampling and by BC using the weight computed from the entries actually observed in the database are almost identical. An apparently surprising exception occurs in the estimation of $p(C = 1|A = 2, B = 2)$ and $p(C = 1|A = 1, B = 2)$: with the 86% of the database but all the entries $B = 1$ removed, for instance, Gibbs Sampling returns in the first case an estimate of

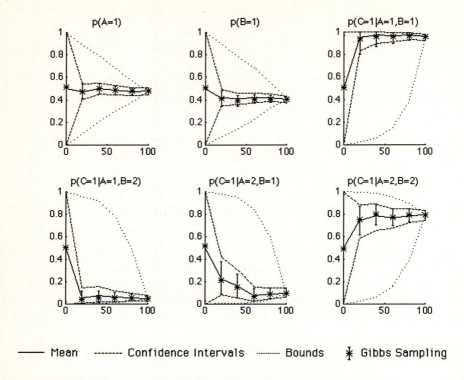

Fig. 2. Estimates against percentage of entries. Entries are randomly missing.

0.4123 and 0.5876, respectively. Both estimates include an error of almost 40% with respect to the estimates computed for the same parameters with the complete databases: 0.0453 and 0.9546 respectively. On the contrary, BC under the same MAR assumption, catches the correct value even when all the observations $B = 1$ are missing. Again, the difference in computation time was significant: our implementation of BC took always less than 20 milliseconds to run to completion, independently of the number of missing entries, while Gibbs Sampler took up to 8 minutes to complete its analysis.

3.3   Discussion

The first important result of our experiment is the overall equivalence between Gibbs Sampling and the BC estimates, which are almost identical when the MAR assumption is fulfilled by the database. When this assumption is violated, BC seems to be significantly more robust than Gibbs Sampling: the little gain in the estimates of $p(B = 1)$ is a small reward for the appalling error in the estimates of $p(C = 1|A = 1, B = 2)$ and $p(C = 1|A = 2, B = 2)$. The strategy of guessing the missing data from the available information leads the Gibbs Sampler to ascribe most of the missing entries to $B = 2$, so that the final

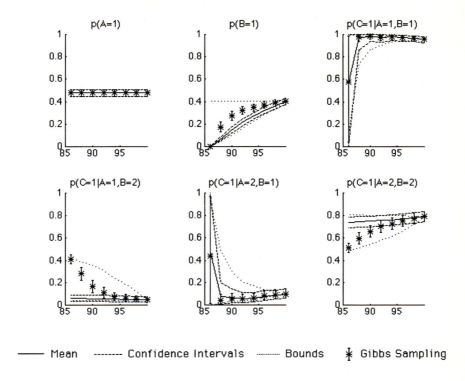

Fig. 3. Estimates against percentage of entries. Entries $B = 1$ are removed.

estimates of $p(C = 1|A = 1, B = 2)$ and $p(C = 1|A = 2, B = 2)$ are based on an inflated sample. On the other hand, the BC estimates computed under the MAR assumption are based on weights which depend only on the actually observed sample. It is worth noting the significant difference in computational cost — milliseconds against minutes — and the remarkable independence of the execution time of BC from the number of missing data.

4 Conclusion

BC provides a method to learn conditional probabilities in a BBN from incomplete databases. Contrary to the current methods, BC does not try to complete the database before learning, but computes the extreme points of the possible distributions consistent with the database and then collapses them into a point estimate using the available information about the pattern of missing data. In this way, BC is able to encompass knowledge external to the database about the pattern of missing data and to subsume the methods using MAR assumption as special cases, providing performances equal or even higher in accuracy. Furthermore, the probability intervals used by BC provide a specific measure of the *quality* of information conveyed by the database and an explicit representation of the impact of the assumption made on the pattern of missing data. From a computational standpoint, BC has all the advantages of a deterministic method: it provides a stopping rule — while Gibbs Sampling is guaranteed to converge only asymptotically — and its computational cost is reduced to the cost of one exact updating and one convex combination for each state of $X_i | \pi_{ij}$. It is therefore not surprising the independence of its execution time from the number of missing data. A final remark is due: although this paper focuses on BBNs, BC is a general method to learn conditional probabilities and may be applied to any data analysis task.

Acknowledgments

Authors thank Greg Cooper, Paul Snow, Zdenek Zdrahal, and two anonymous referees for their helpful suggestions. This research was partially supported by equipment grants from Apple Computers and Sun Microsystems.

References

1. J.O. Berger. *Statistical Decision Theory and Bayesian Analysis*. Springer, New York, NY, 1985.
2. W. L. Buntine. Operations for learning with graphical models. *Journal of Artificial Intelligence Research*, 2:159–225, 1994.
3. G.F. Cooper and E. Herskovitz. A Bayesian method for the induction of probabilistic networks from data. *Machine Learning*, 9:309–347, 1992.
4. R.G. Cowell, A.P. Dawid, and P. Sebastiani. A comparison of sequential learning methods for incomplete data. In *Bayesian Statistics 5*, pages 533–542. Clarendon Press, Oxford, 1996.
5. D. Heckerman. A tutorial on learning Bayesian networks. Technical Report MSR-TR-95-06, Microsoft Corporation, 1995.
6. D. Heckerman, D. Geiger, and D.M. Chickering. Learning Bayesian networks: The combinations of knowledge and statistical data. *Machine Learning*, 20:197–243, 1995.
7. R.J.A. Little and D.B. Rubin. *Statistical Analysis with Missing Data*. Wiley, New York, NY, 1987.
8. M. Ramoni and P. Sebastiani. Learning Bayesian networks from incomplete databases. In *Proceedings of the Thirteenth Conference on Uncertainty in Artificial Intelligence*, San Mateo, CA, 1997. Morgan Kaufmann.
9. M. Ramoni and P. Sebastiani. Robust parameter learning in Bayesian networks with missing data. In *Proceedings of the Sixth Workshop on Artificial Intelligence and Statistics*, pages 339–406, Fort Lauderdale, FL, 1997.
10. D.J. Spiegelhalter and S.L. Lauritzen. Sequential updating of conditional probabilities on directed graphical structures. *Networks*, 20:157–224, 1990.
11. A. Thomas, D.J. Spiegelhalter, and W.R. Gilks. Bugs: A program to perform Bayesian inference using Gibbs Sampling. In *Bayesian Statistics 4*, pages 837–42. Clarendon Press, Oxford, 1992.

Reasoning about Outliers
by Modelling Noisy Data

John X Wu[1], Gongxian Cheng[2] and Xiaohui Liu[2]

[1] Moorfields Eye Hospital, Glaxo Department of Ophthalmic Epidemiology,
Institute of Ophthalmology, Bath Street, London EC1V 9EL, United Kingdom
[2] Department of Computer Science, Birkbeck College, University of London,
Malet Street, London WC1E 7HX, United Kingdom

Abstract. Outliers are difficult to handle because some of them can be
measurement errors, while others may represent *phenomena of interest*,
something "significant" from the viewpoint of the application domain.
Statistical methods for managing outliers do not distinguish between
these two possibilities. In our previous work, we suggested a method for
distinguishing these two possibilities by modelling "real measurements" -
how measurements should be distributed in a domain of interest. In this
paper, we make this distinction by modelling measurement errors instead. The proposed method is better suited to those applications where
it is difficult to obtain relevant knowledge about real measurements. The
test data collected from a recent glaucoma case finding study in a general
practice are used to evaluate the method.

1 Introduction

The development of computational methods to effectively analyse large volumes
of data has long been a challenging task. One of the most important issues
in developing these methods is the management of *outliers*: "outlying" or "un-
representative" data points. Many statistical techniques have been proposed to
manage outliers in the literature [7, 9, 10].

Outliers are difficult to handle because some of them can be severe measurement errors, while others may represent phenomena of interest, something
"significant" from the viewpoint of the application domain [8, 3].

In this situation an outright rejection of outliers based on some statistical
tests may not be a very good idea [1]. To smooth data using robust methods [10]
may also lead to fundamental change in the nature of "phenomena of interest".
Therefore a careful analysis of outliers and the associated data using relevant
background knowledge is desirable.

We have conducted some preliminary research on how to detect and further
analyse outliers using domain knowledge. In particular, a method for distinguishing between phenomena of interest and measurement noise was proposed
and applied to the analysis of a set of visual field test data collected from a
group of glaucoma patients in an eye hospital [13]. That method attempted to

X. Liu, P. Cohen, M. Berthold (Eds.): "Advances in Intelligent Data Analysis" (IDA–97)
LNCS 1280, pp. 549–558, 1997.   © Springer–Verlag Berlin Heidelberg 1997

model "real" measurements, namely how measurements should be distributed in a domain of interest (e.g. how glaucoma manifests itself on visual field data), and rejected values that do not fall within the real measurements. In this paper, however, we attempt to model noise and error processes instead, and accept data outside of the norms if it is not accounted for by a noise model. We describe a program that uses this method which does significantly better at a diagnostic task than an equivalent approach that either utilises all data, or attempts to reject all non-normal values.

2   Reasoning about outliers and the associated data

Our primary objective is to develop computational methods for managing outliers by *explicitly* examining outliers with a view to either reject or welcome them. The methods should not only successfully identify where the outliers are, but also carefully analyse them and their association with the entire data set to distinguish between the truly noisy and "surprising" but useful ones. Here we outline a method for explicitly identifying and analysing outliers in data.

Definitions: Let Ω be a p-dimensional sample space.

Let $X = \{x_1, x_2, ..., x_n\}$ be a set of vectors drawn from Ω.

Let $O = \{o_1, o_2, ..., o_r\}(1 \leq r < n)$ be a set of outliers in X where $O \subset X$.

Let $C = \{noisy, rest\}$ represent two general classes.

Let $F = \{f_1, f_2, ..., f_m\}$ be a set of features extracted from X.

Clustering and Outlier Detection: Given a set of data points, say X, a clustering algorithm is applied to identify the more stable part of the data. The less stable parts of the data then become a set of outliers, O. Note that there is no guarantee that there will be outliers existing in all the data sets. Sometime, there may be no major cluster of data points (e.g. there are several clusters or no cluster at all).

Noise Model Construction: a noise model is constructed which could account for much of recognised measurement noise in a domain of interest. Here we assume such a model is not readily available (things would become much easier if it is), and it needs to be constructed or learned. In particular, we assume that a group of data sets, Xs, can be labeled into two general classes (*noisy, rest*), based on relevant domain knowledge and close examination of data sets. This group of labeled data sets, together with a set of features F which may be extracted from the data sets, is then used to build the classification model (e.g. a set of classification rules) using an inductive learning technique. Those classification rules corresponding to *noisy* then become the noise model, M.

Noise Elimination: Each new data set, say X', is tested against the noise model M generated in the above step. If applicable, then the outliers O' within the data set X' can be rejected (due to known measurement noise).

We make the following observations regarding the proposed method:

1) The clustering algorithm used in the method can be a traditional statistical clustering algorithm [5], a self-organising neural network [11], or other machine learning methods [6].

2) The construction of the noise model requires a set of representative labeled instances on which the "noise model" may be built. There are two mutually exclusive classes that each data set can be assigned to in a domain of interest. Class *noisy* indicates that the corresponding outliers in this data set are noisy data points, therefore can be deleted; while class *rest* says we do not think the outliers in the data set are due to measurement noise, although we cannot say whether they are phenomena of interest either.

3) Note that many classification models may be constructed from a set of labeled instances. The classification model as mentioned in the *Noise Model Construction* step refers to the "best" model in terms of its predictive accuracy, its simplicity or interpretability, misclassification costs, or other appropriate criteria for the problem under investigation [15, 4].

4) The success of the method very much depends on the correctness and completeness of the noise model constructed. The correctness of the model depends largely on the quality of domain knowledge - a set of labelled instances in the proposed method, although the choice of inductive learning algorithm may also matter. On the other hand, if the model does not sufficiently cover all possible types of measurement noise, the data set after cleaning would still contain much noise.

5) The precondition for using the method regarding the availability of relevant knowledge about the distribution of noisy data points (labeled instances) is reasonable in many applications. For example, in time series load forecasting, the understanding of "special irregular events" and their effects on the forecasting results can be used to build the corresponding "noise model" where using the data after removing the effect of the special irregular effects may often increase the forecasting accuracy. In the next section, we will present a detailed case study of analysing behavioural data and show how such noise model can be developed to eliminate the measurement noise.

3   Reasoning about Outliers in Visual Field Data

We have undertaken a major project which aims to develop a software-based test, in which examination of visual field can be performed easily in primary care (e.g. waiting rooms, public halls). Figure 1 demonstrates how this test works. The test, operating on personal computers, examines several locations in the test screen which correspond to crucial positions in the visual field. The test screen consists of a number of objects of the same type (cars in this case) and at any stage of the test, only one of them is moving. The subject, using one of his arms supporting his chin in front of the fixation point (the smiling face on the screen), is asked to click on the mouse to respond to the moving stimulus (car moving).

The test system has been developed on the basis of Computer Controlled Video Perimetry (CCVP) [16], a visual stimuli generating program, but incorporates AI capabilities to overcome problems inherent with any software-based test. The most important problem is that, unlike the conventional visual field

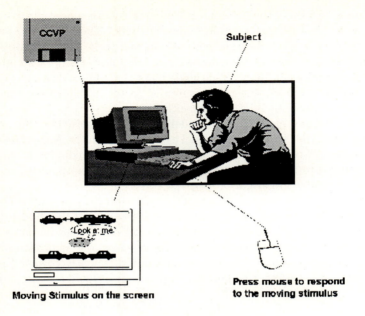

Fig. 1. The Test

test used in eye hospitals, we do not have a specially designed instrument capable of monitoring the behaviour of the subject and providing reliability indicators, and we do not have a standard test environment where many important factors such as light and viewing distance can be strictly controlled. Being flexible and mobile, this type of software-based test is less reliable in that the data collected from subjects will contain much measurement noise.

Because of the concern with the reliability problem, the test was designed to obtain repeated measurements over test locations. For example, 6 test locations could be examined during one particular measurement cycle and this cycle could be repeated 10 times. This is intended to make sure that reliable estimates of subject's visual function can still be obtained even if noise is introduced during one or more measurement cycles.

3.1 Clustering and Outlier Detection

One way of clustering and detecting outliers [12] was developed using self-organising maps [11] and *topographical product* [2]. The self organising map was used to automatically model the features found in the input data and reflect these features on topological maps, while the topographical product was applied to ensure that similar input patterns are mapped onto identical or closely neighbouring *winner* nodes on the output map.

The role of Figure 2 is two-fold. Firstly, it shows one of the physical meanings of the output map: the average sensitivity of all the input vectors associated with each winner node. The top right region of the map is most sensitive, i.e. the subject can see most of the time, and the sensitivity gradually fades towards the bottom left region [12]. Secondly, it demonstrates the results of one particular test. As the selected test locations are examined during one particular measurement cycle and this cycle is repeated ten times, Figure 2 tells us that this subject could see most of the time during the first six measurements, while the nodes of the next four cycles move away from the first node to some nodes with lower sensitivity values, which indicates that the patient cannot see as clearly as early in the test. In this case, the first six measurements form the stable part of the data.

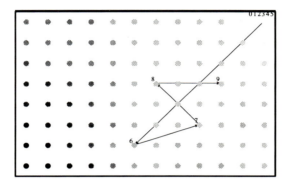

Fig. 2. Meaning of the map and an example of fatigue

One of the advantages of this approach is that it can visualise the subject's behaviour during the test, e.g. whether the patient has demonstrated the signs of fatigue, inattention, and learning effects etc [12]. For example, Figure 2 may have shown a case of fatigue by the subject at the late stage of the test.

3.2   Noise Model I: Noise Definition

In this application, noise in data are defined as those data points typically associated with learning effects, fatigue and inattention.

In this connection, we may define outlying data points due to fatigue on the self-organising maps as follows. If the sensitivities of several initial test cycles are high and similar to each other, and the sensitivities of the remaining cycles are decreasing over time, then the data points corresponding to the remaining cycles are outlying due to fatigue. In this case, the winner nodes of the initial cycles tend to be in a small neighbourhood and the winner nodes of the last few cycles tend to move away from the small neighbourhood to areas where the sensitivities are lower. One example of such cases has already been given in Figure 2.

On the other hand, if the sensitivities of the initial few cycles do not show much regularity but the sensitivities of the remaining cycles gradually become

similar, then the data points corresponding to the initial cycles are outlying due
to learning effects. In this case, the winner nodes of the initial cycles perhaps
are irregular, but are gradually gathered around a small neighbourhood on the
map. One example of such cases can be seen in Figure 3.

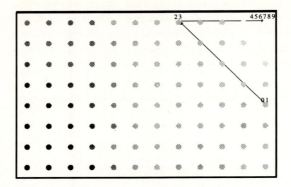

Fig. 3. An example of learning

Figure 4 demonstrates a typical case of inattention. Clearly the subject had
a normal visual field, but was distracted during the 8th measurement cycle.
This results in poor sensitivity values for this particular measurement, leading
to fluctuation in the data. This type of fluctuation, however, should not affect
the overall results of the visual field. Therefore the data collected during this
cycle can be dropped.

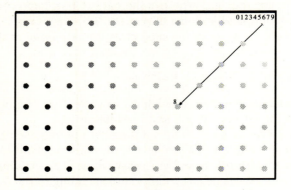

Fig. 4. An example of inattention

In all the above cases, decisions regarding whether to delete certain outlying
data points are relatively easy. For example, the data points corresponding to
measurements 6,7,8 and 9 may be deleted in Figure 2, while those corresponding

to measurements 0,1,2, and 3 in Figure 3 may be cleared. However, things are not always this clear-cut.

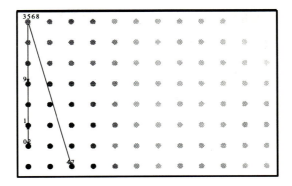

Fig. 5. A glaucoma case (with cluster)

Figures 5 and 6 demonstrate two test results for the same subject who had been confirmed by an ophthalmologist as a glaucoma patient. Figure 5 does seem to show there is a cluster in the top left corner of the map. However, since none of the measurements has shown any high sensitivities (in the top right corner area) and there are six measurements scattered around on the map, there is good reason to believe that these measurements might tell us something about the pathological status of the subject, and should therefore be kept. Meanwhile, Figure 6 does not seem to show any interesting clusters and none of the measurements are very sensitive. In this case, there is no easy way of finding out which measurements are noisy and which are not, and all the measurements will be therefore kept for further analysis.

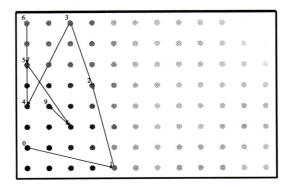

Fig. 6. A glaucoma case (without cluster)

3.3   Noise Model II: Construction

The construction of the noise model in this application is as follows:

1). A set of test records (310 in total) were used for the purpose of building the classification models and each of these records has ten 6-dimensional vectors. Using the visualised data presented by SOMs and relevant knowledge regarding the visual field test, a domain expert labeled each of these records into either *noisy*: corresponding outliers are measurement noise caused by one of the three behavioural factors: learning, fatigue and inattention; or *rest*, the outliers may (or may not) represent useful information.

2). Several features were extracted from the data sets and relevant domain knowledge. These features, together with those labled instances as discussed in the above step, were used to develop the classification models. In particular we have used Quinlan's C4.5 [14] to learn a set of production rules. Experiments were performed to find a set of rules which would minimise the errors on the unseen cases. This includes the division of 310 cases into training and testing cases of various sizes and the use of a more robust method of 10-fold cross validation. It appears that the 10-fold cross validation presents the most promising results for our data set.

3). Those production rules within the *noisy* class now become our "noise model" and can then be used to delete the corresponding outliers for future test data (see below), when applicable.

3.4   Noise Elimination

In this section we present results of applying the noise model generated above to the elimination of those noisy outliers in a set of visual field test data. The corresponding test was recently conducted in a large urban general practice in North London for a glaucoma case findings study. All patients aged 40 years or older who routinely attended the practice for a three-month period during the pilot study were offered the test.

	Detection Rates		
False Alarm	raw data	without outliers	with selected outliers
10%	55%	65%	37%
20%	66%	81%	93%
30%	83%	83%	100%
40%	86%	84%	100%
50%	97%	86%	100%
60%	100%	87%	100%
70%	100%	90%	100%
80%	100%	93%	100%
90%	100%	94%	100%
100%	100%	100%	100%

Tab. 1.: Detection rate by three data sets versus false alarm rate

A total of 925 patients were screened and 78 of them were later assessed clinically in the practice by an ophthalmologist, this sample included all people failing the test and a randomly sampled age matched control group. Among these, 22 eyes were later assessed as glaucoma, 81 were confirmed as normal eyes without any disease, and the rest were diagnosed as other types of ocular abnormalities. The noise model was applied to these 103 test records and Table 1 summarises the results of examining the *discriminating power* of the test in terms of its glaucoma detection rate versus false alarms. The decision threshold used for discriminating between normal and abnormal eyes is the average percentage of positive responses within the test.

Column 2 of Table 1 represents the detection rates corresponding to false alarm rate by using raw test data (with all the outliers); column 3 is the same results by the data without any outliers (all the outlying data points identified using the clustering algorithm are deleted). Finally column 4 describes the same results by the data with selected outliers using the method presented in this paper.

It is clear that the data with selected outliers perform better than the other two in terms of maximising the detection rate and minimising false alarms. For example, this group can achieve 100% detection rate, while the corresponding false alarm rate is 30%. This is equivalent to saying that none of the subjects suffering from glaucoma would have escaped notice and only 30% of those normal subjects would have been unnecessarily referred for further examination. To reach 100% detection rate by using the raw data, however, 60% of normal subjects would receive false alarms. In comparison with the data with selected outliers, this doubles the number of people who will be referred and further examined unnecessarily. This contrast demonstrates the usefulness of deleting noisy outliers from the test data.

It turned out that the data obtained by deleting all the outlying data points (without analysis) performed the worst in this experiment. This appears to suggest two things. Firstly, a great deal of outlying data points among the glaucoma patients were indeed a reflection of pathological status of the patients, and should not be deleted. Secondly, any mathematical similarity criteria used in a clustering algorithm cannot accurately cluster similar data points from the viewpoints of an application domain. Domain-specific knowledge should be applied in this context.

4   Concluding Remarks

Distinguishing between noisy outlying data and noise free outliers is a difficult problem and we have demonstrated how domain knowledge may be used to address this problem. The proposed method for modelling noisy data points is sufficiently general so that it can be applied to other, non-behavioural applications such as data pre-processing in time series forecasting.

An important observation from the work reported in this paper is that due care must be taken when deleting outlying data points. As with the visual field data from a GP clinic, the results obtained using "outlier-free" data are even worse than those from the raw data. Appropriate analysis should be performed using relevant domain knowledge in this regard. We have made initial steps in this direction and we are currently in the process of exploring various ideas of how AI and statistical techniques may be utilised to develop advanced computational methods for reasoning about outliers.

5   Acknowledgements

The work reported in this paper is in part supported by British Council for Prevention of Blindness, International Glaucoma Association, UK's Medical Research Council and the World Health Organisation. We would like to thank referees for their informative review, Sylvie Jami for early C4.5 experiments, Phil Dockings and Claude Gierl for their comments on early drafts.

References

1. Barnet, V., Lewis, T.: Outliers in Statistical Data. John Wiley and Sons, (1994).
2. Bauer, A.U. and Pawelzik, K.R.: Quantifying the Neighbourhood Preservation of Self-Organising Feature Maps. IEEE Trans. on Neural Networks, **3** (1992) 570-9
3. Brachman, A.R.J. and Anand, T.: The Process of Knowledge Discovery in Databases. Advances in Knowledge Discovery and Data Mining (eds. U M Fayyad et al.), (1996) 37-57, AAAI/MIT.
4. Cohen, P.R.: Empirical Methods for Artificial Intelligence (1995), MIT Press
5. Everitt, B.S.: Cluster Analysis. (1993), Gower Publications, London
6. Fisher, D., Pazzani, M. and Langley, P.: Concept Formation: Knowledge and Experience in Unsupervised Learning (1991) Morgan Kaufmann
7. Grubbs, F.E.: Sample Criteria for Testing Outlying Observations. Ann. Math. Statist., **21** (1950) 27-58
8. Guyon, I., Matic, N. and Vapnik, V.: Discovering Informative Patterns and Data Cleaning. Proc. of AAAI-94 Workshop on Knowledge Discovery in Databases, (1994) 143-56.
9. Hawkins, A.: The Detection of Errors in Multivariate Data Using Principal Components. J. Amer. Statist. Assn., **69** (1974) 340-4
10. Huber, P.J.: Robust Statistics. (1981) John Wiley and Sons
11. Kohonen, T.: Self-Organisation and Associative Memory. (1989) Springer-Verlag
12. Liu, X., Cheng, G. and Wu, J. X.: Identifying the Measurement Noise in Glaucomatous Testing: an Artificial Neural Network Approach. Artificial Intelligence in Medicine, **6** (1994) 401-416
13. Liu, X., Cheng, G. and Wu, J.X.: Noise and Uncertainty Management in Intelligent Data Modelling. Proc. of 12th National Conference on Artificial Intelligence (AAAI-94), 263-268
14. Quinlan, J.R.: C4.5: Programs for Machine Learning. (1993) Morgan Kaufmann
15. Weiss, S.M. and Kulikowski, C.A.: Computer Systems that Learn. (1995), Morgan Kaufmann
16. Wu, J.X.: Visual Screening for Blinding Diseases in the Community Using Computer Controlled Video Perimetry, PhD thesis, University of London, (1993).

Section VIII:

Qualitative Models

Reasoning about Sensor Data for Automated System Identification

Elizabeth Bradley and Matthew Easley*

University of Colorado
Department of Computer Science
Boulder, CO 80309-0430
[lizb,easley]@cs.colorado.edu

Abstract. The computer program PRET automatically constructs mathematical models of physical systems. A critical part of this task is automating the processing of sensor data. PRET's intelligent data analyzer uses geometric reasoning to infer qualitative information from quantitative data; if critical variables are either unknown or cannot be measured, it uses delay-coordinate embedding to reconstruct the internal dynamics from the external sensor measurements. Successful modeling results for a sensor-equipped driven pendulum demonstrate the effectiveness of these techniques.

1 Introduction

Constructing a model that a scientist or engineer can use to better understand or control a physical system typically requires analyzing the input/output behavior of that system. Different applications require different types of models; each type requires different measurement and reasoning techniques. In particular, formulating an internal ordinary differential equation (ODE) model from external observations of a system is known as *system identification* (SID). SID has two phases, as shown in figure 1; *structural identification*, in which the general form of the equations that govern the unknown dynamics is determined, and *parameter estimation*, in which coefficient values that match that model to the actual sensor data are found.

One of the aims of qualitative reasoning (QR)[9, 21], a branch of artificial intelligence (AI), is to automate the modeling process by abstracting knowledge and information to a qualitative level. The computer program PRET[3] is an example of a QR modeling tool. It automates the SID process that is diagrammed in figure 1 by building an AI layer on top of a set of traditional SID techniques. This layer automates the high-level stages of the modeling process that are

---

* Supported by NSF NYI #CCR-9357740, NSF #MIP-9403223, ONR #N00014-96-1-0720, and a Packard Fellowship in Science and Engineering from the David and Lucile Packard Foundation.

X. Liu, P. Cohen, M. Berthold (Eds.): "Advances in Intelligent Data Analysis" (IDA–97)
LNCS 1280, pp. 561–572, 1997.   © Springer–Verlag Berlin Heidelberg 1997

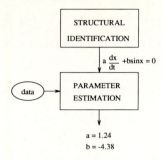

Fig. 1. The System Identification Process. Structural identification yields the general form of the model; in parameter estimation, values for the unknown coefficients in that model are determined. The PRET modeling tool automates this process using artificial intelligence techniques.

normally performed by a human expert. PRET combines several forms of QR, via a special first-order logic inference system[19], to intelligently assess the task at hand; it then reasons from that information to automatically choose, invoke, and interpret the results of appropriate lower-level techniques.

PRET models linear and nonlinear systems by assembling combinations of user-specified and automatically generated model fragments into an ODE system that both fits the domain physics and matches a given set of qualitative and quantitative observations. These user-specified observations guide the modeling process in a fundamental way; properly processed, they allow the inference engine to eliminate large classes of models in a purely qualitative manner, guiding PRET efficiently through the exponentially complex search space of hypothesis combinations. *Qualitative* observations play a straightforward role in this process; for instance, if the target system is known to be damped, PRET uses algebraic reasoning about the divergence of ODEs to quickly discard any candidate model that is conservative. When observations are numeric, PRET uses geometric reasoning to distill qualitative features out of the data — for instance, to recognize that a sensor time series is converging to a fixed point and infer that the system is damped. QR techniques like algebraic and geometric reasoning are far less computationally expensive than techniques that involve the processing of numbers; also, because of their inherent abstraction, they apply to wider problem classes. This application breadth and expense reduction are particularly critical here because of the complexity of PRET's search spaces.

PRET's capabilities have been demonstrated in a variety of simulated and real systems. For example, a commercially acquired radio-controlled (R/C) car used at the University of British Columbia could not be controlled well because no adequate model was available. PRET was applied to this problem; its inputs included a time series of the car's position and heading, the UBC engineer's set of known model fragments, and the mathematical equivalents of a few of his qualitative observations (e.g. "it pulls to the left" or "we assume there is a simple form of frictional damping"). PRET examined different combinations of the model fragments during its structural identification stage, settled on one that matched the qualitative observations, and estimated parameters for this model using its nonlinear parameter estimation reasoner[2]. The resulting model was correct, but it did not match the engineer's intuition because he had omitted a few crucial observations (e.g. "it started from rest"). The discrepancy was useful in a very powerful way: it allowed a human analyst to identify what was missing from his mental model of a problem.

This paper describes a collection of automatic data analysis methods, such as the geometric and algebraic reasoning techniques mentioned two paragraphs above, that are specifically tailored to generate the kinds of information that PRET's inference engine can exploit to build effective models of systems like the R/C car. PRET's intelligent data analysis module, which instantiates these methods, combines ideas from a variety of very different fields — AI, control theory, nonlinear dynamics, and numerical analysis. An important and difficult part of its task is to effectively automate decisions about when to use which techniques and in what sequence, how to set up the invocations of the appropriate code modules, and how to interpret the results. The core of this intelligent data analyzer is based on Hsu's cell-to-cell mapping paradigm for dynamics analysis[11]. This method requires a full state-space trajectory, however, and fully *observable* systems[2] are rare in engineering practice. Sensor data are almost always incomplete and/or noisy; worse yet, the true state variables may not even be known to the user. The partial solutions that we propose to this *observer problem* are based on delay-coordinate embedding[1]. Acting together, geometric reasoning and delay-coordinate embedding allow this intelligent data analyzer to infer, from a quantitative data set, exactly the kinds of qualitative information that PRET can exploit to quickly verify or discard models.

The next two sections describe PRET's intelligent data analysis module, first covering the geometric reasoning that is used to distill qualitative information out of a quantitative data set, and then describing how delay-coordinate embedding techniques can be used to infer knowledge about unobserved state variables. Following these background sections, we present an example PRET run on a parametrically driven pendulum, highlighting the role that the intelligent data analysis techniques play in the successful modeling of the system. We then discuss the results and their implications, describe PRET's relationship to other work, and summarize.

2 Distilling Qualitative Information from Quantitative Data

The intelligent data analyzer's geometric reasoner distills qualitative properties from a numeric data set using phase-portrait analysis, asymptote recognition, and other computer-vision techniques. This information is used by PRET's inference engine to verify or discard large classes of candidate models in a purely

---

[2] those whose state variables are all known and measurable

qualitative manner; for example, if sensor measurements of a state variable indicate that it is linear, the inference engine rules out all models whose second derivative in that state variable are non-zero.

This geometric reasoner is based on the cell-to-cell mapping formalism of Hsu[11], which discretizes a set of n-dimensional state vectors onto an n-dimensional mesh of uniform boxes or *cells*. Because multiple trajectory points are mapped into each cell, this coarse-grained representation of the dynamics is significantly more compact than the original series of floating-point numbers and therefore much easier to work with; because of symbolic dynamics theory[16], which establishes that the mapping preserves the significant properties of the original dynamics, conclusions drawn from the discretized trajectory are also true of the real trajectory.

Simple geometric heuristics, applied to such a discretized trajectory, can quickly classify its dynamics at an abstract level. Some of these heuristics are trivial (e.g. determining if the trajectory exits the mesh), but detecting limit cycles or oscillations requires subtler pattern recognition techniques. Below are several of the geometric reasoner's dynamics classifications, the corresponding heuristics, and some associated implications for the ODE model:

- `fixed-cell`: when a trajectory relaxes to a single cell and remains within that cell for a fixed percentage of its total lifetime. This can, for instance, be used to recognize when a second-order system is overdamped. Appropriate mesh geometry choices can extend this method to asymptote recognition.
- `limit-cycle`: when the trajectory contains a finite, repeating sequence of cells. These patterns are identified by ignoring any transients and searching for periodic mapping sequences; they indicate that the system is either conservative or externally driven (nonautonomous).
- `damped-oscillation`: when a trajectory enters a fixed cell via a decaying oscillation. This pattern is detected by recognition of an inward spiral; such dynamics can indicate, for instance, that a system is underdamped and thus that the model's roots must be complex.
- `constant`: when a state variable does not change over the duration of the trajectory. This computation involves a simple serial scan on each mesh axis; its results are particularly useful to PRET because they have wide-ranging implications about the order of the system.
- `sink-cell`: when a trajectory exits the mesh. This information is used to identify unstable trajectories.

Many other classifications are possible (e.g., `chaotic`, etc.); some are less useful to PRET than others — because their implications either are more limited in range or require processing at a less-abstract reasoning level.

The cell size, mesh boundary, and trajectory length affect the validity and efficiency of this technique. Among other things, a small limit cycle may be classified as a fixed point, and behavior outside the mesh will not be classified at all. All of these discretization and boundary effects are not, in fact, problems; rather, they actually allow PRET to represent and work with the abstraction levels implied by the finite range and resolution that are such fundamental features of a modeling hierarchy — e.g., to avoid including saturation and crossover distortion effects when asked to "model the *small-signal* behavior of the op amp *to within 10mV*." Specifically, we use the range and resolution information from user instructions to set up the mesh boundary and cell size, assuring that behavior outside the range or below the resolution is not modeled.

3   Delay-Coordinate Methods for Observer Theory

If all of a system's state variables are identified and measured, the geometric reasoning techniques described in the previous section can be applied directly to the sensor data. A fully *observable* system like this, however, is rare in engineering practice; as a rule, many — often, most — of the state variables either are physically inaccessible or cannot be measured with available sensors. Worse yet, the true state variables may not be known to the user; temperature, for instance, can play an important and often unanticipated role in the modeling of an electronic circuit. This is part of control theory's *observer problem*: how to (1) identify the internal state variables of a system and (2) infer their values from the signals that *can* be observed. The arsenal of time-series analysis methods developed by the nonlinear dynamics community in the past decade[1] provides powerful solutions to both parts of this problem. This section describes the two methods, Pineda-Sommerer (P-S)[18] and false near neighbor (FNN)[14], that PRET uses to infer the dimension of the internal system dynamics from a time series measured by a single output sensor[3].

Both P-S and FNN are based on *delay-coordinate embedding*, wherein one constructs m-dimensional *reconstruction-space* vectors from m time-delayed samples of the sensor data. For example, if the time series

$$(t, x) = ((0.1, 1.2)\ (0.2, 0.8)\ (0.3, 0.2)\ (0.4, -0.4)\ (0.5, -0.1)\ (0.6, 0.3)...)$$

is embedded in three dimensions with a delay of 0.2, the first two points in the reconstruction-space trajectory are (1.2 0.2 -0.1) and (0.8 -0.4 0.3). Sampling a single system state variable is equivalent to projecting a d-dimensional state-space dynamics down onto one axis; embedding is akin to "unfolding" or "re-inflating" such a projection, albeit on different axes. The central theorem relating such embeddings to the underlying dynamics was suggested in [20] and proved in [17]; informally, it states that given enough dimensions (m) and the right delay (τ), the reconstruction-space dynamics and the true (unobserved) state-space dynamics are topologically identical[4]. This is an extremely powerful theorem: it lets us analyze the underlying dynamics using only the output of a single sensor. In particular, many properties of the dynamics, such as dimension (i.e., fixed

---

[3] Techniques like divided differences can, in theory, be used to derive velocities from position data; in practice, however, this is impractical because the associated arithmetic magnifies sensor error.

[4] More formally, the reconstruction-space and state-space trajectories are diffeomorphic iff $m \geq 2d + 1$, where d is the true dimension of the system.

point, limit cycle, chaotic attractor, etc.), are preserved by diffeomorphisms; if they are present in the embedding, they exist in the underlying dynamics as well. There are, of course, some important caveats, and the difficulties that they pose are the source of most of the effort and subtlety in these types of methods. Specifically, in order to embed a data set, one needs m and τ, and neither of these parameters can be measured or derived, either directly or indirectly, so algorithms rely on numeric and geometric heuristics to estimate them.

The Pineda-Sommerer algorithm creates such estimates; it takes a time series and returns the delay τ and a variety of different estimates of the dimension m. The procedure has three major steps: it estimates τ using the mutual information function, uses that estimated value τ_0 to compute a temporary embedding dimension E, and uses E and τ_0 to compute the *generalized dimensions* D_q, also known as "fractal dimensions." The standard algorithm for computing the fractal dimension of a trajectory, loosely described, is to discretize state space into ϵ-boxes, count the number of boxes occupied by the trajectory, and let $\epsilon \to 0$. Generalized dimensions are defined as

$$D_q = \frac{1}{q-1} \limsup_{\epsilon \to 0} \frac{\log \sum_i p_i^q}{\log \epsilon} \tag{1}$$

where p_i is some measure of the trajectory on box i. D_0, D_1, and D_2 are known, respectively, as the capacity, information, and correlation dimensions; all three are useful to PRET as estimates of the number of state variables in the system. The actual details of the P-S algorithm are quite involved; we will only present a qualitative description:

- Construct 1- and 2-embeddings of the data for a range of τs and compute the saturation dimension D_* of each; the first minimum in this function is τ_0. The D_* computation entails:
  - Computing the information dimension D_1 for a range of embedding dimensions E and identifying the saturation point of this curve, which occurs at D_*. The D_1 computation entails:
    - Embedding the data in E-dimensional space, dividing that space into E-cubes that are ϵ on a side, and computing D_1 using equation (1) with $q = 1$.

Ideally, of course, one lets $\epsilon \to 0$ in the bottom step, but floating-point arithmetic and computational complexity place obvious limits on this; instead, one repeats the calculation for a range of ϵs and finds the power-law asymptote in the middle of the dimension-vs.-ϵ curve. P-S incorporates an ingenious complexity-reduction technique: the ϵs are chosen to be of the form 2^{-k} for integers k and the data are integerized; this enables most of the mathematical operations to proceed at the bit level and vastly accelerates the algorithm. To increase the precision of this computation, we have also implemented an arbitrary-length virtual integer package that facilitates the integerization.

The false near neighbor algorithm is far simpler than P-S. It takes a τ and a time series[5] and returns m. FNN is based on the observation that neighboring points may in reality be projections of points that are very far apart, as shown in figure 2. The FNN algorithm starts with $m = 1$. It finds each point's nearest

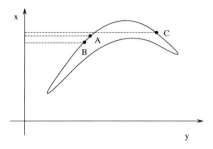

Fig. 2. The geometric basis of the FNN algorithm: the points labeled A and B are true near neighbors in the x-projection, while A and C are false near neighbors.

neighbor, then re-embeds the data with $m = 2$, determines[6] whether any of those neighbors were false (like A and C in the x-projection of figure 2), continues adding dimensions until no more false near neighbors exist, and returns the last m-value as the estimated dimension. We use a K-D tree implementation[10] to reduce the complexity of the nearest-neighbor step from $O(N^2)$ to $O(N \log N)$, where N is the length of the time series.

As both FNN and P-S are based on heuristics, they do not necessarily return the same embedding dimension. Since both algorithms provide conservative estimates, PRET chooses the minimum of the two results as an upper bound for the dimension of the model.

4   Status and Discussion

In order to demonstrate the functions of PRET's intelligent data analysis module, this section presents a real-world modeling example: a PRET run on a parametrically driven pendulum[7]. An actuator controls the pendulum's drive frequency and a sensor (an optical encoder) measures its angular position. The behavior of this apparently simple device is really quite interesting: for low drive frequencies, it has a single stable fixed point, but as the drive frequency is raised, the attractor undergoes a series of bifurcations. In the sensor data, this manifests as interleaved chaotic and periodic regimes[5]. This system is also interesting

---

[5] We run P-S first and then use its τ in FNN; other methods, such as autocorrelation[1] can also be used to estimate τ.

[6] if the point separations change abruptly between the $m = 1$ and $m = 2$ embeddings.

[7] A solid aluminum arm that rotates freely on a standard bearing. The pendulum vertex is driven up and down in a sinusoidal pattern by a motor and a simple linkage.

from a modeling standpoint; at high resolutions, the backlash in the bearings invalidates the standard textbook model. Modeling these effects is critical, for instance, to the accurate deployment of the space shuttle's manipulator arm[12].

```
(find-model
   (domain mechanics)
   (state-variables <theta>)
   (point-coordinates <theta>)
   (hypotheses  (* A1 (sin <theta>))
                (* A2 (deriv <theta>))
                (* A3 (cos (* A4 <time>)))
                (* A5 (deriv (deriv <theta>)))   )
   (observations (nonautonomous)
                 (numeric (<time> <theta>) ((0.00 3.560) ... )) )
   (specifications (<theta> absolute-resolution 0.1)
                   (<theta> range -inf +inf) ))
```

Fig. 3. Instructing PRET to find an ODE model of the driven pendulum. <theta> is the bob angle.

Figure 3 shows how a user instructs PRET to build an ODE model of this system. The details of the syntax are covered elsewhere[3]; briefly, the bulk of the user's input consists of three types of information about the target system: hypotheses, observations, and specifications. The first are ODE fragments from which PRET constructs the model[8] and the third prescribe resolutions and ranges to and over which that model is required to be valid. Observations, which play a central role in this paper, range from the purely qualitative to the purely quantitative. The former are a powerful target for QR techniques; using them, PRET's inference engine can quickly discard broad classes of candidate models. Moreover, to a human engineer, making qualitative observations is a natural part of the modeling process, so this provides a smooth user interface. Quantitative observations, which are the focus of the methods described in this paper, are typically sensor measurements; the source of the numeric observation in this find-model call, for instance, is the optical encoder on the pendulum shaft. Direct inferences from such observations are computationally expensive and of limited utility, but qualitative features extracted from them, as described below, can be leveraged in an abstract and powerful manner by the inference system.

As a first step in the modeling process, PRET's intelligent data analysis reasoner distills a variety of qualitative information from the numeric observation. It first reconstructs the dynamics, beginning by invoking P-S and FNN. The for-

---

[8] PRET also uses power-series techniques to generate ODE fragments from scratch if the user's hypotheses are inadequate.

mer returns $\tau = 15$ and $m = 5$; the latter also returns $m = 5$, corroborating the estimates, so PRET embeds the data set with those parameters. Geometric reasoning classifies this reconstructed trajectory as a limit cycle, so a `limit-cycle` fact is added to the list of `observations` that the model must match[9]. Guided by this augmented observation set, PRET searches the space of `hypothesis` combinations. Inferences drawn from the two qualitative observations, `nonautonomous` and `limit-cycle` — one specified by the user and one inferred automatically by the intelligent data analyzer — play a critical role in search-space reduction. The `nonautonomous` fact lets PRET rule out all models that have no `<time>`-dependent terms; knowledge that the system exhibits a limit cycle is equally powerful, since `limit-cycle` implies `not-constant` and `not-linear`, facts from which the logic system infers that the model must be of at least second order. Moreover, the combination of these two qualitative observations allows PRET to infer that the system is damped. Inferences drawn from all of these facts eventually guide PRET to the model $A_5\ddot{\theta} - A_3\cos[A_4 t] - A_1\sin\theta - A_2\dot{\theta} = 0$. To determine values for the unknown coefficients A_i, PRET invokes its QR-based *nonlinear parameter estimation reasoner* or NPER[2], which returns $A_i = (2.5, 98.0, 80.0, 9.0)$. With these coefficient values, this model meets the requirements in the `find-model` call. This example has been simplified for presentation purposes; for instance, we loosened the resolution so as to avoid modeling the backlash in the bearings. Normally, too, the `find-model` call would contain many more hypotheses and qualitative observations. Most of the models constructed from the former are quickly discarded using quick, inexpensive qualitative reasoning on the latter.

Note that the angular velocity state variable $\dot{\theta} = \omega$ was neither identified nor measured in this example. In the `mechanics` domain, PRET automatically generates a new state variable from the (symbolic) first derivative of each known state variable[10]; here, this step led directly to ω, effectively solving the identification part of the observer problem. PRET's NPER solves the other part of this problem; using a combination of qualitative reasoning and numerical analysis, it automatically synthesizes starting values and initial search directions for a nonlinear least-squares solver routine that performs the coefficient/data regression. If PRET cannot build a successful model with θ and ω, it may be possible to synthesize state variables from scratch using geometric reasoning on the data, as embedded using the P-S and FNN results. This is a current focus of our research effort.

The work described here is only the first half of the full input-output analysis that expert engineers apply to modeling problems; the next step in this research is to incorporate actuators as well as sensors. The overall goal of PRET's automatic sensor/actuator interpretation and control module, of which the intelligent data analyzer described here is a component, is to autonomously generate and refine observations and hypotheses, and possibly even construct and use obser-

---

[9] Note that the results of the automatic data analysis may conflict with the user's observations. If this happens, PRET places a higher confidence level in the former.

[10] more generally, the conjugate momentum for each generalized coordinate

vations that transcend the user's knowledge of physics. There are a variety of potential problems with this; for instance, the type and number of automatically generated hypotheses will have to be limited, lest the search space become unmanageable and/or the models overfit the data. There are two fundamental limits on automated actuator manipulation: *controllability* and *reachability* — the combined actuator and system properties implicitly govern what kinds of experiments can be performed. Automating the experimental observation process gives rise to some interesting AI issues involving representation and reasoning: how to identify and express the appropriate properties, and how to reason from that information in order to determine what experiments are possible and which of those yield the most powerful and useful inferences.

5   Relationship to Related Work

Any type of reasoning about a dynamic system requires a model. The precise form of that model is governed by the form of the reasoning; quantitative analysis requires exact equations, while a qualitative understanding can rest purely on abstract notions like "y is a monotonic function of x." Automated modeling tools reflect this difference in approaches; the models that they produce span the spectrum from precise mathematical descriptions of a system to abstract representations of its physics. Most of the work in the very active AI/QR modeling community, including PRET, focuses on qualitative models; these tools typically generalize a set of descriptions of the state into a higher level abstraction — *qualitative states*[4, 8]. Many QR modeling tools reason about equations at an abstract level by combining model fragments. The abstraction levels and representations of these fragments vary; QSIM[15] models, for instance, are qualitative differential equations (QDEs) like $y = M+(x)$[11], whereas PRET's models are purely mathematical ODEs. Like many of the more physics-based modeling systems (e.g., QPT/QPE[7]), PRET's inputs are expressed in the high-level concepts and terms that are typically used by scientists, and its internal reasoning is guided and governed by the standard ideas of physics. PRET differs from systems like QPT in several important ways; among other things, it does not focus on explanation and causality (e.g. "What happens when a block is dropped from a table"). Perhaps the most significant difference between the work described here and the rest of the QR modeling literature, however, is that PRET takes a practical engineering approach, working with noisy, incomplete sensor data from real-world systems and attempting not to "discover" the underlying physics, but rather to find the simplest ODE that can account for the given observations.

PRET also shares goals, ideas, and tactics with several other fields. In particular, it solves the same problems as traditional system identification[13], but in an automated fashion, and it rests upon many of the standard methods found in basic control theory texts. Finally, it incorporates many of the same ideas that appear in the data analysis literature[6], but it adds a layer of AI techniques,

---

[11] y is a monotonic function of x.

such as symbolic data representation and logical inference, that let it automate many of the higher-level reasoning tasks normally performed by human experts.

6   Conclusion

Geometric reasoning and delay-coordinate embedding allow the automatic data analyzer described in this paper to infer, from a quantitative data set, exactly the kinds of qualitative information that effectively guide the PRET automated system identification tool through the complex search space of ordinary differential equation models. These methods provide a partial solution to the observer problem, allowing PRET to infer some internal system state variable properties from incomplete output sensor data. Importantly, this intelligent data analyzer automates many of the higher-level tasks normally performed by human experts — the choice of what lower-level technique to use in a given phase of the SID process, how to invoke that technique, how to interpret its results, and how to leverage that information later in the process.

The ultimate goal of the PRET project is a tool that can construct internal ODE models of high-dimensional black-box systems in a variety of domains — with minimal human guidance or forethought. Because of this, PRET is designed for easy extension to different domains, and a current focus of our research is to test this extensibility by solving modeling problems in visco-elastic materials and electronic circuits. The initial results have been good, but some subtleties remain: for instance, topology plays a critical role in descriptions of networks of discrete components, and PRET's syntax and reasoning do not yet handle these requirements smoothly. Moreover, human experts in both of these domains — and others — rely heavily on input/output studies such as impulse or frequency response. Intelligent automation of PRET's sensor interaction, the topic of this paper, is only one part of this process; the details of the dynamics cannot be truly exposed without an interactive input-output analysis, which will require intelligent actuator manipulation as well.

Acknowledgments: Joe Iwanski, with some help from Josh Stuart (the virtual integer implementation), coded the algorithms described in section 3. Reinhard Stolle, Apollo Hogan, Brian LaMacchia, and Meenakshy Chakravorty also contributed code and/or ideas to this project, and the IDA-97 reviewers' comments helped focus the content and presentation of this document.

References

1. H. Abarbanel. *Analysis of Observed Chaotic Data*. Springer, 1995.
2. E. Bradley, A. O'Gallagher, and J. Rogers. Global solutions for nonlinear systems using qualitative reasoning. In *Proceedings of the International Workshop on Qualitative Reasoning about Physical Systems*, 1997. Cortona, Italy.
3. E. Bradley and R. Stolle. Automatic construction of accurate models of physical systems. *Annals of Mathematics and Artificial Intelligence*, 17:1–28, 1996.

4. J. de Kleer and J. S. Brown. A qualitative physics based on confluences. *Artificial Intelligence*, 24:7–83, 1984.

5. D. D'Humieres, M. R. Beasley, B. Huberman, and A. Libchaber. Chaotic states and routes to chaos in the forced pendulum. *Physical Review A*, 26:3483–3496, 1982.

6. A. Famili, W.-M. Shen, R. Weber, and E. Simoudis. Data preprocessing and intelligent data analysis. *Intelligent Data Analysis*, 1(1), 1997.

7. K. D. Forbus. Qualitative process theory. *Artificial Intelligence*, 24:85–168, 1984.

8. K. D. Forbus. Interpreting observations of physical systems. *IEEE Transactions on Systems, Man, and Cybernetics*, 17(3):350–359, 1987.

9. K. D. Forbus. Qualitative reasoning. In J. A. Tucker, editor, *CRC Computer Science and Engineering Handbook*. CRC Press, Boca Raton, FL, 1996.

10. J. Friedman, J. Bentley, and R. Finkel. An algorithm for finding best matches in logarithmic expected time. *ACM Transactions on Mathematical Software*, 3:209–226, 1977.

11. C. S. Hsu. A theory of cell-to-cell mapping dynamical systems. *Journal of Applied Mechanics*, 47:931–939, 1980.

12. J. Iwanski and E. Bradley. Modeling nonlinear/chaotic phenomenon: Comparing models. In *Proceedings of the Fourth Experimental Chaos Conference*, 1997. Submitted.

13. J.-N. Juang. *Applied system identification*. Prentice Hall, Englewood Cliffs, N.J., 1994.

14. M. Kennel, R. Brown, and H. Abarbanel. Determining minimum embedding dimension using a geometrical construction. *Physical Review A*, 45:3403–3411, 1992.

15. B. J. Kuipers. Qualitative simulation. *Artificial Intelligence*, 29(3):289–338, 1986.

16. D. Lind and B. Marcus. *Symbolic Dynamics and Coding*. Cambridge University Press, Cambridge, 1995.

17. N. Packard, J. Crutchfield, J. Farmer, and R. Shaw. Geometry from a time series. *Physical Review Letters*, 45:712, 1980.

18. F. J. Pineda and J. C. Sommerer. Estimating generalized dimensions and choosing time delays: A fast algorithm. In *Time Series Prediction: Forecasting the Future and Understanding the Past*. Santa Fe Institute Studies in the Sciences of Complexity, Santa Fe, NM, 1993.

19. R. M. Stolle and E. Bradley. A customized logic paradigm for reasoning about models. In *Proceedings of the Tenth International Workshop on Qualitative Reasoning about Physical Systems*, 1996. Stanford Sierra Camp, CA.

20. F. Takens. Detecting strange attractors in fluid turbulence. In D. Rand and L.-S. Young, editors, *Dynamical Systems and Turbulence*, pages 366–381. Springer, Berlin, 1981.

21. L. Travé-Massuyès and R. Milne. Application oriented qualitative reasoning. *Knowledge Engineering Review*, 10(2):181–204, 1995.

Modelling Discrete Event Sequences as State Transition Diagrams

Adele E. Howe and Gabriel Somlo

Computer Science Dept, Colorado State University, Fort Collins CO 80523, USA
email: {howe,somlo}@cs.colostate.edu

Abstract. Discrete event sequences have been modeled with two types of representation: snapshots and overviews. Snapshot models describe the process as a collection of relatively short sequences. Overview models collect key relationships into a single structure, providing an integrated but abstract view. This paper describes a new algorithm for constructing one type of overview model: state transition diagrams. The algorithm, called *State Transition Dependency Detection* (STDD), is the latest in a family of statistics based algorithms for modeling event sequences called *Dependency Detection*. We present accuracy results for the algorithm on synthetic data and data from the execution of two AI systems.

1   Introduction

Many processes can be viewed as long sequences of discrete events over time. Events may capture the state of the process at regular intervals or may encompass changes in the state at irregular intervals. By modeling these sequences, we can predict likely future states and describe what led to the current state. For example, ecologists perform flora/fauna counts regularly within small areas to monitor ecosystems; credit card companies record and track customer transactions to detect fraud and identify opportunities for additional sales [6].

Discrete event sequences have been modeled with two types of representation: snapshots and overviews. Snapshot models describe the process as a collection of relatively short sequences or rules. These rules relate key events to each other. For example, a rule in a grammar indicates that an adjective should be followed by a noun or that the purchase of the first book in a best-selling trilogy is often followed by the purchase of the second and third books [2]. Overview models collect key relationships into a single structure, providing an integrated but abstract view. For example, finite state machines drive some language parsers, and Markov and Bayesian networks structure probabilistic dependencies [14].

---

[1] This research was supported in part by by NSF Career Award IRI-9624058 and by DARPA-AFOSR contract F30602-93-C-0100 and F30602-95-0257. We also wish to thank Larry Pyeatt for his contributions on the previous version of the algorithm and for collecting the RARS data.

X. Liu, P. Cohen, M. Berthold (Eds.): "Advances in Intelligent Data Analysis" (IDA–97)
LNCS 1280, pp. 573–584, 1997.   © Springer–Verlag Berlin Heidelberg 1997

Each representation has its purposes. Snapshots are convenient, easily exploited representations for rule-based systems (e.g., planning [13]) and are well-suited for short sequences over a large set of separate processes (the *market basket* problem [3]) and tend to include even extremely rare relationships or events. Overviews provide a more comprehensive view, clarifying longer term relationships and interaction effects.

This paper presents a new overview modeling method, called *State Transition Dependency Detection* (STDD), that automatically generates state transition models from event sequences. The method heuristically combines statistically determined snapshot patterns into a cohesive whole. Because the method relies on a statistical technique for determining the basis patterns, it is fairly robust to the presence of noise in the discrete event sequences. Because the combination is heuristic, the resulting model may include cycles: a desirable characteristic for our applications. Consequently, we evaluated the accuracy of the results of the algorithm on both synthetic data with varying levels of noise and real data: discrete event sequences from two AI systems.

1.1 Applications and Techniques for Event Sequence Modeling

Our primary application for event sequence modeling is debugging. Our method for generating snapshot models, called *dependency detection* (DD), has been used to support debugging failure recovery in the Phoenix planner [9] and to identify search control problems in UCPOP[16]. Snapshot DD methods discover unusually frequently or infrequently co-occurring sequences of events (called *dependencies*) using contingency table analysis [8] (see Section 2.1). Mannila et al. [12] find serial (i.e., a strict ordering of events, perhaps with intervening events) and parallel (i.e., events occurring within some time period of one another, the exact ordering is irrelevant) patterns of events that occur more than some number of times. They applied their methods to a database of telecommunications network faults covering a 50 day period.

While the snapshot models did help identify some previously unknown interactions, we found that bugs that are due to cycles are hard to detect with the snapshots, and that fixes based on the snapshots may cause problems that could not be predicted from the snapshots. Based on this experience, we developed an overview modeling algorithm that constructs complete, Semi-Markov models of the event sequences [10]. Abrams et al. generate Semi-Markov models to debug interactions of distributed computational processes [1]. Their method constructs Semi-Markov models from the most strongly associated pairs of events in execution traces; CHAID analysis determines which are most strongly associated [5]. A variety of other methods have been developed for generating graphical models, especially Bayesian Networks and Markov models, most of which are for acyclic graphs [4].

Business and medical applications dominate the applications of data mining methods to event sequence modeling. The *Quest* system at IBM has been used in business for attached mailing, add-on sales and customer satisfaction as well as medical diagnosis [2]. The algorithm for finding sequential patterns counts subsequences and then generalizes the frequently occurring ones to include sets of possible values in positions; the algorithm is extremely efficient and has been designed for mining massive databases [3].

Intelligent agents may use operators for deciding what action to take to achieve a desired state. Agents can learn these operators by modeling their perceptions of state changes in the environment using *Multi-Stream Dependency Detection* (MSDD) [13]. MSDD performs a simple best first search for dependencies, where the sequences are composed of a set of parallel event streams.

2   Building Transition Models of System Execution

State transition models are built by iteratively combining frequently occurring sequences. Thus, the two core ideas underlying the generation algorithm are: finding statistically significantly occurring sequences (snapshot dependencies) to alleviate the effect of noise and carefully checking proposed transitions for consistency with all of the discovered sequences.

2.1   Snapshot Dependency Detection

DD finds dependencies by analyzing event sequences. The idea is to see whether particular subsequences stand out from the whole, based on a statistical analysis. First, the event sequences are scanned to count all subsequences of a length specified by the user. Then, subsequences of influencing events are tested for whether they are more or less likely to be followed by each of the target events using contingency table analysis on the counts. The output is a list of sequences (e.g., $AB \to C$ which means that event C depends on the sequence of events AB) and their "strength" as indicated by the probability that the observed predictor appears as often as other predecessors. Due to the contingency table test, dependencies include both frequently and infrequently co-occurring sequences; we call the types *positive* and *negative* dependencies, respectively.

DD has been extended to find more complex patterns[8, 7]. One version collects dependencies up to some length while pruning overlapping dependencies based on an analysis of which is the best descriptor. Another detects partial order sequences, which is robust to the introduction of intervening events. At present, we use only the simplest form for constructing transition diagrams.

2.2   State Transition Generation Algorithm

To run the algorithm, the user provides discrete event sequences (`dataset`) and designates a sequence length (`seq-length`) and a significance `threshold`. The user can also request either a fully connected model or one that is more complete, but that has states without input from the rest of the diagram (`loose-ends?`).

The basic algorithm, which appears in Table 1, searches for subsequences and then iteratively connects together the dependencies, merging states where possible. The resulting diagram often has states composed of multiple events.

STDD(dataset, seq-length, threshold, loose-ends?)

1. Collect significant **sequences**:
   (a) Find sequences of length **seq-length** below **threshold** in **dataset** using DD
   (b) Prune out negative dependencies (infrequently occurring sequences)
   (c) Cluster sequences by common beginning subsequences (precursor)
   (d) Sort **clusters** by overall frequency of occurrence
2. Create initial state from precursor of most frequent cluster
3. Push initial state on **resolution-list**
4. Repeat until no states in **resolution-list**:
   (a) Get cluster (with precursor and successor subsequences) for first state in resolution-list
   (b) Add successors to the model by:
       – Link to an existing state if all the subsequences that result are consistent with all known dependencies,
       – Merge current state with successor if only one successor and the merge is consistent with dependencies,
       – Otherwise create new state as successor.
   (c) Calculate transition probabilities from counts kept with clusters,
   (d) Add new and merged states to resolution list.
5. If **loose-ends?** and not all **clusters** have been included, then:
   (a) Create a state from precursor of most frequent remaining on **clusters**
   (b) Push new state on **resolution-list**
   (c) Restart loop at 4

Table 1. STDD algorithm for constructing state transition diagrams

The rules for when to link, merge or create new states are based on the information available from DD. This is the most complex part of the algorithm because it requires generating all sequences that could result from the state and its existing predecessors and successors in the state transition diagram. Checking these potential sequences against those found in the data reduces the cases in which the diagram produces event sequences that do not actually occur.

The diagram is enhanced until all states have both incoming and outgoing links. However, at this stage, some subsequences may not be present in the transition diagram; these subsequences are not included because some of the events do not appear in both the precursor and successor portions of dependencies. If the user wishes (by setting **loose-ends?** to true), these states can be added in as dangling states: states with outgoing, but not incoming links. We display the diagram using the Dot graphing package[11].

The resulting model is not intended to completely capture the original event sequences; instead, because it incorporates only significant dependencies, it is unlikely to include spurious sequences (e.g., spurious events or noise in perception of events) and its complexity is adjustable by changing the threshold and indicating whether to include loose ends.

To show how the algorithm works, we generated data from the diagram in

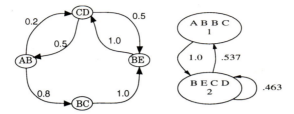

Fig. 1. An idealized five token model (on left) and transition diagram constructed from event data (on right)

the left side of Figure 1 with a small amount of noise in the data (0.05). First, we collected dependencies, as shown in Table 2, and clustered them. The cluster with the most data was $CD \rightarrow A \vee B$. On the first pass through the loop, CD was linked to two new states, A and B, with transition probabilities $\frac{87}{87+75} = .537$ and $\frac{75}{87+75} = .463$, respectively. On the second pass, B is merged with E by finding the dependency $DB \rightarrow E$ and noting that it matches the existing transition $D \rightarrow B$. On the third pass, state A matches the dependency $DA \rightarrow B$; we cannot link to state BE because we can find no $BB \rightarrow E$ dependency so state A is extended to AB. By continuing to add in states from the dependencies, the diagram on the right side of Figure 1 resulted. This diagram collapses the four states into two, making it more compact, but also removing one possible transition ($AB \rightarrow CD$), which did not appear in any dependencies.

(E C → D) (155 1 19 823)	(B E → C) (153 4 95 746)	(C D → A) (87 81 7 823)
(D A → B) (86 2 235 675)	(D B → E) (78 1 83 836)	(C D → B) (75 93 246 584)
(A B → B) (74 18 247 659)	(B B → C) (74 1 174 749)	(B C → B) (74 14 247 663)
(C B → E) (74 1 87 836)		

Table 2. Length 3 dependencies (with contingency table) collected for five event model

To illustrate the effect of adding in loose-ends, we ran the same example with more noise (0.1) and a longer seq-length (three). As Figure 2 shows, the diagram on the right includes two new sequences (e.g., [B C B] and [C B E]), but at the cost of many more false positives and a more complicated diagram.

3   Evaluation of State Transition Model Generation

We emphasize three questions for evaluation: how accurately do the resulting models summarize the datasets, what is the effect of the parameters of the algorithm on accuracy and does accuracy scale-up to realistic datasets. To answer the first two questions, we tested STDD's accuracy on synthetic data because we can control its complexity. We varied the complexity of the underlying models,

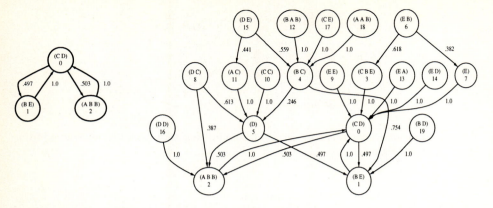

Fig. 2. Diagrams generated with `loose-ends?=nil` (left) and `loose-ends?=T` (right)

the noise introduced in the data and the number of events to determine the effect on accuracy of varying the datasets; we also varied the parameter settings of the STDD algorithm. Then, we tested the accuracy on event sequences from real systems to demonstrate that the method scales up to realistic data sets.

We did not specifically test the computational requirements of STDD. Because it is of concern to the data mining community, we report that "typical" transition diagram generation (as exemplified in Table 3) required about a minute on a SPARC Ultra in Allegro Common Lisp for each synthetic example and up to 15 minutes for the actual program traces.

3.1   Accuracy on Synthetic Data

Testing accuracy is a four step process. First, we developed seven synthetic models in which we varied the number of events (5-13), the number of states (3-10) and the maximum number of events per state (2-7); Figure 1 shows a model with five events, four states and two events per state (abbreviated as M5-4-2). Second, for each of the synthetic models, we generated event sequences of varying lengths (2500 to 70,000 events) by simulating the models and introducing noise (events randomly selected from the set with probabilities of 0.05, 0.1, 0.15, 0.20 and 0.25). Third, we constructed state transition diagrams from the data sets using different parameter settings. Finally, we determined the accuracy of each model for each dataset by comparing the subsequences of varying lengths (2-7) that can be produced by the resulting diagrams to data produced without noise.

We measure accuracy as hit-rate and false positives. Hits are the number of subsequences that both appear in the test data and can be produced by the diagram; Hit-Rate is percentage of hits in the datasets. False positives are the number of distinct subsequences that can be produced with the diagram that do not appear in the data.

	5-4-2	5-4-4	5-10-2	6-6-3	8-4-2	13-3-7	13-10-2	Mean
Ideal Hit-Rate	.75	.58	.16	.89	.91	1.0	.80	.73
len 7 False Pos	3288	3746	4993	407	0	306	75	1831
Ideal Hit-Rate	1.0	1.0	1.0	1.0	1.0	1.0	1.0	1.0
len 2 False Pos	567	0	1254	61	2997	1111	440	918

Table 3. Accuracy of typical transition diagram generation on synthetic models; *len* means the length of sequences checked for accuracy.

Effect of Dataset on Accuracy Table 3 shows typical measures for the seven synthetic models; we report the results for seq-length of four, threshold of .05, loose-ends? off, trace length of 10,000 events and noise of 0.1 as the amount of data and noise level seemed representative of likely data and these parameters appeared to give the best trade-off of hit-rate to false positives overall.

Table 4 summarizes the accuracy across all datasets and parameter settings. The results indicate that most diagrams appear to be capturing most of the dynamics represented by the event sequences. The resulting diagrams have high hit-rates and low false positive rates under most of the tested conditions. A closer examination of the data showed that the measures are best for short sequences (seq-length of 2) with performance degrading as the sequence length increases. Models with fewer types of events tend to be more susceptible to the degradation.

	5-4-2	5-4-4	5-10-2	6-6-3	8-4-2	13-3-7	13-10-2	Median
Hit-Rate								
Best	1.0	1.0	1.0	1.0	1.0	1.0	1.0	1.0
Median	.72	.47	.55	.90	1.0	1.0	.86	.86
Worst	0	.02	0	0	.18	.50	.10	.02
False-Pos								
Best	0	0	0	0	0	0	0	0
Median	998	800	3,484	464	4	0	58	464
Worst	31,482	49,989	69,998	49,996	49,997	1,012	43,342	49,989

Table 4. Accuracy of STDD on synthetic models across all trials

The results also suggest variability in accuracy on different types of models. To determine whether the variability was significant, we ran one way ANOVAs with a model characteristic as the dependent variable and an accuracy measure (hit-rate or false positives) as the dependent. We found that accuracy depends on the model characteristics ($P < .002$ in all cases). More events, less states and more events per state tend to result in more accurate diagrams; this result is not surprising because the extra constraints of more events and more events per state should help separate the true model from the noise.

One-way ANOVAs on noise levels for each model showed a significant effect of noise on accuracy ($P < .0003$) across all trials; examination of the data shows that some parameter settings are more sensitive to the noise than others. For example, the typical parameter settings mentioned earlier were shown to be relatively insensitive to noise (see Figure 3), while longer seq-lengths were more so. We expected this because fewer examples of longer sequences will be found.

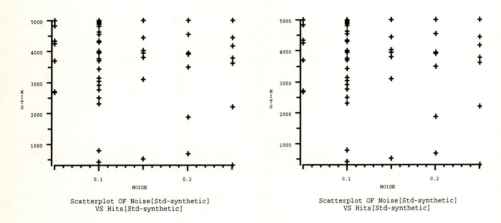

Scatterplot OF Noise[Std-synthetic]
VS Hits[Std-synthetic]

Scatterplot OF Noise[Std-synthetic]
VS Hits[Std-synthetic]

Fig. 3. Noise versus hit-rates and false positives for typical parameter settings

Effect of Algorithm Parameters on Accuracy The algorithm requires the user to make some commitments to parameter settings. We tested the sensitivity of the algorithm to the parameters: `loose-ends?`, `threshold` and `seq-length`. For loose-ends?, we ran t-tests comparing the accuracy of adding the loose ends to not including them for each model for each of the two measures. We found a significant difference ($P < .0001$) for hit-rate on every model; however, for false positives, we found no significant effect for four of the models ($P < .36$) and significant effects ($P < .001$) for models 8-4-2, 6-6-3 and 5-10-2. In general, adding in loose-ends significantly improves the hit-rate while sometimes significantly degrading the false positive rate. As expected, the models are significantly larger when loose ends are included.

Because seq-length and threshold are related through the DD algorithm, we tested the effect of these parameters using two two-way ANOVAs (one for each measure) for each model. In all cases, we found a significant main effect of seq-length ($P < .0001$). In five models, we found a significant main effect of threshold ($P < .01$) on hit-rate; in four models, we found a significant main effect of threshold on false positives ($P < .01$). Only two models exhibited any significant interaction effect (models 5-4-2 and 6-6-3), which suggests that we can set these parameters independently. As expected, the size of the model increases as the threshold becomes more lax.

Both seq-length and threshold affect accuracy. However, the effect is model dependent and is not monotonic. Each model has an ideal parameter setting to minimize false positives and maximize hit-rate. From a scatterplot of the hit-rate versus false positive rate for two models (Figure 4), we see that both models differ significantly on the trade-offs between the two measures and that for the model on the right it is easier to obtain both high hit-rate and low false positives.

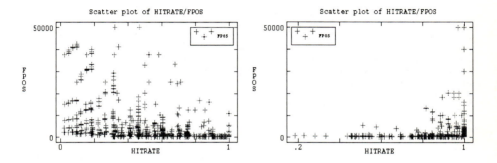

Fig. 4. Scatterplots of false positives and hit-rates for two models: m5-4-4 on left and m8-4-2 on right

3.2 Modeling RARS Controller and Phoenix Planner

We had previously found snapshot dependencies useful for debugging. We have started to build a multi-level, reinforcement learning controller for an agent in a simulated environment (RARS). In this section, we give an example of how we are using transition diagram generation to debug RARS and an assessment of the accuracy of STDD on actual program traces.

Debugging the RARS Controller The RARS (Robot Automobile Racing Simulator) controller must regulate the acceleration and steering of a race car on simulated tracks, which requires that it negotiate the track, avoid crashing into the walls, pass cars and go in for pit stops [15]. We would like to the controller to avoid failures: when the controller loses control of the car and runs off of the track (`crashed` event). For example, Figure 5 shows a fragment of a transition diagram[1] generated from event data from RARS. At this point, the controller has been partially trained to negotiate the track, but is still crashing occasionally. The events are discretized sensor readings: track position (P) and speed relative to a wall (X).

---

[1] The full diagram had 24 states and so would not be readable in this format. We removed low probability transitions ($< .007$) from the diagram to make it fit.

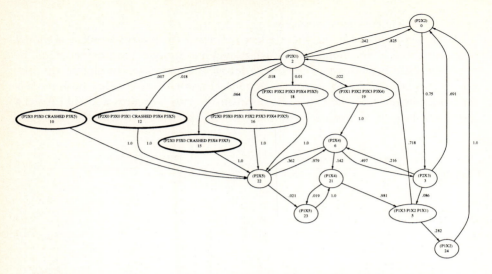

Fig. 5. Portion of transition diagram for RARS reinforcement learning controller

From Figure 5, we identified several problems. First, we observe bottlenecks immediately before and after all states containing the crashed event (highlighted states). These bottlenecks are critical events indicating cases in which training and reinforcement should be focused. Second, the network may have been over-trained for the straightaway, which constitutes roughly 80% of the track. The transition from the bottleneck state away from states containing "crashed" occurs 82% of the time. Thus, we need to tune the reinforcement schedule to prevent overfitting to the common, straight part of the track.

Accuracy on Real Data: RARS and Phoenix We assessed the accuracy of STDD on real data from two systems by comparing the transition diagrams produced from half of the data to the remaining data. Thus, we are testing to see whether a more compact overview model can adequately summarize what was observed.

We generated transition diagrams using two seq-lengths (3 and 4) and two thresholds (.05 and .005) for the two datasets. Table 5 summarizes the results for subsequences of length two and seven. As with the synthetic, the diagrams accurately capture the short sequences. Performance degrades more slowly on RARS than on Phoenix with longer sequences. In fact, as with the synthetic data, the more robust performance was achieved with more events (30 for RARS as opposed to 17 for Phoenix). Finally, the results showed little difference between the diagrams generated with these different parameter settings.

4   Extensions and Conclusions

Given the current performance of the algorithm, we view two issues as paramount for future work: improving the false positive rates through better filtering and

	Phoenix				RARS			
	3		4		3		4	
	.05	.005	.05	.005	.05	.005	.05	.005
Hit-rate	.81	.73	.95	.83	.98	.97	.98	1.0
len 2 False pos	0	0	0	2	16	0	4	11
Hit-rate	.34	.27	.18	.11	.70	.67	.80	.68
len 7 False pos	1944	37	1459	1430	2653	2887	913	1482

Table 5. Accuracy results for Phoenix and RARS event sequences

automatically setting the parameters in the algorithm. At present, we have two mechanisms for filtering noise: reducing the threshold and pruning negative dependencies. We have tested the effect of threshold and found it to be model dependent, further supporting the need to better set the algorithm parameters. As our next study, we intend to focus on testing alternative strategies for pruning negative dependencies.

The algorithm requires three parameters: loose-ends?, seq-length and threshold. Loose-ends? is best set by the user who knows what is most critical to the diagram generation (hit-rate or simplicity). As to the other two, we are currently exploring principled methods of setting these. One option is to use CHAID based analysis to determine seq-length. CHAID constructs an n-step transition matrix, which indicates what earlier point in the event sequences was most predictive of the occurrence of each event[5]. Unfortunately, integrating CHAID based analysis requires significant modification to the underlying code to accommodate multiple dependency lengths simultaneously, and so is yet to be completed.

The significance threshold (probability for the contingency table test) is the primary mechanism for adjusting the complexity of the daigram. Lower probabilities tend to result in less complex diagrams because fewer transitions are recognized as significant. One option for determining how to set the probability threshold is to model its effect. We hope that further analysis of the results presented here will uncover a model of how to set the threshold.

We have proposed a new algorithm for automatically generating state transition diagrams from discrete event sequences. The accuracy of the diagrams generated from this algorithm is quite high for both the synthetic and real data that we have tested and shows gradual degradation as datasets become more difficult to characterize (decreasing dataset sizes with increasing noise levels). Our algorithm extends existing representational options because it does not require that the underlying process be Markov or that the resulting model be acyclic. Finally, ultimately, the utility of the algorithm rests in the utility of the models generated. We view these models as concise descriptions of cyclic processes. We have used the models to help debug the RARS reinforcement learning control system and have started using them to characterize protocols of human programmers and as an internal representation of behavior for a multi-layer simulated robot learning system.

References

1. Marc Abrams, Alan Batongbacal, Randy Ribler, and Devendra Vazirani. CHI-TRA94: A tool to dynamically characterize ensembles of traces for input data modeling and output analysis. Department of Computer Science 94-21, Virginia Polytechnical Institute and State University, June 1994.

2. Rakesh Agrawal, Manish Mehta, John Shafer, and Ramakrishnan Srikant. The Quest data mining system. In *Proceedings of the Second International Conferrence on Knowledge Discovery and Data Mining*, Portland, OR, Aug. 1996.

3. Rakesh Agrawal and Ramakrishnan Srikant. Mining sequential patterns. In *Proceedings of the Int'l Conference on Data Engineering (ICDE)*, Taipei, Taiwan, March 1995.

4. Wray Buntine. Graphical models for discovering knowledge. In U. Fayyad, G. Piatetsky-Shapiro, P. Smyth, and R. Uthurusamy, editors, *Advances in Knowledge Discovery and Data Mining*. AAAI Press, Menlo Park, CA, 1996.

5. Horacio T. Cadiz. The development of a CHAID-based model for CHITRA93. Computer science dept., Virginia Polytechnic Institute, February 1994.

6. Usama Fayyad, Gregory Piatetsky-Shapiro, and Padhraic Smyth. From data mining to knowledge discovery in databases. *AI Magazine*, 17(3):37–54, Fall 1996.

7. Adele E. Howe. Detecting imperfect patterns in event streams using local search. In D. Fisher and H. Lenz, editors, *Learning from Data: Artificial Intelligence and Statistics V*. Springer-Verlag, 1996.

8. Adele E. Howe and Paul R. Cohen. Detecting and explaining dependencies in execution traces. In P. Cheeseman and R.W. Oldford, editors, *Selecting Models from Data; Artificial Intelligence and Statistics IV*, volume 89 of *Lecture Notes in Statistics*, chapter 8, pages 71–78. Springer-Verlag, NY,NY, 1994.

9. Adele E. Howe and Paul R. Cohen. Understanding planner behavior. *Artificial Intelligence*, 76(1-2):125–166, 1995.

10. Adele E. Howe and Larry D. Pyeatt. Constructing transition models of AI planner behavior. In *Proceedings of the 11th Knowledge-Based Software Engineering Conference*, September 1996.

11. Eleftherios Koutsofios and Stephen C. North. *Drawing graphs with dot.* AT&T Bell Laboratories, Murray Hill, NJ, October 1993.

12. Heikki Mannila, Hannu Toivonen, and A. Inkeri Verkamo. Discovering frequent episodes in sequences. In *Proceedings of the International Conference on Knowledge Discovery in Databases and Data Mining (KDD-95)*, Montreal, CA, Aug. 1995.

13. Tim Oates and Paul R. Cohen. Searching for planning operators with context-dependenct and probabilistic effects. In *Proceedings of the Thirteenth National Conference on Artificial Intelligence*, 1996.

14. Judea Pearl. *Probabilistic Reasoning in Intelligent Systems: Networks of Plausible Inference.* Morgan Kaufmann Publishers, Inc., Palo Alto, CA, 1988.

15. Larry D. Pyeatt, Adele E. Howe, and Charles W. Anderson. Learning coordinated behaviors for control of a simulated robot. Technical report, Computer Science Dept, Colorado State University, 1996.

16. Raghavan Srinivasan and Adele E. Howe. Comparison of methods for improving search efficiency in a partial-order planner. In *Proceedings of the 14th International Joint Conference on Artificial Intelligence*, Montreal, CA, August 1995.

Detecting and Describing Patterns in Time–Varying Data Using Wavelets

Sarah Boyd

MRI, Macquarie University
Sydney, Australia 2109
sarahb@mpce.mq.edu.au

Abstract. Reasoning effectively about time-varying data requires so-phisticated pattern detection mechanisms. This paper describes techniques developed for detecting patterns in time-varying data with the ultimate aim of generating textual descriptions of the data. Preliminary experiments are described in which the visually significant features in weather data are extracted and compared against hand-written expert descriptions.

1   Introduction

With the advent of the information explosion huge amounts of data are now available online and much of this data is time-varying in nature; e.g stockmarket prices, patient medical records and meteorological measurements. The time dimension of the data means that besides the standard measures of max, min and mean there are significant patterns and trends observable. An important step in reasoning about time-varying data is being able to detect basic patterns in the data; e.g. as Berndt and Clifford (1996) point out, we must be able to detect the pattern "rising interest rates". Humans have no trouble detecting such patterns but programming computers to do this is difficult. Much research is being undertaken in this area in varied domains such as finance (Berndt and Clifford 1996), process control (Bakshi and Stephanopoulos 1995, Konstantinov and Yoshida 1992), medicine (Haimowitz et al. 1995) and meteorology (Bridges et al. 1995).

Detecting the patterns is sometimes the end goal of the systems, as in Haimowitz et al. (1995), or it may be viewed as the first step towards "mining" the data further, as in Berndt and Clifford (1996). It may be viewed as a way to encode more complex rules in an expert system as in Konstantinov and Yoshida (1992) and Bridges et al. (1995) or alternatively, Bakshi and Stephanopoulos (1995) incorporate pattern detection as part of a more general framework for modelling temporal trends. The research described in this paper examines pattern detection in time-varying data with the aim of generating textual descriptions of the data. It aims to identify and describe visually significant features of time-varying data beyond the standard measures of max, min and mean. The

X. Liu, P. Cohen, M. Berthold (Eds.): "Advances in Intelligent Data Analysis" (IDA–97)
LNCS 1280, pp. 585–596, 1997.   © Springer–Verlag Berlin Heidelberg 1997

intuition is that those features which are visually significant are also those that would be included in a textual description of the data. For example, in the graph of temperature shown in Fig. 1, as well as describing the max, min and mean of this data we would be interested in commenting on the visually noticeable temperature drops from the 2nd to the 5th, the 13th to the 16th and the 25th to the 28th. Automatically generating textual descriptions of time-varying data is

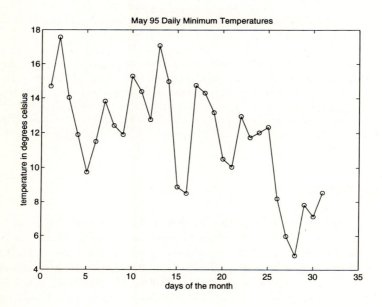

Fig. 1. Graph of May 95 daily minimum temperature

extremely useful when there are large amounts of data involved and descriptions that incorporate trends and patterns in the data should be especially useful. For example, in technical analysis of the stockmarket there are several agreed upon patterns or shapes that indicate possible future movements of the index or underlying stock such as the two-peaked "double top" which may indicate a reversal in the direction of a major upward trend in the market[1] Identifying and describing these patterns in a textual summary would be very useful for technical analysts.

Sophisticated analysis is required to detect patterns or trends in time-varying data and there are three main reasons for this. First, there is a need for an in-exact match; the types of shapes people are interested in are usually vague and hard to specify mathematically. Second, signals are usually noisy and denoising

---

[1] See Murphy (1986) for a full treatment of technical analysis

is a non-trivial task. Third, the patterns or trends may be present at different levels of resolution; e.g either monthly or fortnightly trends. The first two issues of vagueness and smoothing have been addressed using dynamic programming (Berndt and Clifford 1996), fuzzy logic (Bridges et al. 1995) and polynomial approximation (Konstantinov and Yoshida 1992, Haimowitz et al. 1995). However, only Bakshi and Stephanopoulos (1995) have addressed multiscale pattern identification. They applied the signal processing techniques of wavelets and scale-space theory which are easily and efficiently applied to large data sets and can identify patterns at different scales. Wavelets also provide excellent smoothing capabilities and are good at matching vague patterns.

This research provides a framework for identifying patterns in time-varying data and generating textual descriptions applicable to large data sets by applying the techniques of wavelets and scale space theory. Preliminary results performed on monthly weather data are presented. Dramatic changes in monthly temperature data were automatically detected and compared to the dramatic changes included in an expert's written descriptions of the same monthly temperature graphs.

2   Wavelets and Scale Space Theory

2.1   Wavelets and the Fourier Transform

Wavelets are a signal processing tool originating in mathematics which analyse a signal in terms of both time and frequency. They have been succesfully applied to a number of areas including data compression (Mallat and Zhong 1992), image processing (Daubechies 1988) and denoising of noisy data (Donohoe 1993). Wavelet analysis, or the Wavelet Transform, rather like the well known Fourier Transform, decomposes a signal into simpler elements. The Fourier Transform decomposes a signal into individual frequency components consisting of sinusoidal functions giving the relative intensities of each frequency component. However, it does not provide any information about when in time the individual frequencies occur because the trigonometric functions are not localised in time. For example, the Fourier Transform could diagnose a high frequency spike occurring in a dataset but it could not predict where in time the spike occurred. The Short Time Fourier Transform (STFT) was developed by Gabor (1946) to address this shortcoming and is equivalent to a sum of Fourier Transforms of successive windowed segments of the signal. The resulting output is a two-dimensional representation of the signal, one that varies with time and frequency. However, since the time-bandwidth product, or window area has a lower bound according to the Heisenberg uncertainty principle, a potential drawback of the STFT is that once the window width is chosen it is fixed over the entire time-frequency plane. This means that signal components can be analysed with good time resolution or frequency resolution but not both. Many signals are composed of high frequency short duration components and low frequency components of long duration and these two types of signal components cannot be detected by the STFT simultaneously (Rioul and Vetterli 1991). The Wavelet Transform overcomes this drawback by allowing varying window widths at different scales of analysis. Fig. 2 compares the fixed window structure of the STFT with the varying window structure of the Wavelet Transform.

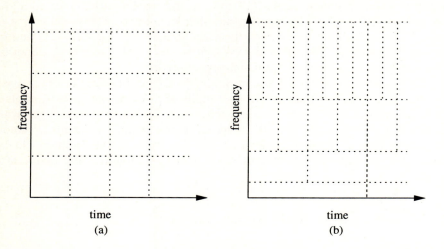

time
(a)

time
(b)

Fig. 2. Coverage of the Time-Frequency Plane. The dimensions of the resolution squares represent minimum time and frequency intervals over which signals can be differentiated: (a) STFT and (b) Wavelet Transform

2.2   The Wavelet Transform

The Wavelet Transform represents the signal as a sum of *wavelets* which are localised in time and frequency. The individual wavelets are obtained from a single prototype wavelet, which must satisfy certain mathematical properties, by the operations of translation and scaling. Each of the wavelets act as a bandpass filter, letting through a certain range of frequencies. In wavelet analysis the term *scale* is used instead of frequency. The relationship between scale a and frequency f is given by:

$$a = f_0/f$$

where f_0 is the centre frequency of the bandpass filter [2]. The local frequency $f = f_0/a$ is now related to the scaling scheme chosen. A common scaling scheme to choose is $a = 2^k$, for integer values of k. This is called *dyadic* as the scales are powers of two. An example wavelet used by Mallat and Zhong (1992) at the dyadic scales of a=4 and a=8 is shown in Fig. 3 next to their corresponding

[2] See Rioul and Vetterli 1991 for a more detailed treatment of the relationship between scale and frequency

frequency spectrum. One can observe their bandpass nature and see that at
higher frequencies, there is finer time resolution and larger frequency bandwidth
than at low frequencies. The transformed signal resulting from applying the

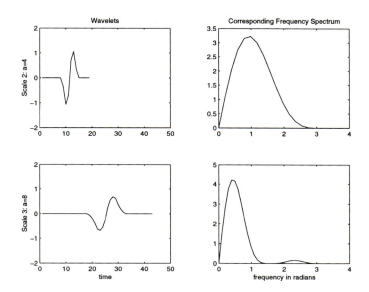

Fig. 3. Wavelets and corresponding frequency spectrum at scales a=4 and a=8

wavelet at each *scale* is often called the *detail signal* because it reflects the details
present at those frequencies in the original signal. The three detail signals for an
artificial input signal are shown in Fig. 4. Since this shows the signal at varying
scales this representation is known as a *scale-space* representation of the signal.

2.3   Wavelets as Edge Detectors

Wavelets have been shown to be effective multiscale edge detectors due to their
time-localisation property. Mallat and Zhong (1992) showed mathematically that
by choosing a particular wavelet they could detect multiple edges of varying
sharpness in the original signal by examining the extrema of the wavelet transform at different scales. This same wavelet is applied to the artificial signal in
Fig. 4 and two observations can be made. First, we can see that the extrema
of the transform match up to the edges in time at multiple scales. Second, the
nature of the transform extrema characterise the edges; e.g. the edge appearing
at time=150 is very steep and thus appears strongest at the first scale.

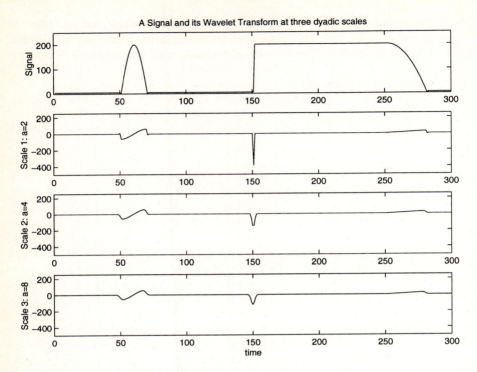

Fig. 4. A signal and its Wavelet Transform at scales a=2, a=4 and a=8. This wavelet tranform is acting as a multiscale edge-detector

2.4   Scale Space Theory

Scale space theory, first introduced by Witkin 1983, encapsulates the idea that there is a "marked correspondence" between the perceptual salience of structures in a signal and the lengths of the intervals during which those structures exist in scale space. This idea has proven empirically to have merit in experiments in extracting significant structures from real images (Lindberg and Eklundh 1991, 1993). When the Wavelet Transform is implemented as a multi-scale edge detector then the wavelet transform extrema can be used to measure the perceptual salience, called "persistence", of the edges in a signal. According to scale space theory, those edges that persist at multiple scales are the perceptually salient edges. These perceptally salient edges are usually described as dramatic changes in the data; e.g. in the weather domain a sharp edge may be described as "sudden temperature increase".

3   Natural Language Generation

The area of research concerned with automatically generating language from some underlying source such as a database or knowledge base so that the language generated sounds "natural"; i.e. as a human would write or speak, is called Natural Language Generation (NLG). There are three recognised steps in Natural Language Generation: text planning, sentence planning and sentence realization. The first of these steps determines the content and overall structure of the text. Content determination, or working out what to say, is an important part of text planning. The second step, sentence planning, assembles the information to be expressed into sentence-sized pieces while the third step realises the content elements as linguistic expressions in the natural language being generated.

There has been much work in NLG in generating textual descriptions of numerical data. The most closely related work is by Mittal et al. (1995) who generated explanatory captions to describe the structure and presentation of complex graphics. They dealt with relational rather than time-varying data, such as geographically distributed real estate figures, and focussed on identifying the perceptual complexity of the graphical presentation. The captions included information about what the graphical elements represent and the relationships between the different variables. They did not attempt to identify significant features, such as trends, in the data itself. Robin and McKeown (1996) examined generating summaries of basketball box scores; however, they focussed on linguistic conciseness of expression rather than content selection. They did not actually implement their "content preselector" and only consider a limited set of features in the data: runs and extrema. The research described here focusses on content determination when automatically generating textual descriptions of time-varying data and seeks to describe more complicated temporal features than previous work.

4   Experimental Methodology

4.1   Generating Weather Reports

This work is a subproject of a larger project which has the aim of automatically generating weather reports similar to those produced by hand each month for a monthly University staff newsletter. Weather data for the University is available online via the Automatic Weather Station [3]. Initially, twenty months of weather reports were analysed to provide motivation for automatic generation of similar reports. In the research described here it was decided to focus on one variable only, temperature - either daily maximum or daily minimum. Since temperature was not always mentioned in the monthly reports, the ten sets of temperature that were described in the monthly reports were then focussed on in more detail. In fact, for evaluation purposes and due to the sparsity of the descriptions of temperature present in the original monthly reports, a weather expert was employed to write separate descriptions of these ten sets of monthly temperature data suitable for inclusion in a weather summary appropriate for the general public.

---

[3] See http://atmos.es.mq.edu.au/aws

4.2   System Overview

Implemented so far is a system written in MATLAB[4] which extracts the four most perceptually salient edges in a time-varying signal using wavelets according to scale space theory. The system was applied to the ten sets of temperature data chosen as described above. Ideally, the system would decide how many edges are actually salient for a particular dataset; indeed in some of the expert descriptions no edges were mentioned. However, it seemed reasonable to start with a system that chose the four most salient edges in each dataset. Input to the system is the time-varying signal while the output of the system is a data structure describing the four most perceptually salient edges in the input signal; adjacent edges are grouped together in the output data structure as a peak or valley. The identification of peaks and valleys is seen as the first step in being able to detect more complex patterns such as those used in stockmarket technical analysis. These patterns are represented as data structures which can then be used to generate textual descriptions.

4.3   System Operation

The multiscale edge detector wavelet described in Mallat and Zhong (1992) is applied to the input data and the detailed signals are obtained at three scales, a=2, a=4, a=8, corresponding to a resolution of half-weekly, weekly and fortnightly. An example of the detail signals corresponding to the data in Fig. 1 can be seen in Fig. 5. The extrema values of the detail signals now correspond to the prominent edges at each scale. According to scale space theory, the edges that persist at multiple scales are the perceptually salient edges. Thus the next step is an algorithm which examines the extrema values of the detail signals and identifies the locations of the four edges that persist at multiple scales. This algorithm returns a single time point. The start and the end of each edge in the original signal are then identified by searching for a monotonic increase and decrease around the edge location. Once the full edge is identified, its magnitude, gradient and sign is also recorded. Significant peaks and valleys are identified by searching for adjacent edges from the four prominent edges. For description purposes, each edge is characterised by five attributes: the *start* and *end* of the edge, the *sign* (which indicates a rise or fall), the *gradient* (which is a measure of the slope) and the *magnitude* (which is the difference between the first and last value).

4.4   Results and Evaluation

In the example given in Fig. 1, four edges were identified automatically as being visually significant edges: the combined fall and rise beginning on the 13th and ending on the 17th and the two temperature drops from the 2nd to the 5th

---

[4] MATLAB is a powerful mathematical package originally developed for matrix calculation which now has the capability to do perform most kinds of data analysis.

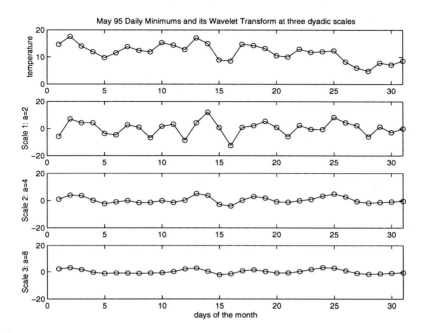

Fig. 5. May 95 Minima and its Wavelet Transform at scales a=2, a=4 and a=8.

Table 1. Four most salient edges and their attributes extracted from May 95 Daily Minima

Edge	Start	End	Sign	Gradient	Magnitude
	13	16	-1	3	9
	16	17	+1	6	6
	2	5	-1	3	8
	25	28	-1	3	8

and from the 25th to the 28th. The attributes of the four edges identified by the algorithm are presented in Table 1. The first two edges were identified as forming a valley by the algorithm and grouped together.

The corresponding expert description matching the temperature data in Fig. 1 and the data structures above is:

> In the first half of the month there was a sharp decrease in temperature followed by a steady recovery and then another sharp decrease. The second half of the month saw a sudden temperature increase followed by a general downward trend. Over the whole month, there was a general decrease in temperature with the month ending 7 degrees cooler.

In order to evaluate the success of the edges extracted automatically it was necessary to study the handwritten descriptions to categorise the features described in the data. The features described by the expert can be broken up into two mutually exclusive categories: *monotonic change* and *non-monotonic change.* We define *monotonic change* as an edge which is made up of each consecutive point being greater than or equal to (or less than or equal to) the previous point, such as the "sharp decrease in temperature" occurring between the 2nd and the 5th in Fig. 1 or the "sudden temperature increase" occurring between the 16th and 17th. We define *non-monotonic change* as any description of change that is not monotonic. We observed that there were different types of non-monotonic change, some easier to identify in the data than others. For example, we identified a simple type of non-monotonic change which we called *two-point change*, which is defined as a difference between two points such as "the month ending 7 degrees cooler" in Fig. 1. More interesting are the harder to identify types of non-monotonic change which require some sort of smoothing, such as wavelets would provide; e.g. the "steady recovery" identified as taking place after the sharp decrease of the 2nd to the 5th. On the graph it can be identified as starting on the 5th and finishing on the 13th May. Although easy for a human to identify it is quite difficult for a computer. Another example of a harder non-monotonic change also commented on in Fig. 1 is the "general downward trend" following the sudden temperature increase, starting on the 17th and extending to the end of the month.

The total number of each type of change was determined by comparing the expert descriptions with the underlying data sets. Each textual phrase describing change was compared to the appropriate temperature graph and then checked to see whether it corresponded to an edge (monotonic change), or included more than one edge (non-monotonic change). Those identified as non-monotonic were then checked to see if they described a difference between two points (two-point change). The results are presented in Table 2. The edges automatically extracted by the system were compared to the number of descriptions of monotonic change mentioned by the expert and the results are presented in Table 3. Only one of the monotonic edges mentioned by the expert was not detected by the system. While these results are promising, it is important to remember that these are only a subset of the changes detected by the system; i.e. the system overgenerated. The scale space algorithm did provide a ranking of the four edges chosen. However, where the expert only mentioned say two of the edges these did not always correspond to the top two ranked edges detected automatically.

5   Conclusions and Future Work

We have demonstrated that, using wavelets and scale space theory, we can detect a subset of visually significant features that humans describe in time-varying data. The ultimate aim of this research is to be able to generate textual descriptions of time-varying data that include all the interesting patterns and trends in the data at different scales of analysis. Efforts so far have only laid the ground

Table 2. Classification of Change Mentioned by Expert

Monthly Description	Monotonic Change	Non-Monotonic Change	
		Two-Point Change	Other
September 95 Daily Min	1	1	2
May 95 Daily Min	3	1	3
May 95 Daily Max	2	1	2
July 95 Daily Max	0	1	1
August 95 Daily Max	1	1	2
November 95 Daily Max	1	1	1
March 96 Daily Max	0	1	1
April 96 Daily Min	4	1	1
April 96 Daily Max	4	1	1
June 96 Daily Min	2	1	2
Total	18	10	16

Table 3. Comparison of Expert and System

Temperature Data	Monotonic Changes Mentioned by Expert	Monotonic Changes Also Detected Automatically
September 95 Daily Min	1	1
May 95 Daily Min	3	3
May 95 Daily Max	2	2
July 95 Daily Max	0	0
August 95 Daily Max	1	1
November 95 Daily Max	1	1
March 96 Daily Max	0	0
April 96 Daily Min	4	4
April 96 Daily Max	4	3
June 96 Daily Min	2	2
Total	18	17

work. An important next step is being able to recognise non-monotonic change and to identify at which scale it exists; it is planned to use the same techiques of wavelets and scale space theory to achieve this. Future work involves identifying more complex patterns in the data such as those shapes used by technical analysts in finance (Murphy 1986).

In order to generate real text, attention will be paid to the sentence planning and text realisation components of the system. To inform this process, a newspaper corpus analysis is underway to identify the types of things described in time-varying data and their linguistic forms.

6 Acknowledgements

Financial support is gratefully acknowledged from the Microsoft Research Institute and the Australian Government.

References

Bakshi, B. R. and G. Stephanopoulos [1995] Reasoning in Time. *Advances in Chemical Engineering*, Vol. 22, pages 485–548. Academic Press, New York.

Berndt, D. J. and J. Clifford [1996] Finding Patterns in Time Series: A Dynamic Programming Approach. *Advances in Knowledge Discovery 1996*, AAAI MIT Press.

Bridges, S. M., C. Higginbotham, J. M. McKinion and J. E. Hodges [1995] Fuzzy Descriptors of Time-Varying Data: Theory and Application. *AI Applications*, Vol. 9, No. 2, 1995.

Bruce, A., D. Donoho, H-Y. Gao [1996] Wavelet Analysis. *IEEE Spectrum*, October 1996, pages 26–35.

I. Daubechies [1988] Orthonormal Bases of Compactly Supported Wavelets. *Communications in Pure and Applied Mathematics*, Vol. 41, no. 7.

Donohoe D. [1993] Nonlinear Wavelet Methods for Recovery of Signals, Densities and Spectra from indirect and Noisy Data In I. Daubechies ed. *Different Perspectives on Wavelets, Proceeding of Symposia in Applied Mathematics* Vol. 47, American Mathematical Society, Providence Rhode Island, pages 173–205.

Gabor D. [1946] Theory of Communication. *Journal of the IEE.*, Vol. 93, pages 429-457.

Haimowitz, I. J. P. Phuc Le, I. S. Kohane [1995] Clinical Monitoring using regression-based trend templates. *Artificial Intelligence in Medicine*, Vol. 7, pages 473–496.

Konstantinov, K. B. and T. Yoshida [1992] Real-Time Qualitative Analysis of the Temporal Shapes of (Bio)process Variables. *AIChe Journal*, Vol. 38, No. 11, pages 1703–1715.

Lindeberg T. [1993] Effective Scale: A Natural Unit for Measuring Scale-Space Lifetime. *IEEE Transactions in Pattern Analysis and Machine Intelligence*, Vol. 15 No. 10, pages 1068–1074.

Lindeberg T. P. and J. O. Eklundh [1991] On the Computation of a Scale Space Primal Sketch. *Journal of Visual Commun. Image Repr.*, Vol. 2, No. 1, pages 55-78.

Lindeberg T.P. and J.O. Eklundh [1993] The Scale-Space Primal Sketch: Construction and Experiments. *Image Vision Computing*, Vol. 10, pages 3–18.

Mallat, S. and S. Zhong [1992] Characterization of Signals from Multiscale Edges. *IEEE Transactions on Pattern Analysis and Machine Intelligence*, Vol. 14, no. 7.

Maybury, M.T. [1995] Generating Summaries From Event Data. *Information Processing and Management*, Vol. 31 No. 5 pages 735–751.

Murphy, J.J. [1986] *Technical analysis of the futures markets: a comprehensive guide to trading methods and applications* New York, N.Y. : New York Institute of Finance.

Rioul O. and M. Vetterli [1991] Wavelets and Signal Processing. *IEEE Signal Processing Magazine*, October 1991, pages 14–38.

Witkin, A. P. [1983] Scale Space Filtering. In *Proceedings Eight International Joint Conference on Artificial Intelligence*.

Diagnosis of Tank Ballast Systems

Björn Schieffer and Günter Hotz

Fachbereich 14 – Informatik, Universität des Saarlandes
Postfach 15 11 50, 66041 Saarbrücken, FRG

URL: http://www-hotz.cs.uni-sb.de/schieffer/
email: schieffer@cs.uni-sb.de

Abstract. The paper deals with the diagnosis problem of hybrid systems. A new two-level approach for that problem is introduced and discussed with the domain example of tank ballast systems. The first level determines the possible defects while the second one calculates their real valued degree. The new approach is shown to be very powerful by 3000 randomly generated single and double faults. In fact, for all of these defects the approach is able to compute the correct diagnoses.

1 Introduction

The control unit of a hybrid system and its environment consisting of mechanical and hydrodynamical components influence each other in a closed loop via sensors and actuators. Such a situation may be found in modern car management units, air planes or manufacturing plants.

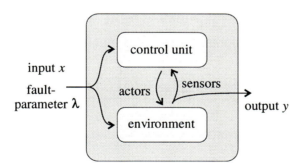

Fig. 1. Scheme of a hybrid system

According to figure 1, the behavior of a hybrid system depends on an input $x \in X$ of the input space X and a fault parameter $\lambda \in C$ of the configuration

X. Liu, P. Cohen, M. Berthold (Eds.): "Advances in Intelligent Data Analysis" (IDA–97)
LNCS 1280, pp. 597–608, 1997. © Springer–Verlag Berlin Heidelberg 1997

space C. The output $y = (y_1, y_2, \ldots, y_r) \in Y$ of the output space Y is measured by the r sensors of the system.

The problem to determine the output $y = sim(x, \lambda)$ with known input x and fault parameter λ is called *simulation*. Reality performs some kind of simulation, it *solves* the underlying differential equations when the system is running. If one is able to control the arising numerical difficultys, he can imitate this natural simulation. Much harder is the problem of *diagnosis* which is an inverse problem to simulation. To compute the diagnoses one has to determine the possible fault parameters λ when the input x and the output y are known. If sim is not injective, a diagnosis does not have to be unique and more than one fault parameters may explain the behavior of the system. Therefore, a *complete diagnosis* is given by

$$diag: \; X \times Y \to \mathcal{P}(C)$$

Much interest has been spent on inverse problems in the linear case [CW84, Ise84, Mas86, EM94]. But the closed loop of the control unit with its environment leads to a non-linear behavior in the case of hybrid systems. Because there exist only domain specific solutions in the case of non-linear inverse problems, we restrict to the diagnosis of tank ballast systems as introduced in [DBMB93].

In [Dav84] it is shown how to obtain a diagnosis of single faults in the case of a discrete configuration space C as for integrated circuits. A generalization to multiple faults is given in [Rei87], improved in [dKMR92] and put into practice in [MNTMQ96]. In [dKW87] and [FS92] strategies to select the sensor positions are introduced. The need of a quantitative system model is shown in [DK89].

The diagnosis of tank ballast systems leads to a continuous configuration space C. This is caused by the fact, that one is not only interested in determining *which* defects are present, but also wants to know the *degree* of these defects in order to decide whether the current operation has to be stopped or not. A defect of a system may occur suddenly or in a smooth way. The determination of the degree of a fault is also necessary to detect smooth defects early.

In this paper, we present a two level approach for the diagnosis problem. At the first level the defects that may cause the measured behavior are determined. In the second level the degrees of these defects are calculated.

Not only single values as a snapshot of the system are considered to perform diagnosis, but functions over a space of time that describe the behavior. Therefore, all available information is used.

The methods are based on two properties of the domain. As the first property, different degrees of the same defect result in the same effects. This is used to combine system configurations of different degrees of the same defect into one fault candidate. The second property is the monotony of the system behavior – with increasing degree of a fault its effect increases too. This property is used to determine the degree with a binary search. In the domain of tank ballast systems these two propertys are not always but generally fulfilled. We will discuss how to treat violations. This may exponentially increase the runtime of the methods. We will show how to slow down the exponential growth with the help of penalty functions.

In section 2, the domain of tank ballast systems is introduced and a clear example is extracted. In section 3, the approach is sketched with some idealizations. In section 4, we discuss the problems if these idealizations do not hold. In the last section, we give some experimental results to demonstrate the applicability of the methods.

Fig. 2. The Brent Spar

2 Domain

Tank ballast systems are used in huge ships or swimming platforms as the *brent spar* shown in figure 2 or the *micopery 7000* which can crane up loads of a weight up to 14 000 tons. To keep the balance it contains 57 ballast tanks that can be filled by sea water with the aid of two strong pumps. It is also possible to change the center of gravity by pumping water from some ballast tanks to others. These operations are directed by an electronical control station. Defects in the ballast tank system may lead to heavy trouble including the sinking of the platform. Therefore, automatic diagnosis tools are needed to assist the control officer.

Tank ballast systems consists of very different components such as tanks, pipes, valves, filters, pumps, vent pipes or float switches. Some typical defects result in a plugging of pipes or filters. Others cause leaks in the pipes, leaking valves or some air in a pump. We want to determine not only which of these defects may be present in the system but the degree of the defect as it may be necessary to know when the control officer has to decide if the current operation has to be stopped or may be continued.

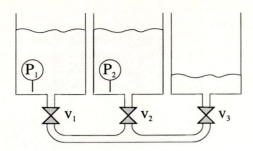

Fig. 3. Running Example

We extract a clear example of that domain. According to figure 3, it contains three tanks that are connected via some pipes and valves. The pressure inside of two of the tanks is measured by two pressure sensors.

An input $x \in X$ consists of the initial water height in the tanks and the positions of the three valves. Therefore, $X := \mathbb{R}^3 \times \{open, closed\}^3$.

Let the possible defects of the system be arbitrary shifts of the valve positions. Therefore, the number n of possible defects is $n = 3$ and $C := [0,1] \times [0,1] \times [0,1]$. The configuration $(0,0,0) \in C$ corresponds to the nominal value when there is no fault in the system.

We observe the two sensors P_1 and P_2 over the interval $[t_{start}, t_{end}]$ of time. Therefore, the sensors result in $y_1, y_2 \in M := \{f : [t_{start}, t_{end}] \to \mathbb{R}\}$ and the output space is $Y := M^2$.

We assume to have a simulator $sim : X \times C \to Y$ that computes the progress of the pressure corresponding to an input x and a fault parameter λ. We denote $(y_1^{x,\lambda}, y_2^{x,\lambda}) := sim(x, \lambda)$, where $y_1^{x,\lambda}, y_2^{x,\lambda} : [t_{start}, t_{end}] \to \mathbb{R}$.

Our task is to determine which valve positions are shifted and to compute the degree of these shifts. We assume that the input $x \in X$ and a measurement $\hat{y} = (\hat{y}_1, \hat{y}_2) \in Y$ of the sensors are known. That means, we try to decide whether $\lambda_i \neq 0$ and we try to compute the value of λ_i for $i \in \{1, 2, 3\}$ and a $\lambda = (\lambda_1, \lambda_2, \lambda_3) \in C$ with

$$(y_1^{x,(\lambda_1, \lambda_2, \lambda_3)}, y_2^{x,(\lambda_1, \lambda_2, \lambda_3)}) = (\hat{y}_1, \hat{y}_2) \tag{1}$$

We restrict to one single input x corresponding to figure 3 and therefore define $x := (1, 1, 0.4, open, open, open)$. Let the product of the liquid density and the gravitational constant be 10000 Pa per meter. Then, in the fault free case the pressure of both sensors goes down from 10000 Pa to 8000 Pa. That is the pressure in the situation that all three tanks have the same water height. According to figure 4, the measured behavior $\hat{y} := (\hat{y}_1, \hat{y}_2)$ differs from the nominal value $(y_1^{x,(0,0,0)}, y_2^{x,(0,0,0)})$. We are looking for a diagnosis to explain this difference.

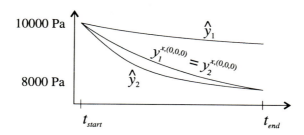

Fig. 4. Nominal value $\left(y_1^{x,(0,0,0)}, y_2^{x,(0,0,0)}\right)$ and measured behavior $\hat{y} = (\hat{y}_1, \hat{y}_2)$

3   Idealized Diagnosis

In this section, we present an approach for the diagnosis problem assuming some idealizations. In the next section, we will discuss how to use it to compute a real diagnosis.

A fault candidate $fc \in Cand := \mathcal{P}(\{1, 2, \ldots, n\})$ is a selection out of the n possible defects. As a simplification one may be interested only in single or double faults. The set of fault candidates is then reduced to $Cand_1 := \{1, 2, \ldots, n\}$ or $Cand_2 := \{\{a, b\} \mid 1 \leq a < b \leq n\}$.

Our interest lies in the set of fault candidates that may explain the measurements as well as in the real valued degree of the defects of these fault candidates. This suggests a two level approach. At the first level, we try to eliminate as many fault candidates as possible with qualitative arguments. In the second level we either determine the degrees of the remaining faults or further eliminate them with quantitative methods.

3.1   First Level

At the first level, we use qualitative attributes over the behavior of the system, that

– are hopefully invariable against the degree of the faults,
– but vary for different fault candidates.

Such attributes may be the sign, the monotony or the curvature of a measurement $m \in M$ or the comparison of the *size* of two measurements $m_1, m_2 \in M$. The size $\|m\|$ of a measurement $m \in M$ is defined by

$$\|m\| := \int_{t_{start}}^{t_{end}} |m| \, dt \tag{2}$$

This implies a partial ordering of M by $m_1 < m_2 \; :\Longleftrightarrow \; \|m_1\| < \|m_2\|$.

A *spot* denotes a sequence of such attributes and a *profile* denotes the evaluation of a spot. Now, we are able to describe the qualitative fault reduction:

1. Choose a spot sp.
2. Determine the profile of each fault candidate with respect to sp.
3. Determine the profile $\hat{p}$ of the measurement $\hat{y}$.
4. Eliminate all fault candidates with a profile unequal to $\hat{p}$.

To give an example, we consider the system introduced in the last section. To keep it simple, we restrict to the fault candidates $\{fc_1, fc_2, fc_3\}$ of the three single faults.

1. Let $Comp(m_1, m_2)$ denote the comparison of the size of the measurements m_1 and m_2. Then, we choose the spot sp by

$$sp := \big(Comp(\hat{y}_1, y_1^{x,(0,0,0)}); \; Comp(\hat{y}_2, y_2^{x,(0,0,0)})\big)$$

2. fc_1: If valve v_1 is shifted, there is a bigger resistance for the water. Therefore, the flow out of the first tank is lower and the measurement of P_1 is greater than expected. As a further consequence the flow out of the second tank is higher and the measurement of P_2 is smaller than expected. To conclude, we get the profile $p_1 := (greater, smaller)$.

   fc_2: If valve v_2 is shifted, we get the profile $p_2 := (smaller, greater)$ because of a symmetry to the fault candidate fc_1.

   fc_3: If valve v_3 is shifted, the water of the first and the second tank can not sink as fast as expected. Thus, we get the profile $p_3 := (greater, greater)$.

3. According to figure 4, the measured behavior $\hat{y}$ implies the profile $\hat{p} := (greater, smaller)$.

4. Because $\hat{p} \neq p_2$ and $\hat{p} \neq p_3$, the only remaining fault candidate is the first one.

3.2   Second Level

There are two motivations for the need of a second level. In the first place the information known from the first level is not enough and one may need the knowledge of the degree of a fault – for example to determine if the current operation has to be stopped or may be continued. And in the second place the qualitative attributes of the first level could not be able to eliminate all but one fault candidate. In that case we need the second level to further reduce the set of fault candidates.

A fault candidate given by the first level corresponds to a subspace $S := [\mu_1, \mu_1'] \times [\mu_2, \mu_2'] \times \cdots \times [\mu_n, \mu_n'] \subset C$ of the configuration space C, where $0 \leq \mu_i \leq \mu_i' \leq 1$ for $1 \leq i \leq n$. A configuration $(\nu_1, \nu_2, \ldots, \nu_n) \in C$ with $\nu_i \in \{\mu_i, \mu_i'\}$ for every $1 \leq i \leq n$ denotes a *corner* of the search space S. The *dimension* $d := \#\{i \mid \mu_i \neq \mu_i'\}$ of a search space S is the number of unfixed parameters.

Now, the task of the second level is to either determine a configuration $\lambda \in S$ for a given search space S that may be responsible for the measurement $\hat{y}$ or to reject S if no such λ exists.

One-dimensional Search Spaces To continue the example, we try to determine the degree of the shift of valve v_1. That means, we search for $\lambda = (\lambda_1, 0, 0) \in S_1$ that fulfills equation (1) where $S_1 := [0, 1] \times [0, 0] \times [0, 0]$.

Considering figure 5 we notice that the effect of a shifted position of valve v_1 increases with an increasing degree of the fault:

$$\mu_1 \leq \mu_1' \;\Rightarrow\; \begin{cases} y_1^{x,(\mu_1,0,0)} \leq y_1^{x,(\mu_1',0,0)} \\ y_2^{x,(\mu_1,0,0)} \geq y_2^{x,(\mu_1',0,0)} \end{cases}$$

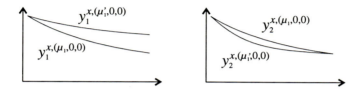

Fig. 5. Monotony of the behavior with $\mu_1 < \mu_1'$

In the case of an existing $\lambda \in S_1$ that fulfills (1), we may derive two bounding conditions:

$$\lambda_1 \in \,]\mu_1, \mu_1'[\quad \Longleftrightarrow \quad y_1^{x,(\mu_1,0,0)} < \hat{y}_1 < y_1^{x,(\mu_1',0,0)} \tag{3}$$

$$\lambda_1 \in \,]\mu_1, \mu_1'[\quad \Longleftrightarrow \quad y_2^{x,(\mu_1,0,0)} > \hat{y}_2 > y_2^{x,(\mu_1',0,0)} \tag{4}$$

These bounding conditions may help to approximate the value of λ_1 with a binary search. Dividing the initial search space S_1 into the two subspaces $S_1^l := [0, \frac{1}{2}] \times [0, 0] \times [0, 0]$ and $S_1^h := [\frac{1}{2}, 1] \times [0, 0] \times [0, 0]$, we are able to decide with the two bounding conditions whether $\lambda \in S_1^l$ or $\lambda \in S_1^h$. Thus, we may choose the one that contains λ and again divide that search space into two subspaces. If we continue that, we get a $\mathbb{R}$-analytic computation in the sense of [HVS95] for which the corners of the required subspaces approximate λ_1.

If there is no $\lambda \in S_1$ with property (1), it is possible that one of the bounding conditions (3) and (4) does not hold for any of the two subspaces or that they hold in different subspaces. In one of theses cases the binary search may be stopped and the corresponding fault candidate may be eliminated. But it is also possible that none of these cases appears and the binary search approximates a value λ' with $(y_1^{x,\lambda'}, y_2^{x,\lambda'}) \neq (\hat{y}_1, \hat{y}_2)$.

To explain this, we consider M and define an equivalence relation $\sim$ of non comparable elements in M by

$$m_1 \sim m_2 :\Longleftrightarrow \int_{t_{start}}^{t_{end}} |m_1|\, dt = \int_{t_{start}}^{t_{end}} |m_2|\, dt$$

This implies a residual class $M_{/\sim}$ that may take over the ordering $<$ of M in a well defined way.

$$[m_1] \quad := \quad \{m_2 \in M \mid m_1 \sim m_2\}$$

$$M_{/\sim} \quad := \quad \{[m_1] \mid m_1 \in M\}$$

$$[m_1] < [m_2] : \iff m_1 < m_2$$

The decisions of the binary search described above are all made by comparisons of the size of the behavior at corners of subspaces with the size of $\hat{y}_1$ and $\hat{y}_2$. Therefore, these decisions are the same for all measurements $([\hat{y}_1], [\hat{y}_2])$ and the binary search approximates a $\lambda' \in S_1$ with

$$(y_1^{x,\lambda'}, y_2^{x,\lambda'}) \in ([\hat{y}_1], [\hat{y}_2])$$

This implies the need to check whether the result of the binary search is a diagnosis of the measurement $\hat{y}$ or not.

Two-dimensional Search Spaces The search space S of a fault candidate corresponding to a double fault is two-dimensional. To determine $\lambda \in S$ we adapt the binary search of the one-dimensional case. Now, the search spaces are divided into four subspaces. For each of these we have to decide whether to continue the search in the subspace or to discard it. Again, these decisions are made by bounding conditions that we conclude from an observed monotonic behavior of the system.

Generalization To avoid restriction to the example we give a general bounding condition for systems with monotonic behavior.

Let $y^{x,\lambda} = (\hat{y}_1, \hat{y}_2, \ldots, \hat{y}_r)$ for $x \in X$ and $\lambda \in C$ and let $S \subseteq C$ be a search space of arbitrary dimension $d \leq n$ with $\lambda \in S$. Then, for each measurement $\hat{y}_i$ with $1 \leq i \leq r$ there are two corners $\nu_{\min}, \nu_{\max} \in S$ of the search space S so that $y^{x,\nu_{\min}} \leq \hat{y}_i \leq y^{x,\nu_{\max}}$.

To see this, one should imagine to reach the corners $\nu_{\min}$ and $\nu_{\max}$ starting from λ by successively moving a single parameter λ_i in the direction of one of its two limits μ_i or μ_i'. In doing so, the behavior is changed monotonical. With the right choice between μ_i and μ_i' it is always possible to increase or decrease the behavior. If done for all n parameters, the corners $\nu_{\min}$ and $\nu_{\max}$ of the search space S are found.

As the bounding condition holds for search spaces of arbitrary dimensions, we may search for arbitrary multiple faults in systems with monotonic behavior.

Concluding, we sketch the analytical algorithm. Its input is a search space $S \subseteq C$, a measurement $\hat{y} = (\hat{y}_1, \hat{y}_2, \ldots, \hat{y}_r) \in Y$ and the system input $x \in X$.

1. Let $\mathcal{S}$ be a set of search spaces, initialized with $\{S\}$

2. If $\mathcal{S} = \emptyset$ then discard the search. No solution $\lambda \in S$ can be found.
3. Delete a search space S' from $\mathcal{S}$ and create the subspaces of S'.
4. For each of these subspaces check the bounding conditions for the measurement $\hat{y}$. Include those subspaces in $\mathcal{S}$ that fulfill the bounding conditions.
5. Continue the calculation with the second step.

If the search is not discarded, the corners of the search spaces in $\mathcal{S}$ converge to a configuration $\lambda \in S$. If the behavior $y^{x,\lambda}$ of the system corresponding to λ is equal to the measurement $\hat{y}$, then a diagnosis is found. As the behavior of the system is monotonical, all diagnoses for $\hat{y}$ build a connected set including λ.

4   Real Diagnosis

In the last section, we simplified the task of diagnosis. What has to be considered if one wants to use the developed ideas for a real diagnosis problem?

1. Sensors do not output the real valued function of behavior of the measured variable in a time interval, but a sequence of single values as points of support of it.
2. – A model of a system is never complete.
    – Due to numerics, the simulation of that model is not an exact but a rough computation.
    – The sensors only have finite precision.
    This leads to an unavoidable difference between the calculated and the measured behavior even in the case of a fault free system. Such differences are called *noise*.
3. In systems bigger than the used example the observed monotony of the behavior may be violated.

These facts lead to some consequences:

From 1: Instead of $M := \{f : [t_{start}, t_{end}] \to \mathbb{R}\}$ we define $M := \mathbb{R}^s$. Now, we need a new definition for the size of a measurement $m = (m^1, m^2, \ldots, m^s) \in \mathbb{R}^s$. Therefore, equation (2) is substituted with some $p \in \mathbb{N}$ by

$$\|m\|_p := \sqrt[p]{\sum_{i=1}^{s} |m^i|^p} \tag{5}$$

From 2: In the presence of noise we are not able to find a configuration λ with a behavior $y^{x,\lambda}$ equal to the measurement $\hat{y}$ but only similar to it. Therefore, equation (1) has to be changed. With a proper definition of the relation $\approx$ we get

$$y^{x,\lambda} \approx \hat{y} \tag{6}$$

From 2: Noise may change the value of the attributes used in the first level and therefore leads to wrong results.

From 2 and 3: Noise and violations of the monotony of the behavior may violate the bounding conditions used in the second level and therefore lead to wrong decisions in the search so that the solution λ is not found.

Remedy To deal with noisy measurements, we need a measure $L \in [0,1]$ of
the precision we use. The extreme value of $L = 0$ stands for absolute precision
without any noise and $L = 1$ means, that there is no precision at all and the
noise dominates.

$$m_1 \approx_0 m_2 \iff m_1 = m_2$$

$$m_1 \approx_1 m_2 \qquad \forall\, m_1, m_2 \in M$$

Now, we define a continuous change from one extreme to the other. Thereby,
we distinguish absolute similarity $\approx^a$ and relative similarity $\approx^r$. The latter one
takes the size of the measurements into account.

$$a \approx_L^a b :\iff (1-L)\,\|a-b\| \leq L$$

$$a \approx_L^r b :\iff \|a-b\| \leq L\,(\|a\| + \|b\|)$$

This similarity implies a robustness against noise.

> An attribute of a measurement m is said to be *robust* against noise of
> level L, iff it holds for all measurements similar to m with level L.

For some attributes, like the monotony or the curveness of a measurement and
for the comparison of the size of two measurements we developed methods to
determine the robustness. From that, we gain robustness of the first level of the
approach.

At the second level, the robust comparison of two measurements prevents
us from wrong decisions whether a subspace should be discarded or not. The
price is a possibly exhaustive search instead of a binary search. For the same
price violations of monotony of the behavior may be handled by a system of
penalties. A penalty $\pi(S, \hat{y}, i)$ of a subspace S expresses the belief that S contains
a diagnosis of the measurement $\hat{y}_i \in M$ of the sensor $i \in \{1, 2, \ldots, r\}$.

For an exhaustive search, the number of search spaces inquired may increase
exponentially. Therefore, we examined different methods to calculate the penalties with the goal to get robustness against noise and violation of monotony
while dealing with a tractable number of search spaces. Best results could be
achieved with a penalty function $\pi(S, \hat{y}, i)$ that uses the distances $\|\hat{y}_i - y_i^{x,\nu}\|$ for
all corners ν of the subspace S in relation to the activity of S. The activity of
a subspace S is defined by the maximum distance $\|y_i^{x,\nu} - y_i^{x,\nu'}\|$ for all corners
ν, ν' of S. As one can see in the next section, the results are nearly as good as
the unreachable optimal search. Thus, the exhaustive search could be avoided.

5   Results

The power of the new approach is demonstrated by some empirical examinations
we made for a system much more complex than the example introduced. It also
consists of three tanks, but in addition there are more pipes and valves, floating

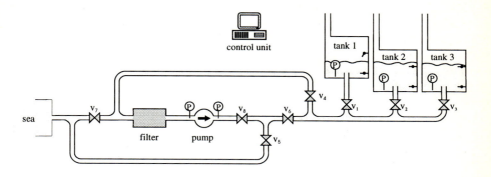

Fig. 6. Scheme of a ballast tank system

switches, bend pipes, a pump, a filter, some additional pressure sensors, a throttle valve and a control unit (figure 6). We designed a fault model that pays attention to the typical defects given by a manufacturer of ballast systems and randomly created 3000 single and double faults with additional noise. An implementation of the two-level approach was able to diagnose all these faults correctly. In about one of three cases the first level leads to a minimal diagnosis in the sense, that for each resulting fault candidate there exists a configuration that may explain the measurements. In the remaining cases the second level was able to eliminate the additional fault candidates for which there exists no such configuration. It appeared that such a elimination is nearly as costly as the computation of a configuration that is a diagnosis. We also omitted the first level and only used the second level. In that case, the runtime roughly doubles.

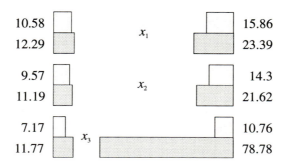

Fig. 7. Mean number of inquired search spaces for three different operations

Figure 7 compares the mean number of inquired search spaces at the second level to the one of a unknown optimal search process. Optimal in the sense that it always knows which subspaces to discard or not. Therefore, it only inquires the minimal number of search spaces needed. The figure distinguishes between three different inputs $\{x_1, x_2, x_3\}$ and the single faults (on the left side) and double faults that are injected. It demonstrates that the approach is very near to the unknown optimal strategy.

The success of the approach for that fixed example encouraged us to start developing a design tool for ballast tanks. With the help of that tool we will test the approach for arbitrary ballast tank systems. We hope that we then will be able to automatically select the sensors in a new system design that are useful to simplify the computation of its diagnoses. The key word of these future plans is *design for diagnosis*.

References

[CW84]   E. Y. Chow and A. S. Willsky. Analytical redundancy and the design of robust failure detection systems. *IEEE Transactions on Automatic Control*, 29:603–614, 1984.

[Dav84]   R. Davis. Diagnostic Reasoning based on structure and behavior. *Artificial Intelligence*, 24:347–410, 1984.

[DBMB93]   O. Dressler, C. Böttcher, M. Montag, and A. Brinkop. Qualitative and Quantitative Models in a Model-based Diagnosis System for Ballast Tank Systems. In *Proc. of the International Conference on Fault Diagnosis TOOLDIAG '93, Toulouse, France*, pages 397–405, 1993.

[DK89]   D. Dvorak and B. Kuipers. Model-Based Monitoring of Dynamic Systems. In *Proceedings IJCAI*, pages 1238–1243, 1989.

[dKMR92]   J. de Kleer, A. K. Mackworth, and R. Reiter. Characterizing diagnosis and systems. *Artificial Intelligence*, 56:197–222, 1992.

[dKW87]   J. de Kleer and B. Williams. Diagnosing multiple faults. *Artificial Intelligence*, 32:97–130, 1987.

[EM94]   H. W. Engl and J. McLaughlin. *Inverse Problems and Optimal Design in Industry*. Teubner press, 1994.

[FS92]   B. Faltings and P. Struss, editors. *Recent Advances in Qualitative Physics*, chapter Sensor Selection in Complex System Monitoring Using Information Quantification and Causal Reasoning. MIT press, 1992.

[HVS95]   G. Hotz, G. Vierke, and B. Schieffer. Analytic machines. Technical Report TR95-025, Electronic Colloquium on Computational Complexity (http://www.eccc.uni-trier.de/eccc), 1995.

[Ise84]   R. Isermann. Process fault detection based on modeling and estimation methods: A survey. *Automatica*, 20:387–404, 1984.

[Mas86]   M. A. Massoumnia. A geometric approach to the synthesis of failure detection filters. *IEEE Transactions on Automatic Control*, 31:839–846, 1986.

[MNTMQ96]   R. Milne, C. Nicol, L. Travé-Massuyès, and J. Quevedo. TIGER: knowledge based gas turbine condition monitoring. *AI Communications*, 9:92–108, 1996.

[Rei87]   R. Reiter. A theory of diagnosis from first principles. *Artificial Intelligence*, 32:57–96, 1987.

Qualitative Uncertainty Models from Random Set Theory

Olaf Wolkenhauer*

Control Systems Centre, UMIST
Manchester M60 1QD, UK
E-mail: olaf@csc.umist.ac.uk
Web: http://www.csc.umist.ac.uk/FUZZY/olaf.html

Abstract. When only incomplete information about the probability distribution of an experiment is available, we may have to admit imprecision in the formulation of an uncertainty model. In this paper Random Set Theory is used to build possibilistic uncertainty models from sampled data. In particular Goodman's one-point coverage function of a class of random sets is estimated from data. Finally, we focus on an example to illustrate how from random sets induced possibility distributions may be used in the detection of changes in time-series data.

1 Introduction

Since its introduction by Zadeh [13], possibility theory has been established as the field that studies formal relationshis between uncertainty techniques. The foundations for a rigorous mathematical treatment and applications have been developed most notably by Dubois & Prade [1].

In the first section of this paper, we describe *random sets* as set-valued random variables or equivalently multi-valued mappings. Random sets are closely related to Dempster's study [4] of probability masses allocated to subsets (*focal elements*) of a given universe rather than to its singletons. Such multi-valued mappings induce imprecise (upper and lower) probabilities. If the focal elements are nested, upper and lower probability measures can be identified with possibility and necessity measures respectively (called *confidence measures*). The possibility or necessity of an event is determined by means of the associated possibility distribution which turns out to be equivalent to the characteristic function of a normalised fuzzy set. Goodman [5] showed that the one-point coverage function of a random set can be identified as the characteristic function of

---

* This work was supported by the UK Engineering and Physical Sciences Research Council (EPSRC) under grant GR/KR09410.

X. Liu, P. Cohen, M. Berthold (Eds.): "Advances in Intelligent Data Analysis" (IDA–97)
LNCS 1280, pp. 609–620, 1997. © Springer–Verlag Berlin Heidelberg 1997

a fuzzy set. This paper will show that if only qualitative knowledge for the uncertainty model is available, a set-valuedisation of the real-valued samples, leading to samples of random sets, induces a possibility distribution. Dubois and Prade [1] demonstrated that such a distribution in fact describes bounds on a class of probability measures and may thus be used in inference systems. The generated fuzzy quantities can easily be integrated in rule-based fuzzy systems for approximate reasoning [10]. Such an interface between *signal processing* (quantitative, numerical data representations) and *knowledge processing* (qualitative, symbolic models) may prove useful in an automated data analysis that interacts with a human analyst or operator. Subjective - vague exogenous knowledge can conveniently be represented by means of fuzzy sets.

2 Random Sets and Set-Valued Statistics

This section introduces basic concepts and definitions used in subsequent sections. Set-valued statistics is related to random set theory as classical statistics is to probability theory. The main difference compared to conventional statistics is, that instead 'variable points', sets are considered.

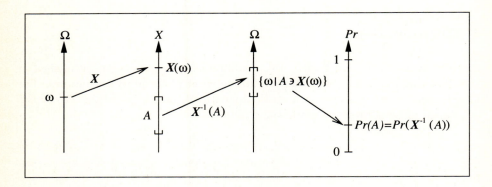

Fig. 1. Ordinary random variables as point-valued mappings.

Before describing a statistical model we derive a probabilistic model for random sets. Reminding us of ordinary point-valued random variables, consider a given probability measure Pr_Ω on $(\Omega, \mathcal{B}_\Omega)$, where $\mathcal{B}_\Omega$ is the collection of subsets of Ω on which Pr_Ω is defined. A random variable $\mathbf{X}$ is then defined as the mapping

$$\mathbf{X}: \quad \Omega \to X \tag{1}$$
$$\omega \mapsto \mathbf{X}(\omega) \in X \ .$$

For any subset A of X, a probability measure Pr_X on X is constructed by means of the inverse image

$$\mathbf{X}^{-1} : \quad \mathcal{P}(X) \to \mathcal{P}(\Omega) \tag{2}$$
$$A \mapsto \mathbf{X}^{-1}(A) = \{\omega \mid \mathbf{X}(\omega) \in A\} \ .$$

as

$$Pr_X(A) = Pr_\Omega \left(\mathbf{X}^{-1}(A) \right) \ . \tag{3}$$

Figure 1 provides an illustration for the one-dimensional case, describing a random variable as a point-valued mapping. Analogously to (1), let $(\Omega, \mathcal{B}_\Omega, Pr_\Omega)$ be a probability field with $\Omega = \{\omega\}$, an associated σ-algebra $\mathcal{B}_\Omega$, and a probability measure Pr. A set-valued mapping from that probability space into the power space $\mathcal{P}(\cdot)$ of the universe of discourse X is called a *random set* :-

$$\Gamma : \quad (\Omega, \mathcal{B}_\Omega, Pr_\Omega) \to (\mathcal{P}(X), \mathcal{P}(\mathcal{P}(X)), g_\Gamma) \tag{4}$$
$$\omega \mapsto \Gamma(\omega) \subseteq X \ .$$

For a subset A in X, the probability measure $g_\Gamma(A)$ is defined from (3) by

$$g_\Gamma(A) = Pr_\Omega \left(\Gamma^{-1}(\tilde{A}) \right) \ . \tag{5}$$

where $\tilde{A} \in \mathcal{P}(\mathcal{P}(X))$ is defined by

$$\tilde{A} = \{\Gamma(\cdot) \mid A \subseteq \Gamma(\cdot) \subseteq X\} \tag{6}$$

and the inverse mapping is analogous to (cf. 2) defined as

$$\Gamma^{-1} : \quad \mathcal{P}(\mathcal{P}(X)) \to \mathcal{P}(\Omega) \tag{7}$$
$$\tilde{A} \mapsto \Gamma^{-1}(\tilde{A}) = \{\omega \in \Omega \mid \Gamma(\omega) \in \tilde{A}\} \ .$$

For the sake of simplicity and to avoid topological complexities, we assume X to be finite and that there is no $\omega \in \Omega$ which maps into an empty set. Figure 2 provides an illustration for the special case of X being one-dimensional.

For a random set, the equivalent of a distribution function for a random variable is called *one-point coverage function* [5] derived for the special case of (5) with A reducing to a singleton element x, that is,

$$g_\Gamma(x \in \Gamma) = Pr_\Omega (\omega \mid x \in \Gamma(\omega)) \ . \tag{8}$$

The coverage function (8), denoted $\mu_\Gamma(\cdot)$, is a generalised distribution function for random sets. Equation (8) in particular, describes the likelihood of a fixed point $x \in X$ being covered by the random set Γ. By using the expectation operator we now show how the coverage function may be estimated using conventional point-valued statistics. For conventional probability theory, with a fixed subset

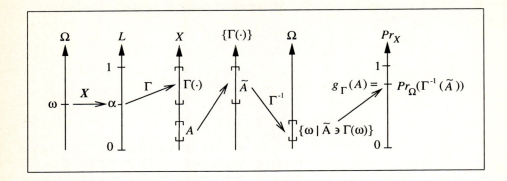

Fig. 2. Random sets as set-valued mappings.

A and variable single-valued outcomes, the probability of event A is defined by the Lebesgue-Stieltjes integral

$$
\begin{aligned}
Pr(A) &= \int_A \mathrm{d}Pr \\
&= \int_X \mathbf{1}_A(x)\,\mathrm{d}Pr \\
&= E[\mathbf{1}_A]\ ,
\end{aligned}
\tag{9}
$$

where $\mathbf{1}_A(\cdot)$ is the characteristic function of the (ordinary) subset $A \in \mathcal{P}(X)$:-

$$
\mathbf{1}_A(x) = \begin{cases} 1 & \text{if } x \in A \\ 0 & \text{otherwise .} \end{cases}
\tag{10}
$$

Similarly we can use $\mathbf{1}_{\Gamma(x)}(\cdot)$ to denote the characteristic function of coverage of x by Γ and as in (9) express the probability of a point being covered by $\Gamma(\omega)$ in terms of the expectation of the characteristic function :-

$$
\begin{aligned}
\mu_\Gamma(x) &= g_\Gamma(x \in \Gamma) \\
&= Pr_\Omega\left(\omega \mid x \in \Gamma(\omega)\right) \\
&= \int_{x \in \Gamma(\omega)} \mathrm{d}Pr_\Omega \\
&= \int_\Omega \mathbf{1}_{\Gamma(\omega)}\,\mathrm{d}Pr_\Omega \\
&= E\left[x \in \Gamma(\omega)\right].
\end{aligned}
\tag{11}
$$

Having introduced the expectation operator, we now can describe how to estimate the coverage probability relation (8) by means of conventional point-valued statistics. Suppose the images of Γ map into a sample $\{C^k\}^n$ of n closed independent identically distributed random intervals : $\Gamma(\omega) \mapsto C^k$, $k = 1,\ldots,n$, drawn from the distribution of Γ, the *covering frequency* of x by Γ is given as

$$\hat{\mu}_{\Gamma^n}(x) = \frac{1}{n} \sum_{k=1}^{n} \mathbf{1}_{C^k}(x) \ . \tag{12}$$

As Wang [8] has shown, eq. (12) is an unbiased and consistent estimator that converges uniformly to the one-point coverage function μ_Γ, i.e.

$$\lim_{n \to \infty} Pr \left(\mid \mu_\Gamma(x) - \hat{\mu}_{\Gamma^n}(x) \mid \, \geqq \epsilon \right) = 0 \ .$$

3   Fuzzy Sets and Confidence Measures

Fuzzy sets, introduced by Zadeh [13], generalise the characteristic function (10) of ordinary or 'crisp' sets by allowing degrees of membership.

$$\forall \ C \in \mathcal{P}(X), \quad \mu_C \colon X \to [0,1] \ , \tag{13}$$

defines the characteristic function μ of a fuzzy set C as a mapping from the reference set X into the unit interval. A fuzzy set may also be represented in terms of a family of *level sets* (or α-cuts) $\{C_\alpha\}$ defined by

$$C_\alpha = \{ \ x \in X \mid \mu(x) \geqq \alpha \ \} \ . \tag{14}$$

The so called *representation theorem* [7] holds that

$$\mu_C(x) = \sup_{\alpha \in (0,1]} \{ \ \min \left(\alpha, \mathbf{1}_{C_\alpha}(x) \right) \} \ , \tag{15}$$

A unimodal fuzzy set with finite support (i.e. $\{x \in X \mid \mu(x) > 0\}$ finite) has nested level sets C_α, that is, for $\alpha_1 > \alpha_2 > \cdots \alpha_i > \cdots > \alpha_l > 0$, $\alpha \in L = (0,1]$, $C_{\alpha_i} \subset C_{\alpha_{i+1}}$ with $\alpha_{l+1} = 0$ by convention. For a *normal* fuzzy set, there exists at least one x for which $\mu(x) = 1$. Viewing μ as a restriction on the values taken by the variable $\mathbf{X}$, a normal fuzzy set is also a possibility distribution [13], denoted by $\pi_\mathbf{X}$. $\pi_\mathbf{X}(x)$ is the degree of possibility that $x \in X$ coincides with an existing, but inaccessible value x_0. For a crisp subset A of X the grade of possibility of the statement "A contains the value of $\mathbf{X}$" is defined by [1]

$$\Pi_\mathbf{X}(A) = \sup_{x \in A} \pi_\mathbf{X}(x) \ . \tag{16}$$

A dual necessity measure of the event A is defined as the impossibility of the complementary event A^c :-

$$Ne_\mathbf{X}(A) = 1 - \Pi_\mathbf{X}(A^c) = \inf_{x \notin A} \{ \ 1 - \pi_\mathbf{X}(x) \ \} \ . \tag{17}$$

4   Level-Set-Valued Statistics and Fuzzy Sets

The statistics for random subsets may be extended by means of confidence values α assigned to the images $\Gamma(\omega)$ then denoted as $\Gamma_\alpha(\omega)$.

Each sampling provides a set $C_\alpha^k = \{x \mid \mu_\Gamma(x) \geq \alpha\}$, $x \in X$ and $\alpha \in L = (0,1]$, such that Γ is monotonic decreasing with respect to set-inclusion, i.e. for each $\omega \in \Omega$ and $\alpha \geq \alpha' \Rightarrow \Gamma_\alpha(\omega) \subseteq \Gamma_{\alpha'}(\omega)$. For each α fixed, set-valued statistics lead to a class of covering frequencies. The class $\{\Gamma_\alpha\}$ is called *random level set* and as with (8)

$$\mu_{\Gamma_\alpha}(x) = Pr_\Gamma (x \in \Gamma_\alpha)$$
$$= Pr_\Omega (\omega \mid x \in \Gamma_\alpha(x)) \ . \tag{18}$$

Following Goodman's analysis [5], given the one-point coverage function μ_Γ, the associated random set Γ may be constructed as follows. Let $\mathbf{Y}$ be a random variable uniformly distributed on the unit interval :

$$\mathbf{Y}(\omega) : \quad \omega \mapsto \alpha \in L = (0,1] \ ,$$

and with

$$\Gamma(\omega) = \{ \ x \mid \mu_\Gamma(x) \geq \mathbf{Y}(\omega) \ \} \ ,$$

then for all $x \in X$, the one-point probability relation (8) holds that

$$Pr_\Omega(\omega \mid x \in \Gamma(\omega)) = Pr_\Omega(\omega \mid \mu_\Gamma(x) \geq \mathbf{Y}(\omega))$$
$$= Pr_\Omega(\omega \mid x \in \Gamma_\alpha(\omega) = C_\alpha)$$
$$= \mu_{\Gamma_\alpha}(x)$$
$$= \mu_C(x) \quad \forall \ \alpha \in L \ . \tag{19}$$

Hence, the one-point coverage function of uniformly randomised level sets is also the characteristic function of a fuzzy set in level set representation.

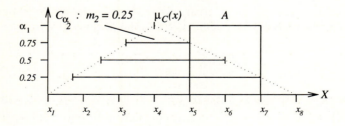

Fig. 3. Equivalence of Pr^* with Π. Note: $\mu_C(x) \equiv \pi(x)$.

The value α describes the minimum degree to which there is certainty that a $\Gamma(\omega)$ represents the outcome of the variable $\mathbf{X}$ on X. In other words, it is the smallest probability that the true value of $\mathbf{X}$ hits C_α. From (17), $Ne(C_\alpha) \geqq 1 - \alpha_{i+1}$. The fact that Ne is a lower probability bound on a whole class of probability measures can be established by defining probabilistic weights (Lebesgue measure) $m(C_{\alpha_i}) = \alpha_i - \alpha_{i+1}$, $\sum m(C_{\alpha_i}) = 1$, on the subsets C_{α_i} of X. Following Dempster's analysis [4], the evidence supporting an event $A \in \mathcal{P}(X)$, is bounded from below by the lower probability

$$Pr_*(A) = \sum_{i: C_{\alpha_i} \subseteq A} m(C_{\alpha_i}) \tag{20}$$
$$= Ne(A) \ .$$

Likewise for (16), the upper bound is given by $Pr^*(A) = 1 - Pr_*(A^c) = \Pi(A)$. The case $\Pi(A) = Pr^*$ is illustrated in fig. 3.

From (15), (19) and (20) the fuzzy quantity C may therefore be associated with a class of probability measures

$$\mathcal{M}(C) = \{Pr \mid \forall A, Ne(A) \leqq Pr(A) \leqq \Pi(A)\} \ , \tag{21}$$

characterised by the upper and lower cumulative distribution functions, respectively:-

$$F^*(u) = Pr^* \left(\mathbf{X} \leqq u\right) \tag{22}$$
$$= \Pi\left((-\infty, u]\right)$$
$$= \sup_{x \leqq u} \mu_C(x)$$
$$F_*(u) = Pr_* \left(\mathbf{X} \leqq u\right) \tag{23}$$
$$= Ne\left((-\infty, u]\right)$$
$$= \inf_{x \leqq u} \left(1 - \mu_C(x)\right) \ .$$

See Dubois and Prade [2] for a comprehensive account of those relationships. The upper and lower distribution functions associated with C are illustrated in fig. 4.

5   Coverage Functions of Closed Random Intervals

This section considers the special case of random closed intervals for which the ends follow a Gaussian - normal distribution law.

Let the images of the random set Γ be closed random intervals, $\Gamma(\omega) = [c_1(\omega), c_2(\omega)]$, with $c_1(\omega), c_2(\omega)$ being independent random variables with distribution functions $F_1(x)$ and $F_2(x)$ respectively, then [8]

$$\mu_\Gamma(x) = F_1(x) \cdot (1 - F_2(x)) \ . \tag{24}$$

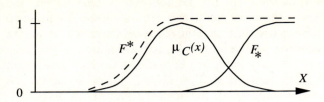

Fig. 4. Upper and lower distribution functions of C.

Let $\mathcal{N}(\eta, \sigma^2)$ denote the normal or Gaussian probablity law with mean η and variance σ^2. The standardised version $\mathcal{N}(0, 1)$ is then defined by the density and cumulative distribution functions

$$g(x) = \frac{1}{\sqrt{2\pi}} \ e^{-\frac{x^2}{2}} \qquad \text{(pdf)} \qquad (25)$$

$$G(u) = \int_{-\infty}^{u} g(x) \ \mathrm{d}x \qquad \text{(cdf)} \ , \qquad (26)$$

and with $F(u) = G(\frac{u-\eta}{\sigma})$ for $\mathcal{N}(\eta, \sigma^2)$ and for (24) we then have

$$\mu_\Gamma(x) = G\left(\frac{x - \eta_1}{\sigma_1}\right) \cdot \left(1 - G\left(\frac{x - \eta_2}{\sigma_2}\right)\right) \ . \qquad (27)$$

Suppose $c_1, c_2 \sim \mathcal{N}(0, 1)$, requiring the normalisation of a fuzzy set for an interpretation as a possibility distribution, the fuzzy set described by the characteristic function (27) would not satisfy this condition. We now consider the case where for **either** the left **or** the right hand end of the random interval, the standard deviation tends to zero. In other words, one end of the interval is fixed to a reference point (here zero). Since $c_1(\omega) \leqq c_2(\omega)$ by definition, we actually have two experiments with normal distributions constrained to the value zero. To satisfy $\int g(x) \ \mathrm{d}x = 1$, we correct the distributions and obtain for the one-point coverage function (8) the following expression:-

$$\mu_\Gamma(x) = \begin{cases} 2 \cdot F_1(x), & \text{when } x \geqq 0 \\ 2 \cdot (1 - F_2(x)), & \text{when } x > 0 \ , \end{cases} \qquad (28)$$

which may be interpreted as a possibility distribution since $\mu_\Gamma(0) = 1$.

6   Local Uncertainty Models for Time-Correlated Data

Suppose the given data are single-valued, $x(k)$, $k = 1, \dots, n$, and subject to uncertainty, i.e. for a given $x(k)$ we cannot be sure that this particular value is the exact value, associated with the variable **X**. The uncertainty may be described by (random) intervals. We then may view the data as drawn from the random set sample $\{C^k\}^n$ of random set Γ, that is, $x(k) \in \Gamma(\omega) = C^k$. We approximate the random set sample C^k by means of a procedure called *set-valuedisation*. For instance, if our knowledge is restricted to, that the values "lie around zero", we may define the following mapping :

$$x(k) \mapsto \begin{cases} [0, x(k)], & \text{if } x(k) > 0 \\ [x(k), 0], & \text{if } x(k) \leq 0 \ . \end{cases} \tag{29}$$

Suppose we are given a set of sampled data $\{x(k)\}^n$, drawn from a normal distribution $\mathcal{N}(0, 1)$ but with only the partial knowledge, that we *expect* the values to be around zero, then with eq. (12) we can estimate a possibility distribution that provides us with bounds on probabilistic evidence about any event under consideration. This is illustrated in fig. 5 with a simulation of a white noise sequence.

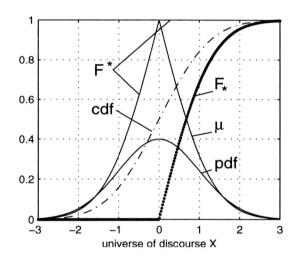

Fig. 5. Random sets inducing a fuzzy quantity and related probability bounds.

In [9], time correlated residual data are monitored for changes in the mean. The uncertainty of the estimate is transformed into a possibility distribution. Applying concepts for ranking fuzzy quantities, fuzzy alarm thresholds have been designed and the algorithms compared to classical control chart techniques. With the present results, such a possibilistic approach to change detection can be put into a more rigorous theoretical framework. Assuming a stationary process, fig. 6 graphically illustrates how a possibility distribution is generated from real-valued data, i.e. random sets. The reference point towards which the intervals are generated, is for change detection problems usually known. However, in principle one may consider the reference point determined, for instance, from a regression line through the given values. Suppose, the alarm-threshold is defined by the value h and the universe of discourse is bounded to $X = [x_1, x_2]$, the crisp set $A = [h, x_2]$ describes the case for which an alarm will be given and the certainty to which there is evidence for this case is determined by (17) :-

$$Ne(A) = \inf_{x \notin A} \{1 - \mu_\Gamma(x)\} \ .$$

If, as in [9], the alarm thresholds are only vaguely defined by fuzzy sets ("alarm if value is very large"), i.e. A is a fuzzy set with characteristic function μ_A, then the necessity to raise an alarm is determined by [1] :-

$$Ne(A) = \inf_{x \in A} \{\mu_A(x) \vee (1 - \mu_\Gamma(x))\} \ .$$

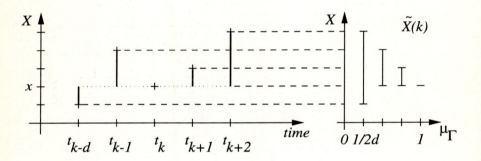

Fig. 6. Possibility distribution induced by correlated data (+).

7   Conclusions

In this paper we have facilitated concepts from random set theory to map sampled data into possibility distribution. The uncertainty model of the experiment from which the data were drawn was only qualitatively described. The close relationship between coverage functions of random sets and Dempster's imprecise probabilities, Shafer's belief functions and confidence measures, opens a wide range of possibilities for *reasoning about data*. The obtained fuzzy quantities can easily be integrated into rule-based fuzzy systems for approximate reasoning [1, 10], combined for data fusion [3] or decision making [6]. The diagram in figure 7 illustrates this concept.

   The application to signal-based data are further discussed in [12]. In [11] the idea of local uncertainty models (inducing possibility distributions from temporal data) is used to generalise classical correlation analysis to a fuzzy similarity sequence. There, the way individual measurements induce random sets is described to model the trend in data. It can be shown, that the possibilistic approach enables us to extract the same qualitative and quantitative information as a conventional correlation sequence does.

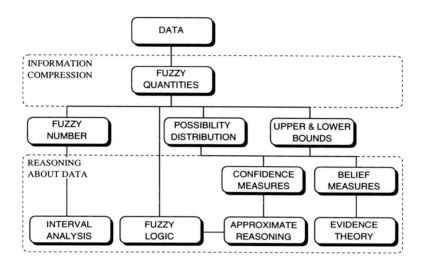

Fig. 7. Reasoning about (fuzzy) data.

References

1. Dubois, D. and Prade, H. : *Possibility Theory.* Plenum Press, New York (1988).
2. Dubois, D. and Prade, H. : *The Mean Value of a Fuzzy Number.* Fuzzy Sets and Systems **24** (1987) 279–300.
3. Dubois, D. and Prade, H. : *Combination of Fuzzy Information in the Framework of Possibility Theory.* In Abidi, M.A. and Gonzales, R.C (eds), 'Data Fusion in Robotics and Machine Intelligence', Academic Press (1992) 481–505.
4. Dempster, A.P. : *Upper and lower probabilities generated by a random closed interval.* The Annals of Mathematical Statistics **39** (1968) 957–966.
5. Goodman, I.R. : *Fuzzy Sets as Equivalence Classes of Random Sets.* In Yager, R.R. (ed.) : Fuzzy Set and Possibility Theory: recent developments. John Wiley & Sons (1982) 327–343.
6. Grabisch, M. et.al. : *Fundamentals of Uncertainty Calculi with Applications to Fuzzy Inference.* Kluwer Academic Publishers (1995).
7. Kruse, R. and Meyer, K.D. : *Statistics with vague data.* D.Reidel Publishing Company, Dordrecht (1987).
8. Wang, P.-Z. : *Knowledge Acquisition by Random Sets.* International Journal of Intelligent Systems **11** (1996) 113–147.

9. Wolkenhauer, O. : *Possibilistic Change Detection: Onset of Oscillations.* Proc. EU-FIT'96, Vol. 3, (1996) 1610–1613.
10. Wolkenhauer, O. and Edmunds, J. : *Dynamic Systems Reliability Evaluation using Uncertainty Techniques for Performance Monitoring.* IEE Proc.-Control Theory Appl. **143** No. 6, November 1996.
11. Wolkenhauer, O. and Edmunds, J. : *Local uncertainty models for the analysis of time series data.* Accepted for EUFIT'97, Aachen, September 1997.
12. Wolkenhauer, O. : *On the Combination of Probabilistic and Fuzzy Concepts in Signal-Based Data Analysis.* Invited Session on Hybrid Methods in the Development of Intelligent Systems, EUFIT'97, Aachen, September 1997.
13. Zadeh, L.A. : *Fuzzy sets as a basis for a theory of possibility.* Fuzzy Sets and Systems **1** (1978) 3–28.

Author Index

Springer
and the
environment

At Springer we firmly believe that an international science publisher has a special obligation to the environment, and our corporate policies consistently reflect this conviction.
We also expect our business partners – paper mills, printers, packaging manufacturers, etc. – to commit themselves to using materials and production processes that do not harm the environment. The paper in this book is made from low- or no-chlorine pulp and is acid free, in conformance with international standards for paper permanency.

Springer

Lecture Notes in Computer Science

For information about Vols. 1–1207

please contact your bookseller or Springer-Verlag